The Impact of Processing Techniques on Communications

NATO ASI Series

Advanced Science Institutes Series

A Series presenting the results of activities sponsored by the NATO Science Committee, which aims at the dissemination of advanced scientific and technological knowledge, with a view to strengthening links between scientific communities.

The Series is published by an international board of publishers in conjunction with the NATO Scientific Affairs Division

A	Life Sciences	Plenum Publishing Corporation
B	Physics	London and New York
C	Mathematical and Physical Sciences	D. Reidel Publishing Company Dordrecht and Boston
D	Behavioural and Social Sciences	Martinus Nijhoff Publishers Dordrecht/Boston/Lancaster
E	Applied Sciences	
F	Computer and Systems Sciences	Springer-Verlag Berlin/Heidelberg/New York
G	Ecological Sciences	

Series E: Applied Sciences – No. 91

The Impact of Processing Techniques on Communications

edited by

J.K. Skwirzynski

Marconi Research Centre
GEC Research Laboratories
Great Baddow, Essex, UK

1985 **Martinus Nijhoff Publishers**
Dordrecht / Boston / Lancaster
Published in cooperation with NATO Scientific Affairs Division

Proceedings of the NATO Advanced Study Institute on The Impact of Processing Techniques on Communications, Chateau de Bonas, (Gers), France, July 11-22, 1983

Library of Congress Cataloging in Publication Data

NATO Advanced Study Institute on "The Impact of
    Processing Techniques on Communications" (1983 :
    Bonas, France)

    (NATO ASI series. Series E, Applied sciences ; no. 91)
    "Published in cooperation with NATO Scientific
Affairs Division."
    1. Telecommunication systems--Congresses.
2. Signal processing--Digital techniques--Congresses.
3. Spread spectrum communications--Congresses.
I. Skwirzynski, J. K. II. North Atlantic Treaty
Organization. Scientific Affairs Division. III. Title.
IV. Series.
TK5101.A1N395 1983   621.38'043   85-4942

ISBN-13: 978-94-010-8760-5   e-ISBN-13: 978-94-009-5113-6
DOI: 10.1007/978-94-009-5113-6

Distributors for the United States and Canada: Kluwer Academic Publishers, 190 Old Derby Street, Hingham, MA 02043, USA

Distributors for the UK and Ireland: Kluwer Academic Publishers, MTP Press Ltd, Falcon House, Queen Square, Lancaster LA1 1RN, UK

Distributors for all other countries: Kluwer Academic Publishers Group, Distribution Center, P.O. Box 322, 3300 AH Dordrecht, The Netherlands

The Editor wishes to express his appreciation to the following officials of NATO's Scientific Affairs Division for their support and co-operation, without which the Advanced Study Institute would not have been possible:

Professor Henry Durand
Dr. Mario di Lullo
Dr. Craig Sinclair

The Institute was also supported by:

National Science Foundation – USA
European Research Office of the US Army – UK
GEC Research Laboratories
    Special thanks are due to its Director, Dr. J.C. Williams, for granting the Editor permission to undertake the time-consuming organisation of the Institute.

LIST OF CONTENTS

PREFACE

This volume contains the full proceedings of the Fourth
Advanced Study Institute organised by myself and my colleagues in
the field of Communication Theory and Allied Subjects[*]. In the
first Institute we associated the subject of signal processing in
communication with that in control engineering. Then we
concentrated on noise and random phenomena by bringing in as well
the subject of stochastic calculus. The third time our subject
was multi-user communication and associated with it, the
important problem of assessing algorithmic complexity. This time
we are concerned with the vast increase of computational power
that is now available in communication systems processors and
controllers. This forces a mathematical, algorithmic and
structural approach to the solution of computational requirements
and design problems, in contrast to previous heuristic and
intuitive methods. We are also concerned with the interactions
and trade-offs between the structure, speed, and complexity of a
process, and between software and hardware implementations.

At the previous Advanced Study Institute in this series, on
Multi-User Communications, there was a session on computational
complexity, applied particularly to network routing problems. It
was the aim of this Institute to expand this topic and to link it
with information theory, random processes, pattern analysis, and
implementation aspects of communication processors.

The first part of these proceedings concentrates on pattern
and structure in communications processing. In organising this
session I was greatly helped and guided by Professor P.G. Farrell
and Professor J.L. Massey. The latter opens our text with his
interesting information-theoretic approach to algorithms, and
uses it to provide both upper and lower bounds on the average
performance of computation algorithms that perform certain tasks
in communication systems. With this aim he introduces the
concept of the Leaf Entropy Theorem and uses it to bound the
average number of essential tests performed by a testing
algorithm. This is followed by two theoretic approaches to the
problem of complexity, inference and coherence of random data
sequences. In the first of these Professor T.M. Cover
investigates the Kolmogorov idea of defining the intrinsic
descriptive complexity of a sequence and shows that this has,

----

* The first three were:-

  "New Directions in Signal Processing in Communication and
   Control"(E12)
  "Communication Systems and Random Process Theory" (E25)
  "New Concepts in Multi-User Communication" (E43)
  (All published in the same series).

with a high probability, the same entropy as that defined by Shannon. These ideas are illustrated by an example of a gambler who gambles on patterns in a sequence. In the second of these papers Professor J. Ziv introduces the normalised complexity of an infinite sequence which provides useful performance criteria for practical data compression algorithms, and can also be used to construct universal encoding algorithms which are asymptotically optimal for all sequences. He also considers the concept of measuring the coherence between two sequences and this new concept can lead to a universal decoding procedure for finite-memory channels. Next Dr. M. Beale investigates the class of composite sequences such as are used in spread-spectrum systems, and in systems using correlation detection or matched filtering. He shows that the structure of composite sequences facilitates rapid synchronisation procedures for correlation receivers.

The second part of our proceedings concentrates on algorithms and processing methods from hardware, firmware and software points of view. This session of the Institute was organised by Professor R.J. McEliece who managed to assemble a powerful team of speakers from the U.S.A. and the U.K. It opens with a tutorial paper by Professor K. Steiglitz on hierarchical, parallel and systolic array processing. He concentrates on methodologies of bit manipulation and provides an important assessment on time, area and architectural bounds of VLSI devices. Then follows Dr. P.J.W. Rayner who reviews the finite field theory as applied to realisation and the exact analysis of digital signal processing systems. He illustrates this with descriptions and realisations of digital filters as finite state machines and of fast transform and convolution devices. Professor D.G. Messerschmitt looks at some trade-offs among alternate means of realisation of signal processing algorithms and applies these to two important examples. The first is a VLSI realisation of a full-duplex data modem with echo cancellation; the second concerns an alternate realisation of digital and switched capacitor digital filters. Both these applications demonstrate the close interaction among architecture, technology and performance of VLSI. Dr. R.E. Blahut concentrates on software - he surveys the state of modern fast algorithms for computations used in digital signal processing and error control, such as linear and cyclic convolutions, discrete Fourier transforms and spectrum estimators. Then he provides an elegant description of Winograd FFT, fast Berlekamp-Massey algorithms and time-domain Reed-Solomon decoders. Professors E.R. Berlekamp and R.J. McEliece discuss the general usefulness of fast and cheap VLSI RAMs in coded telecommunication systems and illustrate this with two detailed examples: an "average-case optimised"

Reed–Solomon decoder, in which a large buffer is used to store
incoming data during times when the decoder is busy decoding an
exceptionally difficult codeword; the second example is of a
family of pseudo-random interleavers, or scramblers, such as are
used in many spread-spectrum communications systems. The last
paper in this part is by Professor P.G. Farrell on code structure
and decoding complexity. He produces an assessment of complexity
for the principal decoding methods for error-correcting codes,
with both hard- and soft-decision demodulation; he then continues
this with decoding using entirely parallel processing, with
decoding by means of simple hardware and very high speed
algorithms, and with decoding algorithms for long and high rate
codes.

This part closes with the verbatim account of a panel
discussion organised by Professor S.W. Golomb* on two-dimensional
signals and codes. After the chairman's opening remarks,
Professor R.J. McEliece describes his recent work on helical
interleavers and Dr. E.T. Cohen describes their realisation. Dr.
R.M.F. Goodman reports on similar situations arising in the
design of a multi-carrier modem, where both frequency and time
are interleaved and yet suffer from fading effects.
Professor P.G. Farrell follows with a clear and thorough
demonstration of two-and three-dimensional codes and
corresponding error correcting methods. Professor J.K. Wolf
reminds us of the old Elspas method for correcting burst errors
using 2D cyclic codes. Then the chairman gives an extensive
explanation of his 'finite-field' construction of safe
frequency-hop patterns for radar or sonar purposes.
Dr. R.M.F. Goodman rejoins the discussion with a description of
multi-dimensional convolutional codes and of his trouble with
them. Professor J.L. Massey argues that convolutional code is
much easier to frame synchronise than is the block code. Finally
Dr. S. Matic gives a short survey of Orchard codes.

We now come to the third part of our proceedings which was
organised by two of my colleagues in this venture,
Professors K.W. Cattermole and J.K. Wolf. This concerns the wide
subject of communication networks and is therefore to some extent
a continuation and addition to contributions in our last
deliberations on multi-user communication. It opens with a
challenging suggestion by Professor J.K. Wolf to apply the
methods of group testing to the design of multi-access protocols.

----

*
  *His natural cheerfulness and ready wit have again contributed
to the atmosphere of the Institute. As at the last Institute, he
composed an 'ad hoc' limerick, read first at the closing banquet
and reproduced here.*

The idea was originally developed for efficient testing of large
groups of blood samples for the presence of a particular disease.
First the method and its history are reviewed, and a
probabilistic model is developed. Then a generalised model is
adapted to our problem, with new criteria of efficiency and
assessment of information theoretic bounds.
Professor K.W. Cattermole offers an extensive paper on network
organisation, control signalling and complexity. The control
organs here are ancilliary channels used for the conveyance of
operating messages. The design of these channels is naturally
related to more complex network topologies and routing
procedures, to the increasing variety of required services,
facilities etc. All of these aspects are considered in detail
and qualitative as well as quantitative limitations are assessed.
Professor A. Ephremides considers the problem of designing
distributed protocols for mobile radio networks, where users,
although adequately synhronised, are moving randomly about, are
subject to failure or jamming and are spread over an area greater
than the communication range of any single user. The distributed
protocols are designed to suit a network architecture of
reconfigurable clusters of nodes which are locally controlled and
are linked to each other via gateway nodes. Jamming protection,
multiple access interference and signalling waveform design are
also discussed. The last paper in this section is by
Dr. S. Harari on seal functions as used in data banks, or in
tranmission of sensitive data. In the proposed method a digital
quantity, called a 'seal', is concatenated to data and shares all
the processing that is done on it. A method of designing these
functions is proposed.

As above, this section contains also a verbatim account of a
panel discussion on 'Local Area Networks and Layered Architecture
and Protocols'. It was organised by Professors A. Ephremides and
J.K. Wolf. Both open the discussion with short accounts of the
subject and an amusing example. Then Mr. M.A. Padlipsky gives a
full account of the ARPANET Reference Model, its architectural
characteristics, design axioms and networking protocols.
Dr. I. Cidon reports on his work on multi-station single-hop
radio networks and describes two forwarding schemes: a fixed one
and a random one. Professor J.W.R. Griffiths describes the
project UNIVERSE, used in the U.K. for extending the local area
technology into a wide area by the use of satellite communication
and other high speed links. Dr. A.B. Cooper concentrates on
land-mobile computer networks and provides a short account of
re-synchronisation protocols. Finally Professor J.L. Massey
reports on a recently proposed ETHERRING system which combines a
ring topology with a multi-access protocol in which there is a
'natural' resolution of collisions, independent of the stability
of the system. The discussion closes with an argument on merits
or otherwise of ring structures.

The fourth part of proceedings contains a set of presentations on different and particular aspects of processing techniques. This set can be divided into five sections.

In the first of these we offer papers on various communication systems or devices and on their implementations. Dr. G.R. Cooper presents a short review on the role of processing techniques in generation and detection of spread spectrum signals for both commercial and military systems. This includes means reducing acquisition time, increasing both bandwidth and jamming resistance, decreasing the probability of intercept and improving security. Professor L.E. Franks considers VLSI implementation of signal processing devices which exhibit a high degree of modularity and a low degree of interconnectivity between modules. He also investigates the use of a ternary-coefficient transversal structure as an adaptive equaliser for synchronous data communication. Dr. A. Morgul and Dr. P.M. Grant discuss a new realisation of frequency adaptive filters based on 100 point SAW chip transform processors which give a 4 MHz real bandwidth capability.

Then follows another verbatim account of a panel discussion on fault-tolerant computers, organised by Professor R. McEliece. This opens with an interesting account by Professor D.G. Messerschmitt on methods (including redundancy) which ensure fault-tolerance in the 4 ESS toll switching machine designed by the Bell North American Telephone in 1976. Mr. E.A. Palo relates the complexity problem of modern and future VLSI technology, as predicted by Gordon Moore, to the obvious needs to ensure continuous testing and high reliability. He provides an example of his argument, namely a 16-bit data bus for Reed-Solomon coding, and discusses the role of redundancy in function processing. Professor K. Steiglitz reports his and his Department's work on providing production testing facilities of VLSI chips in order to ensure that a chip is still functional at a certain time of operation. Dr. C. Heegard discusses the use of algebraic codes for optical data storage and produces detailed examples of implementation. Finally, Dr. R.M.F. Goodman considers the problem of the prediction of reliability (e.g. MTBF) of memory coding for VLSI RAM's.

There are two papers on hardware implementations of Reed-Solomon decoders: Dr. E.T. Cohen uses this as an example of design methodology for 'special purpose microprocessors' and Dr. B.L. Johnson stresses the need for an electronically reconfigurable structure in order to separate extension-field operations from normal binary ones; this results in a highly repetitive architecture, well suited to the VLSI technology. Both designs are described in detail.

We now come to three special subjects which were discussed thoroughly at the Institute: cryptography, image processing and text generation.

In the first of these Dr. R.M.F. Goodman presents an approach to implement fast hardware for public key cryptosystems, particularly the RSA scheme. He concludes with an example developed in the GEC Research Laboratories, at the Hirst Research Centre, based on efficient use of systolic arrays. This is followed by the last verbatim text of a panel discussion on cryptography and security, organised by Professor J.L. Massey. He has attempted, with some success to interlace speakers who are either 'contra' or 'pro' the PKC system. The first 'contra' is Professor I. Ingermarsson who gives a short review of available encryption algorithms and reports on recent work on factoring large products of primes. Dr. S. Harari describes a toll television encyphering systems and defends the knapsack method. Dr. R. Johannesson presents an elegant, though intuitive, method for determining the logarithm of a polynomial in a finite field. Mr. P.G. Wright describes the encryption method and implementation used in the project Universe. (See above.) Dr. R. Blom introduces a PKC system for security in large networks, where he proposes the use of a server which handles the verification of users' keys and thus attempts to defeat a wire-tapper. Professor J. Ziv gives an account of the search for 'one-way' functions. Finally, Dr. R.M.F. Goodman describes a Public Key Broadcast System. The discussion closes with a lively argument on the merits or otherwise of different crypto-systems.

Image processing and pattern recognition are subjects which in recent years have made great advances and have found many fields of application; they are also of great interest to the editor of these proceedings. We have therefore invited Professor K.S. Fu to present an account of his work on the semantic-syntactic approach to these subjects. He starts with the basic formulation of the problem and discusses an important result in syntax-semantics trade-off. This is illustrated by an example of the characterisation of three kinds of chromosone patterns. Then Professor E. Panayirci discusses tendencies in the inappropriate use of clustering algorithms and proposes an introduction of measures of clustering tendencies. Finally, Dr. M.C. Fairhurst and Professor P.G. Farrell discuss parallels betweem pattern recognition problems and those underlaying error-control coding. They propose a fruitful interaction between these two fields.

This part of proceedings closes with a paper by Dr. F. Jelinek on Markov source modelling of text generation. This is an attempt to direct a hypothesis search in a language model for a sentence that was spoken. It can provide the

statistical basis for efficient encoding and text storage during
transmission. Here an algorithm is presented which leads to the
discovery of word string equivalance classes and for the
automatic extraction of statistics for the production of rules in
ambiguous context-free grammars.

It has been my privilege to introduce at each Institute
organised by me a presentation by one of the considered leading
establishments in the field. After DFVLR from Germany in 1974,
the SHAPE Technical Centre from the Netherlands, in 1977, and
COMSAT Communications Laboratory from the U.S.A. in 1980, this
time we have invited representatives from two establishments in
the U.K.: the GEC Research Laboratories and the National
Physical Laboratory.

The National Physical Laboratory (NPL) session was organised
and chaired by Dr. E.L. Albasiny. It opens with an article by
Dr. D. Rayner on testing protocol implementation. He reports on
work at NPL on the objective and independent testing of the
conformance of protocols with the Open Systems Interconnection
(OSI) standards, and relates this work to that done in other
countries in Europe and America. Dr. D. Schofield reports on the
research at NPL on speech recognition, in particular, recognition
of continuous speech from different speakers. The pre-processing
hardware and front-end software are described and so is their use
for the identification and classification of features.
Dr. W.L. Price reviews the development of standards for data
encipherment, when a large community of data users wishes to
exchange data with security over a public data network.

The GEC session was organised and chaired by
Dr. P.V. Collins. The first three contributions concern image
processing activities. Mrs. C.J. Oddy and Mr. R.J.M. Mason
review image processing algorithms, in particular algorithms for
automatic segmentation, their assessment and optimisation.
Dr. P.V. Collins describes image processing systems used at the
GEC Hirst Laboratory. Mr. A.G. Cory et al. report on special
purpose VLSI hardware developed for image processing activities,
such as a grid processor chip or spatial filter. Finally,
Dr. T.W. Chong reviews speech processing prospects in
telecommunications applications, such as voice announcement
systems.

It is hoped that this volume will be of interest to anyone
who wishes to be acquainted with recent trends in processing
techniques and facilities in the domain of communications. We
may follow this Institute with another one in 1986, concentrating
this time on the limitations on speed and efficiency of

processing imposed by forever denser packing on chips, by quantal limits in optical communications and by difficulties of modelling or estimating stochastic channels.

It took me two years to prepare this Institute, and this could not have been done without the help and advice from my colleagues: Professor K.W. Cattermole, Professor P.G. Farrell, Professor R.J. McEliece and Professor J.K. Wolf. It is my privilege to thank them here, particularly for their counsel ensuring a coherent technical programme.

I would also with to thank my colleagues Mr. B.G. West and Mr. P.G. Wright, who were responsible for the tape recording of panel discussions and then for editing these tapes, distributing text to speakers for approval and finally for reconstituting the present version. It has been my aim to reproduce these discussions in an exact verbatim form and I compliment all who were involved in this task.

I wish to acknowledge the help of Miss M. Sadler and Mrs. G. Cooper for administrative assistance during the Institute, as well as for organising the successful social programme.

The typing and editing of word processor files of panel discussions was done by Mrs. S.J. Gibson, who showed remarkable patience with frequent and unforeseen changes in these texts.

August 1984
Great Baddow.

P.S.

This preface is followed by two "poems". The first is the 'Ode of Bonas', sung to the tune of Clementine at our evening barbecue. I take the opportunity of inserting this here, to show the pleasant atmosphere of the Conference in Bonas, which was only occasionally spoiled by the need to attend our evening panel discussions.

Then follows yet another limerick by my old friend Professor Sol Golomb. This one was composed at my request.

ODE OF BONAS

(Tune : Clementine)

Refrain: Dear Director, dear director, dear director Skwirzynski
         We are lost without your guidance at Bonas in eighty-
         three.

1.   In a chateau, on a hill top, in the land of Armagnac
     There had Joseph had his conference, held us stretched upon
     the rack
     REFRAIN

2.   Every morning Joseph drove us to the lecture just at nine
     Wish he'd left us, soundly sleeping, after all the cheap red
     wine
     REFRAIN

3.   Golomb codes, and Galois fields are convolving in our minds
     On-chip layout sounds so way out; all we want is cheap red
     wine
     REFRAIN

4.   Now it's lunch time, prisons over, all head for the foaming
     brine
     Standing up to necks in water, talking codes and cheap red
     wine
     REFRAIN

5.   After lunch when Madame fed us, now it comes siesta time
     All are snoring, very boring, sleeping off the cheap red
     wine
     REFRAIN

6.   Afternoons are very weary listening to the same old line
     Complex theory very dreary, longing now for cheap red wine
     REFRAIN

7.   Time for dinner, Madame is waiting, punching tickets, mighty
     fine
     Quiche Lorraine & soup du jour and yet more jugs of cheap
     red wine
     REFRAIN

8. Evening sessions is upon us passed another day in time
    Orchard codes are 'a la mode' not helped along by cheap red
    wine
    REFRAIN

9. End is near so au revoir now, see you in a few years time
    Friendship made here will continue, helped along by cheap
    red wine
    REFRAIN

Ballad of the 1983 NATO ASI

Château de Bonas in July

- Sol Golomb

I.   All the Wheelings and Dealings infernal
Are recorded and stored in this Journal:
       How we went to Château Bonas
       Where we spent our Franks and Kronas

And gained Knowledge of Technology eternal.

II.   We'll remember in December at our Ski-drills
That we mentioned high-dimensioned Polyhedrals -
       That the Local Loops were ringing
       While the Vocal Groups were singing -
And those Tours that went to scores of French Cathedrals.

III.   (Was it noted if we quoted Aristotle,
Being driven, God-forgiven, at full Throttle?)
       At this Institute of NATO
       We ate Mincemeat with Potato
And drank Booze from old Toulouse within a Bottle.

IV.   From this Fare we were rarely exempted,
And though meek, we were frequently tempted
       To evade Skwirzynski's Tyrannies;
       But, afraid to scale the Pyranees
That Endeavour was never attempted.

V.   With arboreal tutorial Precision
Dr. Massey has a classy rustic Vision
       Where each rooted Tree was tortured
       As it fruited in its Orchard
Whilst its Edges felt the Sledges of excision.

VI.   Dare you talk of those awkward Occasions
When you've struggled to juggle Equations?
       Watch your Mastery increase
       Till you're faster than McEliece
With his stereospheric Evasions.

VII.   Do you cry out when you try out your Decoding
With Expressions of Depression and Foreboding?
       Do you suffer much from knowing
       That your Buffer's overflowing
And the Hope that it can cope is fast eroding?

VIII. The Arrival of Archival Data Bases
      As reflected in projective Metric Spaces
            Makes us state a Description
            Of Data Encryption
      Where the easier Keys leave no Traces.

 IX. But before all the Snoring was over,
     Came the Word that we heard from Tom Cover;
            Of those Monkeys, all in Tandem,
            As they clunk keys, quite at Random,
     While one Rhesus types her Thesis, back in Dover.

  X. There's a Lemus, quite a Schemer, name of Thomas
     Who types Myriads of Periods and Commas,
            While a Gibbon types Othello
            With a Ribbon frayed and yellow -
     That's no climate for a Primate with such Promise!*

 XI. Want to digitize the Fidgest of Enrico
     Or to scramble how you ramble as you speak? Oh,
            One who frolics and plays
            With Systolic Arrays
     Claims it's reckoned in a Second that is Pico -.

XII. There's a sedentary Pedant, I should warn ya
     Who knows Runes and foreign Tunes like Ochy Chornya;
            Who learns Curses from Roumania
            And writes Verses t' entertain ya,
     Then adjourns and now returns to California.

XIII. From this Preview I will leave you with my Greeting:
      "May the Lectures and Conjectures from this Meeting
            Make you zealous in your Research."
            On the Trellis of your Tree-Search
      Are they mentioned for Retention or Deleting?

---

*It was another Gibbon who wrote The Decline and Fall of the Roman Empire.

Part 1.

PATTERN AND STRUCTURE IN COMMUNICATIONS PROCESSING

Organised by:

P.G. Farrell
*Electrical Engineering Department,
The University, Manchester, U.K.*

and

James L. Massey
*Institute of Telecommunications,
Zürich, Switzerland*

AN INFORMATION-THEORETIC APPROACH TO ALGORITHMS

James L. Massey

Professor for Digital Systems Engineering
Institute of Telecommunications
Swiss Federal Institute of Technology
Zurich, Switzerland

ABSTRACT

A quite general result, called the Leaf Entropy Theorem, is proved. This theorem gives the relation between the entropy of the leaves and the branching entropies of the nodes in a rooted tree with probabilities. It is shown that this theorem can be used to bound the average number of essential tests performed by a testing algorithm for a discrete random variable. The approach is illustrated by two examples, namely, the sorting of a list and the polling of the stations in a multiple-access communications system.

1. INTRODUCTION

This paper aims to demonstrate that information theory can be a useful tool in the analysis and design of a rather large class of algorithms that are used in communications and elsewhere. Analogous to the manner in which information theory provides both upper and lower bounds on the average data rate of reliable communications systems, so can it also provide upper and lower bounds on the average "computation" (appropriately defined) of algorithms that can perform certain tasks. This presumes, of course, that one can place a probability distribution over the possible values of the "data" that the algorithm must process -- but, as we shall see, such a probability distribution is often suggested by the nature of the problem. The utility of this information-theoretic approach depends both on the reasonableness of the incorporated definition of computation as well as on the appropriateness of an average (rather than worst-case) computational criterion.

The techniques used in this paper are a generalization and extension of those in [1]. The techniques in [1] have also been generalized in a different way by Hartmann et al. [2] to whose paper the reader is referred for an up-to-date bibliography of other work in this area. The approach given here differs from that in [2] principally in the fact that the latter admits a more general complexity measure but requires more detailed information about the algorithm(s) being analyzed. Here, we seek to make useful statements about the average computation of algorithms after only a superficial analysis of their properties.

In the next section, we introduce the terminology and notation for rooted trees with probabilities and prove the Path Length Lemma. In Section 3, we prove the Leaf Entropy Theorem for such trees, which forms the basis of our information-theoretic approach to algorithms. Section 4 introduces the notion of a testing algorithm. Section 5 shows how to use the Leaf Entropy Theorem to obtain upper and lower bounds on the average computation of testing algorithms. Section 6 provides an example of the application of these bounds to a polling problem. Finally, in Section 7, we discuss the capabilities and limitations of this approach to algorithms.

2. ROOTED TREES WITH PROBABILITIES

A tree is a connected undirected graph with no loops. A rooted tree is a tree in which one of the vertices is designated as the root. (In our figures, we will designate the root by attaching the symbol for an electrical ground to that vertex.) If two vertices are adjacent in a rooted tree, the one farther from the root is called a child of the other, which in turn is called the parent of that child. Note that a child has only one parent and that the root is the only vertex with no parent. A vertex with no children is called a leaf. The depth of a leaf is the number of branches on the path from the root to that leaf; the other vertices on this path are called the ancestors of this leaf. Notice that the depth of a leaf equals the number of its ancestors. The non-leaf vertices (including the root) are called nodes.

By a rooted tree with probabilities, we mean a rooted tree to which a nonnegative real number has been assigned to each leaf (and which is called the probability of that leaf) such that the sum over all leaves of these numbers is 1. We then assign to each node a probability equal to the sum of the probabilities of all leaves having this node as an ancestor. It follows that the root has probability 1 and that the probability of a node equals the sum of the probabilities of its children.

In a rooted tree with probabilities, we shall write P_1, P_2,...,
P_L for the probabilities of the L leaves and shall write p_1, p_2,...,
p_N for the probabilities of the N nodes; we do not exclude the case
$L = N = \infty$ when the number of leaves and nodes is countably infinite,
but we do assume that there is a finite maximum for the number of
children with the same parent. We shall always suppose that the first
node is the root so that $P_1 = 1$.

The following simple result, which is (more or less) well known,
will be extremely useful in the sequel.

Path Length Lemma: The average depth $E[W]$ of the leaves of a
rooted tree with probabilities equals the sum of the node probabilities, i.e.,

$$\sum_{i=1}^{L} P_i w_i = \sum_{j=1}^{N} P_j,$$

where w_i is the depth of the i-th leaf.

Proof: Recall that w_i equals the number of nodes that are ancestors
of the i-th leaf. Hence, in the sum of the node probabilities where
each term is expressed as the sum of the probabilities of those leaves
having that node as an ancestor, the i-th leaf will contribute its
probability P_i exactly w_i terms. This proves the lemma.

It is illuminating to interpret the leaf probability P_i as the
probability that a one-way journey from the root will terminate on
the i-th leaf. Under this interpretation, the node probability P_j is
the probability that the j-th node will be visited on the journey
and $E[W]$ is the average length of the journey in branches. [To avoid
annoying trivialities, we assume hereafter that all nodes (but not
necessarily all leaves) have non-zero probability; were this not so,
we could prune the tree at the nodes of zero probability to produce
an equivalent (for our purposes) tree with all nodes having non-zero
probability.] Let q_{jk} denote the probability of that vertex which is
the k-th child of node j. [We have chosen q here rather than p or P
to denote the vertex probability since this vertex could be either
a leaf or another node.] It follows that

$$\sum_{k} q_{jk} = P_j. \tag{1}$$

It follows also that

$$q_{k|j} = q_{jk} / p_j \qquad (2)$$

is the probability that the k-th child of node j will be visited on the journey, given that node j is visited.

3.   ENTROPIES IN ROOTED TREES WITH PROBABILITIES

In the above interpretation of a rooted tree with probabilities, the leaf entropy, defined as

$$H_{LEAF} = - \sum_{i=1}^{L} p_j \log p_j , \qquad (3)$$

is just the uncertainty about where the journey will terminate or, equivalently, about which path from the root to a leaf will be followed. [For examples, we shall always use base 2 for our logarithms so that entropies will be in "bits".] The branching entropy at node j, defined as

$$H_j = - \sum_{k} q_{k|j} \log q_{k|j} , \qquad (4)$$

is just the uncertainty about which branch will be chosen next, given that the journey has reached node j.

Example 1: For the rooted tree with probabilities of Fig. 1 (where here and hereafter nodes are indicated by squares and leaves by circles), we have $q_{1|1} = q_{2|1} = 1/2$ so that, for the root node, (4) gives

$$H_1 = 1 \text{ bit.}$$

Similarly, we have $q_{1|2} = 1/2$, $q_{2|2} = 1/4$ and $q_{3|2} = 1/4$ so that

$$H_2 = 3/2 \text{ bits.}$$

Because $p_1 = 1/2$, $p_2 = 1/4$, and $p_3 = p_4 = 1/8$, (3) gives

$$H_{LEAF} = 7/4 \text{ bits.}$$

Because $P_1 = 1$ and $P_2 = 1/2$, the path length lemma gives

$$E[W] = P_1 + P_2 = 3/2.$$

This agrees, of course, with the direct calculation of $E[W]$ which, because $w_1 = 1$, $w_2 = w_3 = w_4 = 2$, yields

$$E[W] = (\frac{1}{2})1 + (\frac{1}{4})2 + (\frac{1}{8})2 + (\frac{1}{8})2 = 3/2.$$

It follows from (2) and from our definition (4) of the branching entropy at node j that

$$P_j H_j = -\sum_k q_{jk} \log (q_{jk}/P_j)$$

$$= P_j \log P_j - \sum_k q_{jk} \log q_{jk}. \tag{5}$$

Summing over j now gives

$$\sum_{j=1}^{N} P_j H_j = \sum_{j=1}^{N} P_j \log P_j - \sum_{n=1}^{N} \sum_k q_{nk} \log q_{nk} \tag{6}$$

where our reason for changing the last index of summation will soon be apparent. We now note that every vertex, except the root, will be the k-th child of the n-th node for exactly one pair (n,k). Thus,

$$\sum_{n=1}^{N} \sum_k q_{nk} \log q_{nk} = \sum_{j=2}^{N} P_j \log P_j + \sum_{i=1}^{L} P_i \log p_i. \tag{7}$$

Using (7) in (6) and noting that $P_1 \log P_1 = 1 \log 1 = 0$, we obtain the following result.

LEAF ENTROPY THEOREM: The leaf entropy, H_{LEAF}, in a rooted tree with probabilities equals the sum of the node branching entropies weighted by the node probabilities, i.e.,

$$-\sum_{i=1}^{L} p_i \log p_i = \sum_{j=1}^{N} P_j H_j.$$

Example 1 (concluded):

$$\sum_{j=1}^{2} P_j H_j = (1)1 + (\frac{1}{2})\frac{3}{2} = \frac{7}{4} \text{ bits}$$

which, of course, agrees with our previous direct calculation of H_{LEAF}.

The above theorem, which is the main result of this paper, is a reformulation and minor generalization of previous work [1]. The generalization consists in removal of the previous restriction that the rooted tree be a so-called D-ary tree, i.e., that exactly D branches stem from each node in the direction away from the root. The reformulation in tree rather than sequence terminology suggested the above proof, which is a major simplification of the previous proof.

In the sequel, we shall use the above theorem as the link between algorithms and information theory. We note here in passing, however, that this theorem implies one of the fundamental source coding results of information theory [3,p.71], namely, that the average number of letters, E[W], required to code a discrete random variable U into D-ary letters such that no codeword is the prefix of another codeword satisfies

$$E[W] \geq \frac{H(U)}{\log D}. \tag{8}$$

This can be seen as follows. Any such "prefix condition" code can be used to label the branches of a D-ary rooted tree whose leaf probabilities are the probabilities of the codewords and hence of the corresponding values of U. Thus, $H_{LEAF} = H(U)$. But, since each node has D children, $H_i \leq \log D$. Thus, the Leaf Entropy Theorem gives

$$H(U) = H_{LEAF} \leq \log D \sum_{i=1}^{N} P_j = E[W] \log D$$

where, for the last equality, we have invoked the Path Length Lemma. This derivation of (8) completely bypasses the Kraft Inequality [3,p.67], which has been the basis of previous derivations of (8), and could thus be regarded as a more fundamental proof.

4. TESTING ALGORITHMS

Suppose that the input data to some algorithm A can be described as a discrete random variable V. Suppose further that U is a random variable uniquely determined by V. Then V induces a probability distribution on U, which, in particular, permits calculation of the entropy H(U) as

$$H(U) = - \sum_u P(u) \log P(u).$$

Example 2: Suppose that V is a list of N distinct integers and that U is an N-tuple whose i-th component is the location in the list of the i-th smallest number. Suppose further that the probability distribution over V is such that U is equally likely to be any of the N! permutations of [1,2,..,N]. Then

$$H(U) = \log (N!).$$

We may suppose that the algorithm A is described by a flowchart. The "branching points" or "test points" in the flowchart may be of two types, which we shall call essential and inessential, respectively. An essential branching point (or test point) is one in which the branch followed after that point is reached cannot be determined without having recourse to the input data; an inessential branching point is one where the branch followed does not depend on the input data. Clearly, an algorithm could always be described by a flowchart with no inessential branching points, although inessential branching points (such as those that determine whether some loop has been performed some prescribed number of times) are often convenient to use in flowcharts.

We may also suppose that the outcome of the essential tests of the algorithm are coded in some convenient fashion, e.g., by the integers 0,1,..,D-1 if the test has D possible outcomes. Let $X_1, X_2, .., X_W$ denote the sequence of essential test outcomes when the algorithm A is applied to the input data V. Note that, in general, W will be a random variable. We shall say that A is a testing algorithm for U if (a) the value of U uniquely determines the outcome of every essential test performed and also determines which (if any) essential test will next be performed, and (b) no further essential tests are performed if and only if the sequence of essential test outcomes to that point uniquely determines the value of U.

It follows from this definition of a testing algorithm that the possible executions of the algorithm define a rooted tree in which each leaf corresponds to a possible value u of U. We now assign to

each leaf the probability P(u) specified by the probability distribution for U to obtain a rooted tree with probabilities such that

$$H_{LEAF} = H(U). \tag{9}$$

This assignment of probabilities to the leaves induces a probability
assignment on the nodes such that the node probability P_j is the
probability that this node will be reached in the execution of the
algorithm; the branching entropy H_j is just the uncertainty about
the corresponding essential test performed when this node is reached,
and the average depth of the leaves E[W] is the average number of
essential tests performed.

Example 3: Consider the special case N = 3 of Example 2 when
$V = [n_1, n_2, n_3]$ is a list of 3 distinct integers. Fig. 2 shows a
testing algorithm for the random variable U (where U = [2,1,3] means,
for instance that n_2 is the smallest integer, that n_1 is the second
smallest and that n_3 is the largest) in which the only type of test
used is a comparison of two integers to see which is greatest. Because
the 3! = 6 values of U are equally likely, the corresponding rooted
tree with probabilities is that shown in Fig. 3. We see that the
branching entropies are

$$H_1 = H_4 = H_5 = 1 \text{ bit}$$

and

$$H_2 = H_3 = h(1/3) = .918 \text{ bits},$$

where we have introduced the binary entropy function

$$h(x) = - x \log x - (1-x) \log (1-x).$$

By the Path Length Lemma, we see that

$$E[W] = P_1 + P_2 + P_3 + P_4 + P_5 = \frac{8}{3} = 2.667$$

is the average number of essential tests used. Finally, we note that

$$H(U) = H_{LEAF} = \log (3!) = 2.585 \text{ bits}.$$

5.   BOUNDING COMPUTATION OF TESTING ALGORITHMS

In our study of testing algorithms, we shall equate a computation with the performance of an essential test, and we shall measure the goodness of an algorithm by the smallness of the average number of essential tests that it uses. Our goal here is two-fold:

(i) to underbound $E[W]$ for all testing algorithms of a specified kind, and

(ii) to overbound $E[W]$ for particular testing algorithms after only a rudimentary analysis of their properties.

A.   The Basic Bounds

The following simple result turns out to be surprisingly useful.

Theorem 1: The average depth of the leaves in a rooted tree with probabilities satisfies

$$H_{LEAF}/H_{max} \leqslant E[W] \leqslant H_{LEAF}/H_{min}$$

where H_{LEAF} is the leaf entropy, and where

$H_{max} = \sup_j H_j$ and $H_{min} = \inf_j H_j$ are the maximum and minimum node branching entropies, respectively.

Proof: The definitions of H_{min} and H_{max} imply

$$H_{min} \sum_{j=1}^{N} P_j \leqslant \sum_{j=1}^{N} P_j H_j \leqslant H_{max} \sum_{j=1}^{N} P_j .$$

By the Path Length Lemma, the leftmost and rightmost members of this inequality are $E[W] H_{min}$ and $E[W] H_{max}$, respectively. By the Leaf Entropy Theorem, the central member is H_{LEAF}. This proves the theorem.

This theorem was given previously in [1] but with the restriction, removed here, that the tree be D-ary.

We now have our first bounds on computation.

Corollary 1: If all testing algorithms for U of a specified type have $H_{max} < H_o$, then the average computation of any such testing algorithm satisfies

$$E[W] \geqslant H(U)/H_o .$$

Corollary 2: If a testing algorithm for U has minimum branching entropy H_{min}, then the average computation of this testing algorithm satisfies

$$E[W] \leqslant H(U)/H_{min}$$

Example 4: Consider again the general sorting problem of Example 2 for a specified, but arbitrary, value of N. Consider all testing algorithms for U that use only tests which compare two integers to see which is greater. Because there are only 2 possible outcomes for any test, $H_{max} \leqslant H_o = 1$ bit for all such algorithms. Thus, Corollary 1 specifies that for any such sorting algorithm

$$E[W] \geqslant \log (N!). \tag{10}$$

For instance, with N = 64, the bound is

$$E[W] \geqslant 296.0 \text{ comparisons.}$$

The bound (10) has long been familiar in sorting theory where it is often referred to as the "information-theoretic bound".

Example 5: For an arbitrary but fixed N, $N \geqslant 3$, consider the following recursive sorting algorithm (of which the algorithm in Fig. 2 is the N = 3 special case): After the first i numbers have been sorted, determine where the (i+1)-st number should be inserted in this list, first by comparing it to the middlemost number of the sorted list (say, that in position $\lceil i/2 \rceil$ where $\lceil x \rceil$ denotes the smallest integer equal to or greater than x), then to the middlemost number in the reduced by half list to which it belongs, etc., until its proper place has been found. When there are j ($j \geqslant 1$) numbers in the sublist whose middlemost member is used for comparison, the test uncertainty will be $h\big(\lceil j/2 \rceil/(j+1)\big)$, which achieves its minimum for j = 2, namely $h(1/3) = .918$ bits. Thus,

$$H_{min} = .918 \text{ bits}$$

for this algorithm. Corollary 2 now gives

$$E[W] \leqslant 1.09 \log (N!). \tag{11}$$

From (10), we see that this sorting algorithm is inferior by at most 9% to the best sorting algorithm, whatever that might be. For N = 64, (11) gives

$$E[W] \leqslant 322.4 \text{ comparisons.}$$

The sorting bound (11) was given in [1]. In the next section, we sharpen this bound by a refinement of Corollary 2.

B. Sharpened Bounds

Suppose now that the nodes in a rooted tree with probabilities have been divided into two sets S_1 and S_2, respectively. Suppose moreover that

$$H_{1min} = \inf_{j \in S_1} H_j > H_{2min} = \inf_{j \in S_2} H_j \tag{12}$$

and let

$$\Delta H_{2min} = H_{1min} - H_{2min}. \tag{13}$$

It then follows that

$$\sum_{j=1}^{N} P_j H_j \geqslant \sum_{j=1}^{N} P_j \left[H_{1min} - \theta_2 \Delta H_{2min} \right] \tag{14}$$

where

$$\theta_2 = (\sum_{j \in S_2} P_j) / (\sum_{j=1}^{N} P_j). \tag{15}$$

Using the Leaf Entropy Theorem and Path Length Lemma in (14) now gives the following result.

Theorem 2: For the partition of the nodes of a rooted tree with probabilities into two classes and with the defintions given in (12), (13) and (15),

$$E[W] \leqslant \frac{H_{LEAF}}{H_{1min} - \theta_2 \Delta H_{2min}}.$$

Example 6: For the tree in Fig. 3 and the choice $S_1 = \{1,4,5\}$ and $S_2 = \{2,3\}$, we have $H_{1min} = 1$ and $H_{2min} = h(1/3) = .918$ bits. Thus, $\Delta H_{2min} = .082$ bits. Also $\theta_2 = (1/2 + 1/2)/(8/3) = 3/8$. Recall from Example 3 that $H_{LEAF} = 2.585$ bits. Theorem 2 now gives the bound

$$E[W] \leqslant 2.667,$$

which is actually an equality as can be seen from Example 3.

In order to apply the bound of Theorem 2 to algorithms, we need a convenient way to find (or to overbound) θ_2. It is instructive first to get a physical interpretation of θ_2 as defined by (15) when applied to the tree of a testing algorithm. We claim that θ_2 <u>is the fraction of time that the essential test performed corresponds to a node in S_2</u> in many executions of the testing algorithm for independent samples of U. To see this, note that P_j is just the probability that node j is reached in one execution of the algorithm. Thus, in a large number M of executions of the algorithm, this node would be reached about MP_j times, and nodes in S_2 would be reached about

$$M \sum_{j \varepsilon S_2} P_j$$

times. The total number of nodes reached would be about

$$M \sum_{j=1}^{N} P_j .$$

Thus, the fraction of nodes reached that are in S_2 would be about θ_2 as given by (15), and this becomes a law-of-large-numbers "certainty" as $M \to \infty$. Actually, we do not need to be more precise about θ_2 here, as we intend only to argue that if $\tilde{\theta}_2$ is <u>any upper bound on the fraction of times that a test in S_2 is used on every particular execution of the algorithm</u>, then surely

$$\theta_2 \leqslant \tilde{\theta}_2$$

and we obtain the following result.

<u>Corollary</u>: If $\tilde{\theta}_2$ overbounds the fraction of essential tests that are in S_2 on every execution of a testing algorithm for U, then the average computation of this algorithm satisfies

$$E[W] \leqslant \frac{H(U)}{H_{1min} - \tilde{\theta}_2 \Delta H_{2min}}$$

where H_{1min} and ΔH_{2min} are defined by (12) and (13), respectively.

<u>Example 7</u>: Consider the sorting algorithm of Example 5 and let S_2 consist of all nodes j where $H_{2min} = H_j = h(1/3) = .918$ bits. The branching entropy of all nodes in S_1 is at least $h(2/5) = .971$ bits, as the worst case in S_1 occurs when a number is to be inserted into a sublist of four numbers. Thus, $H_{2min} = .053$. It is readily checked that, for $N \geqslant 8$, the fraction of time that a test in S_2 will be used [i.e., the fraction of times that a number is to be inserted into

a sublist of two numbers⌋ is less than 1/3, i.e.,

$$\tilde{\theta}_2 < 1/3.$$

Using this bound in the Corollary gives, for $N \geq 8$,

$$E[W] \leq 1.05 \log (N!) \tag{16}$$

so that we have substantially improved our upper bound (11) on $E[W]$ for this algorithm. This sorting algorithm is inferior by less than 5% to any other!

It should be obvious that Theorem 2 can be generalized by partitioning the nodes into any number of subsets rather than two. We leave the details to the reader, but remark that bounding the fraction of time that an algorithm spends in each subset generally becomes more difficult as more subsets are used and our motivation is to avoid tedious analysis. We content ourselves in the remainder of this paper to applying the above techniques to an example more in the realm of communications than the sorting example used above.

6. POLLING ALGORITHMS

Consider a multi-access communications system with N sending stations. Let U be the binary N-tuple whose i-th component is a 1 if and only if station i has a message to transmit. We assume that each station independently has probability p of having such a message so that

$$H(U) = Nh(p). \tag{17}$$

A _polling algorithm_ is a testing algorithm of the following type. At each step, some subset of the stations is "enabled" by the central controller, i.e., given permission to send some signal if they have a message to send. The central controller learns from this feedback only whether no station in the enabled set has a message to send (absence of signal)or whether at least one station in the enabled set has a message to send (presence of signal).The central controller continues to poll in this fashion until he has determined precisely which stations have messages to send,i.e.,until he has determined the value of U. Since each test has only two outcomes, $H_o = 1$ bit is an upperbound on the maximum branching entropy of every essential test performed by the central controller. It follows then from (17) and Corollary 1 that the average number of polls made by the central controller satisfies

$$E[W] \geq Nh(p) \tag{18}$$

for any such polling algorithm.

Example 8: With N = 30 stations and p = .110 which gives h(p) = .500 bits , the bound (18) gives

$$E[W] \geqslant 15.0 \tag{19}$$

for any polling algorithm. Note that the average number of stations that have messages to send is only Np = 3.30.

We shall say that a station belongs to the <u>fully unknown set</u> of stations at some time if that station still has probability p of having a message to send, given the results of all polls conducted up to that time by the central controller. Clearly a station is in the fully unknown set if it has not yet been enabled. Suppose that some subset B of the fully unknown set is enabled, that the response indicates one or more stations have messages (which removes the subset B from the fully unknown set), and that a subset E of B is then enabled. If the response again indicates one or more busy stations, it is easy to see that those stations in B but not in E return to the fully unknown set.

We shall say that a station belongs to a <u>one-plus busy set</u> at some time if the only information about this set of stations derivable from the results of all polls conducted up to that time is that there is at least one station in the set with a message to send. Clearly an enabled subset E of the fully unknown set becomes a one-plus busy set if the response indicates that one or more stations have messages. Suppose that B is a one-plus busy set and that a subset E of B is enabled. If the response indicates that no station has a message, then it is easy to see that those stations in B but not in E now form a one-plus busy set.

These considerations motivate the following polling algorithm that depends on a single positive integer parameter n and that, after each response, has all stations whose status is in doubt either in the fully unknown set F or in a single one-plus busy set B. For convenience, let M be the set of stations found to have a message to send, let E be the set of stations enabled in a poll, and let X denote the response to a poll in the manner that X = 1 indicates that one or more of the enabled stations have messages to send, while X = 0 indicates that none do.

Step 0: (Initialization) F ←{1,2,..,N}, B← ∅, M ← φ.

Step 1: If # (F) < n, go to step 2. Otherwise, enable the first n stations in F, and F ← F - E. If X = 0, return to step 1. If X = 1, then B ← E and go to step 3.

Step 2: If F ≠ ∅, stop. Otherwise enable all stations in F, and
F ← ∅. If X = 0, stop. If X = 1, then B ← E and go to step 3.

Step 3: If # (B) = 1, then M ← M ∪ B, B ← ∅, and go to step 1.
Otherwise, go to step 4.

Step 4: Enable the first $\lfloor i/2 \rfloor$ stations in B where i = # (B) [and
where $\lfloor x \rfloor$ denotes the integer part of x]. If X = 0, then
B ← B - E and to to step 3. If X = 1, then F ← F ∪ (B - E),
B ← E and go to step 3.

Let q = 1 - p be the initial probability that a station is idle.
Then, for the above algorithm, we have

$$P(X = 0 \mid \text{step 1}) = q^n \tag{20}$$

and

$$P(X = 0 \mid \text{\# (E)} = m, \text{step 2}) = q^m, \quad 2 \leq m < n \tag{21}$$

as follows from the fact that E is a subset of the fully unknown
set in both step 1 and step 2. Similarly, we find that

$$P(X = 0 \mid \text{\# (B)} = i, \text{step 4}) = \frac{q^{\lfloor i/2 \rfloor} - q^i}{1 - q^i}, \quad 2 \leq i \leq n \tag{22}$$

as follows from the facts that # (E) = $\lfloor i/2 \rfloor$ and that the original
probabilities of events are now increased by the factor $1/(1-q^i)$ because of the conditioning that not all i stations in B are idle.

We now choose the parameter n so that the above polling algorithm will give a small average number of polls, E[W], when N is
large. We observe first that step 2 is reached only in the "end
game" so that the number of tests there will have little influence
on E[W]. Thus, we choose to ignore step 2 for the moment. To make
step 1 efficient, i.e., to maximize the branching entropy of the
test performed there, it follows from (20) that we should choose
n so that q^n is as close to 1/2 as possible.

Example 8 (continued): For q = 1 - p = .890, the choice n = 6
gives the best approximation to 1/2, namely

$$q^6 = .497.$$

It follows that all tests performed in step 1 of the polling algorithm have branching entropy

$$H_j = h(.497) = 1.000 \text{ bits} \tag{23}$$

which is very good indeed.

We now turn our attention to step 4 of the algorithm, and we discover that we have been lucky. The fact that

$$q^n \approx \frac{1}{2} \tag{24}$$

ensures that the right side of (22) will also be reasonably close to 1/2 when $p = 1 - q$ is small, as we would expect in a multiple access system appropriate for polling.

Example 8 (continued): For $q = .890$ and $n = 6$, the right side of (22) becomes .471, .373, .442, .529, and .413 for $i = 2,3,4,5$ and 6, respectively. The corresponding branching entropies are .998, .953, .990, .998 and .978 bits, respectively, for these step 4 tests.

Finally, we must face up to those tests performed in step 2 as our polling algorithm goes into its "end game".

Example 8 (continued): For $q = .890$ and $n = 6$, the right side of (21) becomes .890, .792, .705, .627, and .558 for $m = 1,2,3,4$ and 5, respectively. The corresponding branching entropies are .500, .737, .875, .953, and .990 bits, respectively, for these step 2 tests.

We now partition our test nodes into the sets S_1 and S_2 as described in Section 5B, putting into S_1 all test nodes whose branching entropies are at least as great as the smallest branching entropy of a test performed in step 4 of the polling algorithm. It remains only to determine an upper bound $\tilde{\theta}_2$ on the fraction of time that the algorithm will perform tests in S_2 on every execution of the algorithm in order to be able to use the Corollary of Theorem 2 to overbound the average number of polls performed by our algorithm.

Example 8 (concluded): For the specified choice of S_1 and S_2, we see that S_2 contains only those step 2 tests for which the size m of the enabled set is 2 or 3. Moreover, $H_{1min} = .953$, $H_{2min} = .500$ and $\Delta H_{2min} = .453$ bits. It is easy to check that, for $N = 30$, the maximum fraction of tests in S_2 occurs on that execution of the algorithm in which the first four tests are in step 1 and give $X = 0$, the fifth is in step 1 and gives $X = 1$, the sixth in step 4 gives $X = 1$, the seventh in step 2 with $m = 3$ gives $X = 1$, the eighth in step 4 gives $X = 1$, the ninth in step 2 with $m = 2$ gives $X = 1$, the tenth in step 4 gives $X=1$, and the eleventh and last in step 2 with $m = 1$ gives $X = 0$. Thus, $\tilde{\theta}_2 = 3/11$. Using these values in the Corollary of Theorem 2 now gives the upper bound

$$E[W] \leqslant \frac{15.0}{.953 - (3/11)(.453)} = 18.1 \tag{25}$$

on the average number of polls performed by our polling algorithm.

Comparison of the bound (25) to the lower bound (19) on any polling algorithm shows that our simple polling algorithm cannot be far from optimal.

The reader may notice that in fact the polling algorithm analyzed here is an example of "nested group testing" as described elsewhere in this volume [4].

7. DISCUSSION

We have described an information-theoretic approach to algorithms that seeks to obtain useful upper and lower bounds on the performance of an algorithm from only a cursory analysis of the algorithm itself. As our polling example illustrated, the nature of these bounds can sometimes suggest an efficient algorithm for the problem at hand. The avoidance of the need to analyze in detail the computation performed by the algorithm is perhaps the chief virtue of our approach.

There are four limitations to our approach that should be mentioned. The first is that we require a probability distribution over the input data for the algorithm. This is hardly inconvenient as most problems suggest a natural probability distribution. The second limitation is that our computational measure is an average rather than worst-case measure; in most cases, this would seem to be an advantage rather than a limitation, particularly if the algorithm is to be executed many times. The third limitation is more serious, namely, that we identify a computation with a test, and all tests are thus given equal weight regardless of their "complexity". This can sometimes be compensated for by splitting a complex test into several smaller tests, or by combining simple tests performed consecutively into one test, but it remains a drawback of our approach. The fourth and most serious limitation is that the algorithm must be a testing algorithm for a random variable U determined by the data. We have found it rather surprising that algorithms for dividing two integers (when U is taken as their quotient) and for finding the greatest common divisor of two integers (when U is taken as the pair of integers after division by the greatest common divisor) turn out to be testing algorithms. But algorithms for such problems as finding the maximum in a list of N unequal integers are not testing algorithms for any U determined from the data. Although testing algorithms turn up in more places than one might expect, our approach to algorithms will not be very satisfying until it can be generalized to a much broader class of algorithms. We would be delighted if this paper would stimulate someone to make such a generalization.

ACKNOWLEDGEMENT

I am grateful to my doctoral student, R. Rueppel, for the observation that the sum over incomplete codewords used in [1] to compute entropy could be more insightfully considered as a sum over nodes in a rooted tree. This observation stimulated our search for a simple proof of the Leaf Entropy Theorem.

REFERENCES

1.   Massey, J.L., An Information-Theoretic Approach to Data Processing Algorithms, Abstracts of Papers, IEEE Int. Symp. Info. Th., IEEE Cat. No. 74 CHO 883-9 IT, 1974.

2.   Fano, R.M., Transmission of Information, M.I.T. Press, Cambridge, Massachusetts, 1961.

3.   Hartmann, C.R.P., Varshney, P.K., Mehrotra, K.G., and Gerberich, C.I., Application of Information Theory to the Construction of Efficient Decision Trees, IEEE Transactions on Information Theory, 28, 565, 1982.

4.   Wolf, J.K., Principles of Group Testing and an Application to the Design and Analysis of Multi-Access Protocols, this volume, 1983.

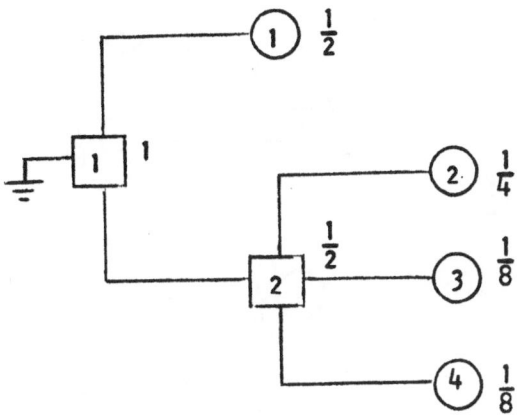

Fig. 1   A rooted tree with probabilities

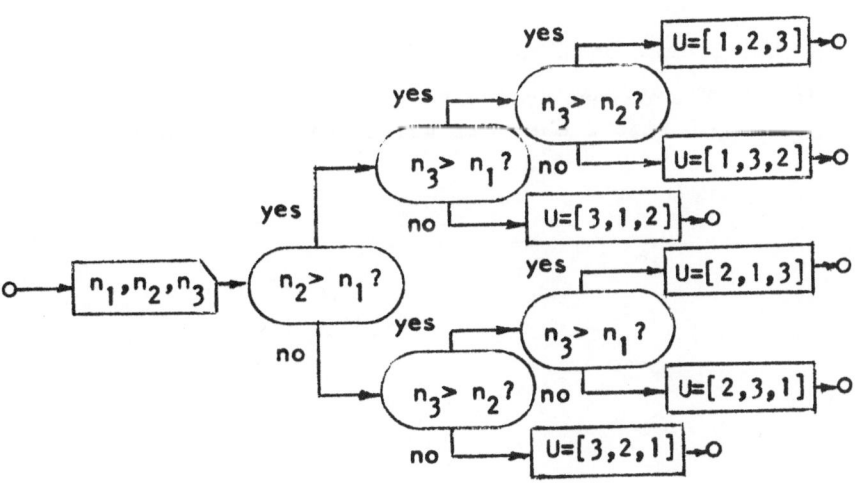

Fig. 2   An algorithm for sorting three distinct integers

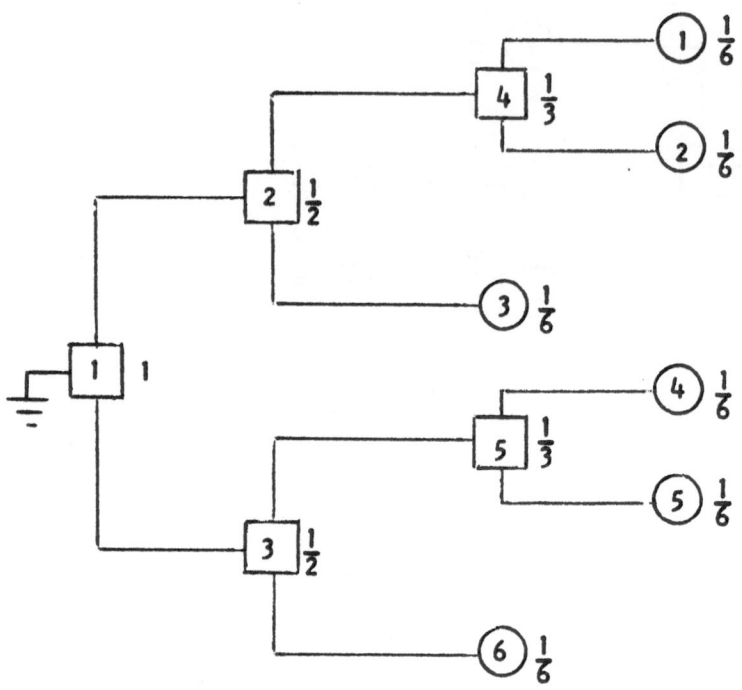

Fig. 3   The rooted tree with probabilities for the algorithm
of Fig. 2 when all values of U are equally likely.

Kolmogorov Complexity, Data Compression, and Inference

Thomas M. Cover
Stanford University

Abstract

If a sequence of random variables has Shannon entropy H, it is well known that there exists an efficient description of this sequence which requires only H bits. But the entropy H of a sequence also has to do with inference. Low entropy sequences allow good guesses of their next terms. This is best illustrated by allowing a gambler to gamble at fair odds on such a sequence. The amount of money that one can make is essentially the complement of the entropy with respect to the length of the sequence.

Now suppose that the sequence is not random. Although the entropy of such a sequence is not defined, there is a notion of its intrinsic descriptive complexity. This idea, put forth by Kolmogorov, Chaitin, and Solomonoff, says that the intrinsic complexity of a sequence is the length of its shortest description. Here too there is a tradeoff between complexity and inference. Low complexity sequences allow a high degree of inference. Again there is a gambling tradeoff.

Finally, it will be shown that if a sequence is random and has entropy H, then with high probability its Kolmogorov complexity will also be H.

Special attention will be given to the so-called Kolmogorov H function, a function that has not yet made its appearance in the literature. We argue that it plays the role of a minimal sufficient statistic. Thus, we can assert that there is a sufficient statistic for the Mona Lisa. This idea will capture the fundamental structure of geometrical patterns, probability distributions and the laws of nature.

1. Kolmogorov Complexity.

Let N denote the natural numbers $\{0,1,2,...\}$. Let $x \in \{0,1\}^\infty$ denote an infinite binary sequence $x = (x_1, x_2,...)$ and let $x(n) = (x_1, x_2, \ldots, x_n)$ denote the first n terms. Let $\{0,1\}^*$ denote all binary sequences of finite length. Let A be a partial recursive function $A: \{0,1\}^* \times N \to \{0,1\}^*$. We restrict A to have a prefix free domain, i.e., no program p accepted by A is the prefix of another. Let $l(x)$ denote the length of the sequence x. Then

$$K_A(x(n)\,|\,n) = \min_{A(p,n)=x(n)} l(p) \tag{1}$$

is defined to be the complexity of $x(n)$ with respect to the algorithm A, given the length n of the sequence $x(n)$. Similarly, let

$$K_A(x) = \min_{A(p,0)=x} l(p) \tag{2}$$

If A is a universal partial recursive function, then K_A, or simply K, is called the Kolmogorov complexity [1,2,3,4,5,9,10,11]. We know that

i) $K(x(n)\,|\,n) \leq K_B(x(n)\,|\,n) + c_B$ for all $n \in N$, $\forall$ x $\qquad$ (3)

ii) $|\{x \in \{0,1\}^* : K(x) < k\}| \leq 2^k$, $\forall$ $k \in N$. $\qquad$ (4)

Now we define a complexity measure for functions $f : D \to \{0,1\}$, where the domain D is some finite set. Let A be a universal partial recursive function

Definition (Function complexity)

$$K_A(f\,|\,D) = \min_{\substack{A(p,x)=f(x)\\ \forall\, x \in D}} l(p) \tag{5}$$

Thus the complexity of f given the domain D is the minimum length program p such that a Turing machine A, or equivalently a mechanical algorithm A, can compute $f(x)$ in finite time, for all $x \in D$.

2. Some More Properties of the Kolmogorov Complexity K.

First some examples. Let all sequences $x \in \{0,1\}^n$. Let n be known to the computer. Let 0^n denote a sequence of n 0's. Examples:

1. $K(0^n \,|\, n) = c$ (some constant independent of n). $\qquad$ (6)

2. $K(\pi_1 \pi_2 \cdots \pi_n \,|\, n) = c$, where $\pi_1 \cdots \pi_n$ are the first n bits of π . $\qquad$ (7)

3. $K(1st \ n \ \text{bits of Shakespeare} \,|\, n) \approx n/4$. $\qquad$ (8)

4. $K(\alpha_1 \alpha_2 \cdots \alpha_n \,|\, n) = ?$,where α_i is the ith bit in the binary

    expansion of the fine structure constant $\alpha = e^2 \,/\, hc$. $\qquad$ (9)

5. $K(x_1, x_2 \cdots x_n \,|\, n) \le nh(\frac{1}{n} \sum_{i=1}^{n} x_i) + \log n + c$.

    If $X_i \sim \text{Bernoulli} \ (p)$, $\qquad$ (10)

    then $Pr\{| \frac{1}{n} K(X_1 X_2 \cdots X_n \,|\, n) - h(p) \,| > \epsilon\} \to 0$ $\qquad$ (11)

We investigate some additional properties of K. Again, we assume n known to the computer.

Proposition 1:

$$K(x(n)\,|\,n) \le n + c \,, \quad \text{for all} \quad x \in \{\,0,1\,\}^n \,.$$ $\qquad$ (12)

Proof: The program "Print $x_1 x_2 \cdots x_n$ " achieves the bound.

Proposition 2:

$$K(x) \le K(x \,|\, l(x)) + 2 \log l(x) + c \,, \quad \text{for all} \quad x \in \{\,0,1\,\}^* \,.$$ $\qquad$ (13)

Theorem: (Complexity version of law of large numbers) (Fine [6])

$$K(x_1 x_2 \cdots x_n \mid n) \geq n(1 - \epsilon) => \mid \frac{1}{n} \sum_{i=1}^{n} x_i - \frac{1}{2} \mid \leq \epsilon', \qquad (14)$$

where $\epsilon' \to 0$ as $\epsilon \to 0$. Thus high complexity sequences satisfy the most frequently used test for randomness.

3. Gambling on Patterns.

We shall now develop some properties of the function complexity defined in Equation (2). This section follows the development in Cover [7].

Given a domain D of patterns $D = \{x_1, x_2, \ldots, x_n\}$ and an unknown classification function $f : D \to \{0,1\}$ assigning the patterns to two classes, we ask for an intelligent way to learn f as the correctly classified elements in D are presented one by one. We ask this question in a gambling context in which a gambler, starting with one unit, sequentially bets a portion of his current capital on the classification of the new pattern. We find the optimal gambling system when f is known a priori to belong to some family F. We also exhibit a universal optimal learning scheme achieving $\exp_2(n - K(f \mid D) - log(n + 1))$ units for each f, where $K(f \mid D)$ is the length of the shortest binary computer program that calculates f on its domain D. In particular it can be shown that a gambler can double his money aproximately $n(1 - H(d/n))$ times, where $H(p) = -p \log p - (1-p) log(1-p)$, if f turns out to be a linear threshold function on n patterns in d-space.

Let F denote a set of (classification) functions $f : D \to \{0,1\}$. For example, F might be the set of all linear threshold functions. Let $|F|$ denote the number of elements in F.

The interpretation will be that D is the set of patterns, and $f(x)$ is the classification of the pattern x in D.

Consider the following gambling situation. The elements of D are presented in any order. A gambler starts with one dollar. The first pattern $x_1 \in D$ is exhibited. The gambler then announces amounts b_1 and b_0 that he bets on the true class being $f(x_1) = 1$ and $f(x_1) = 0$, respectively. Without loss of generality we can set $b_1 + b_0 = 1$. The true value $f(x_1)$ is then announced, and the gambler loses the incorrect bet and is paid fair odds (2 for 1) on the correct bet. Thus his new capital is

$$S_1 = \begin{cases} 2b_1, & f(x_1) = 1 \\ 2b_0, & f(x_1) = 0. \end{cases}$$

Now a new pattern element $x_2 \in D$ is exhibited. Again, the gambler announces proportions b_1 and b_0 of his current capital that he bets on $f(x_2) = 1$ and $f(x_2) = 0$ respectively. Without loss of generality, let $b_0 + b_1 = 1$. Thus the bet sizes are $b_1 S_1$ and $b_0 S_1$. Then $f(x)$ is announced and the gambler's new capital is

$$S_2 = \begin{cases} 2b_1 S_1, & f(x_2) = 1 \\ 2b_0 S_2, & f(x_2) = 0. \end{cases}$$

Continuing in this fashion, we define

$$b_1^{(k)}[\ x_k\,|\,(x_1, f(x_1)), \ \ldots, (x_{k-1}, f(x_{k-1}))]\ ,\quad x_k \in D\ ,$$

and

$$b_0^{(k)} = 1 - b_1^{(k)}, b_0^{(k)} \geq 0,\ b_1^{(k)} \geq 0,$$

as a gambling scheme that depends only on the previously observed properly classified (training) set.

The accrued capital after all patterns $x_1, x_2, \ldots, x_n$, $n = |D|$, have been observed is

$$S_k = \begin{cases} 2b_1^{(k)}S_{k-1}, & f(x_k) = 1 \\ 2b_0^{(k)}S_{k-1}, & f(x_k) = 0, \end{cases}$$

for $k = 1,2,...,n$ and $S_0 = 1$. Let

$$b = (\ (b_0^{(1)}, b_1^{(1)}\ ), (\ (b_0^{(2)}, b_1^{(2)}\ ), \ldots, (\ (b_0^{(n)}, b_1^{(n)}\ ),$$

denote a sequence of gambling functions.

Theorem 1: For any $F \subseteq D^{\{0,1\}}$, there exists a gambling scheme b^* achieving $S_n(f) = S^* = 2^{n-log|F|}$ units, for all f in F and for all orders of presentation of the elements $x \in D$. Moreover, there exists no b that dominates b^* for all f; thus, b^* is minimax. This gambling scheme is given by the expression

$$b_1^{(k)*}(x) = \frac{|\{g \in F : g(x_1) = f(x_1),\ i = 1,2,...,k-1,\ \text{and}\ g(x) = 1\}|}{|\{g \in F : g(x_1) = f(x_i),\ i = 1,2,...,k-1\}|}$$

Remark: This gambling scheme simply asserts at time k, "Bet all of the current capital on the hypotheses $f(x_k) = 1$ and $f(x_k) = 0$ in proportion to the number of functions g in F that *agree* on the training set and assign the new pattern x_k to classes $g(x_k) = 1$ and $g(x_k) = 0$ respectively."

The proof will not be given here but can be found in [1].

Applications and Examples:

1. Let F be all 2^n functions $f : D \rightarrow \{0,1\}$, where $n = |D|$. Then $log|F| = n$, and $S^* = 1$. No money can be gained. The training set gives no information about future pattern classifications. This is the worst case.

2. Let D denote a set of n vectors in Euclidean d-space R^d. Let us also assume that $\{x_1, x_2, \ldots, x_n\} = D$ is in *general position* in the sense that every d-element subset of D is linearly independent. Let F be the set of all

linear threshold functions on D; i.e., $f \in F$ implies there exists $w \in R^d$, $T \in R$, such that

$$f(x) = sgn(w^t x - T), \quad \forall \ x \in D ,$$

where

$$sgn(t) = \begin{cases} 1, & t \geq 0 \\ 0, & t < 0 . \end{cases}$$

Then from Cover [8], we have

$$|F| = 2 \sum_{k=0}^{d} \binom{n-1}{k} , \quad \forall \ d, n .$$

Using bounds derived from Stirling's approximation, it can be shown that

$$log\left(2 \sum_{k=0}^{d} \binom{n-1}{k} \right) \approx nH\left(\frac{d}{n}\right), \quad for \quad n \geq 2d ,$$

where $H(p) = -p \log p - (1-p)\log(1-p)$ is the Shannon entropy function. Thus we conclude, for $n \geq 2d$, that an amount $S_n = 2^{n(1 - H(d/n))}$ can be won if in fact the n patterns are linearly separable in R^d . Note also that $H(d/n)$ is the Kolmogorov complexity of most of the linear threshold function $f \in F$. Finally, we observe that S_n is not much greater than 1 until $n \geq 2d$, at which point the behavior of S_n is exponential. This is yet more evidence that $n = 2d$ is a natural definition of the capacity of a linear threshold pattern recognition device with d variable weights.

3.   Let F be the set of all functions $f : D \rightarrow \{0,1\}$ that can be represented by rth degree polynomial discriminant functions:

$$f(x) = sgn \left[\sum_{i_1, i_2, \ldots, i_r} w_{i_1 i_2 \cdots i_r} \ x_{i_1} x_{i_2} \cdots x_{i_r} - T \right]$$

If the elements of D are in general position with respect to rth degree polynomials, we see [8] that there are precisely $2 \sum_{k=0}^{d-1} \binom{n-1}{k}$ elements in F

where d' is the number of coefficients in an arbitrary rth degree polynomial in d variables. For example, for $r = 2$, $d = 2$, we have

$$f(x) = a_{11}x_1^2 + a_{22}x_2^2 + a_{12}x_1x_2 + a_1x_1 + a_2x_2 + a_0,$$

and $d' = 6$.

The point is that d' is the number of degrees of freedom of the manifold $\{x : f(x) = 0\}$. Again by the theorem, we have $S_n \geq 2^{n(1 - H(d'/n))}$, where now d' is the number of degrees of freedom of the family of separating surfaces F.

4. Suppose it is not known what degree polynomial is needed to classify D correctly. Since the degree r need take on only $(n + 1)$ values before the degree is sufficient to make an arbitrary assignment f, we merely invest an initial amount $1/(n + 1)$ in the betting system for each degree $r = 0,1,...,n$. Then the theorem becomes

$$S(f) > 2^{n(1 - H(d(f)/n)) - log(n + 1)}, \quad \text{for all} \quad f : D \rightarrow \{0,1\}$$

where $d(f)$ is the number of degrees of freedom of an rth degree polynomial, and r is the minimal degree necessary to yield f.

Theorem: These results are special cases of the following theorem:

Theorem: There exists a betting scheme b^* such that the total accumulated capital satisfies

$$S(f) \geq 2^{n - K(f \mid D) - log(n + 1)}.$$

Comment: If f is a linear threshold function, then

$$K(f \mid D) \leq log \, 2 \left(\sum_{k=0}^{d-1} \binom{n-1}{k} \right) + c.$$

Simply write a program saying "f is the ith function in the lexicographically ordered list of linear threshold functions on D^n. Thus i requires $\log 2 \left(\sum_{k=0}^{d-1} \binom{n-1}{k} \right)$ bits and c is the length of the rest of the program specified above.

Similarly, the polynomial threshold functions can be seen to be special cases of this theorem.

4. Kolmogorov's H_k Function.

Consider the function $H_k : \{0,1\}^n \rightarrow N$, $H_k(x) = \min_{p:l(p) \leq k} \log |S|$, where the minimum is taken over all subsets $S \subseteq \{0,1\}^n$, such that $x \in S$, $U(p) = S$, $l(p) \leq k$. This definition was introduced by Kolmogorov in a talk at the Information Theory Symposium, Tallin, Estonia, in 1974. Thus $H_k(x)$ is the log of the size of the smallest set containing x over all sets specifiable by a program of k or fewer bits. Of special interest is the value

$$k^*(x) = \min\{k : H_k(x) + k = K(x)\} .$$

Note that $\log |S|$ is the maximal number of bits necessary to describe an arbitrary element $x \in S$. Thus a program for x could be written in two stages: "Use p to print the indicator function for S; the desired sequence is the ith sequence in a lexicographic ordering of the elements of this set." This program has length $l(p) + \log |S|$, and $k^*(x)$ is the length of the shortest program p for which this 2-stage description is as short as the best 1-stage description p^*. We observe that x must be maximally random with respect to S -- otherwise the 2-stage description could be improved, contradicting the minimality of $K(x)$. Thus $k^*(x)$ and its associated program p constitute a minimal sufficient description for x.

Example: Let $x \in \{0,1\}^a$, $\sum_{i=1}^{n} x_i = k$. Then $k^*(x) \approx log(n+1)$, and the associated program is "S is the set of all $x \in \{0,1\}^a$ such that $\sum x_i = k$."

Arguments can be provided to establish that $k^*(x)$ and its associated set S^* describe all of the "structure" of x. The remaining details about x are conditionally maximally complex. Thus pp^{**}, the program for S^*, plays the role of a sufficient statistic.

References

[1] A.N. Kolmogorov, "Three Approaches to the Concept of the Amount of Information," *Problemy Peredachi Informatsii*, 1, (1965), pp. 3-11.

[2] A.K. Zhvonkin and L.A. Levin, "The Complexity of Finite Objects and the Development of the Concepts of Information and Randomness by Means of the Theory of Algorithms," Russian Mathematical Surveys 25, (1970), pp. 83-124.

[3] C.P. Schnorr, "A Unified Approach to the Definition of Random Sequences," *Math. Systems Theory*, 5, No. 3, (1971), pp. 246-258.

[4] R.J. Solomonoff, "A Formal Theory of Inductive Inference, Part I," *Information and Control*, 7, (1964), pp. 1-22.

[5] R.J. Solomonoff, "A Formal Theory of Inductive Inference, Part II," *Information and Control*, 7, (1964), pp. 224-254.

[6] T. Fine, Theories of Probability, 1974.

[7] T. Cover, "Generalization on Patterns Using Kolmogorov Complexity," *Proc. 1st Internatinal Joint Conference on Pattern Recognition*, Washington, D.C. (1973).

[8] T. Cover, "Geometrical and Statistical Properties of Linear Threshold Func-
     tions with Applications in Pattern Recognition," *IEEE Trans. Elec. Comp.*,
     (1965).

[9] T. Cover and S.K. Leung-Yan-Cheong, "Some Equivalences between Shan-
     non Entropy and Kolmogorov Complexity," *IEEE Trans. on Information
     Theory*, Vol. IT-24, No. 3, May 1978, pp. 331-338.

[10] T. Cover, "Universal Gambling Schemes and the complexity Measures of
     Kolmogorov and Chaitin," Technical Report No. 12, (1974) Dept. of Statis-
     tics, Stanford University.

[12] G. Chaitin, "A Theory of Program Size Formally Identical to Information
     Theory," *J. of the Assoc. for Computing Machinery* Vol. 22, No. 3, July 1975,
     pp. 329-341.

COMPLEXITY AND COHERENCE OF SEQUENCES

J. Ziv

Dept. of Electrical Engineering, Technion - Israel Institute of
Technology, Haifa, Israel.

Abstract

For every individual infinite sequence x, a quantity $\rho(x)$
is defined, called the normalized complexity (or compressibility)
of x, which is shown to be the asymptotically attainable lower
bound on the compression ratio (i.e., normalized encoded length)
that can be achieved for x by any finite-state information lossless encoder. This is demonstrated by a constructive coding theorem
and its converse that, apart from their asymptotic significance,
also provide a useful performance criteria for finite practical data
compression algorithms, and leads to a universal encoding algorithm
which is asymptotically optimal for all sequences.

This concept of complexity is shown to play a role analogous
to that of entropy in classical information theory where one deals
with probabilistic ensembles of sequences rather than with individual sequence.

The notion of compressibility of an individual sequence is
then generalized to the case where an additional side-information
sequence is available. For every individual infinite sequence x,
and given an infinite side information y, we define a quantity
$\rho(x(y))$ called the conditional compressibility of x given y.
It is shown to be the asymptotically attainable lower bound on the
compression ratio which can be achieved for x by any information
lossless finite state encoder, given y.

The conditional compressibility (conditional normalized complexity) which is a measure of the coherence between the two sequences is shown to play a role similar to that of conditional

entropy in classical information theory. Here again, the constructive coding theorem leads to a universal algorithm that is asymptotically optimal for all sequences.

This new concept leads to a universal decoding procedure for finite-memory channels. Although the channel statistics are unknown, universal decoding can achieve an error probability with an error exponent which (for large block length) is equal to the random-coding error exponent that is associated with the optimal maximum-likelihood procedure for the given channel.

1. INTRODUCTION

In this lecture, compressibility of individual sequences is investigated with respect to a broad class of encoders that can operate in a variable-rate mode as well as in a fixed-rate one and that allow for any finite-state scheme of variable-length-to-variable-length coding. No distortion is allowed, and the original data must be fully recoverable from the compressed image. This class of encoders can be modeled by the class of finite-state information-lossless generalized automata [1]. In our model, an encoder E is defined by a quintuple (S,A,B,g,f) where S is a finite set of states, A is a finite input alphabet, B is a finite set of output *words* over a finite output alphabet, g is the "next-state" function that maps $S \times A$ into S, and f is the output function that maps $S \times A$ into B.

By allowing the words in B to be of different finite lengths, we allow for block-to-variable coding, and by including in B the null-word λ (i.e., the "word" of length zero), we allow for any finite-state scheme of variable-to-variable coding.

When an infinite sequence $x = x_1 x_2 \ldots$, $x_i \in A$ is fed into $E = (S,A,B,g,f)$, the encoder emits an infinite sequence $y = y_1 y_2 \ldots$, $y_i \in B$ while going through an infinite sequence of states $z = z_1 z_2 \ldots$, $z_i \in S$ according to

$$y_i = f(z_i, x_i)$$

$$z_{i+1} = g(z_i, x_i), \quad i = 1, 2, \ldots$$

where z_i is the state of E when it "sees" the input symbol x_i.

A finite segment $x_i x_{i+1} \ldots x_j$, $1 \le i \le j$, of x will be denoted by x_i^j; similar notation will naturally apply to finite segments of other sequences. Following conventional practice, we shall extend the use of the encoder functions f and g to indicate the output sequence as well as the final state, which results from a given initial state and a finite sequence of input letters. For instance, we shall occasionally write $f(z_1,x_1^n)$ for y_1^n and $g(z_1,x_1^n)$ for z_{n+1}.

An encoder E is said to be *information lossless* (IL) if for all $z_1 \in S$ and all $x_1^n \in A^n$, $n \geq 1$, the triple $\{z_1, f(z_1,x_1^n), g(z_1,x_1^n)\}$ uniquely determines x_1^n. E is said to be *information lossless of finite order* (ILF) if there exists a finite positive integer n such that for all $z_1 \in S$ and all $x_1^m \in A^m$, the pair $z_1, f(z_1,x_1^m)$ uniquely determines the first input letter x_1. It is easy to verify that if E is ILF then it is also IL, but there exist IL encoders that are not ILF.

In the sequel, we assume the IL or the ILF property depending on which leads to a stronger result. For example, we prove a coding theorem (Theorem 2) by means of an ILF construction, while its converse (Theorem 1) is proved under the broader IL assumption.

To simplify the discussion without any real loss of generality, we also assume throughout that the output alphabet of the encoder is binary and that the initial state z_1 is a prescribed fixed member of the state-set S.

Given an encoder $E = (S,A,B.g.f)$ and an input string x_1^n, the *compression ratio* for x_1^n with respect to E is defined by

$$\rho_E(x_1^n) \stackrel{\Delta}{=} \frac{L(y_1^n)}{n \log \alpha} \tag{1}$$

where $\alpha = |A|$, $y_1^n = f(z_1,x_1^n)$, $L(y_1^n) = \sum_{i=1}^n \ell(y_i)$, and $\ell(y_i)$ is the length in bits of $y_i \in B$. (Note that when $y_i = \lambda$, $\ell(y_i) = 0$.) Here and elsewhere in the sequel, $\log K = \log_2 K$.

The minimum of $\rho_E(x_1^n)$ over the class $E(s)$ of all finite-state IL encoders with $|A| = \alpha$ and $|S| \leq s$ is denoted by $\rho_{E(s)}(x_1^n)$. That is

$$\rho_{E(s)}(x_1^n) \stackrel{\Delta}{=} \min_{E \in E(s)} \{\rho_E(x_1^n)\} . \tag{2}$$

Furthermore, let

$$\rho_{E(s)}(x) \stackrel{\Delta}{=} \limsup_{n \to \infty} \rho_{E(s)}(x_1^n) \tag{3}$$

and

$$\rho(x) \stackrel{\Delta}{=} \lim_{s \to \infty} \rho_{E(s)}(x) . \tag{4}$$

It is clear that for every individual sequence x, $0 \leq \rho(x) \leq 1$. This normalized quantity $\rho(x)$ that depends solely on x will be referred to as the (finite-state) *compressibility* (or *complexity*) of x. Theorem 1 (the converse-to-coding theorem) gives a lower bound on $\rho_{E(s)}(x_1^n)$. Theorem 2 (the coding theorem) demonstrates the existence of an asymptotically optimal universal ILF encoding scheme under which the compression ratio attained for x tends in the limit to the compressibility $\rho(x)$ of x for every x. It is important to note that apart from their asymptotic significance, the results of Theorems 1 and 2 also provide useful performance criteria for finite (and practical) data-compression tasks.

The concept of complexity as defined here, like the quantity H(·) in [2], plays a role analogous to that of entropy in classical information theory where one deals with probabilistic ensembles of sequences rather than with individual sequences. This analogy is reinforced by Theorems 3 and 4 in [1].

The results of [1] can be generalized to include the case where a side-information is available to the encoder and the decoder as follows:

The input sequence x and the side-information $u = u_1 u_2 \ldots$, $u_i \in D$ are fed into $E = (S,A,B,D,g,f)$, the encoder emits an infinite sequence y while going through an infinite sequence of states $z = z_1 z_2 \ldots$, $z_i \in S$, according to

$$y_i = f(z_i, x_i, u_i)$$

$$z_{i+1} = g(z_i, x_i, u_i), \quad i = 1, 2, \ldots \quad .$$

An encoder with side-information is said to be information lossless (IL) if for all $z_1 \in S$ and all $x_1^n \in A^n$ and any given $u_1^n \in D^n$ the quadruple $\{z_1, f(z_1, x_1^n, u_1^n), g(z_1, x_1^n, u_1^n), y_1^n\}$ uniquely determines x_1^n.

We assume again that the output of the encoder is binary and that the initial state z_i is a prescribed member of the state set S.

Given an encoder $E(S,A,B,D,g,f)$, an input string x_1^n and side information u_1^n the *conditional compression ratio* for x_1^n given u_1^n is defined by

$$\rho_E(x_1^n | u_1^n) \triangleq \frac{L(y_1^n)}{n \log \alpha} \qquad \text{(see Eq. (1))} \quad . \tag{5}$$

The minimum of $\rho_E(x_1^n | u_1^n)$ over the class $E(s)$ of all IL finite-state encoders is denoted by $\rho_{E(s)}(x_1^n | u_1^n)$. That is,

$$\rho_{E(s)}(x_1^n | u_1^n) = \min_{E \in E(s)} \{\rho_E(x_1^n)\} \quad . \tag{6}$$

Furthermore, let

$$\rho_{E(s)}(x|u) = \lim_{n\to\infty} \sup \rho_{E(s)}(x_1^n|u_1^n) \quad , \tag{7}$$

and let

$$\rho(x|u) = \lim_{s\to\infty} \rho_{E(s)}(x|u) \quad . \tag{8}$$

This quantity, which depends solely on x and u is called the conditional complexity (conditional compressibility) of x given u.

In Theorem 3 (converse to coding theorem), a lower bound on $\rho_{E(s)}(x_1^n|u_1^n)$ is derived. In Theorem 4 it is demonstrated that there exists an asymptotically optimal universal IL finite-state encoding scheme under which the compression ratio attained for x given u tends in the limit to $\rho(x|u)$ for every x and u. Clearly, the concept of conditional complexity plays a role analogous to that of conditional entropy in classical information theory where one deals with probabilistic ensembles of sequences rather than individual sequences.

Aside from the direct applications to universal data compression algorithms for sequences with side-information, this new concept leads to a universal decoding procedure for finite-memory channels. Although the channel statistics are unknown, universal decoding can achieve an error probability with an error exponent which (for large block-length) is equal to the random-coding error exponent that is associated with the optimal maximum-likelihood procedure for the given channel. This application is discussed in Section 3.

2.   STATEMENT AND DISCUSSION OF RESULTS

In the first part of this section, some of the results which first appeared in [1] are stated in the form of Theorem 1 and Theorem 2.

The first result is a lower bound on the compression ratio attainable by any encoder E, from the class E(s) of IL encoders with no more than s states, for any finite input string x_1^n over an alphabet A of α letters.

Theorem 1 (Converse-to-Coding Theorem): For every $x_1^n \in A^n$

$$\rho_{E(s)}(x_1^n) \geq \frac{c(x_1^n)+s^2}{n\log\alpha} \log \frac{c(x_1^n)+s^2}{4s^2} + \frac{2s^2}{n\log\alpha} \quad , \tag{9}$$

where $c(x_1^n)$ is the largest number of *distinct* strings (or "phrases") whose concatenation forms x_i^n. (The proof of this theorem is given in [1].

Corollary 1 (Compressibility Bounds):

$$\rho(x) = \lim_{s \to \infty} \rho_{E(s)}(x) \geq \limsup_{n \to \infty} \frac{c(x_1^n) \log c(x_1^n)}{n \log \alpha} \tag{10a}$$

$$\rho(x) \geq \limsup_{n \to \infty} \limsup_{k \to \infty} \frac{1}{kn \log \alpha} \cdot \sum_{i=0}^{K-1} c(x_{in+1}^{(i+1)n}) \log c(x_{in+1}^{(i+1)n}) . \tag{10b}$$

An interesting application of the compressibility bound is the use of (11) to identify certain infinite sequences x that, while being rather easy to describe, satisfy $\rho(x) = 1$ and thus are incompressible by any finite-state IL encoder. To illustrate this point, let $u(k)$ denote the binary sequences of length $k2^k$ that lists, for example, in lexicographic order, all the 2^k binary words of length k, and let

$$u_1^{n(k)} = u(1)u(2)\ldots u(k)$$

where

$$n(k) = \sum_{i=1}^{k} i 2^i = (k-1)2^{k+1} + 2 .$$

It is easy to verify that when each $u(i)$ is parsed into its 2^i distinct i-tuples, we obtain a parsing of $u_1^{n(k)}$ into a maximum number of distinct phrases, namely,

$$c(u_i^{u(k)}) = \sum_{i=1}^{k} 2^i = 2^{k+1} - 2 . \tag{11}$$

For example, $u_1^{n(3)}$ is parsed as follows:

$$u_1^{n(3)} = 0,1,00,01,10,11,000,001,010,011,100,101,110,111 , \tag{12}$$

For this particular sequence u, inequality (10-a) implies

$$\rho(u) \geq \lim_{k \to \infty} \frac{(2^{k+1}-2) \log(2^{k+1}-2)}{(k-1)2^{k+1} + 2} = 1 .$$

In our next result we employ a universal compression algorithm 1 to demonstrate the existence of an ILF compression scheme under which, for every x, the compression ratio attained for x_1^n tends to $\rho(x)$ as n tends to infinity.

Theorem 2 (Coding Theorem): For every $n > 0$ there exists a finite-state ILF encoder $\mathcal{E}$ with $s(n)$ states that implements a block-to-variable code with the following performance characteristics.

i) For any given block length n and every input block x_1^n, the compression-ratio attained by $\mathcal{E}$ satisfies

$$\rho_{\mathcal{E}}(x_1^n) \leq \frac{c(x_1^n) + 1}{n \log \alpha} \log(2\alpha(c(x_1^n) + 1)) \; ; \tag{13}$$

the compression ratio attained for successive blocks $x_{(i-1)n+1}^{in}$, $i = 1, 2, \ldots,$ satisfies the same inequality with $x_{(i-1)n+1}^{in}$ replacing x_1^n .

ii) For every finite s,

$$\rho_{\mathcal{E}}(x_1^n) \leq \rho_{E(s)}(x_1^n) + \delta_s(n) \tag{14}$$

where

$$\lim_{n \to \infty} \delta_s(n) = 0 \; .$$

iii) Given an infinite input sequence x, let $\rho_{\mathcal{E}}(x, n)$ denote the compression ratio attained for x by $\mathcal{E}$ while feeding $\mathcal{E}$ with successive input blocks of length n. Then for any $\varepsilon > 0$

$$\rho_{\mathcal{E}}(x, n) \leq \rho(x) + \delta_{\varepsilon}(x, n) \tag{15}$$

where

$$\lim_{n \to \infty} \delta_{\varepsilon}(x, n) = \varepsilon \; .$$

Proof: The proof is constructive, and before going into computational details, we present a short outline of the construction. For the encoder $\mathcal{E}$, we employ an ILF finite-state machine that realizes a concatenated coding scheme by combining a fixed block-to-variable inner code. The inner code is used to encode sequentially and state dependently growing segments of an input block of relatively large length n. Upon completion of a block the machine returns to its initial state, thus "forgetting" all past history before starting to encode the next input block.

The segments of a block that serve as input words of the inner code are determined according to a so-called *incremental parsing* procedure. This procedure is sequential, and it creates a new phrase as soon as a prefix of the still unparsed part of the string differs from all preceding phrases. The parsing is indicated as

$$x_1^n = x_{n_0+1}^{n_1} x_{n_1+1}^{n_2} x_{n_2+1}^{n_3} \cdots x_{n_p+1}^{n_{p+1}}, \quad n_o \overset{\Delta}{=} 0, \; n_{p+1} \overset{\Delta}{=} n \,, \tag{16}$$

and is called incremental if the first p words $x_{n_{j-1}+1}^{n_j}$, $1 \le j \le p$ are all distinct and if for all $j = 1,2,\ldots,p+1$ when $n_j - n_{j-1} > 1$ there exists a positive integer $i < j$ such that $x_{n_{i-1}+1}^{n_i} = x_{n_{j-1}+1}^{n_j-1}$.

It is clear from this definition that if (16) is an incremental parsing of x_1^n, then $n_1 = 1$. The last word $x_{n_p+1}^n$ may or may not be distinct from the first p words, and for every word of length $\ell > 1$, its prefix of length $\ell - 1$ can be found as an earlier word of the parsing. For example, (12) is an incremental parsing of $u_1^{n(3)}$.

Now let $x_{n_{-1}+1}^{n_o} \overset{\Delta}{=} \lambda$, the word of length zero, and (since $x_1^n = \lambda x_1^n$) let us adopt the convention that $x_{n_{-1}+1}^{n_o}$ is always the initial word of an incremental parsing. Also given a word w, let $d(w)$ denote the word obtained by deleting the last letter of w. It follows that for every $j = 1,2,\ldots,p+1$ there exists a unique nonnegative integer $\pi(j) = i < j$ such that $d(x_{n_{j-1}+1}^{n_j}) = x_{n_{i-1}+1}^{n_i}$.

The incremental parsing of a given block x_1^n and the coding of the words determined by it are executed sequentially as follows. To determine the j-th word, $k \le j \le p+1$, we take n_j to be the largest integer, not exceeding n, for which $d(x_{n_{j-1}+1}^{n_j})$ equals some earlier word, for example, $x_{n_{i-1}+1}^{n_i}$, and we set $\pi(j) = i$ (e.g., for $j = 1$, $n_1 = 1$, $x_1^1 = x_1$, $d(x_1) = \lambda$, and $\pi(1) = 0$). The information which is transmitted to the decoder is the value of $\pi(j)$ and the value of the last letter of the j-th word, namely x_{n_j}. It follows that [1]:

$$\rho_{\mathscr{E}}(x_1^n) \le \frac{c(x_1^n) + 1}{n \log \alpha} \log[2\alpha(c(x_1^n) + 1)] \,,$$

which proves (13).

From (9) and (13), after some manipulation, we obtain [1]

$$\rho_{\mathscr{E}}(x_1^n) \le \rho_{\mathscr{E}(s)}(x_1^n) + \delta_s(n) \,. \tag{17}$$

where

$$\lim_{n \to \infty} \delta_s(n) = 0 \,, \tag{18}$$

which proves (14).

Finally, the compression attained by our encoder $\mathcal{E}$ for an infinite sequence x is by definition:

$$\rho_{\mathcal{E}}(x,n) = \limsup_{k\to\infty} \frac{1}{kn \log \alpha} \sum_{i=1}^{k} L_i \qquad (19)$$

where L_i is the number of bits used to encode the i-th block $x_{(i-1)n+1}^{in}$ of x. Since (15) and (16) hold for any input block of length n, we can write

$$\frac{L_i}{n \log \alpha} = \rho_{\mathcal{E}}(x_{(i-1)n+1}^{in}) \leq \rho_{E(s)}(x_{(i-1)n+1}^{in}) + \delta_s(n) . \qquad (20)$$

From (19) and (20) we obtain

$$\rho_{\mathcal{E}}(x,n) \leq \delta_s(n) + \limsup_{k\to\infty} \frac{1}{K} \sum_{i=1}^{k} \rho_{E(s)}(x_{(i-1)n+1}^{in}) ,$$

which reduces to

$$\rho_{\mathcal{E}}(x,n) \leq \delta_s(n) + \rho_{E(s)}(x) . \qquad (21)$$

where $\lim_{n\to\infty}\delta_s(n) = 0$. By (4), we can write $\rho_{E(s)}(x) = \rho(x) + \delta_s^*(x)$, where $\lim_{s\to\infty}\delta_s^*(x) = 0$ for all x. Hence given x and any $\varepsilon > 0$, there always exists a sufficiently large finite s for which $\delta_s^*(x) \leq \varepsilon$. Since (21) holds for every finite s, it follows that for any $\varepsilon > 0$, we can write

$$\rho_{\mathcal{E}}(x,n) \leq \rho(x) + \varepsilon + \delta_s(n) ,$$

which proves (5) with $\delta_{\mathcal{E}}(x,n) \overset{\Delta}{=} \varepsilon + \delta_s(n)$ and completes the proof of the theorem

Q.E.D.

It is easy to verify that inequality (10-b) and the proof of Theorem 2 also imply the following corollary.

Corollary 2: Let $p(x_i^j)$ denote the number of phrases in the incremental parsing of x_i^j. Then for every infinite sequence x,

$$\rho(x) = \limsup_{n\to\infty} \limsup_{k\to\infty} \frac{1}{kn \log \alpha} \cdot \sum_{i=0}^{k} p(x_{in+1}^{(i+1)n}) \log p(x_{in+1}^{(i+1)n}) .$$

We proceed now to generalize the results of [1] and to discuss the case where a side-information sequence u is available to the finite-state IL encoder and the decoder.

Consider the sequence $w = w_1 w_2 \ldots$, $w_i \in A \times D$ of ordered pairs

$$w_i = x_i, u_i; \quad i = 1, 2, \ldots$$

and let $c(w_1^n)$ be the number of phrases generated by a parsing of w_1^n into distinct phrases.

For example if:

$$x_1^n = 00101100$$

and

$$u_1^n = 00000100$$

an *incremental* parsing of w_1^n yields:

$$\begin{array}{|c|c|c|c|} \hline 0 & 01 & 011 & 00 \\ \hline 0 & 00 & 001 & 00 \\ \hline \end{array} \tag{22}$$

and $c(w_1^n) = 4$.

Let $c(u_1^n)$ be the number of distinct phrases in the parsed u_1^n and let $u(\ell)$ denote the ℓ-th distinct phrase of u_1^n $(1 \le \ell \le c(u_1^n))$. In the above example,

$$c(u_1^n) = 3; \quad u(1) = 0; \quad u(2) = 00; \quad u(3) = 001 \quad .$$

Let $c_\ell(x_1^n | u_1^n)$ be the number of times that the phrase $u(\ell)$ appears in the parsed u_1^n. In the above example, $c_1(x_1^n | u_1^n) = 1$, $c_2(x_1^n | u_1^n) = 2$; $c_3(x_1^n | u_1^n) = 1$. Clearly $c_\ell(x_1^n | u_1^n)$ is the number of distinct phrases of x_1^n that appears jointly with $u(\ell)$ and

$$\sum_{\ell=1}^{c(u_1^n)} c_\ell(x_1^n | u_1^n) = c(w_1^n) = c(x_1^n, u_1^n) \quad .$$

Theorem 3 (Converse-to-Coding Theorem): For every $x_1^n \in A^n$, every side-information sequence $u_1^n \in D^n$, and every parsing of w_1^n into distinct phrases

$$\rho_{E(s)}(x_1^n | u_1^n) \ge \sum_{\ell=1}^{c(u_1^n)} \left[\frac{c_\ell(x_1^n | u_1^n) + s^2}{n \log \alpha} \log \frac{c_\ell(x_1^n | u_1^n) + s^2}{4s^2} \right] + \frac{2s^2 c(u_1^n)}{n \log \alpha} \quad . \tag{23}$$

Proof: All the $c_\ell(x_1^n | u_1^n)$ that correspond to $u(\ell)$ are distinct. Thus, by Theorem 1, the total length $L(\ell)$ of the encoded output strings that corresponds to the $C_\ell(x_1^n | u_1^n)$ input phrases is lower bounded by

$$L(\ell) \ge (c_\ell(x_1^n | n_1^n) + s^2) \log \frac{(c_\ell(x_1^n | u_1^n) + s^2)}{4s^2} + 2s^2 \quad .$$

Therefore,

$$L(y_1^n) = \sum_{\ell=1}^{c(u_1^n)} L(\ell) \geq \sum_{\ell=1}^{c(u_1^n)} (c_\ell(x_1^n|u_1^n) + s^2) \log \frac{c_\ell(x_1^n|u_1^n) + s^2}{4s^2}$$

$$+ 2s^2 c(u_i^n) \quad . \qquad \square$$

Corollary 3 (Conditional Compressability Bounds):

$$\rho(x|u) \geq \limsup_{n\to\infty} \sum_{\ell=1}^{c(u_1^n)} \frac{c_\ell(x_1^n|u_1^n)}{n \log \alpha} \log c_\ell(x_1^n|u_1^n) \qquad (24\text{-a})$$

$$\rho(x|u) \geq \limsup_{n\to\infty} \limsup_{k\to\infty} \frac{1}{kn \log \alpha} \sum_{i=o}^{k-1} \sum_{\ell=1}^{c(u_{in+1}^{(i+1)n})} c_\ell\left(x_{in+1}^{(i+1)n}\big|u_{in+1}^{(i+1)n}\right) \cdot$$

$$\cdot \log c_\ell\left(x_{in+1}^{(i+1)n}\big|k_{in+k}^{(i+1)n}\right) \quad . \qquad (24\text{-b})$$

In the next result, we employ a variant of the Lempel-Ziv encoding algorithm, denoted by $\mathcal{E}$, to demonstrate the existence of an information-lossless, finite-state compression scheme which operates on successive n-blocks of x under which, for every x and u, the conditional compression ratio attained for x tends to $\rho(x|u)$ as the block-length n tends to infinity.

The encoding algorithm $\mathcal{E}$ based on an incremental parsing of $w_1^n = (x_1^n, u_1^n)$ yielding $c(w_1^n)$ *distinct* phrases (the last phrase of w_1^n generated by incremental parsing might not be distinct): The side-information u_1^n is available to the decoder and we assume that all the earlier phrases have already been decoded.

Therefore, the prefix $w_{n_i+1}^{n(i+1)}{}^{-1}$ of the current phrase $w_{n_i+1}^{n(i+1)} = {}^{n_{i+1}}$ (resulting by the deletion of the last symbol $w_{n(i+1)}$ = $x_{n(i+1)}$, $u_{n(i+1)}$ of the current phrase) can be regenerated at the decoder by pointing to the *one* earlier phrase which is identical to this prefix. This is done by first informing the decoder of the length of this prefix, denoted by $L(u(\ell))$, where $u(\ell)$ is the u-phrase which is associated with the prefix $w_{n_i+1}^{n(i+1)}{}^{-1}$ namely $u_{n_i+1}^{n(i+1)}{}^{-1}$ Once the $L(u(\ell))$ is known, $u(\ell)$ itself is known (since u_1^n is available at the receiver!). The decoder is then informed of the serial number of the *one earlier phrase* which is identical with the prefix $w_{n_i+1}^{n(i+1)}$ among the $c_\ell(x_1^n|u_1^n)$ phrases which are characterized by the same associated u-phrase $u(\ell)$.

Once the prefix of the current phrase is regenerated at the decoder, the last x-letter x_{n_i+1} of the phrase $w^n(i+1)$ is regenerated by informing the decoder of its value. n_i+1

The total length of the codeword for the current phrase is thus upper-bounded by

$$\lceil \log L(u(\ell)) \rceil + \lceil \log(c_\ell(x_1^n|u_1^n) + 1) \rceil + \lceil \log \alpha \rceil \ .$$

In the example (22) we have for the last phrase that $u(\ell) = u(1) = 0, \quad L(u(1)) = 1, \quad C_1(x_1^n|u_1^n) = 1, \quad \alpha = 2.$

Thus, the length of the codeword for x_1^n, given u_1^n, is upper-bounded by:

$$L(y_1^n) \leq \sum_{\ell=1}^{c(u_1^n)} \left\{ (c_\ell(x_1^n|u_1^n) + 1)\log(c\ (x_1^n|u_1^n) + 1) + \log L(u(\ell)) + 3 \right\}$$

$$(25)$$

Now, by the convexity of the logarithmic function,

$$\sum_{\ell=1}^{c(u_1^n)} c_\ell(x_1^n|u_1^n)\log L(u(\ell)) \leq c(w_1^n)\log \frac{n}{c(w_1^u)} \ .$$

Also, by Eq. (6) in [1],

$$c(w_1^n) \leq O\left(\frac{n}{\log n}\right) \ .$$

Thus,

$$\frac{1}{n} \sum_{\ell=1}^{c(u_1^n)} (c_\ell(x_1^n|u_1^n) + 1)\log L(u(\ell)) \leq O\left(\frac{\log \log n}{\log n}\right) \ . \qquad (26)$$

Therefore, by Eqs. (7), (25), (24-b) and (26), the following theorem can be stated:

Theorem 4: For any given length $n > 0$, there exists a finite-state IL encoder $\mathcal{E}$ with $s(n)$ states that implements a block-to-variable code with the following characteristics.

i) For any given block length n and every input block x_1^n and side-information u_1^n, the compression-ratio attained by by $\mathcal{E}$ satisfies, for any finite s,

$$\rho_{\mathcal{E}}(x_1^n|u_1^n) \leq \rho_{E(s)}(x_1^n|u_1^n) + \delta_s(n) \qquad (27)$$

where

$$\lim_{n\to\infty} \delta_s(n) = 0 .$$

ii) Given an infinite input sequence x, let $\rho_{\mathscr{E}}(x,n|u)$ denote the compression ratio attained for x by $\mathscr{E}$ while feeding $\mathscr{E}$ with successive input blocks of length n. Then, for any $\varepsilon < 0$,

$$\rho_{\mathscr{E}}(x,n|u) \leq \rho(x|u) + \delta_\varepsilon(x,u,n) , \qquad (28)$$

where

$$\lim_{n\to\infty} \delta_\varepsilon(x,u,n) = \varepsilon .$$

3. APPLICATIONS: UNIVERSAL DECODING

Consider the class P_K of finite alphabet channels with finite memory which are characterized by a transitional probability distribution of the form

$$W(y_i|x_1^i) \in P_K \triangleq \{W(y_i|x_1^i): W(y_i|x_1^i) = W(y_1|x_{i-K}^i\} \quad \text{for every } i > K,$$

where

1) x_i is the input to the channel at the i-th instant
   $x_i \in X; \quad |X| = \alpha.$

2) $x_i^j = x_i, x_{i+1}, \ldots, x_j.$

3) y_i is the output of the channel at the i-th instant
   $y_i \in Y; \quad |Y| = \beta.$

Here K is the channel memory. A channel is memoryless if $K = o$.

Let $\underline{x} = x_1^n$ and consider a code $C = \{\underline{x}^1, \underline{x}^2, \ldots, \underline{x}^\ell, \ldots \underline{x}^M\}$ for $M = 2^{nR}$ equiprobable messages where $\underline{x}^\ell \in \bar{X}$; $\ell = 1,2,\ldots,M$. It is well known that the optimal decoding procedure that minimizes the probability of error is the maximum likelihood decoding procedure where the decoded code word is the one $\underline{x}' \in C$ for which

$$f(\underline{x}',\underline{y}) = \min_{\hat{\underline{x}} \in C} f(\hat{\underline{x}},\underline{y}) , \qquad (29)$$

where the function $f(\underline{x},\underline{y})$ is given by:

$$f(\underline{x},\underline{y}) = -\log(W(\underline{y}|\underline{x})) . \qquad (30)$$

The probability of error which is associated with this optimal decoding procedure is denoted by $P_{e,o}(C,R,n)$.

In general, for large $M = 2^{nR}$, it is very hard to find the optimal code which minimizes $P_{e,o}(C,R,n)$ over all codes with 2^{nR} code words. However, it has been demonstrated [2] that the expectation of $P_{e,o}(C,R,n)$ over the ensemble of randomly selected codes where each code-word is generated independently at random, governed by some probability distribution $q(\underline{x})$, has the following interesting properties:

Let the expectation of $P_{e,o}(C,R,n)$ with respect to $q(\cdot)$ be denoted by $\bar{P}_e(q,R,n)$. Then,

a) For memoryless channels $(K = 0)$, there exists an $R(q)$ such that

$$-\frac{1}{n} \log \bar{P}_e(q,R,n) \geq E(q,R) > 0; \quad R < R(q) \quad . \tag{31}$$

Hence, the expectation of the probability of error decreased exponentially with n for $R < R(q)$. Furthermore

$$\max_{q(\cdot)} \left[-\frac{1}{n} \log \bar{P}_e(q,R,n) \right] = -\lim_{n \to \infty} \frac{1}{n} \log \{ \min_{C} P_{e,o}(C,R,n) \}$$

$$\text{for } R_o \leq R \leq R_{max}, \tag{32}$$

where R_{max} is the channel capacity.

b) For finite-memory channels (i.e. $K \geq 1$)

$$-\frac{1}{n} \log \bar{P}_e(q,R,n) \geq E(q,R) > 0 \quad \text{for} \quad R < R(q) \quad , \tag{33}$$

where now $\max_q R(q)$ is not necessarily equal to R_{max}

In some cases, the channel's transition probability distribution $W(\underline{y}|\underline{x})$ is not available. In such cases, maximum likelihood decoding cannot be utilized and the optimizing channel-input distribution $q(\cdot)$ is not known.

It is therefore assumed that $q(\cdot)$ is a uniform probability distribution, i.e. $q(\underline{x}) = \frac{1}{|B|}$ for every input vector $\underline{x} \in X^n$, and therefore $q(\underline{x}) = \prod_{i=1}^{n} Q(x_i); \quad \underline{x} = x_1, x_2, \ldots, x_n$ and where $Q(x) = \frac{1}{|X|} = \frac{1}{\alpha}$ for every $x \in X$.

In the following, a universal decoding procedure is described which yields essentially the same random-coding error exponent as that which is associated with the optimal, maximum likelihood procedure for *any finite-memory* channel. (See [3] for universal decoding for *memoryless* channels only.)

The function $f(x,y)$ in Eq. (30) is now replaced by a universal function $u(\underline{x},\underline{y})$ (which does not depend on the channel statistics) as follows: Let

$$u(\underline{x},\underline{y}) = \frac{1}{n} \sum_{\ell=1}^{C(\underline{y})} C_\ell(\underline{x}|\underline{y}) \log(C\ (\underline{x}|\underline{y})) \ , \tag{34}$$

where $C_\ell(\underline{x}|\underline{y})$ and $C(\underline{y})$ are associated with incremental parsing of $(\underline{x},\underline{y})$, as in Theorem 4, and let $\bar{P}_{e,u}(R,n)$ denote the probability of error which is associated with this universal decoding procedure and random coding with $Q(x) = 1/\alpha$.

Then,

Theorem 5: For any finite-memory channel (finite K),

$$\lim_{n \to \infty} \frac{1}{n} \log \bar{P}_{e,u}(R,n) = \lim_{n \to \infty} \frac{1}{n} \log \bar{P}_e(q,R,n); \quad Q(x) = \frac{1}{\alpha} \ .$$

The proof of Theorem 1 appears in [4].

REFERENCES

1. Ziv, J. and Lempel, A. Compression of Individual Sequences via Variable-Rate Coding. IEEE Trans. Inform. Theory, 25, 5 (1978), 530-536.

2. Gallager, R.G. Information Theory and Reliable Communication. (Wiley, 1968).

3. Csiszar, I. and Körner, J. Information Theory: Coding Theorems for Discrete Memoryless Systems. (Academic Press, 1981).

4. Ziv. J. Universal Decoding for Channels with Finite Memory. (Final version in preparation.)

A CLASS OF COMPOSITE SEQUENCES AND ITS IMPLICATIONS FOR MATCHED FILTER SIGNAL PROCESSING

Maurice Beale

Electrical Engineering Laboratories, University of Manchester, M13 9PL, U.K..

1   INTRODUCTION

In direct-sequence spread-spectrum multiple-access (SSMA) communication systems, a unique code sequence is assigned to each user (transmitter-receiver pair). A transmitter modulates its code sequence with the data to be transmitted, usually by sequence-inversion keying (SIK), and then procedes with RF carrier modulation (e.g., using PSK or MSK). The corresponding receiver recovers the data by either matched filtering (using a filter matched to the code sequence) or active correlation (using a locally-generated synchronous replica of the code sequence). The performance of such a system depends upon the correlation properties of the set of sequences used. Specifically, the out-of-phase values of the autocorrelation function (ACF) of each sequence are required to be small in magnitude to ease problems associated with the synchronisation circuits at a receiver and, in some applications, to provide protection against multipath interference. Also, the cross-correlation function (CCF) between each pair of sequences in the set must be small in magnitude, for all relative phase-shifts, to minimise interference between users in the face of unpredictable (often time-varying) delays in the communication channel.

Certain SSMA applications call for long code sequences to ensure sufficient protection against external interference and enable an adequate number of users to be accommodated. However, long codes usually imply a long synchronisation delay at a receiver, at the start of each transmitted message, and such long delays may not be tolerable. The class of Composite sequences discussed here may find application to such systems by virtue of their rapid-acquisition structure (see section 6). The correlation properties of these sequences (see sections 3-5) are such that no performance penalty, in terms of average data error-rates, need be incurred. Indeed, by careful signal design, a Composite sequence SSMA system can outperform a system using conventional sequences (e.g., m-sequences).

Finally, the same structural properties which enable fast synchronisation procedures to be adopted can be used to advantage in the design of matched filters for Composite sequences, as discussed in section 7. These results imply that Composite sequences may be useful for a much wider range of applications (e.g., measuring the impulse response of a communication channel, system identification for control systems, etc.), since they enable matched filters to be implemented for time-bandwidth (TB) products significantly greater than would be technologically feasible if conventional sequences were used.

2   THE COMPOSITE SEQUENCE STRUCTURE

Let $\{A\}$ be a repetitive sequence of binary elements (chips) $A(j) \in \{+1, -1\}$, of period n_a, such that $A(j) = A(j + Kn_a)$ for all integers K and for all $j \bmod n_a$. Similarly, let $\{B\}$ be another repetitive sequence of binary elements $B(k) \in \{+1, -1\}$, of period n_b, with $B(k + Kn_b) = B(k)$ for all integers K and for all $k \bmod n_b$.

A Composite sequence $\{R\}$, with elements $R(i) \in \{+1, -1\}$, is constructed by sequence-inversion keying the 'inner' component sequence $\{A\}$ with the 'outer' component sequence $\{B\}$. The chip-rate of $\{B\}$ is made n_a times slower than that of $\{A\}$, so that one chip of $\{B\}$ has the same duration as one complete period of $\{A\}$. Note that $\{R\}$ has the same chip-rate as $\{A\}$, but that one period of $\{R\}$ comprises $n_r = n_a n_b$ chips and has the same duration as one period of $\{B\}$. The i<u>th</u> chip of $\{R\}$, for $0 \leqslant i \leqslant n_r - 1$, can be expressed as:-

$$R(i) = A(j).B(k) \quad \text{for } 0 \leqslant j \leqslant n_a - 1, \ 0 \leqslant k \leqslant n_b - 1$$

$$\text{with } i = kn_a + j. \tag{1}$$

Since this construction is equivalent to forming the Kronecker product of $\{A\}$ and $\{B\}$, Stark and Sarwate [3] have referred to sequences formed in this way as Kronecker sequences.

We can generalise the construction in two ways: firstly, the component sequences $\{A\}$ and $\{B\}$ need not necessarily be binary, and most generally could be complex-valued; secondly, we need not restrict the process to just two component sequences, but could go on to SIK the Composite sequence $\{R\}$ with a third component sequence, and so on.

3   CORRELATION FUNCTIONS FOR COMPOSITE SEQUENCES

As discussed earlier, for SSMA system applications, we are
interested in the ACF and CCF behaviour of sets of sequences. The
relevant correlation functions are first defined.

The _periodic_ (or even) CCF between two, generally complex-
valued sequences {x} and {y}, of common period P chips, can be
defined as:

$$\chi_{x,y}(\tau) = \sum_{i=0}^{P-1} x(i).[y(i + \tau)]^*, \tag{2}$$

where * represents complex conjugation. The periodic ACF for {x},
$\chi_x(\tau)$, is defined by (2) with y = x. The periodic CCF specifies
the interference due to sequence {y} on a receiver matched (and
synchronised) to sequence {x}. However, in SSMA systems, the
sequences are modulated by data, and the interference is only given
by the periodic CCF when there is no data transition (i.e.,
sequence inversion) during the correlation interval. When such a
data transition occurs, the interference is determined by the _odd_
CCF [2, 4-7], defined below. It is convenient first to define the
aperiodic CCF:-

$$\rho_{x,y}(\tau) = \begin{cases} \sum_{i=0}^{P-1-\tau} x(i).[y(i + \tau)]^* & \text{for } 0 \leqslant \tau \leqslant P-1 \\ \sum_{i=0}^{P-1+\tau} x(i - \tau).[y(i)]^* & \text{for } 1-P \leqslant \tau < 0 \\ 0 & \text{for } |\tau| \geqslant P. \end{cases} \tag{3}$$

It can be seen that the aperiodic CCF is evaluated over only a part
of the sequence period. In terms of this function, the _odd_ CCF is
defined as:-

$$\hat{\chi}_{x,y}(\tau) = \rho_{x,y}(\tau) - \rho_{x,y}(\tau - P), \quad \text{for } 0 \leqslant \tau \leqslant P. \tag{4}$$

Since this is the _difference_ between two partial correlation
functions, it corresponds to the interference experienced by a

receiver matched to $\{x\}$, due to sequence $\{y\}$, when part of the latter sequence is inverted by a data transition. The aperiodic and odd ACFs for sequence $\{x\}$ are given by (3) and (4), respectively, with $y = x$.

Now, let $\{R_s\}$ be a Composite sequence constructed as described in section 2 from an 'inner' component sequence $\{A_s\}$ of period n_a and an 'outer' component sequence $\{B_s\}$ of period n_b. We consider a pair of such Composite sequences by letting $s = 1,2$. The structure of these Composite sequences is such that their correlation behaviour can be expressed entirely in terms of that of their component sequences. For example, in references 1 and 2 it is shown that the periodic CCF between Composite sequences $\{R_1\}$ and $\{R_2\}$ can be expressed as:-

$$\chi_{R_1,R_2}(\tau) = \rho_{A_1,A_2}(\tau_a)\chi_{B_1,B_2}(\tau_b) + \rho_{A_1,A_2}(\tau_a - n_a)\chi_{B_1,B_2}(\tau_b + 1),$$

$$(5)$$

where $0 \leqslant \tau_a \leqslant n_a - 1$, $0 \leqslant \tau_b \leqslant n_b - 1$ and $\tau = \tau_b n_a + \tau_a$.

Similarly, the aperiodic CCF between $\{R_1\}$ and $\{R_2\}$ can be expressed as follows [2,3]:-

$$\rho_{R_1,R_2}(\tau) = \rho_{A_1,A_2}(\tau_a)\rho_{B_1,B_2}(\tau_b) + \rho_{A_1,A_2}(\tau_a - n_a)\rho_{B_1,B_2}(\tau_b + 1),$$

$$(6)$$

in which $\tau = \tau_b n_a + \tau_a$, with $0 \leqslant \tau_a \leqslant n_a$ and $-n_b \leqslant \tau_b \leqslant n_b - 1$, so that $-n_a n_b \leqslant \tau \leqslant n_a n_b$.

We can also express the odd CCF between $\{R_1\}$ and $\{R_2\}$ in terms of the aperiodic CCF for their 'inner' component sequences and the odd CCF for their 'outer' component sequences [2,3], as follows:

$$\hat{\chi}_{R_1,R_2}(\tau) = \rho_{A_1.A_2}(\tau_a)\hat{\chi}_{B_1,B_2}(\tau_b) + \rho_{A_1,A_2}(\tau_a - n_a)\hat{\chi}_{B_1,B_2}(\tau_b + 1),$$

$$(7)$$

for $0 \leqslant \tau_a \leqslant n_a - 1$, $0 \leqslant \tau_b \leqslant n_b - 1$, with $\tau = \tau_b n_a + \tau_a$.

The periodic, aperiodic and odd ACFs for a Composite sequence $\{R_1\}$ can be obtained by putting $R_2 = R_1$, $A_2 = A_1$ and $B_2 = B_1$ in equations (5)-(7).

All the above results are valid for the general class of complex-valued Composite sequences, when each is constructed from two component sequences. The extension to the case of Composite sequences constructed from more than two component sequences is straightforward. For example, consider a pair of Composite sequences $\{Q_s\}$, $s = 1,2$, each of period $n_q = n_a n_b n_c$ and constructed from three component sequences: $\{A_s\}$ of period n_a, $\{B_s\}$ of period n_b and $\{C_s\}$ of period n_c, in the following way. First, $\{A_s\}$ is SIK'd by $\{B_s\}$ to give a Composite sequence $\{R_s\}$, of period $n_r = n_a n_b$, as before. Then $\{R_s\}$ is SIK'd by $\{C_s\}$ to give $\{Q_s\}$. The periodic CCF between $\{Q_1\}$ and $\{Q_2\}$ can be expressed via equation (5) as:-

$$\chi_{Q_1,Q_2}(\tau) = \rho_{R_1,R_2}(\tau_r)\chi_{C_1,C_2}(\tau_c) + \rho_{R_1,R_2}(\tau_r - n_r)\chi_{C_1,C_2}(\tau_c+1),$$

(8)

where $0 \leqslant \tau_c \leqslant n_c - 1$, $0 \leqslant \tau_r \leqslant n_r - 1$ and $\tau = \tau_c n_r + \tau_r$. Next, we can express the aperiodic CCFs between $\{R_1\}$ and $\{R_2\}$, in terms of those between $\{A_1\}$ and $\{A_2\}$, and $\{B_1\}$ and $\{B_2\}$, using equation (6). In equation (8), we require $\rho_{R_1,R_2}(\tau_r)$ for $0 \leqslant \tau_r \leqslant n_a n_b - 1$, which can be obtained from (6) with $0 \leqslant \tau_a \leqslant n_a - 1$, $0 \leqslant \tau_b \leqslant n_b - 1$ and $\tau_r = \tau_b n_a + \tau_a$. (Note that this is only part of the range of values for which (6) is valid). We also require $\rho_{R_1,R_2}(\tau_r - n_r)$ in equation (8), again for $0 \leqslant \tau_r \leqslant n_a n_b - 1$. Since $\tau_r = \tau_b n_a + \tau_a$ and $n_r = n_a n_b$, this can be obtained from equation (6) by subtracting n_b from the value of τ_b:-

$$\rho_{R_1,R_2}(\tau_r - n_r) = \rho_{A_1,A_2}(\tau_a)\rho_{B_1,B_2}(\tau_b - n_b) + \rho_{A_1,A_2}(\tau_a - n_a)\rho_{B_1,B_2}(\tau_b - n_b + 1)$$

(9)

Thus, using equations (6) and (9) in (8), we obtain:-

$$\chi_{Q_1,Q_2}(\tau) = [\rho_{A_1,A_2}(\tau_a)\rho_{B_1,B_2}(\tau_b) +$$

$$\rho_{A_1,A_2}(\tau_a - n_a)\rho_{B_1,B_2}(\tau_b+1)]\,\chi_{C_1,C_2}(\tau_c)$$

$$+ [\rho_{A_1,A_2}(\tau_a)\rho_{B_1,B_2}(\tau_b - n_b) +$$

$$\rho_{A_1,A_2}(\tau_a - n_a)\rho_{B_1,B_2}(\tau_b - n_b + 1)]\,\chi_{C_1,C_2}(\tau_c+1),$$

(10)

where $0 \leqslant \tau_a \leqslant n_a - 1$, $0 \leqslant \tau_b \leqslant n_b - 1$, $0 \leqslant \tau_c \leqslant n_c - 1$ and $\tau = \tau_c n_a n_b + \tau_b n_a + \tau_a$. This allows the periodic CCF for $\{Q_1\}$ and $\{Q_2\}$ to be evaluated for all relative phase-shifts τ mod n_q, from the aperiodic and periodic CCFs of the three pairs of component sequences.

In the same way, the aperiodic CCF between the three-component Composite sequences $\{Q_1\}$ and $\{Q_2\}$ can be derived from equation (6), as follows:-

$$\rho_{Q_1,Q_2}(\tau) = [\rho_{A_1,A_2}(\tau_a)\rho_{B_1,B_2}(\tau_b) +$$

$$\rho_{A_1,A_2}(\tau_a-n_a)\rho_{B_1,B_2}(\tau_b+1)] \; \rho_{C_1,C_2}(\tau_c)$$

$$+ [\rho_{A_1,A_2}(\tau_a)\rho_{B_1,B_2}(\tau_b-n_b) +$$

$$\rho_{A_1,A_2}(\tau_a-n_a)\rho_{B_1,B_2}(\tau_b-n_b+1)] \; \rho_{C_1,C_2}(\tau_c+1), \qquad (11)$$

and since equation (7) is identical to equation (5) except that all χs are replaced by χ̂s, the odd CCF for $\{Q_1\}$ and $\{Q_2\}$ is given by equation (10) with the same modification.

This process can be applied recursively to obtain the correlation functions for Composite sequences constructed from any number of component sequences. Although the expressions become rather cumbersome, they are nevertheless straightforward and computationally-efficient to evaluate. For example, consider equation (10). We require the aperiodic CCF for $\{A_1\}$ and $\{A_2\}$ for $-n_a \leqslant \tau_a \leqslant n_a - 1$, which can be evaluated by computing $2n_a$ correlation values. Similarly, the terms involving $\{B_1\}$ and $\{B_2\}$ in (10) require $2n_b$ correlation values, while the periodic CCF for $\{C_1\}$ and $\{C_2\}$ can be obtained from n_c correlation values. Thus, a total of only $2(n_a + n_b) + n_c$ correlation values are required to completely specify all $n_a n_b n_c$ values of the periodic CCF for $\{Q_1\}$ and $\{Q_2\}$, by virtue of equation (10). Moreover, each of these correlation values is computed for one of the three pairs of much shorter component sequences (compared with the Composite sequence length), giving further computational savings.

4   AVERAGE INTERFERENCE PARAMETERS FOR COMPOSITE SEQUENCES

For many SSMA systems, the most useful measure of the system performance is the average data-bit error probability [5]. The averaging here is carried out over all relative time-delays and carrier phases among the set of received signals, and over all combinations of data modulation on the interfering signals. Unfortunately, an exact calculation of this average error probability consumes intolerable quantities of computer time [8-10].

One alternative approach which appears to give accurate results [8,9] is to evaluate the average signal-to-noise ratio (SNR) at the output of a matched filter (or synchronised correlator) at a receiver, and to use this to obtain an approximation to the average error probability [5]. This average SNR at the jth receiver in an N-user SSMA system, in which codes of period P are employed by all users, is given by [5]:-

$$SNR_j = \left(\frac{\eta}{2E_b} + \frac{1}{6P^3} \sum_{\substack{i=1 \\ i \neq j}}^{N} r_{i,j} \right)^{-1/2} , \qquad (12)$$

where E_b is the signal energy per data bit, $\eta/2$ is the two-sided power spectral density of the background noise and $r_{i,j}$ is the average interference parameter (AIP), which specifies the effect of the ith received signal on SNR_j. For real-valued sequences, the AIP can be expressed [5] as:-

$$r_{i,j} = 2\mu_{i,j}(0) + \mu_{i,j}(1), \qquad (13)$$

where $$\mu_{i,j}(n) = \sum_{\tau=1-P}^{P-1} \rho_{i,j}(\tau) \, \rho_{i,j}(\tau+n) \qquad (14)$$

The AIP can therefore be evaluated for any such pair of sequences from their aperiodic CCF. In fact, by means of a fundamental correlation identity [7], $r_{i,j}$ can be expressed entirely in terms of the aperiodic autocorrelation functions [6].

Using this approach, Stark and Sarwate [3] have shown that the AIP for any pair of real-valued, two-component Composite sequences can be expressed in terms of the parameters r and μ for their component sequences, as follows:-

$$\left. \begin{array}{l} r_{R_1,R_2} = \mu_{B_1,B_2}(0) \cdot r_{A_1,A_2} + \mu_{B_1,B_2}(1) \cdot f_{A_1,A_2} \\[2mm] \text{where } f_{A_1,A_2} = \mu_{A_1,A_2}(n_a-1) + 4\mu_{A_1,A_2}(n_a) + \mu_{A_1,A_2}(n_a+1). \end{array} \right\} \qquad (15)$$

As before, $\{A_s\}$ is the 'inner' and $\{B_s\}$ the 'outer' component sequence used to construct the Composite sequence $\{R_s\}$, for s = 1,2. Although equation (15) applies only to real-valued Composite sequences, the generalisation to the complex-valued case is straightforward [3].

Another extension of the above result is that to the case of multiple-component Composite sequences. For example, the AIP for real-valued, three-component Composite sequences $\{Q_1\}$ and $\{Q_2\}$, constructed as described in section 3, is easily seen from equation (15) to be:-

$$r_{Q_1,Q_2} = \mu_{C_1,C_2}(0) \cdot r_{R_1,R_2} + \mu_{C_1,C_2}(1) \cdot f_{R_1,R_2},$$

$$\text{where } f_{R_1,R_2} = \mu_{R_1,R_2}(n_r-1) + 4\mu_{R_1,R_2}(n_r) + \mu_{R_1,R_2}(n_r+1). \qquad \Bigg\} \qquad (16)$$

Using equation (15) and some further manipulation, it is then possible to express the AIP for $\{Q_1\}$ and $\{Q_2\}$ entirely in terms of the parameters r and μ for the three pairs of component sequences. (The rather long expressions are omitted here for brevity).

5   COMPARISONS BETWEEN COMPOSITE AND OTHER SEQUENCES

First, the bad news...

5.1 Peak Correlation Magnitudes

If, in a particular SSMA application, we wish to optimise the worst-case error-rate performance, then Composite sequences are a poor choice. This is because in worst-case analyses, the relative time-delays among the set of received signals are assumed to be those which maximise the interference between users, and the maximum magnitudes of the correlation functions for Composite sequences are larger than for many other sequences. For example, consider the out-of-phase values of the periodic ACF for a two-component Composite sequence, as given by equation (5) with $A_2 = A_1$, $B_2 = B_1$ and $R_2 = R_1$. Whenever τ is an integer multiple τ_b of the period n_a of the 'inner' component sequence $\{A_1\}$, we have $\tau_a = 0$ in equation (5), which then reduces to:-

$$\chi_{R_1}(\tau = \tau_b n_a) = \rho_{A_1}(0)\, \chi_{B_1}(\tau_b),$$

because $\rho_{A_1}(-n_a) = 0$ from the definition (3). Also, from (3) and (2), note that $\rho_{A_1}(0) = \chi_{A_1}(0) = n_a$, provided the sequence $\{A_1\}$ is normalised with a mean chip amplitude of unity. Hence:-

$$\chi_{R_1}(\tau = \tau_b n_b) = n_a \cdot \chi_{B_1}(\tau_b)$$

and even if the 'outer' component $\{B_1\}$ is an m-sequence, with $\chi_{B_1}(\tau_b) = -1$ for all $\tau_b \not\equiv 0 \bmod n_b$, the out-of-phase values of the ACF for $\{R_1\}$ include peaks of magnitude n_a, which represents a fraction $1/n_b$ of the complete Composite sequence length $n_r = n_a n_b$. This may well be a large fraction, since we often require $n_b \ll n_a$ to minimise acquisition time (see section 6).

These relatively large peak values of the correlation functions for Composite sequencs may not be a significant disadvantage, however. This is because the peak values occur for only a small fraction of the possible relative phase-shifts between sequences, and therefore only rarely influence performance. A more meaningful approach for many applications is to compare systems on the basis of their <u>average</u>, rather than worst-case performance. We are therefore interested in how Composite sequences compare with others from the standpoint of their average interference parameters, discussed in section 4 (and this brings us to the good news...)

5.2 Average Interference Parameters

One way to investigate the effect of the Composite sequence structure on the AIP is to consider <u>random</u> binary sequences of length n_r chips and Composite sequences of the same length constructed from independent pairs of random binary component sequences. In both cases, the expected value of the AIP can be shown [3] to be $2n_r^2$, so in this respect there is no difference between binary Composite sequences and any other binary sequences.

Now consider the general case of a pair of Composite sequences $\{Q_1\}$ and $\{Q_2\}$, of common period n_q, each constructed from an <u>arbitrary</u> number, t, of component sequences. If all the component sequences are mutually-independent, random, binary sequences, it is easily seen that equation (16) still holds, provided that $\{C_2\}$ and $\{C_1\}$ are interpreted as the 'outermost' component sequences, while $\{R_1\}$ and $\{R_2\}$ are now a pair of (t-1)-component Composite sequences. The expected value of the AIP for $\{Q_1\}$ and $\{Q_2\}$ is therefore

$$E(r_{Q_1,Q_2}) = E[\mu_{C_1,C_2}(0)] . E(r_{R_1,R_2}) + E[\mu_{C_1,C_2}(1)] . E(f_{R_1,R_2})$$

and, from equations (14) and (3), we note that $E[\mu_{C_1,C_2}(1)] = 0$ and $E[\mu_{C_1,C_2}(0)] = n_c^2$. Hence, we obtain:-

$$E(r_{Q_1,Q_2}) = n_c^2 . E(r_{R_1,R_2}) . \tag{17}$$

Now, suppose that the (t-1)-component Composite sequences $\{R_1\}$ and $\{R_2\}$ are such that $E(r_{R_1,R_2}) = 2n_r^2$. Then equation (17) implies that $E(r_{Q_1,Q_2}) = 2n_c^2 n_r^2 = 2n_q^2$. Thus, if the expected value of the AIP for (t-1)-component Composite sequences is twice the square of their period, the same holds for t-component Composite sequences. But we know that $E(r_{R_1,R_2}) = 2n_r^2$ when $\{R_1\}$ and $\{R_2\}$ are each constructed from <u>two</u> component sequences, so the same result must hold for 3-component Composite sequences, and so on. This proves (by induction) that:-

$$E(r_{Q_1, Q_2}) = 2n_q^2, \tag{18}$$

for any pair of Composite sequences $\{Q_1\}$ and $\{Q_2\}$, each of period n_q and constructed from an arbitrary number, t, of component sequences.

Thus, by going to a Composite sequence structure, with any number of levels, we should not expect to incur any degradation in the SNR at the correlator output of a receiver.

The above results for Composite random sequences suggest that it may be possible, by careful choice of component sequences, to construct sets of Composite sequences for which the AIPs are smaller than for other sets of sequences (such as m-sequences). This is indeed the case, as shown by the following examples.

First, consider binary Composite sequences of period $n_r = 124$, constructed using m-sequences of period 31 (as 'inner' components) and a Barker sequence [11] of period 4 (as the 'outer' component in each case). Typical values of the AIP for these Composite sequences were shown by Stark and Sarwate [3] to vary from 22768 to 26464. All of these values are smaller than the AIPs for m-sequences of period 127, which were shown by Roefs [12] to range from 29070 to 42894. Note also that the AIP for random and Composite random sequences has an expected value of $2n_r^2 = 30752$ for $n_r = 124$, which is larger than any of the values for the Composite sequences of this length constructed from Barker and m-sequence components.

Other numerical results demonstrate that the above is not an exceptional case. For instance, computing the AIPs for Composite sequences of period 155 constructed from m-sequences of period 31 and a Barker code of period 5, we obtain a range of values from 45762 to 46226, compared with the expected value for random sequences of 48050. Also, for a pair of Composite sequences of period 195 constructed from a pair of m-sequences of period 15 and a Barker code of period 13, the value of the AIP is 68418, which is again smaller than the expected value of 76050 for random sequences.

We therefore conclude that as far as the AIP is concerned, Composite sequences are likely to perform about as well as, if not better than, other sequences of the same (or similar) period. Since the AIP determines the average SNR at the correlator output of a receiver in a SSMA system, it follows that Composite sequences are likely to provide similar, if not smaller average data error-rates than other sequences with a comparable period.

6   RAPID ACQUISITION OF COMPOSITE SEQUENCES

Consider a Composite sequence $\{R\}$ with a period $n_r = n_a n_b$ chips and a chip-rate $1/T_c$ chips/sec, constructed from an inner component sequence $\{A\}$, with a period n_a and chip-rate $1/T_c$, and an outer component sequence $\{B\}$, with a period n_b and a chip-rate of $1/n_a T_c$. A receiver can be designed to synchronise rapidly to $\{R\}$ by separately acquiring its component sequences $\{A\}$ and $\{B\}$, using a correlator for each. Assuming that the received SNR is large enough (see below), the maximum total acquisition time, $T_a(\max)$, for this scheme can be shown [2,3] to be:-

$$T_a(\max) = n_a n_b T_c \left(\frac{n_a}{n_b} + n_b\right), \tag{19}$$

while the average total acquisition time $T_a(av) = \frac{1}{2}T_a(\max)$. This result implies that we can choose the component sequence periods n_a and n_b so as to $\underline{\text{minimise}}$ the acquisition time, for a given Composite sequence period $n_r = n_a n_b$. By differentiating (19) with respect to n_a and setting the result equal to zero, we find that the acquisition time is minimised if we choose

$$n_a = (n_r^2/2)^{1/3} \quad \text{and} \quad n_b = (2n_r)^{1/3}, \tag{20}$$

which yields:-

$$T_a(\max) = (2^{1/3} + 2^{-2/3})\, n_r^{4/3} T_c. \tag{21}$$

The success of this rapid acquisition scheme for Composite sequences depends upon the received SNR being large enough to permit reliable detection of the in-phase ACF peak (of height n_a) for the inner component sequence $\{A\}$, by performing correlation integrals over $n_a T_c$ seconds. To provide a reasonable comparison, we now suppose that the same SNR exists at the input of a receiver for a conventional (non-Composite) sequence of period n_r and chip-rate $1/T_c$. Such a receiver can synchronise with the same reliability as the Composite sequence receiver, when using an integration time of $n_a T_c$ (rather than its full-period integration time of $n_r T_c$). However, the maximum number of correlations required to bring the local replica of the conventional sequence into synchronism is n_r, so its maximum acquisition time becomes:-

$$T'_a(\max) = n_a n_r T_c$$

If n_a and n_b are chosen according to (20), the use of Composite sequences can be seen to provide a reduction in acquisition time by a factor of

$$\frac{T'_a \text{ (max)}}{T_a \text{ (max)}} = \frac{n_r^{1/3}}{2^{2/3} + 2^{-1/3}} \approx (0.420) n_r^{1/3}. \tag{22}$$

Examples of this improvement factor are 4.2 for a code period of $n_r = 1000$, rising to 17 for $n_r = 65000$ and 42 for $n_r = 10^6$.

7 MATCHED FILTERS FOR COMPOSITE SEQUENCES

From the structure of the rapid-acquisition receiver discussed above, it follows that a matched filter (MF) receiver for a Composite sequence can be implemented using separate MFs for its component sequences. The critical parameters in the design of a MF for any code are the bandwidth and time-delay required, these being given by the chip-rate and the duration of one period, respectively, of the code. For a conventional code of period n_r chips and chip duration T_C, we require a MF with a delay of $n_r T_C$ and a bandwidth of the order of $1/T_C$, i.e., a time-bandwidth (TB) product of n_r. However, for a Composite sequence of period $n_r = n_a n_b$ and chip duration T_C, we can use a MF for each of the component sequences, with TB products of n_a and n_b. The following example shows that this structure enables a MF receiver to be implemented using surface-acoustic wave (SAW) technology, for TB products significantly greater than if conventional sequences were used.

Suppose that $T_C = 6.67$ns so that $1/T_C = 150$MHz, and $n_a = n_b = 127$ so that $n_r = 16129$. We first note that a MF for a conventional code of period n_r and chip duration T_C would require a bandwidth of 150MHz and a delay of $n_r T_C = 107\mu s$, which is beyond the capabilities of current SAW technology [13]. For a Composite sequence, however, a MF for the inner component needs to provide a delay of only $n_a T_C = 0.85\mu s$ with a bandwidth of 150MHz, while the delay required of a MF for the outer component is $107\mu s$ over a reduced bandwidth of $1/n_a T_C = 1.18$ MHz. Both of these MFs can be realised using SAW devices, and samples of such devices, with delay/bandwidth characteristics similar to those required in this example, have been produced by a number of manufacturers [14].

8 CONCLUSIONS

Various properties of a class of Composite sequences have been investigated. The structure of these sequences can be used to advantage in the design of direct-sequence SSMA systems, and also in other systems using correlation detection or matched filtering. The results on the correlation behaviour of these sequences are encouraging and show that by careful choice of component sequences, we can construct Composite sequences which outperform conventional (e.g., m-) sequences, in terms of average data error-rates. The Composite sequence structure also enables rapid synchronisation procedures to be adopted at a receiver, providing significant reductions in acquisition time. Finally, the same structure also allows matched filter to be implemented in cases where, due to technological limitations, this would not be feasible if conventional sequences were used.

REFERENCES

[1] BEALE, M. & TOZER, T. C. 'A class of Composite sequences for spread-spectrum communications', IEE J. Comput. & Digital Tech., Vol. 2, 1979, pp. 87-92.

[2] BEALE, M. 'Direct-sequence spread-spectrum multiple-access systems', Ph.D. Thesis, Electronics Labs., University of Kent at Canterbury, U.K., 1982.

[3] STARK, W. E. & SARWATE, D.V. 'Kronecker sequences for spread-spectrum communication', IEE Proc., Vol. 128, Pt. F, No. 2, April 1981, pp. 104-109.

[4] MASSEY, J. L. & UHRAN, J. J. 'Sub-baud coding', Proc. 13th Allerton Conf. on Cct. & System Theory, Oct. 1975, pp. 539-547.

[5] PURSLEY, M. B. 'Performance evaluation for phase-coded spread-spectrum multiple-access communication - Part I: system anaylsis', IEEE Trans., Vol. COM-25, 1977, pp. 795-799.

[6] PURSLEY, M. B. & SARWATE, D. V. 'As ref. 5 - Part II: code sequence analysis', ibid, Vol. COM-25, 1977, pp. 800-803.

[7] SARWATE, D. V. & PURSLEY, M. B. 'Crosscorrelation properties of pseudorandom and related sequences', Proc. IEEE, Vol. 68, 1980, pp. 593-619.

[8] BORTH, D. E. PURSLEY, M. B. SARWATE, D. V. & STARK, W. E. 'Bounds on error probability for direct-sequence spread-spectrum multiple-access communications', MIDCON Proc. Chicago, Nov. 1979, paper No. 15/1, pp. 1-14.

[9] PURSLEY, M. B., SARWATE, D. V. & STARK, W. E. 'On the average probability of error for direct-sequence spread-spectrum multiple-access systems', Proc. 1980 Conf. on Information Sciences & Systems, Mar. 1980, pp. 320-325.

[10] PURSLEY, M. B. SARWATE, W. E.& STARK, W.E. 'Error probability for direct-sequence spread-spectrum communications - Part I: Upper & Lower bounds', IEEE Trans., Vol. COM-30, 1982, pp. 975-984.

[11] BARKER, R. H. 'Group synchronising of binary digital systems', in JACKSON, W. (Ed.), 'Communication Theory', Butterworth, 1953, pp. 273-287.

[12] ROEFS, H. F. A. 'Binary sequences for spread-spectrum multiple-access communication', Ph.D. Thesis, Coordinated Science Lab., University of Illinois, Urbana, IL, 1977.

[13] STIGLITZ, M. R. & SETHARES, J. C. 'Magnetostatic waves take
     over where SAWs leave off', Microwave Journal, Feb. 1982,
     pp. 18-38 and p. 111.
[14] BRISTOL, T. W. 'SAW devices and applications - a status report',
     ibid, Dec. 1977, pp. 16-22.

Part 2.

ALGORITHMS AND PROCESSING METHODS

Organised by:

R.J. McEliece
*California Institute of Technology,
Pasadena, CA, U.S.A.*

HIERARCHICAL, PARALLEL AND SYSTOLIC ARRAY PROCESSING

Kenneth Steiglitz

Electrical Engineering and
    Computer Science Department
Princeton University
Princeton, New Jersey 08544

1. SIGNAL PROCESSING AND VLSI

Many signal processing algorithms are highly regular, data-independent, and access the data in fixed patterns. For these reasons the current technological advances in very large scale integrated circuits hold especially great promise for signal processing, and in fact we now see the development of many highly integrated processors of a more or less specialized nature. At one end of the spectrum, we see programmable signal processing chips that are really microprocessors, with program, memory and logic separated as in a general-purpose machine. At the other extreme, we see highly-specialized, custom chips that perform fixed tasks; typically the data moves through the chip along fixed, regular paths, the arithmetic logic is distributed in space, and the "program" is really "hard-wired" into the topology. This talk is devoted to a study of this latter, custom variety of architecture.

The range of algorithms that are commonly used for digital signal processing is not very great; a few very important algorithms are used intensively. They fall roughly into four categories: convolution and filtering; Fourier transforms; matrix calculations (see, for example, [36]); and iterative algorithms for adaptive filtering. All these applications are characterized by two important characteristics that make special-purpose, highly dense hardware very attractive:

- *high-volume, real- or nearly real-time data-flow require-ments*, and
- *effective algorithms with regular patterns of data access and fixed operation sequences.*

There are direct architectural consequences of these characteristics. The regularity of the patterns of data access and operation sequences makes possible a high degree of pipe-lining and multiplexing. This, in turn, makes possible a high-volume data flow. Furthermore, the regularity of the algo-rithms is reflected in a regularity of VLSI circuit structure, so that a hierarchical approach to layout design and specification becomes possible, and that greatly simplifies the design and development of large-scale, custom VLSI circuits for digital signal processing. The rest of this talk is devoted to these architectural consequences: In the next section we will discuss some general aspects of parallelism, and why the need for parallelism justifies the development of custom, single-purpose chips. Section 3 is devoted to highly-pipelined and systolic structures, using filtering as an example, with a review of some useful topologies. Section 4 deals with how some of the important structures can be combined in a hierarchical way . We will review mathematical models of VLSI computation and available lower and upper performance bounds in Section 5. Section 6 will deal with the practical matter of maximizing throughput by appropriate choice of latching density.

2. PARALLELISM AND THE CUSTOM/PROGRAMMABLE DECISION

There is a general tradeoff between specialization and pro-grammability in digital signal processing chips. The obvious advantages of flexibility afforded by programmability must be weighed against the higher potential throughput of a custom chip. The choice between the two is dictated by the cost in time and money of designing and testing a chip, and by the need for very high-throughput real-time processing. This tra-deoff changes with time and technology: as chip design becomes a highly automated process, and as more real-time, high-volume applications arise (such as in the fields of com-munications and robotics), we are likely to see the prolifera-tion of very specialized signal processing chips.

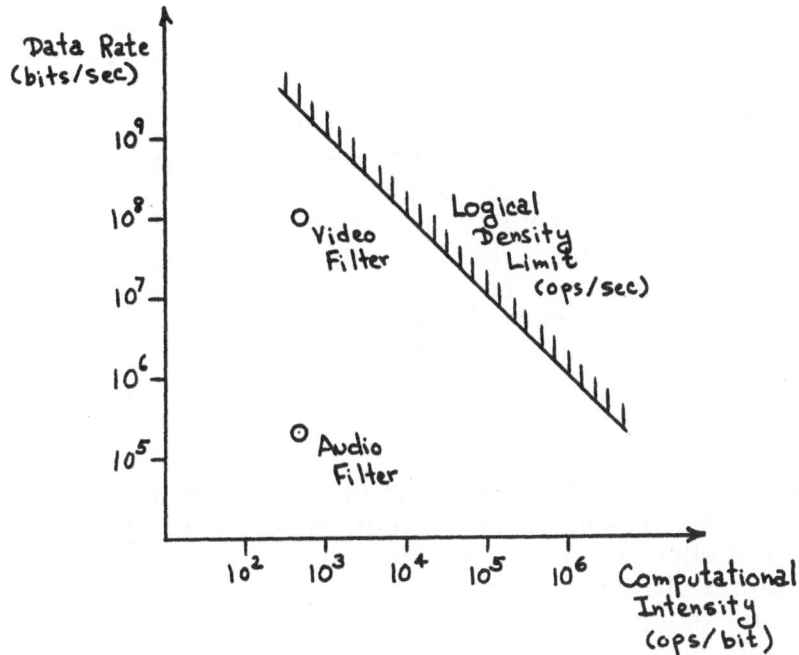

Figure 1. Signal Processing Tasks in the intensity-rate plane (after [7,9]).

Figure 1 shows one way of looking at the range of signal processing applications. We plot the data rate in bits/sec as ordinate, and the computational intensity requirement of a given task in operations/bit as abscissa. For example, a low-order FIR filter has a low computational intensity, whereas a high-order filter has a correspondingly high intensity. A 50th-order FIR filter at a sampling rate of 20,000 words/sec (a typical audio rate) and 10 bits/word, with say 100 logical operations (at the gate level) per fixed-point multiply-by-constant, corresponds to a data rate of 2×10^5 bits/sec, and a computational intensity of 5×10^2 operations/bit. For a chip of fixed size and for a given technology, the product of the two coordinates in operations/sec is bounded from above by some constant: the total number of operations possible if every piece of the chip were doing useful work all the time. For a chip with 10^5 gates and a clock period of 100 nsec, this is 10^{12} operations/sec. This boundary is shown by a hyperbola in Fig. 1 (a straight line in log-log coordinates). At the same time, there is an upper limit on the data rate, determined by the number of I/O pins and the speed of the I/O drivers.

We are thus constrained to work within the region shown. Whenever an operation is carried out that does not contribute directly to the processing of the signal, as counted by the measure of computational intensity, we move away from the boundary. Consider, for example, the operation of a pro-grammable signal processor with a stored program. Every instruction fetch or store, every instruction decoding, and every test for branching, is wasted in the sense that a part of the chip is doing work that is not essential. Also wasted, of course, is any part of the chip that remains idle during any particular clock cycle.

We are lead to the conclusion that the most efficient use of chip area, the dearest resource at present, should avoid pro-grammability, and should make concurrent use of as much of the chip as possible. When demands on performance are very high, at the limits of applications technology, we are lead to the design of custom, single-purpose chips with fixed data-flow paths. Thus, some filtering tasks at audio bandwidths may be best implemented now with programmable chips, but applica-tions at video rates, like robot vision, demand custom designs.

In this talk we will use the operation of convolution for our examples. It is no doubt the most widely-used of all the digital signal operations, and is also representative in terms complex-ity and throughput requirements. We write it as

$$y = w \otimes x = \sum_k w_k x_{n-k} = \sum_k x_k w_{n-k} \tag{1}$$

The function w_k will be called the *weight sequence*, and will usually be of finite duration, so that the limits of the summa-tions in (1) will be finite. We will distinguish two situations in which convolution is usually implemented: *general convolu-tion*, where the weights w_k are variable on a short-term basis, as fast as the signal x_k; and *filtering*, where the weights w_k are fixed (or at least infrequently changed). We will make no dis-tinction between convolution and *correlation*, which is simply convolution with one of the signals time-reversed.

Convolution can be applied in many ways. At the bit level, with Boolean product, and Boolean sum with carry, it means binary multiplication. At the signal level it means filtering or correlation. At the logical level it means pattern matching. This observation allows us to develop highly regular VLSI topologies by first developing a structure at the *word* level for convolution. A similar structure is then used recursively to build a multiplier at the *bit* level. The result is a hierarchical structure that is highly regular, being uniform in topology all the way down to the bit level. In the next two sections we carry out just this plan.

3. SYSTOLIC AND COMPLETELY-PIPELINED STRUCTURES

Highly concurrent VLSI circuits can be characterized by the following desirable properties [7,9]:

- *Local-Connectedness* : This means that computational elements are connected only to nearby neighbors.
- *Flow-Simplicity* : This means that each element is used only once per elementary computation.
- *Cell-Simplicity* : This means that each element takes only constant time for its computation; that is, its computation time does not depend on the such parameters as the number of bits in a word, or the number of coefficients in a filter.

Systolic arrays [21,22,27] can be characterized as those that are both *locally-connected* and *flow-simple*. Wires are short and the data flows through the structure in a smooth way. However, each "cell" may be very complex (a multiplier, possibly), and may take time dependent on the problem parameters.

Completely-pipelined circuits [7,9] are another class of highly concurrent, pipelined circuits, characterized by the properties of being *flow-simple* and *cell-simple*. These circuits are more general than systolic ones in that long wires are allowed, but more restricted in the sense that the computational cells operate in time independent of the problem parameters (such as word-size).

In what follows we will concentrate on the simplicity and regularity of some computational structures and ignore some problems that are important at a practical level, but which would obscure the presentation. For example, the question of efficient use of area will be ignored for now, but will be discussed in Section 5. The problem of distributing power, ground, and clock lines will likewise be ignored; some discussion of these points in the present context can be found in [9]. We will also not worry about the signs of numbers in describing multiplication, but assume that extension bits are added to two's-complement numbers so that the answer is always in range (see [13,31,33] for some discussion of this issue).

3.1 *Word-Serial Filtering*

Figure 2a shows the conventional signal flow graph F for FIR filtering (see [37], for example): the input signal x_k is delayed along a chain of registers, and during each clock period the appropriate samples are multiplied by the corresponding weights w_{n-k} and summed. At the right we see the computation that must be performed every clock cycle. Notice that for this example of a 4-coefficient filter, not only do we need to perform 4 multiplications and 3 additions between clock pulses, but the input of the second addition depends on the result of the first, and the input of the third depends on the result of the second. This means that we cannot perform these additions in parallel, and therefore that the throughput rate is limited by the time for three additions

Figure 2b shows another signal flow graph F^t, the *transpose* (see [34], for example) of signal flow graph F (called B1 in [21]). The transformation of transposition entails reversing the direction of every arc, replacing summing nodes by branching nodes, replacing branching nodes by summing nodes, and interchanging input and output. Here the sequence of sums is replaced by a **broadcast** of the input signal x_k at any time; the computation during each cycle is again shown at the right. This broadcasting, or *fanout*, of a signal carries with it a certain penalty in terms of delay, but is generally much faster than sequential add operations, so that this signal flow graph can be implemented with a much higher throughput if the three additions are implemented with three adders operating in parallel. The commercial chip described in [47] uses this transposed structure.

The fanout problem of graph F^t and the sequential delay of graph F can be avoided by using the graph F^2, shown in Fig. 2c. Here on every clock pulse the input signal moves to the right, and the output signal moves to the left. The transfer function of F^2 is $H(z^2)$ if the original graph F has transfer function $H(z)$, so that meaningful output is obtained only every other clock cycle (this is the structure W1 of [21]). Thus the input and

73

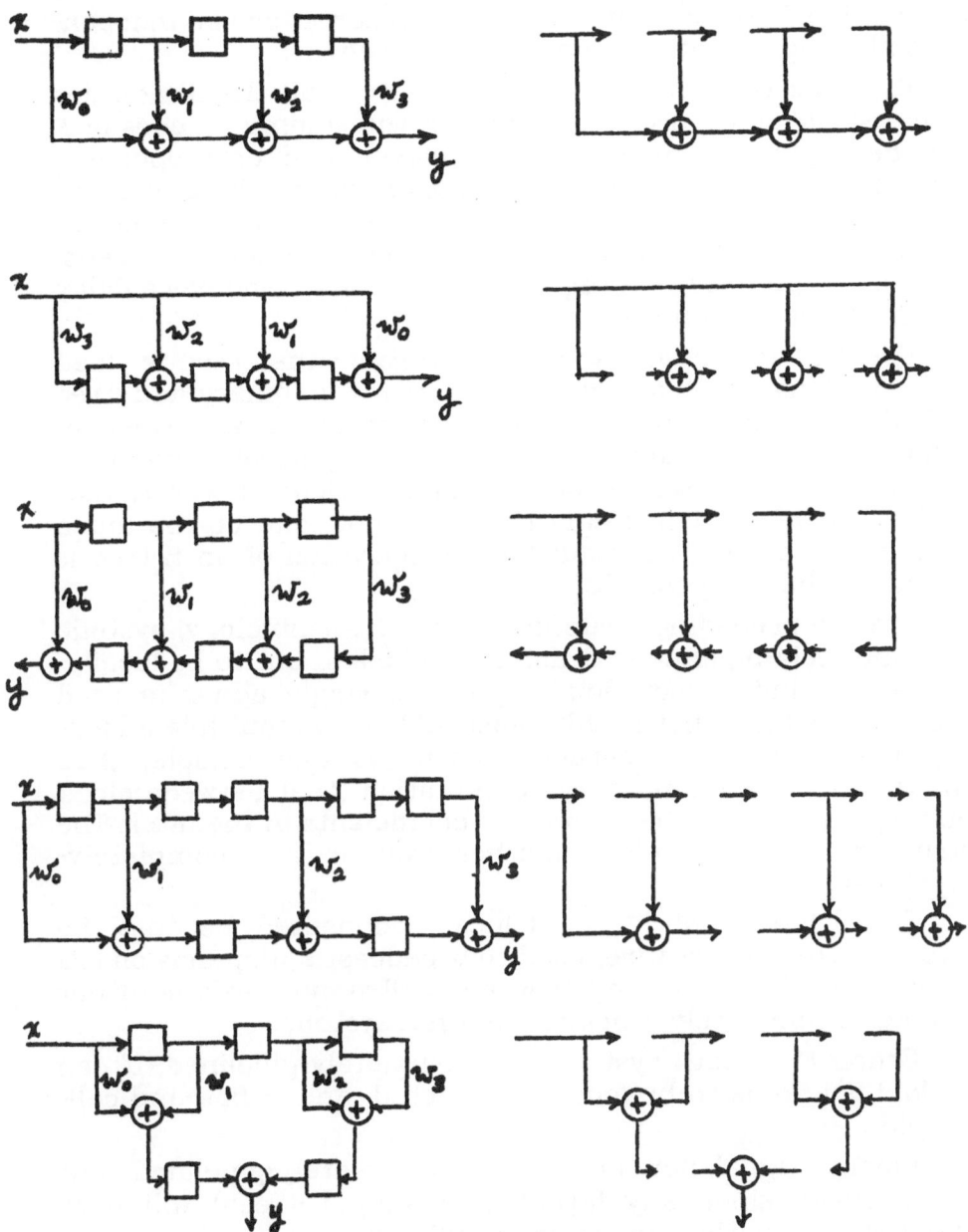

Figure 2. Structures for word-serial filtering. The boxes are delay elements. The computation during each clock period is shown at the right. From the top: a) F, b) F^t, c) F^e, d) F^d, e) F^Δ.

output signals must be interleaved with zeros (or two independent filtering operations can be interleaved).

It seems that there is no way to avoid all the difficulties with a single structure. For example, the summing nodes of F can be separated by registers (delays), and corresponding extra delays inserted between inputs, producing the circuit F^d shown in Fig. 2d (called W2 in [21]). Here the input and output signals move in the same direction, but at different speeds. But this graph has more registers than F or F^t, and has a delay before the output appears.

Finally, Fig. 2e shows the structure F^Δ that results when the additions in F are performed using a binary-add tree [7,9,37]. This is a convenient structure for visualizing the convolution operation, and may be useful for general convolution (as opposed to filtering). Care must be taken, however, that the tree is laid out in a way that does not take up too much area on the chip. The recursive configuration of an H-tree is useful for that purpose [32].

Are the preceding structures, by our terminology, systolic or completely-pipelined? Graph F can be laid out to be locally-connected, but is not flow-simple if a single adder is used sequentially (and that is only reasonable since multiple adders would not be usable in parallel). Neither is it cell-simple, since the time for the elementary computation (that between clock pulses) depends on the number of coefficients in the filter. The structure F is therefore neither systolic nor completely-pipelined.

Signal flow graph F^t is not locally-connected, because the length of the longest wire, used to broadcast x, depends on the filter order. It is, however, flow- and cell-simple, so it is by our definition completely-pipelined, but not systolic.

Graph F^2 is both systolic and completely-pipelined; it can be laid out so as to be locally-connected, and is flow-and cell-simple.

Finally, signal flow graph F^Δ is completely-pipelined, but not systolic, since any layout (including H-trees) will have wires whose lengths depend on the filter order.

3.2 Bit-Serial Multiplication

The same structures used for word-serial filtering can be used for bit-serial multiplication by a constant, with the difference that each summing node is a full adder with three

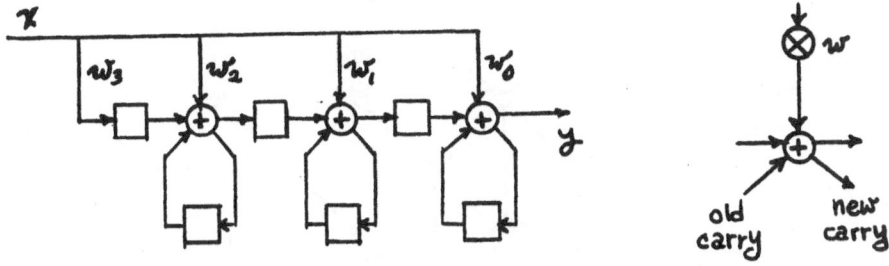

Figure 3. M^t: The structure F^t adapted for bit-serial multiplication.

inputs and two outputs. The inputs to each full adder are the two addend bits and the preceding carry bit, the outputs are the sum bit and the carry bit. Figure 3 shows the multiplier corresponding to graph F^t; it is really no more than a simple implementation of the ordinary shift-and-add elementary-school multiplication algorithm. We will denote by M, M^t, M^2 and M^Δ the multipliers corresponding to F, F^t, F^2, and F^Δ, respectively.

3.3 *Word-Parallel Filtering*

We now consider word-parallel filtering, and, in the next section, the corresponding operation of bit-parallel multiplication. Figure 4 shows a diamond array with the signal x entering from the top left and the filter weights from the top right (when the weights w_k are fixed, they need not be transmitted through the array as shown but can be stored in place). Notice that the signal values x_k corresponding to a given signal are arranged on a horizontal line, and hence skewed in time so that successive values enter the diamond array at successive clock pulses. The next horizontal line will have another signal in it, and blocks of filtered signals emerge from the bottom of the array at successive clock pulses. The sides of the diamond array have length proportional to the filter order.

Figure 5 shows the detail of a node of the array: Each node in Fig. 4 contains a multiplication by a weight and an adder, and each arc has a delay element (a latched register). We will call the overall structure F^A (for *Array Filter*).

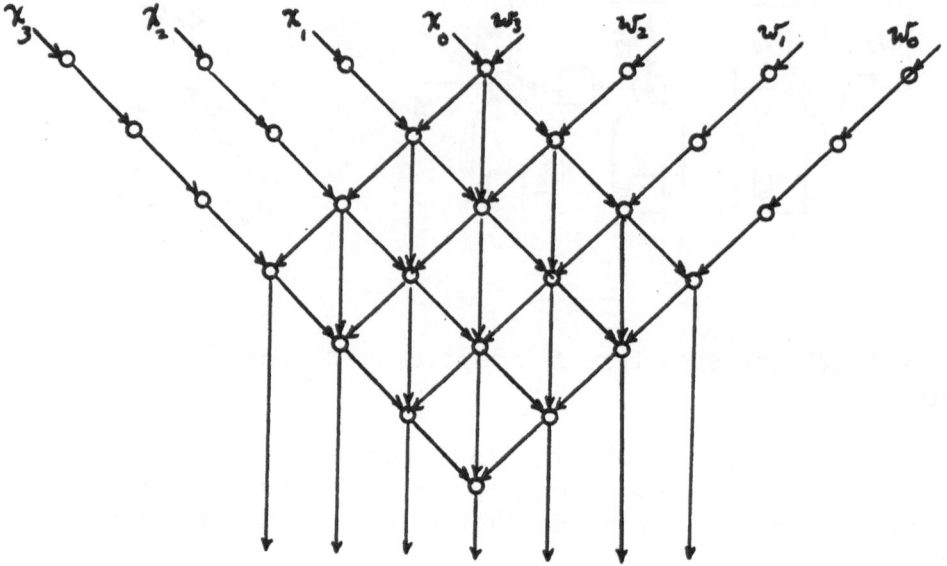

Figure 4. A structure for word-parallel filtering, F^4.

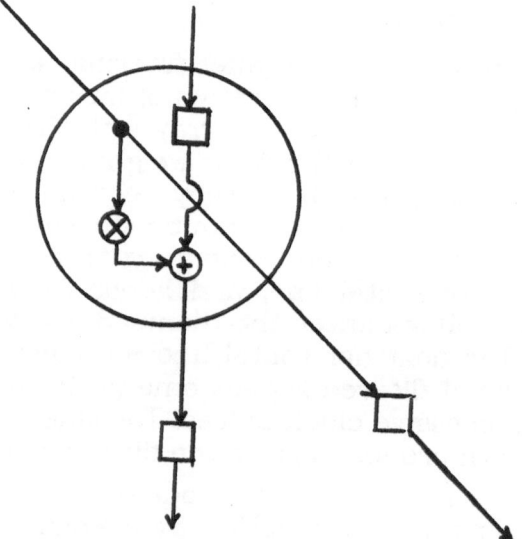

Figure 5. Detail of a node in F^4.

3.4 *Bit-Parallel Multiplication*

As before, the filtering structure becomes a multiplication structure when the multipliers are replaced by Boolean product and the summers by full adders with carries. In this case the carries propagate down and to the left, which direction corresponds to the next-higher bit of the product. An extra triangle is needed at the lower left so that the carry bits can propagate all the way to the left (see [31], for example). The reader will recognize this signal flow graph as nothing more than an array multiplier (we will call it M^A), with every value latched between clock pulses (see [31,37]). This parallel multiplication structure has the property that the full adders at the top of the diamond can accept inputs without extra logic, so that the multiplier can function as an accumulator as well, and this fact is useful in FIR filters and other applications [10,15].

This hexagonally-connected array multiplier is locally-connected, cell-simple, and flow-simple, and is therefore both systolic and completely pipelined by our definition.

3.5 *Other Useful Structures*

We have already seen the *linearly-connected array* (in all the serial examples), the hexagonally-connected array (in F^A and M^A), and the *leaf-connected tree* [7,9] (in F^A and M^A). We now mention some of the other regular topologies that have proven useful in constructing computational networks for VLSI. A structure called *cube-connected cycles* is used in [35] for bit-parallel multiplication. A tree-like structure can be used to shorten the delay (latency) of a parallel multiplier, and the resulting structure, called a *mesh-of-trees* can be found in [8,25]. Leighton also discusses an analogous topology called a *tree-of-meshes* [25].

We have seen above a variety of different topologies, all of which perform similar computational tasks. Some work has been done towards developing a unified treatment of computational structures of this type, and showing how they can be expressed conveniently and derived from each other. For more about such mathematical representations see [12,14,20,45,46].

4. HIERARCHICAL METHODOLOGY

An important feature of the regular topologies exemplified in the preceding section is that they can be combined in a recursive, or hierarchical, way. The most obvious application of this idea is to use bit-serial multiplication within a word-serial filter, yielding a bit-serial, word-serial filter that is fully-pipelined. On each clock pulse, every bit moves, every piece of hardware (silicon) is used, and one output bit appears. Bit-serial adders are needed at the summing nodes. Such structures have been discussed widely in the literature recently [6,13,16,23,30], and are attractive at this time because a reasonably high-order filter can fit on one chip, and the interconnection problems caused by high pin counts are greatly alleviated by the bit-serial nature of the computation.

Suppose for illustration that the multiplier M^i is used within the filter F^i, resulting in what we will call $F^i(M^i)$. Figure 6 shows a schematic representation of this filter, which is similar to those described in [6,13]. In theory, then, we have the ingredients for $5 \times 5 = 25$ different bit-serial, word-serial filters, all of which have slightly different timing and layout details.

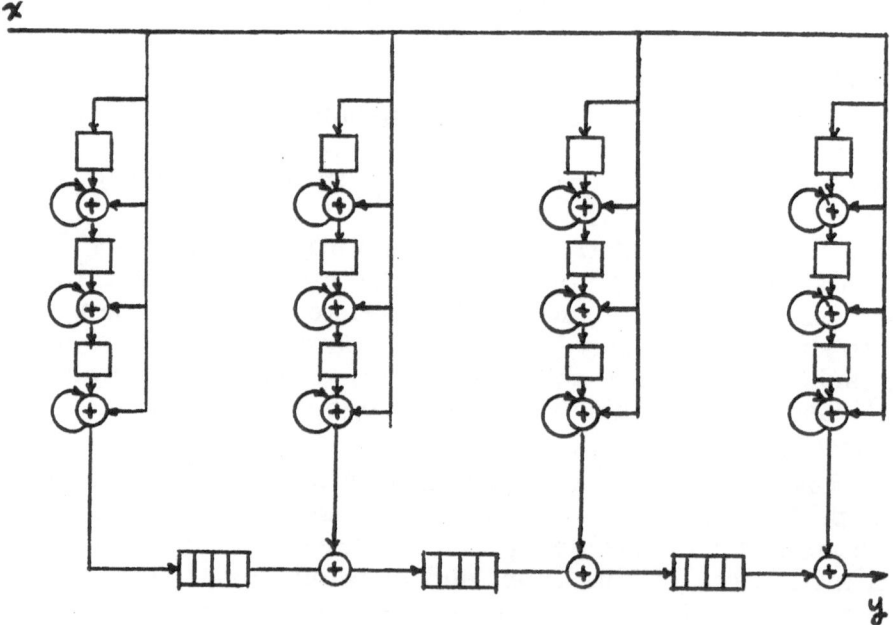

Figure 6. Recursive use of the structure F^i: $F^i(M^i)$.

To go one step further, we can combine serial and parallel structures. For example, at the other extreme from the completely bit-serial filter just described, we can assemble $F^A(M^A)$, producing a bit-parallel, word-parallel filter — one that produces a completely filtered block of signal samples once every clock pulse. (Now we need bit-parallel adders at the summing nodes.) Of course, the amount of area is greatly increased over the bit-serial filter, but so is the throughput.

In the same way, we could use a bit-serial multiplier within a word-parallel filter (yielding $F^A(M^s)$, for example), which produces a bit-serial, word-parallel filter, which for B-bit words, produces a complete block of filtered samples every B clock pulses. With only the 6 different structures discussed here, there are 36 possibilities, each having its own characteristics in terms of layout area and throughput.

An important advantage of this approach to VLSI layout is that some of the problems associated with design and layout are greatly simplified, since the overall problem is broken down into natural pieces, each of which can be handled in relative isolation. Such a design methodology is well-suited to the use of high-level layout languages and silicon compilers (see, for example, [17,18,29,38]).

Another advantage of the hierarchical approach is in the crucial but often neglected area of testing. Because the complexity of testing arbitrary circuits can grow exponentially with the size of the circuit, it is a great advantage to be able to break a circuit into blocks whose function can be tested independently of the other blocks. Much the same approach has proven very valuable in the design of large software systems. Some recent results in the testing of regular bilateral arrays can be found in [19,40,42].

5. MODELS AND BOUNDS FOR VLSI

The signal processor has heretofore been concerned mainly with speed of operation. High throughput on a general-purpose computer is achieved by managing *time*. But now the designer of systems has a new resource to manage: *area*. It is no longer sufficient to specify a sequence of instructions for data processing. We must now specify a geometric layout. Of course the requirements of high speed and small area are mutually conflicting. Consider, for example, the multipliers discussed above. A B-bit-serial multiplier like M^s will generally have a throughput rate of one product every B clock pulses.

before the answer is ready, and generally takes area proportional to B. On the other hand, a B-bit-parallel multiplier such as M^A has a throughput rate of one product every clock pulse (once the pipeline is full), but area proportional to B^2. Thus there appears to be a conservation law at work, and we expect that bounds can be obtained on such quantities as *throughput per unit area*. We will describe some such bounds below.

5.1 *Some Terminology*

We need to define some important terms precisely. First, the *delay* or *latency* T of a signal processing device is the time between the arrival of the first bit of the input signal at the input port, and the time that the last bit of the answer appears at the output port. This is the usual usage of the term "computation time." But in many signal processing applications we are concerned more with the throughput rate than with the delay. We define the time between successive outputs with pipelines full as the *period* P of a chip, and the reciprocal of the period as the *throughput*.

If a quantity is bounded from above by a constant multiple of $f(B)$ for sufficiently large B, where B is any parameter of interest (often the number of bits), we say the quantity is $O(f(B))$. So, for example, the array multiplier M^A requires area $O(B^2)$. A corresponding *lower* bound is written $\Omega(B)$.

5.2 *A VLSI Model*

We will next describe a mathematical model for a VLSI chip, one that is abstract and simple enough so that results can be proved about it, but one that is also realistic enough so that the results provide some guidelines, or at least hints, about reality. The model we describe is attributed by Vuillemin to the three sources [4,32,41]; this and similar models have been used by many others. There is a fairly large literature on models and bounds that we will not attempt to survey completely here (see, for example, [1,2,4,5,24,26,28,35,39,41,43]).

The basic premise is that there is a minimum feature size λ_w, and minimum delays τ_w and τ_g, dictated by the technology. The important assumptions are then that a) no two wires can have their midpoints closer together than λ_w, b) every logical unit (such as a gate) must have area at least λ_w^2, c) passing a signal through a wire entails a delay of at least τ_w, *independent of the wire's length*, and d) passing a signal through a gate entails a delay of at least τ_g.

As discussed in [2], the assumption that the delay is independent of wire length is true only in certain regimes. Depending on the technology, the time for propagation of signals may be independent of wire length, as we assume here (the *synchronous* model [4]), or proportional to the logarithm of the wire length (using repeaters), or proportional to the square of the wire length (diffusion case).

5.3 *Lower Bounds*

The essential result is expressed in a nicely general form by Vuillemin [43]. He defines a wide class of functions called *transitive* functions, which includes integer product, convolution, linear transform, and matrix product.

Theorem [43]: Any circuit that computes a transitive function has wire area

$$A = \Omega(D^2) \tag{2}$$

where D is the data rate in bits/sec.

The period P is related to the data rate D in a simple way: $P = N/D$, where N is the number of input bits. Therefore the bound above can also be written

$$AP^2 = \Omega(N^2) \tag{3}$$

We can also observe that if any one of N input bits can affect the output, and if there is a constant bound on the allowed fan-in, then the circuit must have at least $\log N$ stages. This implies the following lower bound on the latency T:

$$T = \Omega(\log N) \tag{4}$$

A good example to illustrate the use of these bounds is B-bit multiplication. The bounds above tell us that $AP^2 = \Omega(B^2)$, and $T = \Omega(\log B)$. The array multiplier described above has $A = O(B^2)$, $P = O(1)$, and $T = O(B)$. It is therefore asymptotically optimal under the measure AP^2, but possibly has more delay than necessary. In fact, the delay can be reduced to the asymptotically optimal $O(\log B)$ with the area increasing only from $O(B^2)$ to $O(B^2 \log B)$ [8] (see also [44]).

An interesting measure of goodness, that takes into account both period and latency, is $AP^2 T^2$. By the arguments above the lower bound on multiplication is then $AP^2 T^2 = \Omega(B^2 \log^2 B)$. The array multiplier mentioned above [8] has the upper bound $AP^2T^2 = O(B^2\log^3 B)$, which is therefore no more than one log-factor away from asymptotic optimality, by this measure at least.

Compare this result with the bit-serial multiplier M^t, for example. That structure has area $A = O(B)$, period $P = O(B)$, and latency $T = O(B)$, so that $AP^2T^2 = O(B^5)$. This is an indication that the overall efficiency of silicon utilization is not as good asymptotically as that of the parallel array multipliers, but one should not conclude too much from this argument. For one thing, we may need a multiplier with small area simply because an array multiplier will not fit on a chip, in which case we must settle for the possible instead of the asymptotically good. It is also quite likely that the constants of proportionality favor the simple linearly-arranged designs such as M^t, so that for reasonably-sized B the measures above may be misleading. The asymptotic measures give us useful guidelines for comparing designs that are similar, but there are so many factors in choosing a multiplier for a particular application at a particular point in technological development, that mathematical analysis should be interpreted with caution.

6. PIPELINING AND LATCHING FOR THROUGHPUT

In the designs discussed above it was assumed that every signal value was latched (that is, held in a register) at every stage in the signal flow graph. So, for example, the array multiplier M^4 has a register after every full adder (this was stressed in [31]). This means that every part of the circuit can be used for holding intermediate results — that every part can function as a pipeline. This approach leads to high throughput at the expense of delay. In contrast, array multipliers that are commercially produced on packaged chips do not generally have a high degree of pipelining in this sense of the term; the answer is usually produced in one or two clock cycles, and the carry signals ripple through the structure, settling in time for each new clock pulse. In [3], for example, an array multiplier with combinational logic that is 113 gates deep is mentioned. Thus, commercial single-package multipliers are optimized for latency and not throughput, and are therefore not necessarily "fast" for custom chip designs for signal processing applications.

However, latching at every possible stage of a circuit does not necessarily lead to the highest throughput. First, the latches themselves take time to operate; their input stages

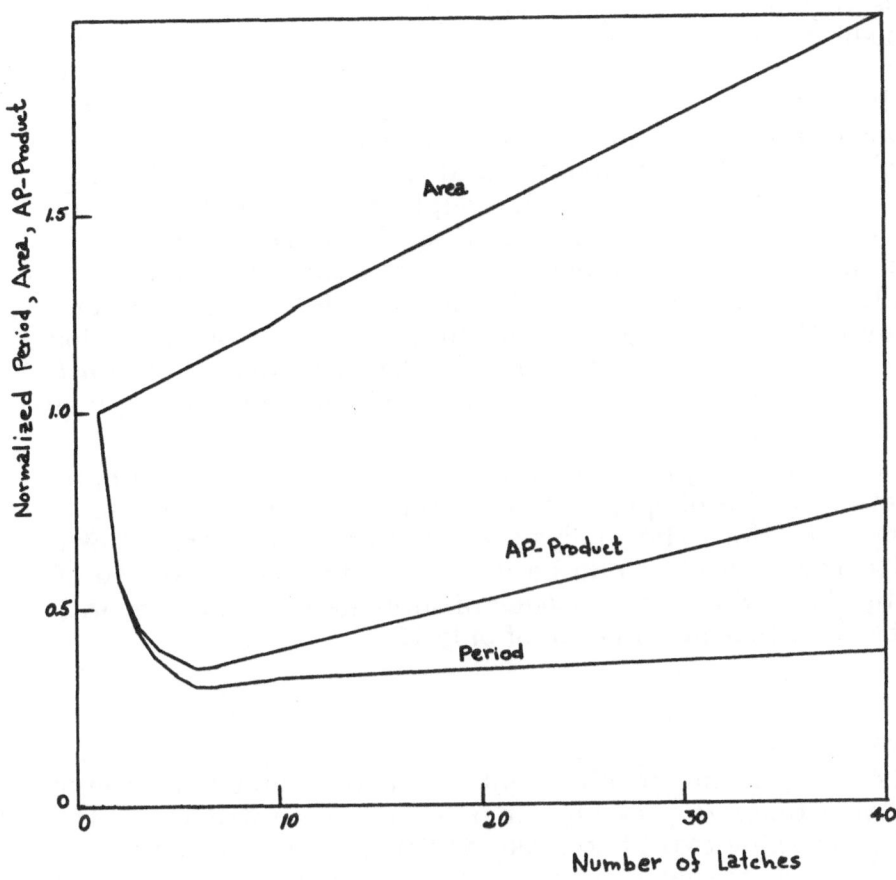

Figure 7. Area, period, and area-period product, as functions of the number of intermediate latching stages m, in a typical pipelined array multiplier (after [11]).

must be charged, and they must charge the input stages of the next layer of logic. Second, the clock driver must drive the additional capacitance of the latches, and for a given driver this lengthens the clock rise- and fall-times and decreases the possible clock rate.

If we start with one stage of a circuit that has combinational logic that is many gates deep, and we introduce m intermediate stages of latching, we decrease the period by a factor of roughly $1/m$, up to the point that the latching and clock-driving time becomes comparable to the propagation time of signals through one stage of logic. After that point there are diminishing returns to the addition of more latching. In [11] the generic situation of a block of combinational logic is modeled mathematically, and the optimal choice of the number of additional latches, m, was studied. Figure 7 shows typical curves of area, period, and area-period product as a function of m for a circuit with a depth of 100 gates. As can be seen, the period as a function of m decreases sharply to a minimum and stays almost constant, the area increases steadily with m, and the product has a well-defined minimum. We may wish to minimize this product AP instead of the period P; $1/AP$ can be written as $(1/P)/A$ — the *throughput-per-unit-area*. In any case, the optimal values of m for minimizing P or AP will be close to each other.

A typical example of such a situation occurs in the implementation of the bit-parallel array multiplier M^4 discussed in Section 3.4. Here the analysis predicts that the period of a 16-bit array multiplier can be decreased from 210 nsec to 66 nsec by the addition of 5 stages of intermediate latches, with an attendant increase in area of only 13%.

7. CONCLUSIONS

The design and development of custom chips for signal processing tasks is very challenging, calling as it does on the signal processing expert to make decisions at many design levels: he must manage overall system architecture, circuit topology, timing, area utilization, and layout. At the same time, making good use of such resources can lead to reliable low-cost devices that have very high throughput in many signal processing applications.

The key to effective design is a high degree of pipelining using regular, repetitive structures and fixed data-flow paths. Such structures can be hierarchically organized, making the design and layout problems manageable.

8. ACKNOWLEDGEMENTS

I want to thank Prof. Peter R. Cappello of the Computer Science Department, University of California, Santa Barbara,

California. Some of the work discussed here is due originally to him, and forms part of his Ph. D. dissertation (Princeton University, 1982).

This work was supported in part by NSF Grant ECS-8120037, U. S. Army Office-Durham Grant DAAG29-82-K-0095, and DARPA Contact N00014-82-K-0549.

REFERENCES

1. Abelson, H. and P. Andreae, "Information Transfer and Area-Time Tradeoffs for VLSI Multiplication," *CACM*, Vol. 23, Jan. 1980, pp.20-23.

2. Bilardi, G., M. Pracchi, and F. P. Preparata, "A Critique and an Appraisal of VLSI Models of Computation," in *VLSI Systems and Computation*, H. T. Kung, Bob Sproull, and Guy Steele (eds.), Computer Science Press, Rockville, Md., 1981.

3. Bötcher, K., A. Lacroix, M. Talmi, D. Wesseling, "Integrated Floating Point Signal Processor," *Proc. 1982 IEEE Int. Conf. on Acoustics, Speech, and Signal Processing*, Paris, May 1982, pp. 1088-91.

4. Brent, R. P. and H. T. Kung, "The Chip Complexity of Binary Arithmetic," *Proc. 12th Annual ACM Symposium on the Theory of Computing*, Los Angeles, Ca., April 1980, pp. 190-200.

5. Brent, R. P. and H. T. Kung, "The Area-Time Complexity of Binary Multiplication," *JACM*, Vol. 28, No. 3, July 1981, pp. 521-534.

6. Cappello, P. R., and K. Steiglitz, "Digital Signal Processing Applications of Systolic Algorithms," in *VLSI Systems and Computations*, H.T. Kung, Bob Sproull, and Guy Steele (eds.), Computer Science Press, Rockville, Md., 1981.

7. Cappello, P. R. and K. Steiglitz, "Bit-Level Fixed-Flow Architectures for Signal Processing," *Proc. 1982 IEEE Int. Conf. on Circuits and Computers*, New York, N. Y., Sept. 29 - Oct. 1, 1982.

8. Cappello, P. R. and K. Steiglitz, "A VLSI Layout for a Pipelined Dadda Multiplier," *ACM Trans. on Computer Systems*, Vol. 1, No. 2, May 1983, pp. 157-174.

9. Cappello, P. R. and K. Steiglitz, "Completely Pipelined Architectures for Digital Signal Processing," *IEEE Trans. on Acoustics, Speech, and Signal Processing*, Vol. ASSP-31, No. 4, August 1983, in press.

10. Cappello, P. R. and K. Steiglitz, "A Note on 'Free' Accumulation in VLSI Filter Architectures,' submitted for publication.

11. Cappello, P. R., A. S. LaPaugh, and K. Steiglitz, "Optimal Choice of Intermediate Latching to Maximize Throughput in VLSI Circuits," *Proc. 1983 IEEE Int. Conf. Acoustics, Speech, and Signal Processing*, Boston, Mass., April 14-16, 1983, pp. 935-938. (Also *IEEE Trans. on Acoustics, Speech, and Signal Processing*, in press.)

12. Cappello, P. R. and K. Steiglitz, "Unifying VLSI Array Designs with Geometric Transformations," 1983 IEEE Int. Conf. on Parallel Processing, Aug. 1983.

13. Caraiscos, C. and B. Liu, "Bit Serial VLSI Implementations of FIR and IIR Digital Filters," *Proc. 1983 Int. Symp. on Circuits and Systems*, May 1983.

14. Culik II, K. and J. Pachl, "Folding and Unrolling Systolic Arrays," Research Report CS-82-11, Faculty of Mathematics, University of Waterloo, Waterloo, Ontario, Canada, April 1982.

15. Denyer, P. B. and D. J. Myers, "Carry-Save Arrays for VLSI Signal Processing," in *VLSI 81: Very Large Scale Integration*, John P. Gray (ed.), Academic Press, London, 1981.

16. Denyer, P. B., "An Introduction to Bit-Serial Architectures for VLSI Signal Processing," Draft of a paper presented at Advanced Course on VLSI Architecture, University of Bristol, U.K., July 1982.

17. Denyer, P. B. and D. Renshaw, "Case Studies in VLSI Signal Processing using a Silicon Complier," *Proc. 1983 IEEE Int. Conf. on Acoustics, Speech, and Signal Processing*, Boston, Mass., 1983, pp. 939-942.

18. DeMan, H., J. Van Ginderdeuren, and N. Gonçalves, "Custom Design of Hardware Digital Filters on I.C.'s," *Proc. Custom Integrated Circuits Conf.*, Rochester, N. Y., 1982.

19. Gray, F. G. and R. A. Thompson, "Fault Detection in Bilateral Arrays of Combinational Cells," *IEEE Trans. on Computers*", Vol. C-27, 1978, pp. 1206-1213.

20. Johnsson, L. and D. Cohen, "A Mathematical Approach to Modeling the Flow of Data and Control in Computational Networks," in *VLSI Systems and Computation*, H. T. Kung, Bob Sproull, and Guy Steele (eds.), Computer Science Press, Rockville, Md., 1981.

21. Kung, H. T., "Why Systolic Architectures?" Carnegie-Mellon Univ., Dept. of Computer Science, CMU-CS-81-148, Nov. 1981.

22. Kung, S. Y., and D. V. Bhaskar Rao, "Highly Parallel Architectures for Solving Linear Equations," *Proc. 1981 Int. Conf. on Acoustic, Speech, and Signal Processing*, Atlanta, Ga., 1981, pp. 39-42.

23. Kung, H. T., L. M. Ruane, and D. W. L. Yen, "A Two-Level Pipelined Systolic Array for Convolutions," in *VLSI Systems and Computations*, H. T. Kung, Bob Sproull, and Guy Steele (eds.), Computer Science Press, Rockville, Md., 1981.

24. Leighton, F. T., "New Lower Bound Techniques for VLSI," *Proc. 22nd Annual Symposium on Foundations of Computer Science*, Nashville, Tenn., Oct. 1981.

25. Leighton, F. T., "A Layout Strategy for VLSI Which is Provably Good," *Proc. 14th Annual ACM Symposium on the Theory of Computing*, San Francisco, Ca., May 1982.

26. Leiserson C. E., "Area-Efficient Graph Layouts (for VLSI)," *Proc. 21st Annual Symposium on Foundations of Computer Science*, Syracuse, N.Y., 1980.

27. Leiserson, C. E. and H. T. Kung, "Algorithms for VLSI Processor Arrays," Section 8.3 of *Introduction to VLSI Systems*, C. Mead and L. Conway, Addison-Wesley Publishing Co., Menlo Park, Ca., 1980.

28. Lipton, R. J. and R. Sedgewick, "Lower Bounds for VLSI," *Proc. 13th Annual ACM Symposium on the Theory of Computing*, May 1981, pp. 300-307.

29. Lipton, R.J., J. Valdes, R. Sedgewick, "Programming Aspects of VLSI," *Proc. 9th Annual ACM Symposium on Principles of Programming Languages*, Albuquerque, N.M., Jan. 1982.

30. Lyon, R. F., "A Bit-Serial VLSI Architecture Methodology for Signal Processing," in *VLSI 81: Very Large Scale Integration*, John P. Gray (ed.), Academic Press, London, 1981. (*Proceedings of the First International Conference on Very Large Scale Integration*, University of Edinburgh, August 18-21, 1981.)

31. McCanny, J. V., J.G. McWhirter, J. B. G. Roberts, D. J. Day, T. L. Thorp, "Bit Level Systolic Arrays," *Proc. 15th Asilomar Conf. on Circuits, Systems, and Computers*, Nov. 1981.

32. Mead, C. and L. Conway, *Introduction to VLSI Systems*, Addison-Wesley, Menlo Park, Ca., 1980.

33. Myers, D. J., "Multipliers for LSI and VLSI Signal Processing Applications," M. Sc. Project report MSP5, University of Edinburgh, U.K., Sept., 1981.

34. Oppenheim, A. V., and R. W. Schafer, *Digital Signal Processing*, Prentice-Hall, Englewood Cliffs, N. J., 1975.

35. Preparata, F. P. and J. E. Vuillemin, "Area-Time optimal VLSI Networks Based on the Cube Connected Cycles," Rapport INRIA #13, Rocquencourt, France, 1980.

36. Priester, R. W., H. J. Whithouse, K. Bromley, J. B. Clary, "Signal Processing with Systolic Arrays," *Proc. 1981 IEEE Int. Conf. on Parallel Processing*, 1981, pp. 207-215.

37. Rabiner, L. R. and B. Gold, *Theory and application of digital signal processing*, Prentice-Hall, Inc., Englewood Cliffs, N.J., 1975.

38. Sastry, S. and S. Klein, "PLATES: A Metric-Free VLSI Layout Language," *Proc. MIT Conf. on Advanced Research in VLSI*, Cambridge, Mass., 1982.

39. Savage, J. E., "Area-Time Tradeoffs for Matrix Multiplication and Related Problems in VLSI Models," *J. Computer and Systems Science*, April 1981.

40. Sung, C. H., "Testable Sequential Cellular Arrays," *IEEE Trans. on Computers*, Vol. C-25, Jan. 1976, pp. 11-18.

41. Thompson, C. D., "Area-Time Complexity for VLSI," *Proc. 11th Annual ACM Symposium on the Theory of Computing*, April 1979, pp. 81-88.

42. Vergis, A. and K. Steiglitz, "Testability Conditions for Bilateral Arrays of Combinational Cells," 1983 IEEE International Conference on Computer Design: VLSI in Computers, New York, Oct. 31 - Nov. 3, 1983.

43. Vuillemin, J., "A Combinatorial Limit to the Computing Power of VLSI Circuits," *Proc. 21st Annual Symposium on the Foundations of Computer Science*, 1980, pp. 294-300.

44. Vuillemin, J., "A Very Fast Multiplication Algorithm for VLSI Implementation," *Integration*, Vol. 1, 1983, pp. 39-52.

45. Weiser, U. and A. L. Davis, "Mathematical Representation for VLSI Arrays," Technical Report UUCS-80-111, Dept. of Computer Science, University of Utah, Salt Lake City, Utah, Sept. 1980.

46. Weiser, U. and A. L. Davis, "A Wavefront Notation for VLSI Array Design," in *VLSI Systems and Computations*, H. T. Kung, Bob Sproull, and Guy Steele (eds.),Computer Science Press, Rockville, Md., 1981.

47. Williams, F. A., "An Expandable Single-IC Digital Filter/Correlator," *Proc. 1982 IEEE Int. Conf. on Acoustics, Speech, and Signal Processing*, Paris, May 1982, pp. 1077-80.

NUMBER THEORETIC & FINITE FIELD PROCESSING

P J W Rayner

Department of Engineering Cambridge University, Cambridge, U.K.

1   INTRODUCTION

The processing of continuous signals by means of a digital system requires that the signal be quantised in both time and amplitude. Quantisation in time is well-understood in terms of the Sampling Theorem. However amplitude quantisation is less well understood and in general the errors induced by amplitude quantisation are approximated by a white noise sources placed at appropriate positions in the system being considered. For many systems this model produces results in accordance with measured results on real systems, but in some classes of system, particulary those with feedback, the model does not predict accurately the behaviour of a real system.

The usual method of realising a digital signal processing system is to describe the required system in terms of equations which are discrete in time but continuous in amplitude (ie Difference equations) and then approximate the continuous arithmetic operations by means of finite wordlength binary arithmetic. In general the product of two binary numbers, each of length r bits, is a number of length 2r bits and some means must be used to approximate the product to r bits (eg. truncation or rounding). Such approximation leads to errors and in some systems can lead to instability known as limit cycle oscillation.

2   DIGITAL SIGNAL PROCESSING SYSTEM AS FINITE STATE MACHINES

Any digital signal processing system may be viewed as a Finite State Machine (F.S.M.) in that the system output at any clock instant is some function of past and present values of input and machine state. A model is shown in figure 1.

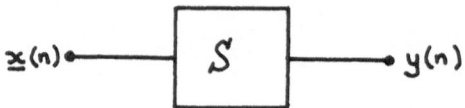

Fig.1. A Finite State Machine

$\underline{x}$(n) and $\underline{y}$(n) are respectively the discrete input and output
vectors. In general the system has memory and this is represented
by internal states $\underline{s}$(n). At each clock instant the operation of
the machine is to map from the set of states and inputs into a
new set of states and outputs. This operation may be represented
as:

$$\underline{s}(n+1)=f_1[\underline{s}(n), \underline{x}(n)]$$
$$\underline{y}(n+1)=f_2[\underline{s}(n), \underline{x}(n)] \tag{1}$$

This formulation is usually referred to as the Mealy model of a
F.S.M. It will be noted that the normal binary arithmetic
realisation of a digital signal processing system is a special
case of eqns.1. One must now ask the question whether the binary
arithmetic realisation is necessarily the optimum form of eqns.1
to achieve some "real word" specification such as frequency
response, impulse response, signal to noise ratio etc. Indeed,
in view of the presence of limit cycle oscillations in finite
wordlength realisations of Infinite Impulse Response (I.I.R)
digital filters, it seems likely that the normal arithmetic realisation is certainly not optimal. Moreover the arithmetic type
of realisation is restricted largely to linearly encoded signals.
Thus there are two problem areas to be considered. The first is
to decide on the optimal form of the finite mappings, given by
eqns.1, to achieve a "real world" specification; this will be
termed the System Representation problem. The second is to
determine the optimal realisation of the mappings in terms of
hardware complexity, speed of operation etc., this will be termed
the System Realisation problem.

2.1 System Representation

Attention will be focussed on I.I.R. digital filters for
two reasons. Firstly, I.I.R. filters exhibit limit cycle oscillations when realised in the normal way and secondly the complexity
of I.I.R. filters is reasonable, when measured in terms of the
number of variables, compared with other signal processing systems
such as Finite Impulse Response filters and transform computation.

One method of approach is to evaluate the system output
for all possible combinations of finite input and system state
using the continuous algebraic equation description of the system.

The outputs calculated in this manner will not, of course, be in the same finite set as the input and state so that some form of approximation is needed to achieve this. However the system outputs determined in this way can be within $+\frac{1}{2}$ least significant bit over the representable range of outputs. The process is best illustrated by a simple example.

Consider the single pole filter described by the difference equation:

$$y(n) = K.y(n-1) + x(n) \qquad (2)$$

which may be represented as shown in fig.2.

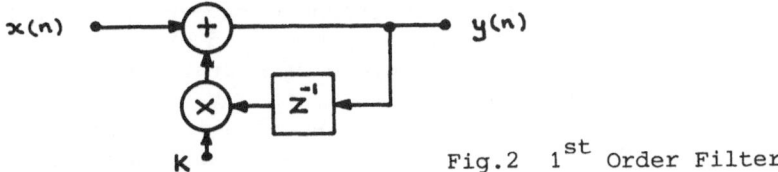

Fig.2   1st Order Filter

The normal realisation of this would use finite wordlength binary multiplication and addition with truncation or rounding at the multiplier output. Below is a table showing the filter output y(n) calculated from eqn.2 with K=0.6, input x(n)=0. Also shown in the table is the finite output $\bar{y}(n)$ obtained by rounding y(n) assuming 3 bit representation. Such a table may be set up for every possible input thus giving a tabular form of the finite mapping of eqns.1.

y(n-1)	y(n)	$\bar{y}$(n)
3	1.8	2
2	1.2	1
1	0.6	1
0	0	0
-1	-0.6	-1
-2	-1.2	-1
-3	-1.8	-2
-4	-2.4	-2

$$y(n) = 0.6.y(n-1)$$

In this simple example the outputs obtained are no different than would be obtained from the arithmetic realisation. However, in a system containing more than one multiplier the results obtained from the proposed method would more accurately match the results from an infinite precision filter.

The method proposed above is based on the "best" representation of the static characteristics of the system and gives no
consideration to the dynamic behaviour.   It is instructive to
draw the next-state graph for the system shown in the table. We
see immediately that the state graph gives a clear pictorial
representation of the dynamical behaviour of the system.   If the
system starts in any state other than O then the system can never
reach the O state.   This behaviour is limit cycle oscillation and
the system shown exhibits 3 limit cycles of length zero, including the required cycle on the zero state.   For a higher order
filter limit cycles of length greater than zero are exhibited.
Clearly then the method proposed for determining the system state
table is not adequate in that limit cycles exist.   However
modifications may be made to the graph to remove limit cycles and
in the simple example considered the modified graph would obviously
be as shown in Fig.4.   A limit cycle of n states can be broken at
n different points and, in general, it is not obvious how to break
cycles in an optimal way.   The next section introduces a technique,
developed by I. Proudler of Cambridge University Engineering
Department, for solving this problem.

The next-state graph of Fig.4 is known as an Arborescence
(Directed Tree) in graph theory and it shows just one way of
breaking the cycles in the graph of Fig.3.   Such an arborescence

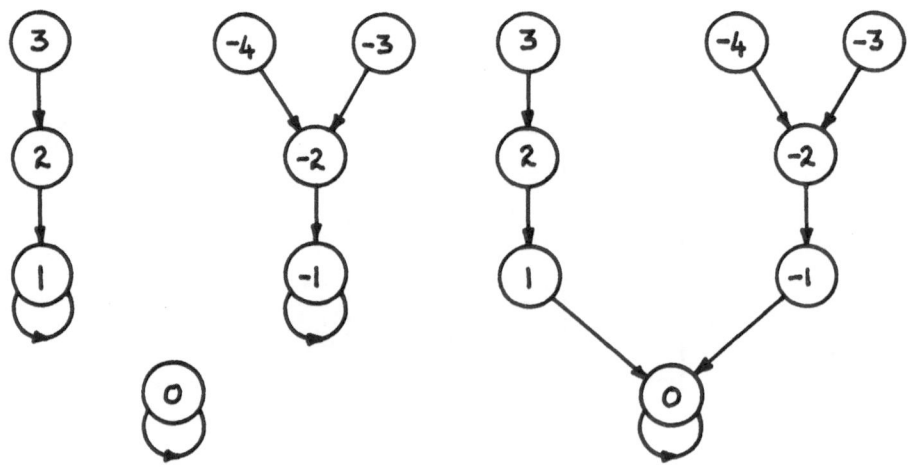

Fig.3.Next-state Graph   Fig.4.Corrected Next-state
                                          Graph

is said to be a Spanning Arborescence since it includes all vertices of the next-state graph. However there are a number of
ways of breaking the limit cycles and some means is needed to
choose the "optimum" spanning arborescence to yield the "optimum"
next state graph without limit cycles.

 If a "value" is allocated to each arc in such a way that
the "value" is a measure of the error in making the state transition along the arc then the optimum system is given by the
spanning arborescence in which the sum of the values along the
arcs is minimum; this is termed the Minimum Spanning Arborescence
Fig.5 shows the labelled graph with all possible transitions from
state 3. The arcs have been labelled with the error measure
given by the square of the difference between the system output
calculated from the system difference equation (eqn.2) and the
output given as the next state on the graph. For example, if the
system state, y(n-1), is 3 then the output calculated from the
difference equation is:

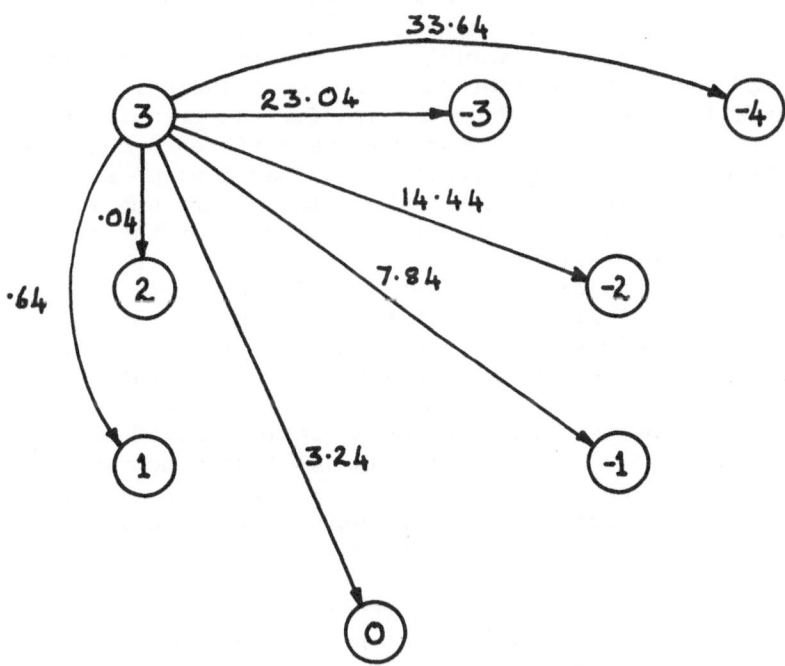

Fig.5,Possible Transitions from State 3 with associated
errors.

$$y(n)=0.6.y(n-1)=1.8$$

and the error E in making the state transition 3 to 1 is:

$$E=(1.8-1)^2=0.64$$

For the sake of clarity, fig.5 shows only the possible transitions from state 3; similar sets of transitions must be considered for each machine state. Efficient algorithims for generating minimum spanning arborescences have been developed(1).

A number of filters have been designed using this technique and fig.6 shows some results for a 2-pole filter,

$$y(n)=0.5y(n-1)-0.6y(n-1)+x(n)$$

using 6 bit variables. The graph shows the output signal to noise ratio as a function of the input amplitude for three different filter realisations. The "Quantised Output" filter is designed by calculating the filter output from the difference equation for every possible combination of input and the two system states $y(n-1)$ and $y(n-2)$; the calculated output is then rounded to produce a number representable by 6 bits. The filter designed by this method exhibits limit cycles which are removed by the minimum spanning arborescence method to give the "optimal filter. The "Arithmetic" filter is a normal twos complement arithmetic realisation of the filter difference equation. The criterion used for generating the "value" of the arc for the transition from state A to state B is the square of the difference between the output from the arithmetic realisation operating on state A and the next state B in the graph. The "Clamped Arithmetic" filter is an arithmetic realisation with the outputs of the adders clamped to avoid overflow.

A number of points are clear from fig.6 First, the "Quantised Output" filter has significantly better signal to quantising noise ratio performance than the arithmetic realisation. Note that the decrease in signal to noise ratio after the input amplitude has reached a particular level is due to overload. Second, it can be seen that the optimisation procedure for removing the limit cycle oscillations causes little degradation of the signal to noise ratio.Fig.7 shows a typical limit cycle oscillation and the corresponding response for the "optimised" filter.

2.2 System Realisation

The technique introduced in the previous section allows an optimal state transition table to be formed. Consideration must now be given to methods whereby this transition table may be realised in hardware.

A general model of the system is shown in fig.8 where the two main system blocks are seen to be state storage and a block representing the mapping of eqn.1. The most simple method, in

Fig.6. Performance of 2ⁿᵈ Order Filters

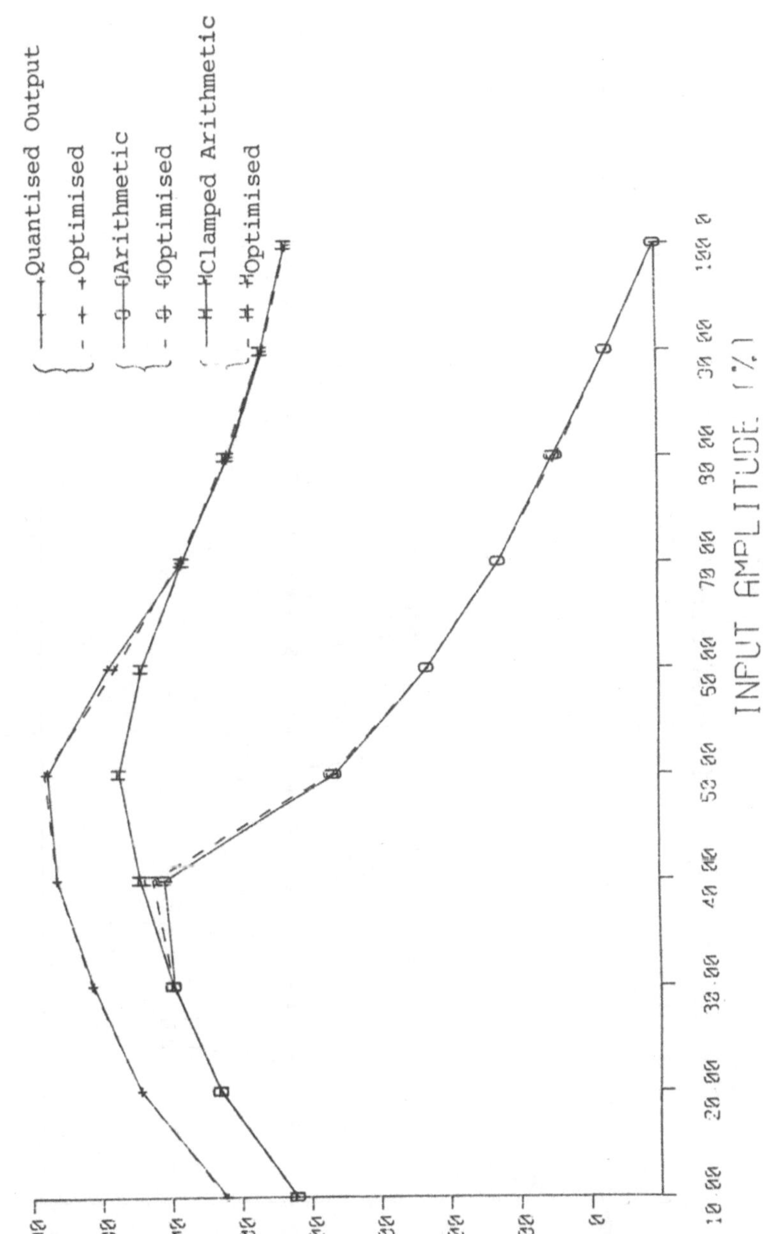

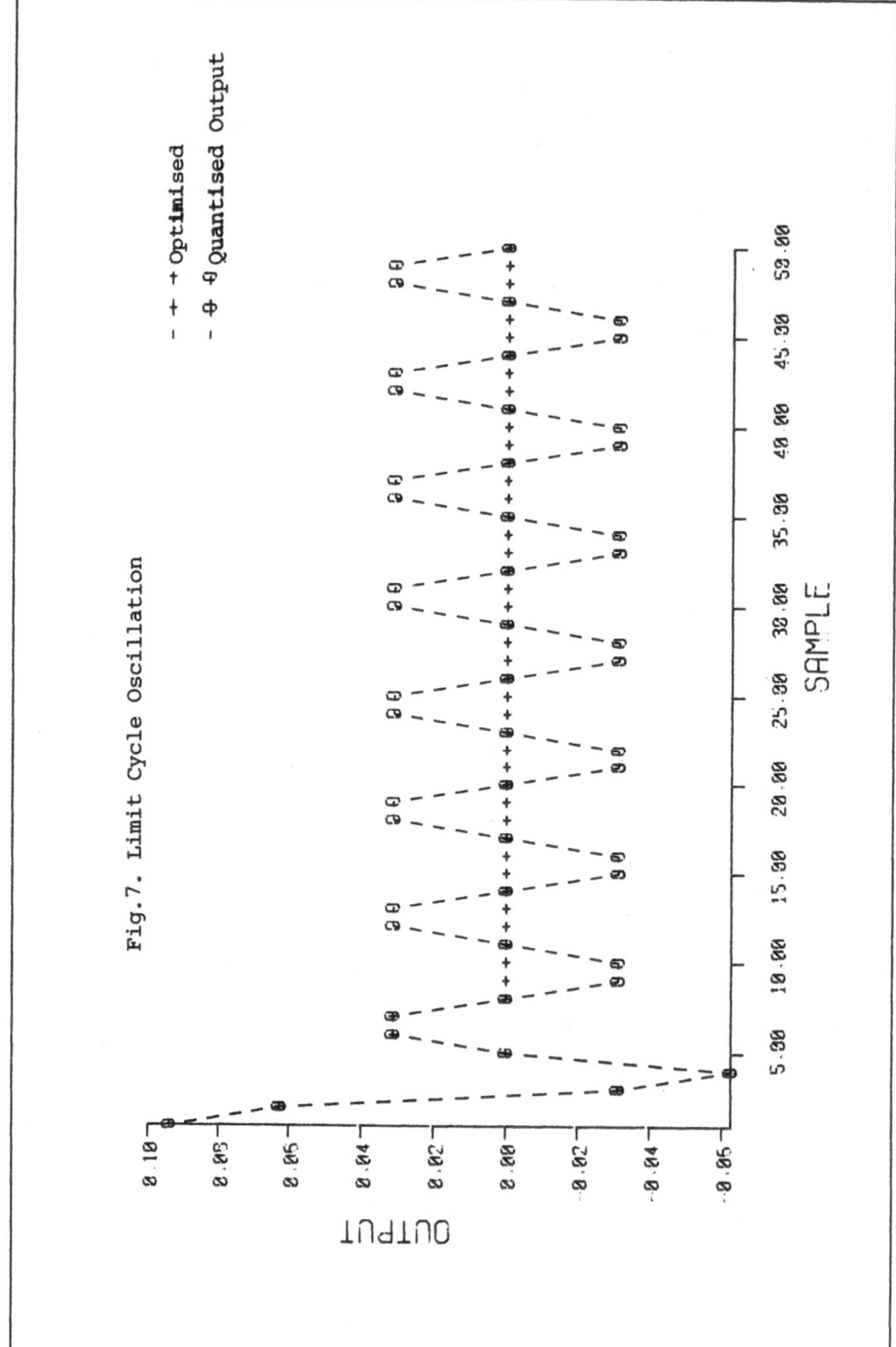

Fig.7. Limit Cycle Oscillation

- + + Optimised
- ⊕ ⊕ Quantised Output

concept, of realising the mapping operation is to store the state transition table in a Read Only Memory (ROM). The memory is addressed by a word which is the concatenation of the present states and the inputs. Stored at that address in the ROM is a word representing the next states. This realisation is conceptually very simple and potentially very fast. However the major disadvantage is the size of ROM required for any practical system. For example, even a simple system such as a two pole filter with variables defined with a precision of 10 bits would require a ROM having 2^{30} storage locations, each containing 10 bit data.

A modification of this method has been suggested by I. Proudler who proposes that a normal arithmetic system realisation be used together with a ROM containing only those state transitions in the optimised system which are different from the transitions generated by the arithmetic realisation. This technique, although not producing the global optimum filter, requires typically only a few hundred storage locations for a two pole filter.

Since the mapping of eqn.1 is finite, it may be represented as a set of multinomials over a finite field of order N where N is sufficiently large that each of the finite values in the mapping may be represented by a unique field element (2).
Consider the state mapping:

$$\underline{s}(n+1) = f_1[\underline{s}(n),\underline{x}(n)]$$

where

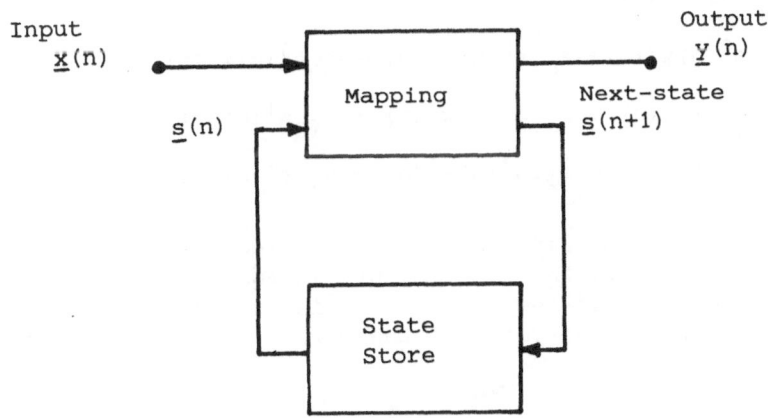

Fig.8. Finite State Machine

$$\underline{S}(n) = \begin{bmatrix} S_o(n) \\ S_1(n) \\ \vdots \\ \vdots \\ S_{M-1}(n) \end{bmatrix} \qquad\qquad \underline{x}(n) = \begin{bmatrix} x_o(n) \\ x_1(n) \\ \vdots \\ \vdots \\ x_{P-1}(n) \end{bmatrix}$$

$x_i(n), s_i(n+1) \in S = \{$Set of elements of finite field F$\}$

The ith element in the state vector $\underline{s}(n)$ may be expressed as:

$$s_i(n+1) = f_i[s_o(n), --, s_{M-1}(n), x_o(n), --, x_{P-1}(n)] \tag{3}$$

which in turn may be expressed as a multinomial:

$$S_i(n+1) = \sum_{q_o=0}^{N-1} \cdots \sum_{q_{M-1}=0}^{N-1} \sum_{j_o=0}^{N-1} \cdots \sum_{j_{P-1}=0}^{N-1} a(i)_{q_o \cdots q_{M-1} j_o \cdots j_{M-1}} \left\{ \prod_{k=0}^{M-1} s_k^{q_k}(n) \right\} \left\{ \prod_{m=0}^{P-1} x_m^{j_m}(n) \right\} \tag{4}$$

where the arithmetic operations are those of the field F.

Some insight is obtained by writing out eqn.4 for the simple system having a single input variable and a single state variable over the field GF(3)

$$s(n+1) = a_{00}s^0(n)x^0(n) + a_{01}s^0(n)x^1(n) + a_{02}s^0(n)x^2(n)$$

$$+a_{10}s^1(n)x^0(n) + a_{11}s^1(n)x^1(n) + a_{12}s^1(n)x^2(n)$$

$$+a_{20}s^2(n)x^0(n) + a_{21}s^2(n)x^1(n) + a_{22}s^2(n)x^2(n) \tag{5}$$

If the system is to be realised in the multinomial form then it is necessary to calculate the coefficients in eqn.4 from the specified state behaviour of the system. It is convenient to consider first a single variable system of the form:

$$s(n+1) = \sum_{q=0}^{N-1} a_q s^q(n) \tag{6}$$

where $s(n+1), s(n), a_q \in S$

The coefficients a_i in the polynomial representation of eqn.6 may be determined by allowing the present state $s(n)$ to be each field element in turn and writing the corresponding value of the next state s(n+1).

Set s(n) = r and the let the corresponding value of $s(n+1) = y_r$ then eqn.6 becomes:

$$y_r = \sum_{q=0}^{N-1} a_q r^q \qquad \forall r \in S \qquad (7)$$

Multiply each side of eqn.7 by r^j and sum over $\forall r \in S$

$$\sum_{r \in S} y_r r^j = \sum_{r \in S} \sum_{q=0}^{N-1} a_q r^{q+j}$$

$$= \sum_{q=0}^{N-1} a_q \sum_{r \in S} r^{q+j}$$

Now $\qquad \sum_{r \in S} r^{q+j} = \begin{cases} 0, & q+j \neq 0 \\ -1, & q+j = 0 \end{cases}, \quad 0^0 \equiv 0$

where r^{-i} is the multiplicative inverse of r^i over F.

$$\therefore \quad a_q = \sum_{r \in S} y_r r^{-q} \qquad (8)$$

Eqn.8 enables the polynomial coefficients a_q to be calculated from the specified state transitions of the system. The extension to a multivariable system is straightforward. For example, for the two-variable system

$$S_o(n+1) = \sum_{q=0}^{N-1} \sum_{j=0}^{N-1} a_{qj}(o) \, s_o^q(n) \, s_i^j(n) \qquad (9)$$

$$S_i(n+1) = \sum_{q=0}^{N-1} \sum_{j=0}^{N-1} a_{qj}(1) \, s_o^q(n) \, s_i^j(n) \qquad (10)$$

Then $\qquad a_{ij}(o) = \sum_{r \in S} \sum_{t \in S} y_{rt}(o) \, r^i \, t^j \qquad (11)$

$$a_{ij}(l) = \sum_{r \in S} \sum_{t \in S} y_{rt}(l) \, r^i \, t^j \tag{12}$$

Where

$$y_{rt}(0) = S_0(n+1) \Big|_{\substack{S_0(n) = r \\ S_1(n) = t}}$$

$$y_{rt}(1) = S_1(n+1) \Big|_{\substack{S_0(n) = r \\ S_1(n) = t}}$$

Thus a general F.S.M. may be represented by the general multi-nomial of eqn.4 and the multinomial coefficients easily calculated from the specified state transitions of the system.

For the special case of a linear F.S.M. the general non-linear system mapping, give by eqn.2, reduces to:

$$\underline{s}(n+1) = \underline{A} \ \underline{s}(n) + \underline{B} \ \underline{x}(n)$$

$$y(n+1) = \underline{C} \ \underline{s}(n) + \underline{D} \ \underline{x}(n)$$

Where $\underline{A}$, $\underline{B}$, $\underline{C}$ and $\underline{D}$ are matrices of appropriate dimension. The stability of such a machine may be investigated by considering the autonomous machine.

$$\underline{s}(n+1) = \underline{A} \ \underline{s}(n) \tag{13}$$

Application of a similarity transformation to matrix $\underline{A}$ enables any linear machine to be represented as a linear feedback shift register having a cycle structure isomorphic to the original machine(3) and system stability is easily investigated. It would be convenient if a similar formulation could be found for the general non-linear autonomous machine:

$$\underline{s}(n+1) = f \ [\underline{s}(n)]$$

Consider the polynomial representation of eqn.4 for a single state variable machine

$$s(n+1) = \sum_{j=0}^{N-1} a_j \, s^j(n) \tag{14}$$

where $s(n+1), s(n), a_i \in S$

Since the field F is finite, powers of the next state $s(n+1)$ may be expressed as polynomials and in general:

$$s^i(n+1) = \sum_{j=0}^{N-1} a_{ij} \, s^j(n) \tag{15}$$

It should be noted that the coefficients a_{ij} in eqn.15 are functions of the coefficients a_j in eqn.14 and should not be considered as independent variables. Eqn.15 may be written in matrix form as:

$$
\begin{bmatrix}
1 \\
s(n+1) \\
s^2(n+1) \\
\vdots \\
\vdots \\
s^{N-1}(n+1)
\end{bmatrix}
\begin{bmatrix}
1 & 0 & 0 \text{------} 0 \\
a_{10} & a_{11} & a_{12} \text{-----} a_{1(N-1)} \\
a_{20} & a_{21} & a_{22} \text{-----} a_{2(N-1)} \\
\vdots & \vdots & \vdots \\
\vdots & \vdots & \vdots \\
a_{(N-1)0} & a_{(N-1)1} & a_{(N-1)2} \text{----} a_{(N-1)(N-1)}
\end{bmatrix}
\begin{bmatrix}
1 \\
s(n) \\
s^2(n) \\
\vdots \\
\vdots \\
s^{N-1}(n)
\end{bmatrix}
$$
$$\tag{16}$$

or $\quad \underline{s}(n+1) = \underline{M} \, \underline{s}(n) \tag{17}$

Note that eqn.17 has the same form as eqn.13 for the autonomous linear machine. However there are some important differences. The vectors in eqn.13 have elements which are the individual state variables in the multivariable linear machine whereas the elements of the vectors in the formulation of the non-linear machine are not independent variables. In spite of this the form of eqn.17 allows the non-linear machine to be studied as a linear machine embedded in an extended state space. A major disadvantage of this method is that the polynomial coefficients have to be calculated for each power of $s(n+1)$. However this computation may be included in the state transition matrix formulation in the following way.

Let the elements of the field F be denoted by the integers $0, 1, 2, --N-1$. It should be noted that the integers are merely labels for the abstract field elements although it will be assumed that 0 is the identity element under addition and 1 is the identity under multiplication. In eqn.16 let the present state $s(n)$ take successively the value of each field element, then:

$$[\underline{y}_0 \ \underline{y}_1 \qquad \underline{y}_{N-1}] = \underline{M} \cdot \begin{bmatrix} 1 & 1^0 & 2^0 & - - - - (N-1)^0 \\ 0 & 1^1 & 2^1 & - - - -(N-1)^1 \\ 0 & 1^2 & 2^2 & - - - -(N-1)^2 \\ \vdots & \vdots & \vdots & \vdots \\ 0 & 1^{N-1} & 2^{N-1} & - - -(N-1)^{N-1} \end{bmatrix} \qquad (18)$$

where $\underline{y}_r = [1 \ s(n+1) \ s^2(n+1) - - - - - s^{N-1}(n+1)]^t$ is the next state column vector corresponding to the present state column vector $\underline{s}(n)$ when $s(n) = r$. Now $\underline{y}_r \in S$, so that $\underline{y}_r$ will be identical to one of the columns of the post-multiplying matrix on the right hand side of eqn.18. Therefore eqn.18 may be written as:

$$\begin{bmatrix} 1 & 1^0 & 2^0 & - - - (N-1)^0 \\ 0 & 1^1 & 2^1 & - - - (N-1)^1 \\ 0 & 1^2 & 2^2 & - - - (N-1)^2 \\ \vdots & \vdots & \vdots & \vdots \\ 0 & 1^{N-1} & 2^{N-1} & - - - (N-1)^{N-1} \end{bmatrix} \begin{bmatrix} \underline{c}_0 & \underline{c}_1 & - - - \underline{c}_{N-1} \end{bmatrix} = \underline{M} \begin{bmatrix} 1 & 1^0 & 2^0 & - - - (N-1)^0 \\ 0 & 1^1 & 2^1 & - - - (N-1)^1 \\ 0 & 1^2 & 2^2 & - - - (N-1)^2 \\ \vdots & \vdots & \vdots & \vdots \\ 0 & 1^{N-1} & 2^{N-1} & - - - (N-1)^{N-1} \end{bmatrix}$$

$$(19)$$

where $\underline{c}_r$ is a column vector having one element equal to unity and all other elements equal to zero. The position of the unity element is such that the vector $\underline{c}_r$ selects that vector from the pre-multiplying matrix, in eqn.18, which is identical to $\underline{y}_r$.

An example may make this step more clear. Consider an F.S.M. defined over field GF(4) with the following state transition table.

s(n)	s(n+1)
0	0
1	3
2	1
3	2

Eqn.17 becomes

$$\begin{bmatrix} 1 & 3^0 & 1^0 & 2^0 \\ 0 & 3^1 & 1^1 & 2^1 \\ 0 & 3^2 & 1^2 & 2^2 \\ 0 & 3^3 & 1^3 & 2^3 \end{bmatrix} = \underline{M} \begin{bmatrix} 1 & 1^0 & 2^0 & 3^0 \\ 0 & 1^1 & 2^1 & 3^1 \\ 0 & 1^2 & 2^2 & 3^2 \\ 0 & 1^3 & 2^3 & 3^3 \end{bmatrix}$$

It is clear that the above expression may be written in the form of eqn.19 as:

$$\begin{bmatrix} 1 & 1^0 & 2^0 & 3^0 \\ 0 & 1^1 & 2^1 & 3^1 \\ 0 & 1^2 & 2^2 & 3^2 \\ 0 & 1^3 & 2^3 & 3^3 \end{bmatrix} \begin{bmatrix} 1 & 0 & 0 & 0 \\ 0 & 0 & 1 & 0 \\ 0 & 0 & 0 & 1 \\ 0 & 1 & 0 & 0 \end{bmatrix} = M \begin{bmatrix} 1 & 1^0 & 2^0 & 3^0 \\ 0 & 1^1 & 2^1 & 3^1 \\ 0 & 1^2 & 2^2 & 3^2 \\ 0 & 1^3 & 2^3 & 3^3 \end{bmatrix}$$

Finally eqn.19 leads to an expression for the state transition matrix $\underline{M}$ as:

$$\underline{M} = \begin{bmatrix} 1 & 1^0 & 2^0 & \cdots & (N-1)^0 \\ 0 & 1^1 & 2^1 & \cdots & (N-1)^1 \\ 0 & 1^2 & 2^2 & \cdots & (N-1)^2 \\ \vdots & & & & \vdots \\ 0 & 1^{N-1} & 2^{N-1} & \cdots & (N-1)^{N-1} \end{bmatrix} \begin{bmatrix} \underline{c}_0 & \underline{c}_1 & \cdots & \underline{c}_{N-1} \end{bmatrix} \begin{bmatrix} 1 & 1^0 & 2^0 & \cdots & (N-1)^0 \\ 0 & 1^1 & 2^1 & \cdots & (N-1)^1 \\ 0 & 1^2 & 2^2 & \cdots & (N-1)^2 \\ \vdots & & & & \vdots \\ 0 & 1^{N-1} & 2^{N-1} & \cdots & (N-1)^{N-1} \end{bmatrix}^{-1} \quad (20)$$

The inverse matrix may be determined by making use of the following identities over the finite field GF(N).

$$u^{N-1} = 1 \qquad \forall u \in S, \ u \neq 0$$

$$\sum_{u \in S} u^i = 0 \qquad i = 1, 2, \ldots, N-2$$

$$\sum_{i=1}^{N-1} u^i = 0 \qquad \forall u \in S, \ u \neq 1 \qquad (21)$$

$$\sum_{i=1}^{N-1} 1 = -1$$

$$\underline{M} = \begin{bmatrix} 1 & 1^0 & 2^0 & \cdots & (N-1)^0 \\ 0 & 1^1 & 2^1 & \cdots & (N-1)^1 \\ 0 & 1^2 & 2^2 & \cdots & (N-1)^2 \\ \vdots & & & & \vdots \\ 0 & 1^{N-1} & 2^{N-1} & \cdots & (N-1)^{N-1} \end{bmatrix} \begin{bmatrix} \underline{c}_0 & \underline{c}_1 & \cdots & \underline{c}_{N-1} \end{bmatrix} \begin{bmatrix} 1 & 0 & 0 & \cdots & -1 \\ 0 & -1^{-1} & -1^{-2} & \cdots & -1^{-(N-1)} \\ 0 & -2^{-1} & -2^{-2} & \cdots & -2^{-(N-1)} \\ \vdots & & & & \vdots \\ 0 & -(N-1)^{-1} & -(N-1)^{-2} & \cdots & -(N-1)^{-(N-1)} \end{bmatrix} \quad (22)$$

The stability of a single variable non-linear F.S.M. may now be studied in terms of the simple permutation matrix in the above equation.

Although consideration has been given to only single state variable systems the formulation developed can be extended readily to multivariable systems in the following way. Consider a system having q variables each of which may be represented by elements from the finite field GF(F). The system can take on any one of F^q combinations of states so that the total system state can be represented as a single variable in the extension field GF(F^q).

3. DISCUSSION

A method, termed the minimum spanning arborescence method, has been introduced for the optimal representation of digital signal processing systems. The representation obtained is that of a state-transition table which must be realised in terms of hardware. In general the state transition table is simply an abstract mapping in which the labelling of a state need not have any particular physical significance. In other words the particular label adopted for a state need not be interpreted as a binary number equal to the quantised signal level. This being the case it is no longer necessary to constrain the signal quantisation to be linear (or some simple law). The quantisation may be non-linear and may be chosen to optimise some aspect of the system performance in the "real world" eg. signal to noise ratio(4).

There are many aspects of the system realisation problem which must be investigated before the finite field system is of general use. The most pressing problem is efficient realisation of the general multinomial form given by eqn.4. There are two approaches to this which bear promise. As mentioned above the state labels are abstract and may, therefore, be assigned a binary representation (assuming a binary logic realisation) arbitrarily and one would wish to choose a binary representation which minimises the hardware required in the realisation. To an extent this is a problem which has been studied by many people with a fairly marked lack of success in the general case. However it is hoped that restricting attention to particular classes of system (eg. stable I.I.R. filters) could be more tractable than the general problem. The second approach to efficient realisation of the multinomial form is to consider factorisation of the multi-nomial. As a simple example consider the single state variable system defined by:

$$s(n+1) = s^6(n) + 4s^5(n) + 2s^4(n) + s^3(n) + 4s^2(n) + 2s(n) + 6$$

over the field GF(7). This may be expressed in the form:

$$s(n+1) = [s(n)+3]^6 + 5$$

which is obviously much simpler to realise than the previous expression.

Some rather special classes of system have very simple realisations over a finite field. Such a class is that of circular

convolution where the general state mapping of eqn.1 becomes a linear operator on a finite field. The linear operators are, of course, the Number Theoretic or Finite Field Transforms.

4. References
1. Even,S.Graph Algorithms.(Computer Science Press)
2. Rayner,P.J.W.The Application of Finite Arithmetic Structures to the Design of Digital Processing Systems.(Digital Signal Processing,eds.Cappellini and Constanides,Academic Press,1981)
3. Stone,H.S.Discrete Mathematical Structures and their Applications (Science Research Associates Inc.1973)
4. Albinson,L.J.and Rayner,P.J.W.Finite Automata Representation of Digital Filters.Proceedings International Conference on Digital Signal Processing.Florence,Italy,1981.pp432-437.

VLSI IMPLEMENTED SIGNAL PROCESSING ALGORITHMS

David G. Messerschmitt

Department of Electrical Engineering and Computer Sciences
University of California
Berkeley, California 94720
U.S.A.

1. INTRODUCTION

The sophistication and complexity of signal processing algorithms which can be implemented at reasonable cost is increasing dramatically with time. This is due to the rapid advances in integrated circuit technology. If, however, the capabilities of the technology are to be fully exploited, certain inherent characteristics must be taken into account in the design of the signal processing system. Conversely, basic architectural decisions, such as whether the implementation is to include programmable or special-purpose chips, is to be analog or digital, etc., affect the choice of signal processing algorithms.

This purpose of this article is to briefly summarize, in Section 2, the characteristics of the evolving integrated circuit technology as they relate to signal processing and communications applications, and then to illustrate the rather dramatic impact that the implementation approach can have on the algorithms which are chosen by way of a couple of examples. These examples are drawn from our recent research at Berkeley, and include the implementation of fast digital filters in Section 3 and full-duplex data modems in Section 4.

2. GENERAL CHARACTERISTICS OF VLSI

The integrated circuit technology is rapidly evolving toward improved performance, as measured by speed, complexity, power consumption, and cost. The basic driving force in this performance improvement is the continual scaling down of the dimensions of the devices on the chip in the MOS-derived technologies. All of the performance parameters, including speed, available complexity, and power consumption, improve as this scaling proceeds.

In this Section, the characteristics of this technology both today and as they are projected to change in the future are briefly summarized.

2.1. Digital Integrated Circuits

The most complex digital chips today have several hundred thousand devices. Somewhere in the range of fifty thousand devices is the transition between large scale integration (LSI) and very large scale integration (VLSI). The transition between LSI and VLSI is a part of the general evolutionary advance in device complexity. The term VLSI is usually associated with MOS digital IC's, although analog MOS IC technology is also advancing rapidly as discussed in the next Section. The primary driving force behind the advance in digital IC's is the computer industry.

The most rapidly advancing technology is MOS, and in particular CMOS. Bipolar technologies offer inherently higher speed, since speed is controlled by the vertical base thickness rather than lateral channel length which can be controlled less accurately by lithography. However, bipolar technologies also have a lower device density, due to the need for device isolation. Thus, VLSI is an MOS technology, and due to the major research and development effort increasingly a CMOS technology.

If the relatively low speed MOS VLSI technology is to be fully exploited in performance-dominated applications, ways must be found to utilize its complexity without relying on speed. This implies an inherent advantage for processing techniques which employ concurrency or parallelism.

To get more specific, MOS device density can be expected to follow the present trend of doubling approximately every 18 months [1] for about ten years until some fundamental limits are reached [2] and the largest chips have approximately 10 million devices (two orders of magnitude greater than today). The power supply voltage will have to be reduced to approximately two volts or less (to maintain reliability), and the device speed will be approximately one order of magnitude greater than today [3]. These devices will be no faster than the fastest bipolar circuits available today.

The conclusion is that the performance of VLSI as measured by the product of speed and complexity will be approximately a factor of 1000 greater than today's MOS circuits. However, perhaps only a factor of 250 to so can be exploited due to the inefficiencies of using the greater complexity. VLSI will offer no functional capability that is not available with today's fastest bipolar technologies, but it will offer dramatically lower cost, size, and power consumption.

One characteristic of VLSI is its complexity, which results in a high design cost. To some extent this can be reduced with automated design aids, but in systems where performance is not critical other alternatives which exchange lower performance for reduced design cost will continue to be used. These include using off-the-shelf programmable chips (such as microprocessors and programmable digital signal processors), gate arrays, standard cells, and logic arrays [4, 5]. Innovation in design approaches can be expected, a recent example being the idea of restructurable VLSI [6].

While in low performance applications there are many design alternatives, in applications where the capabilities of the technology are being stretched it is our thesis that a fairly rigid and structured design approach in which the capabilities and limitations of the technology are carefully taken into account must be followed. In the case of digital systems, high performance can usually be taken to mean high processing or computational speed. The nature of the constraints will be illustrated in Section 3 for an example of digital filter design.

2.2. Analog Integrated Circuits

In the literature of VLSI, digital integrated circuits are usually emphasized, and indeed for a broad class of applications including the computer industry this is appropriate. However, analog integrated circuits will be critical to the development of communication systems, and conversely communications will continue to be the driving force behind the development of analog technologies as it has been in the past.

The first reason for the importance of analog IC's is that most of the interfaces to the external world in a communication system are analog in nature. This includes the signals being transmitted, including voice and video but excluding data, as well as the interface to the actual transmission media. Of course, this is obvious in the case of an analog transmission system, but it is perhaps ironic that most of the design of even a digital transmission system (whether the medium is cable or radio) is inherently analog. This fact will be illustrated in Section 4, where the design of a data modem will be considered.

The second reason for the importance of analog IC's is that many of the signal processing functions in a communication system can be most efficiently realized with analog circuitry. This will be discussed further in the remainder of this Section.

The most important analog circuits are the interface circuits, which perform the analog-to-digital and digital-to-analog conversions at the interfaces to the system. Because many interfaces in a communications system (to both the signal source and the channel) are inherently analog, interface circuits will always be a critical and often performance limiting (where performance is measured by accuracy as well as speed). The speed and accuracy of a digital signal processing system are limited by the interface circuits.

Aside from the interfaces, analog circuits are well suited to certain types of signal processing functions, and principally filtering. This is particularly true as the required speed is increased. The fundamental reason is that in a digital implementation, efficiency is based on the multiplexing of hardware, and therefore as the sampling rate is increased the amount of digital circuitry must be increased almost in direct proportion (this will be illustrated in the digital filtering approach described in Section 3). In an analog realization, on the other hand, an increase in speed comes for "free", in the sense that it does not usually result in more circuitry. The speedup is achieved by a most careful circuit design, within the fundamental limitations of the technology used.

The tradeoff between analog and digital realizations is affected directly by the cost, speed, and accuracy of the interface analog circuits in a system which interfaces an analog world. Digital realizations are penalized when the filtering function is simple by the relatively high cost of the data converters. However, as the data converters and other analog interface functions continue to improve in cost and performance, digital implementation of signal processing functions looks more attractive. Of course, improvements in analog interface circuitry will continue to be accompanied by improvements in analog circuits for functions such as filtering, so that the trend toward increased digital implementation of signal processing functions such as filtering will not occur as rapidly as might otherwise be expected.

Analog integrated circuits can benefit from the same scaling phenomenon as digital circuits, and in fact some performance properties such as speed and power consumption improve with the scaling. However, scaling and the attendant reduction in power supply voltage have the undesirable effect of reducing the dynamic range of the signals (or equivalently increasing the level of the circuit noise in relation to the signal level). Analog circuits are of course more adversely affected by this effect than digital circuits, and hence benefit less from scaling.

The accuracy of analog circuits such as filters are also dependent on the component tolerances of the circuit elements (often as manifested by the ratio of component values), and these tolerances also must loosen as scaling occurs since they are often dependent on the accuracy of the component dimensions.

As in digital circuitry, MOS and CMOS is likely to be the dominant technology. This is because of the availability of almost ideal capacitors and switches in MOS, and because analog and digital circuitry can be combined on the same chip [7]. Combinations of switches, capacitors, and operational amplifiers have been shown to be applicable to many signal processing functions, the most predominant being filtering. Switched capacitor techniques are now used prominently for both data converters and for the low pass filtering function required to prevent aliasing in the sampling operation at voiceband frequencies [8]. Current research at Berkeley is focusing on realization of inherently accurate data converters which do not depend on component tolerances for accuracy, and on extending switched capacitor filtering techniques to frequencies as high as 10 to 20 MHz.

It can be anticipated that digital filtering techniques will make gradual inroads at voiceband frequencies. Switched capacitor techniques will dominate at high frequencies up to approximately 70 MHz. Higher frequencies will continue to be beyond the capabilities of a totally monolithic filtering solution for some time to come (although other technologies such as GaAs and SAW devices will be applicable).

Digital circuit techniques are capable of much more complicated control algorithms, particularly in conjunction with programmable processors. For this reason, digital techniques will continue to dominate functions which require processing more complicated than filtering (for example adaptive algorithms), as well as at voiceband and lower frequencies.

3. FAST RECURSIVE FILTERING

This Section will describe some techniques for fast digital filtering. While it is true as pointed out in Section 2.1 that filtering functions, particularly at high frequencies, are most efficiently implemented by analog techniques, this will serve to illustrate some of the constraints inherent in digital VLSI implementation of some digital signal processing functions. In particular, our point is that in the attempt to achieve the highest speed, the characteristics of the technology will influence the choice of a filter realization in a major way.

The most fundamental characteristic of VLSI is that computation and control functions are relatively plentiful, and communications is expensive. When we say communications, we mean communications between different parts of the same chip, as well as communications between chips. The reason for this is that as the devices are scaled down and the power supply voltage is reduced, there is less capability to drive the high capacitance of a metalization wire on a chip or even more so a wire between chips. The result is that the devices which drive these transmission lines cannot be scaled as effectively, and an increasing portion of a chip is devoted to transmission line drivers. These transmission line drivers are also a major consumer of power on the chip, and the total chip power consumption is limited by the ability of the packaging to dissipate the heat.

An additional factor is that the increase in complexity which can be reasonably implemented with VLSI cannot be accompanied by an increased requirement for communications between parts of the system because of the physical wiring problem. There is a limit to the complexity of the interconnection on a chip, because of the constraint of a small number of layers of metalization. Similarly, at the system level, a major expense is the physical wiring on both the printed circuit board and the backplane.

It was previously mentioned that in VLSI a premium is put on exploiting parallelism rather than speed for increases in processing throughput. This consideration coupled with the high cost of communication in a VLSI system implies the following: The most desirable algorithms for achieving maximum performance in a VLSI system are those with achieve increased parallelism with a minimum of communication between the parts of the algorithm. Where communication is necessary, it is advantageous that it be as localized as possible. The high cost of design also places a premium on structures which are regular; that is, have a replication of identical computational units. This is the source of interest in regular computational structures such as systolic arrays which have only localized interconnection [9].

As an illustration of these ideas, consider the problem of digital filtering. Assume in particular that it is desired to achieve the maximum performance, as measured by sampling rate, in the VLSI implementation of a particular digital filter transfer function. An implementation architecture similar to a systolic array which is capable of high performance will now be shown.

Suppose the Nth order filter we want to implement has the following transfer function:

$$y_n = \sum_{i=1}^{N} b_i y_{n-i} + \sum_{i=0}^{M} a_i x_{n-i} \tag{1}$$

where there are two interesting cases. One, the finite impulse response (FIR) filter corresponds to N=0. The other, the infinite impulse response (IIR) filter, corresponds to $N \geqslant M$ and $N > 0$. The tradeoff between these two cases is roughly that to approximate a given desired response, the FIR filter requires a much larger order M and hence a larger computation rate (as measured by multiplies and adds per output sample). However, when the sampling rate at which a filter can be implemented for a given speed of hardware is considered, the FIR filter appears at first examination to be faster because of the natural way in which parallelism can be exploited.

The method for achieving speed with an FIR filter is simply to calculate a vector of L successive output samples in parallel. To see this, define output and input vectors of L successive samples as

$$Y_n = [y_{nL}, y_{nL+1}, \ldots, y_{(n+1)L-1}]^T \tag{2a}$$

$$X_n = [x_{nL}, x_{nL+1}, \ldots, x_{(n+1)L-1}]^T \tag{2b}$$

where n denotes the block number. If we take $L \geq M$ for an FIR filter, then it is easy to see that the filter can be represented as

$$Y_n = AX_n + BX_{n-1} \tag{3}$$

where A and B are appropriate $L \times L$ band matrices. The form of the filter given by (1) we refer to as SISO (single-input single-output) and (2-3) is referred to as MIMO (multiple-input multiple-output). The MIMO system is shown schematically in Figure 1.

Since there are no feedback terms in (3), an implementation can be completely pipelined. Each stage of the pipeline can perform an elementary operation such as multiplication or addition. The rate at which new Y_n vectors can be generated is determined by the slowest of these elementary operations (usually the multiplication time). Since each vector of samples represents L samples at the original sampling rate, the sampling rate which is achieved for a given speed of hardware is proportional to L. Hence, by simply increasing L, of course at the expense of additional hardware, the sampling rate of the filter can be increased without increasing the required speed of the hardware.

The FIR filter implementation thus addresses the problem of achieving increased speed through the application of parallelism. The additional constraint of VLSI is to minimize the communications, and keep the communications localized as much as possible. This problem is addressed neatly for the FIR type of computation by the use of systolic arrays [9, 10, 11]. An example of a systolic architecture will be seen shortly in conjunction with the IIR filter realization.

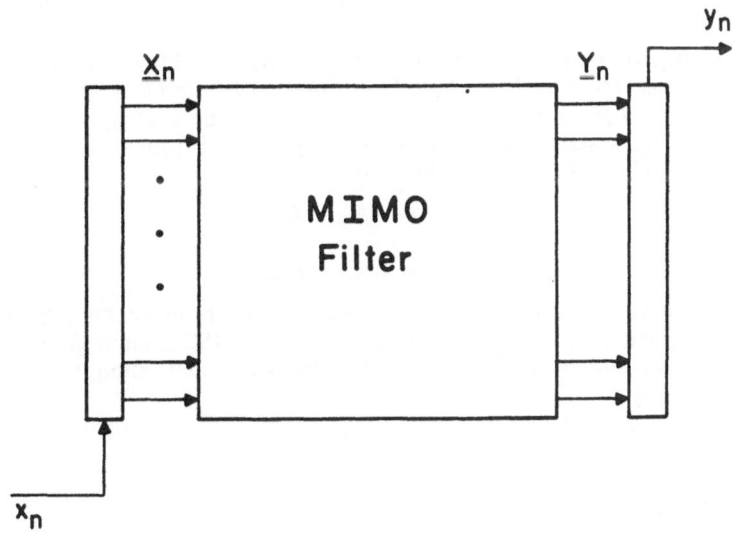

Fig. 1 MIMO Filter Structure

Conventional wisdom would say that unlike the FIR filter, for a given speed of hardware the sampling rate of the IIR filter is limited by the feedback inherent in the recursion. It will now be shown that this is not the case, and specifically that an architecture with unlimited speed (similar to the FIR filter) can be defined. This architecture of course exploits parallelism, and can be implemented in a fashion which requires only local communication.

The representation we will describe is based on a block state realization of the IIR filter. Consider first the realization of the filter given as in (1), namely a single-input single-output form. For an Nth order filter, the minimal state space representation can be written as

$$r_{n+1} = Ar_n + bx_n \qquad (4a)$$

$$y_n = cr_n + dx_n \qquad (4b)$$

where A, b, c and d are, respectively, $N \times N$, $N \times 1$, $1 \times N$ and 1×1 constant matrices. The r_n, x_n and y_n are the N-component state vector, the input sample and the output sample at time n respectively. Taking the z-transform of equations (4), we get:

$$H(z) = \frac{Y(z)}{X(z)} = c(zI - A)^{-1}b + d \qquad (5)$$

Hence, a state space realization can be represented by a quartuple (A, b, c, d).

State equations can be applied to filters of any order N, as long as the state matrix A is of size $N \times N$. A can vary from a full matrix to a very simple matrix, such as a diagonal one. A block diagonal representation can always be obtained as follows. First, the filter is realized as a cascade of filters, each of which has no multiple poles. Then each of these transfer functions with only simple poles is expanded in partial fractions to yield a parallel realization. Finally, this parallel realization results in straightforward fashion in a state representation as in (4) in which the state matrix A is block diagonal. It consists of the form

$$A = diag(A_1, A_2, \ldots, A_m) \qquad (6)$$

where each of the A_i is a 2×2 matrix corresponding to a state representation of one complex pole pair.

The key to finding an architecture suitable for a fast IIR filter is to define a block state realization of the filter which realizes the relationship between a block of input samples and a block of output samples given in (2). This is analogous to the strategy which was followed in the FIR case. The block state equation can be easily derived from single-input single-output state equations[12]. The state equations become

$$R_{n+1} = AR_n + BX_n \qquad (7a)$$

$$Y_n = CR_n + DX_n \qquad (7b)$$

where the new state vector is $R_n = r_{nL}$ and where X_n and Y_n are defined by (2). The state matrix A can be determined from the state matrix of the SISO system given by (6). Let us represent the SISO system by $(\hat{A}, b, c, d)$. We can get a realization (A, B, C, D) in the MIMO system given by

$$A = \hat{A}^L \qquad\qquad B = [\hat{A}^{L-1}b \quad \hat{A}^{L-2}b \ \ldots \ b]$$

$$C = \begin{bmatrix} c \\ c\hat{A} \\ \cdot \\ \cdot \\ \cdot \\ c\hat{A}^{L-1} \end{bmatrix} \qquad D_{ij} = \begin{cases} 0, & i<j \\ d, & i=j \\ c\hat{A}^{i-j-1}b, & i>j \end{cases} \qquad (8)$$

Taking a close look at equations (7a,b), we find that matrices B, C and D can be pipelined without affecting the overall speed, while matrix A has to finish in one block period LT_s where L is the block size and T_s is the input sampling period.

Since A is block diagonal, we can assign each submatrix to one computational element without requiring any communication links among these submatrices. In order to satisfy the real time condition, the block size should be the smallest integer greater than or equal to the operation time of one submatrix divided by the input sampling period. The only constraint for the other three matrices is that the operation time for any computational element must not be greater than the block period.

The multiplications by matrices B, C and D can be achieved by a systolic array approach. Define the operation of one cell as in Figure 2 and the connection of this cell to implement a 3×4 matrix-vector multiplication as in Figure 3. The operation of this matrix on a vector at time n can be represented as:

$$\begin{bmatrix} a_{11} & a_{12} & a_{13} & a_{14} \\ a_{21} & a_{22} & a_{23} & a_{24} \\ a_{31} & a_{32} & a_{33} & a_{34} \end{bmatrix} \begin{bmatrix} X_{n,1} \\ X_{n,2} \\ X_{n,3} \\ X_{n,4} \end{bmatrix} = \begin{bmatrix} Y_{n,1} \\ Y_{n,2} \\ Y_{n,3} \end{bmatrix}$$

In Fig. 3, the coefficient stored in each cell is equivalent to the value in the corresponding position of the matrix. This two dimensional array receives input vectors from the top and outputs vectors to its right. Each cell sends its upper input to the cell below in the same column. The cells in the same row calculate the corresponding output sample. Each cell updates the output sample by adding the product ax to its left input. It then transmits the result to its right cell. The rightmost cells do the final updating and send out the results.

Suppose this array starts working on the nth input vector X_n. The left input y_{in}'s of the cells in the leftmost column are initialized to 0. In the first cycle, the upper left cell works on the first input sample of $X_{n,1}$. At the end of this cycle, the input $X_{n,1}$ will be sent to the cell

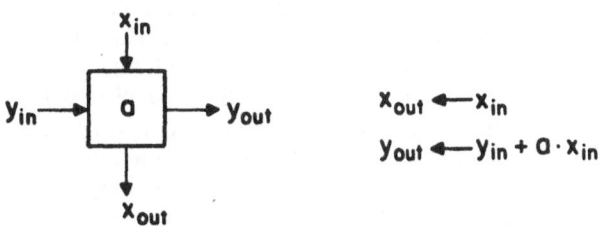

$$x_{out} \leftarrow x_{in}$$
$$y_{out} \leftarrow y_{in} + a \cdot x_{in}$$

Fig. 2 One Cell of a Systolic Array

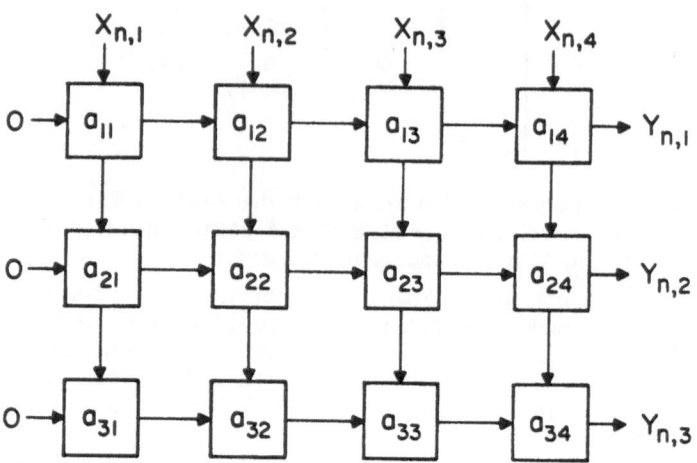

Fig. 3 A Systolic Array for Matrix Multiplication

below and the updated output $Y_{n,1} = a_{11} \times X_{n,1}$ will be sent to the right cell. After the upper left cell has completed its calculation, it will fetch the first input samples of the following input vector X_{n+1}. Thus, the time between two adjacent output samples on the same row will always be the execution time of one cell. In other words, the time between two adjacent vectors equals the execution time of one cell.

The block diagram of the block state filter using systolic arrays is drawn as in Figure 4. The three big rectangles represent the two dimensional arrays for implementing the multiplication of matrices B, C and D respectively. Each small square contains the operations of one 2×2 submatrix in matrix A. Notice that there are no downward links connecting these squares. Since each small square takes the time of four multiplications and four additions, each cell in any one of the three big rectangles should comprise four basic cells. The sizes of the three big rectangles depend, obviously, on the number of poles of the filters and on the block length of the input samples.

The block state filter architecture of Figure 4 is capable of very high speed IIR filtering. If the speed of the systolic cells remains fixed, the rate at which blocks of input samples are processed remains fixed. Thus, as the block size L increases, the overall effective input sampling rate increases in proportion to L. The only sampling rate limitation will be the speed of the hardware which does the data distribution and collection, since there remains a "serial to parallel conversion" function which must run at the rate of input samples. In addition to achieving concurrency, this structure has the desirable property of only local communications.

This structure illustrates the point made earlier; namely, that the hardware complexity of a digitally implemented filter inherently increases with the sampling rate. In this case, the number of cells increases in direct proportion to L, as does the input sampling rate. This property should be contrasted to analog filters, where the basic complexity is largely independent of bandwidth (although the difficulty of design increases with bandwidth).

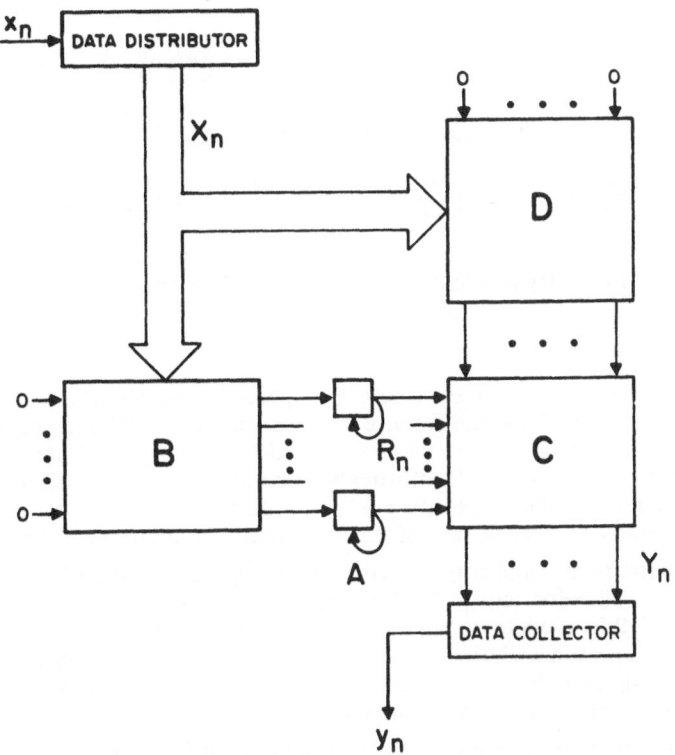

Fig. 4 Block State Filter Implemented with Systolic Arrays

4. FULL-DUPLEX DATA TRANSMISSION

The second application to be discussed in this article will illustrate how it is frequently impossible in communications applications to avoid analog circuitry, and the fact that it is often necessary to modify the algorithms to overcome the accuracy limitations of that analog circuitry.

The point was made earlier that in a communications application the interfaces to both the user and the transmission medium are inherently analog. This implies that at minimum data converters are required at these interfaces, and implementation of all or a part of the processing circuitry using analog circuitry can be considered.

Figure 5 shows a block diagram of one such application. This is a modem for transmitting a data stream over a single pair of wires simultaneously in both directions in the same bandwidth. The objective in an application of this technique to subscriber loop data transmission is to achieve full-duplex data at a rate of 144 kb/s over a distance on the order of 5 km maximum. The problem which must be faced is obtaining adequate isolation between the two directions of transmission, since there is a strong tendency for the local transmitter to interfere with the local receiver. The first defense is the hybrid, a bridge circuit which depends on knowledge of the loop impedance to obtain an isolation which is guaranteed to be no better than about 10 dB. Since the line attenuation can be up to 40-45 dB for a 5 km subscriber

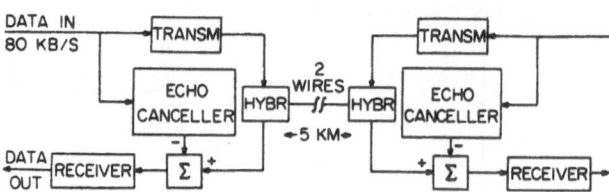

Fig. 5 Hybrid Method Full Duplex Data Transmission

loop (measured at half the data rate, where the spectral density peaks for the bipolar coded signal), the resulting signal to interference ratio is about -35 dB for the worst case. The second source of isolation is the echo canceler, which is an adaptive transversal filter which learns the response of the hybrid feedthrough, generates a replica of that feedthrough, and cancels it. The required echo cancellation is 50-55 dB for a 20 dB signal to noise ratio after cancellation. Even greater cancellation of the echo would be desirable if it could be achieved.

The implementation complexity of hybrid mode digital subscriber loops would result in a high manufacturing cost for any realization using off-the-shelf components. A monolithic implementation of this modem is clearly required to achieve a sufficiently low cost for widespread application on a majority of subscriber loops, and it would be very desirable if the modem could be implemented in standard integrated circuit technologies without trimming, adjustments, or external precision components.

Although this application is typical in the requirement for an analog interface, it requires somewhat greater accuracy than most applications. It is therefore an interesting case study in how the accuracy requirement in conjunction with the constraints of the implementation technology almost totally constrains what signal processing algorithms and implementation approach can be used. In fact, the accuracy requirements will now be shown to dictate either a largely analog implementation of the echo canceler or a largely digital implementation in which the algorithms are specifically modified to overcome the limitations of the implementation technology.

Four alternative implementation approaches will be considered, starting with the completely analog echo canceler using switched-capacitor techniques illustrated in Figure 6a. Further detail on these implementation approaches can be found in earlier papers [13, 14]. In the analog approach, the coefficients of the transversal filter are stored in analog integrators, and the binary weighting is done using switched capacitors. A summing amplifier performs the convolution sum. The adaptation algorithm is also implemented by a switched capacitor circuit that adds a correction term to the coefficients stored in the integrators. The desired correction term is obtained by multiplying the cancellation error coming from the output of the summing amplifier by either +1 or -1, corresponding to the data bit at the corresponding tap. The difficulty with this approach is in obtaining the required time constant for the adaptation circuit [13]. This is virtually impossible in the face of inevitable offset currents, etc.

In the fully digital approach, shown in Fig. 6b, the canceler is implemented entirely with digital logic, and the received signal is sampled and A/D converted prior to the cancellation. For someone oriented toward digital implementation, this is the most natural approach. Unfortunately, it too is very difficult to implement at present due to the required linearity and accuracy for the data converter. To achieve the required cancellation, the A/D converter must have the performance of a 12 or 13 bit ideal converter. Such performance may be

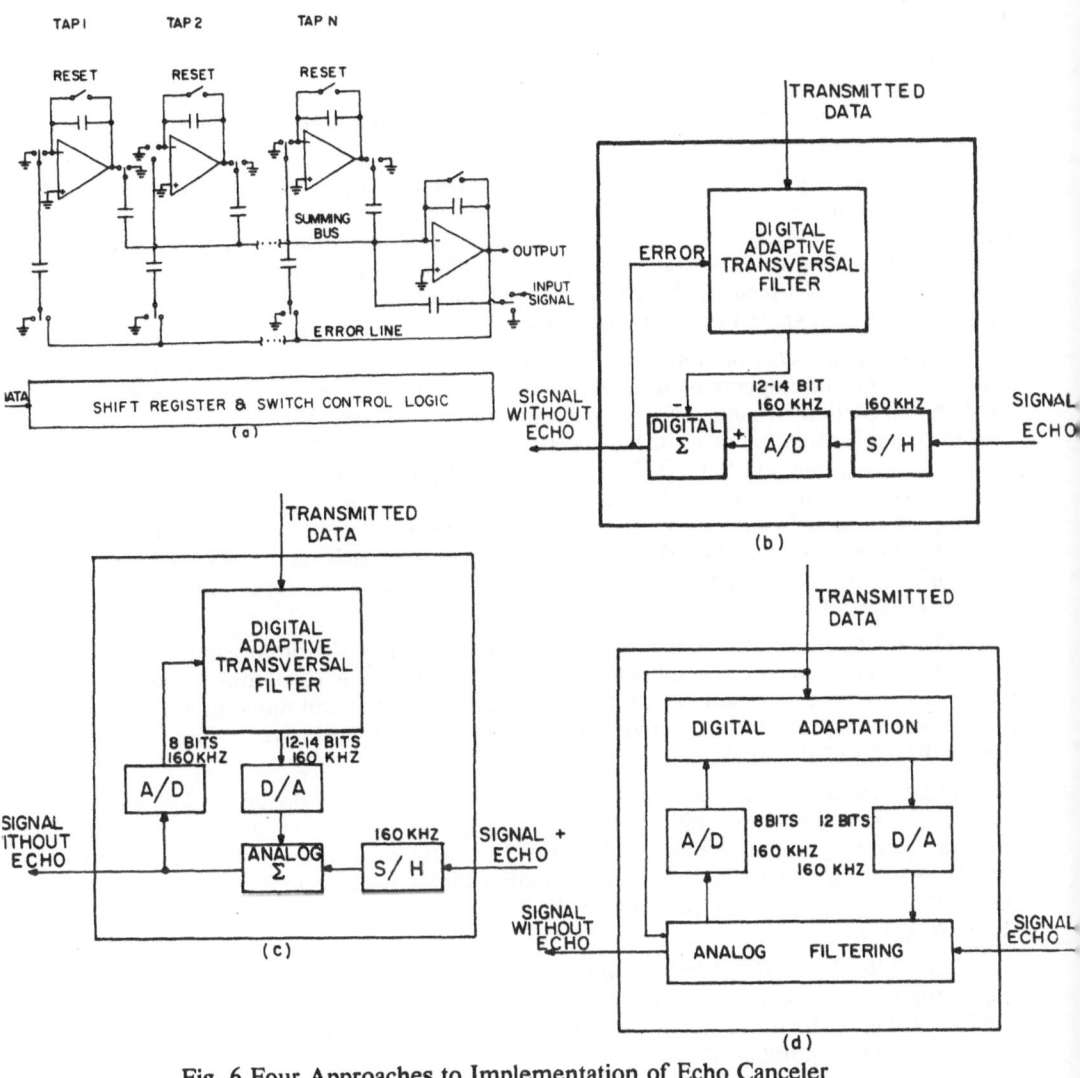

Fig. 6 Four Approaches to Implementation of Echo Canceler

available at some future date, but cannot be achieved today in a monolithic realization at the required sampling rate of 144 or 288 kHz.

The digital echo canceler with analog cancellation is shown in Fig. 6c. It uses an all-digital implementation of the canceler and converts the echo replica to analog using a D/A converter prior to cancellation in the analog domain. Resolution requirements for the D/A converter are similar to those for the A/D converter in Figure 6b, so at first it appears that this technique has no advantage. Although this accuracy would be easier to achieve in a D/A converter than an A/D converter, it is still very difficult. However, for this case if the D/A converter meets the relaxed condition of monotonicity, which is easily achieved, it turns out that the digital transversal filter can be modified so that it can adaptively overcome the limit tions of the data converter. This technique will be described momentarily.

Since the A/D is used only to generate a feedback signal for the adaptation algorithm, must be monotonic, but not necessarily linear. Furthermore, a resolution of only 8 bits required [2]. The MOS monolithic implementation of these converters poses no major pro lems.

An architecture in which the convolution sum and the echo cancellation are perform with analog techniques is shown in Fig. 6d. The adaptation algorithm is performed digitall thereby taking advantage of the long time constants that can be achieved. The tap weights a converted to analog using a single time-shared D/A converter. The multiplications and add tions are then performed in the charge domain, where the nonlinear distortion can be ke very small without special circuit design effort.

Since a large dynamic range and slow time-constant are required in the adaptation alg rithm, it cannot be implemented by analog means. Thus the use of a digital processor to pe form the adaptation, while a switched capacitor analog transversal filter computes the convol tion sum and cancels the echo in the received signal. The coefficients of the transversal filt are stored in sample and hold (S/H) capacitors and refreshed periodically by the digital ada tation processor. In order to reduce the required speed of the D/A converter, only one of th S/H capacitors is refreshed each pulse period; however, all are updated in the digital proce sor. The possibility of a slow D/A converter in this approach represents a significant adva tage, since slow operation could be traded off for increased resolution and reduced die ar and power consumption.

The major advantage of Fig. 6d is that the adaptation algorithm can adapt to overcon the effect of D/A converter nonlinearity as long as the D/A is monotonic. The reason is th the error is in the tap weight, which the adaptation algorithm will move in a direction so as reduce the overall cancellation error. This is therefore an attractive architecture for mono ithic realization of the modem.

Most designers would prefer to design a chip which is as digital as possible, and add tionally digital technologies will benefit more from scaling as mentioned previously. Ther fore, the primarily digital approach of Figure 6c is still intriguing. It has been shown that modification to the digital canceler can be made which enables it to compensate adaptively fc the nonlinearity of the D/A converter [14]. This modification is based on the following nor linear expansion.

Let $f(B_0, \cdots, B_{N-1})$ be an arbitrary (nonlinear) function of N bits, where B_i assume one of the values 0 or 1. Over all combinations of N bits this function assumes a total of 2 possible values (which are not necessarily distinct). This function can be represented as series with a finite number of terms,

$$f(B_0, \cdots, B_{N-1}) = f_0 + \sum_{k=0}^{N-1} f_1(k) B_k$$

$$+ \sum_{k_1 \neq k_2} f_2(k_1, k_2) B_{k_1} B_{k_2}$$

$$+ \cdots$$

$$+ \sum_{k_1 \neq k_2 \neq \cdots \neq k_L} f_L(k_1, k_2, \cdots, k_L) B_{k_1} B_{k_2} \cdots B_{k_L} \qquad (9)$$

$$+ \cdots$$

$$+ \sum_{k=0}^{N-1} f_{N-1}(k) B_0 B_1 \cdots B_{k-1} B_{k+1} \cdots B_{N-1}$$

$$+ f_N B_0 B_1 \cdots B_{N-1} .$$

The general $L-th$ order sum is over all combinations of L of the N indexes.

This expansion can be used as the basis of an adaptive nonlinear filter which compensates for a general time-invariant nonlinearity in the echo path as well as D/A converter. The coefficients of the expansion are simply adapted in a fashion nearly identical to the linear adaptive transversal filter. The details are given elsewhere [14]. For practical situations, only a subset of the 2^N total terms are required. Experience suggests that about $2N$ terms, consisting of a subset of the second and third order terms, usually suffice to give the approximately 55 dB of cancellation required. The nonlinear digital echo canceler has an additional advantage of being able to cancel nonlinear echo mechanisms such as transformer saturation and transmitted pulse asymmetry [14].

In summary, two viable approaches to the canceler implementation have been described. The first is based on modification of the architecture (and in particular the partitioning between analog and digital) so that the existing adaptation algorithms can compensate for the shortcomings of the analog circuitry. The second is based on a modification to the adaptive filtering algorithms to more directly compensate for the nonidealities of the technology. As scaling of the analog integrated circuits proceeds, we can expect to see many examples of the use of adaptation to overcome implementation constraints.

5. SUMMARY AND CONCLUSIONS

This article has addressed the implementation of signal processing and communications components where the goal is to achieve maximum performance using a custom integrated circuit technology. Where the performance requirements are relaxed, the algorithmic and architectural choices are generally non-critical, and the system can be successfully implemented. Where the performance will stretch the limits of the technology, however, the algorithmic and architectural choices are critical to success. In particular, many of the inherent characteristics of the technology must be taken into account, as the examples have illustrated.

6. Acknowledgements

The author has benefited greatly from discussion of these issues with Profs. David Hodges and Paul Gray. Much of the work on digital filtering reported here was carried out by Mr. Hui-Hung Lu, and the work on full-duplex data transmission was performed in part by Dr. Oscar Agazzi.

REFERENCES

1. Gordon Moore, "VLSI: Some Fundamental Challenges," *IEEE Spectrum*, pp. 30-37 (April, 1979).

2. B. Hoeneisen and C. A. Mead, "Fundamental Limitations in Microelectronics -- MOS Technology," *Solid State Technology* 15 pp. 819-829 (1972).

3. Robert Brodersen, "VLSI for Signal Processing," *Trends and Perspectives in Signal Processing* 1(1) (Jan. 1981).

4. John G. Posa, "Gate Arrays -- A Special Report," *Electronics*, p. 145 (Sept. 25, 1980).

5. Stephen Trimberger, "Automating Chip Layout," *IEEE Spectrum*, p. 38 (June 1982).

6. Rob Budzinski et. al., "A Restructurable Integrated Circuit for Implementing Programmable Digital Systems," *Computer*, p. 43 (March 1982).

7. D.A. Hodges, P.R. Gray, and R. W. Brodersen, "Potential of MOS Technologies for Analog Integrated Circuits," *IEEE J. Solid State Circuits* SC-13 pp. 285-294 (June, 1978).

8. R.W. Brodersen, P.R. Gray, and D.A. Hodges, "MOS Switched-Capacitor Ladder Filters," *Proc. of the IEEE* 67(No. 1) pp. 61-75 (Jan., 1979).

9. H. T. Kung and C. E. Leiserson, "Systolic Arrays (for VLSI)," pp. 256-282 in *Sparse Matrix Proceedings 1978*, ed. I. S. Duff and G. W. Stewart,Society for Industrial and Applied Mathematics (1979).

10. H. T. Kung, "Special-Purpose Devices for Signal and Image Processing: an Opportunity in Very Large Scale Integration (VLSI)," pp. 76-84 in *Proceedings of the Society of Photo-Optical Instrumentation Engineers*, ed. Tien F. Tao, (July 1980).

11. H. T. Kung, "Why Systolic Architectures?," *IEEE Computer* 15(1) pp. 37-46 (Jan. 1982).

12. Casper W. Barnes and S. Shinnaka, "Block Shift Invariance and Block Implementation of Discrete-Time Filters," *IEEE Trans. Circuits and Systems* CAS-27 pp. 667-672 (Aug. 1980).

13. O. Agazzi, D.A. Hodges, and D.G. Messerschmitt, "Large Scale Integration of Hybrid Method Digital Subscriber Loops," *IEEE Transactions on Communications* COM-30(9) (September 1982).

14. O. Agazzi, D. Hodges, and D. Messerschmitt, "Nonlinear Echo Cancellation for Full-Duplex Data Signals," *IEEE Transactions on Communications* COM-30(11) (November 1982).

FAST ALGORITHMS FOR SIGNAL PROCESSING AND ERROR CONTROL

Richard E. Blahut

IBM Owego

ABSTRACT

Algorithms for computation are a central part of digital signal processing and of decoders for error-control codes. When restricted to the study of their computational algorithms, there is not much to distinguish those two subjects. Only the arithmetic field is different; in one case the real or complex field, and a Galois field in the other. Even this distinction is hard to defend; signal processing problems may use Galois fields, and error-control codes in the real or complex field are now under study.

We will survey the state of modern fast algorithms for certain types of computations used in digital processing and error-control including tasks such as linear convolution, cyclic convolution, discrete Fourier transforms, and spectral estimation. In particular we will discuss the decoding of error-control codes as a problem in spectral estimation, and see how this viewpoint leads to the development of efficient decoder algorithms. The Winograd FFT, a fast Berlekamp-Massey algorithm, and a time-domain Reed-Solomon decoder will be described.

INTRODUCTION

Digital signal processing and error-control codes are engineering subjects that have grown rapidly in the last decade. Modern communication systems depend heavily on developments in these subjects. Many of the methods of digital signal processing and error-control codes are strikingly similar, especially in the area of fast computational algorithms.

In this article we survey the common areas of the two fields, treating the subject of error-control codes as a branch of the subject of digital signal processing[1]. The rewards are of three kinds. First, the subject of error-control coding becomes more easily accessible to the many engineers familiar with digital signal processing. Second, the many fast algorithms of digital signal processing can be passed over to the field of error-control codes. Third, the techniques of error-control can be passed over to the field of digital signal processing to protect against impulsive noise. Wolf[2] and Marshall[3] have discussed such methods.

THE FOURIER TRANSFORM

The Fourier transform is familiar in digital signal processing. A vector of complex numbers $\underline{v}$ of length n has a spectrum $\underline{V}$ given by

$$V_j = \sum_{i=0}^{n-1} \omega^{ij} v_i \qquad j = 0 \ldots, n-1$$

where $\omega = e^{-\sqrt{-1}\, 2\pi/n}$ is an nth root of unity in the complex field. The Fourier transform in a Galois field plays a role in error-control codes. Let $\underline{v}$ be a vector of length n of symbols from a Galois field GF(q). Then the Fourier transform is the spectrum vector $\underline{V}$ given by

$$V_j = \sum_{i=0}^{n-1} \omega^{ij} v_i \qquad j = 0 \ldots, n-1$$

where now ω is an nth root of unity in the Galois field GF(q). This formula looks exactly the same as before, but the additions and multiplications it expresses are in the Galois field GF(q). The inverse Fourier transform and the convolution theorem hold because the proofs only use the formal structure of a field. There is one important difference here however; in a Galois field, an nth root of unity does not exist for every n, so a Fourier transform does not exist for every n. This is why error-control codes are usually limited in the choice of blocklength.

A Fourier transform in a Galois field is familiar to the field of digital signal processing under the name of a number-theoretic transform. In this case GF(q) has a number of elements q equal to a prime number, and the arithmetic operations are modulo q arithmetic operations. For example if q=5, then GF(q) = $\{0,1,2,3,4\}$.

It is easily verified that the element 2 is a fourth root of unity; $2^4 = 1$; hence there is a Fourier transform of blocklength 4 given by

$$V_j = \sum_{i=0}^{3} 2^{ij} v_i \qquad j = 0, \ldots, 3$$

Usually, in practical error control codes, one wants q to be a power of two, so GF(5) would not be desirable. Similarly, GF(5) is too small for applications in digital signal processing. However large prime fields such as $GF(2^{16} + 1)$ have been used both for applications in digital signal processing and in error-control.

If the Galois field GF(q) does not contain an element of order n for some desired n, one can sometimes find such a root of unity in the extension field $GF(q^m)$.

SPECTRAL PROPERTIES OF ERROR CONTROL CODES

We give a spectral description of Reed-Solomon codes. These codes of blocklength n are defined over a field F whenever F contains an nth root of unity. Reed-Solomon codes are normally defined over GF(q) but, as Wolf has suggested, all of the ideas hold over any field. Let c be a vector of length n over the field F with spectrum C in F. The t-error-correcting Reed-Solomon code of blocklength n with symbols in F is the set of all vectors c whose spectrum satisfies $C_j = 0$ for $j = 1, \ldots, 2t$. One way to find these codewords is to encode in the frequency domain. This means to set $C_j = 0$ for $j = 1, \ldots, 2t$ and set the remaining n-2t components of C equal to the n-2t information symbols. An inverse Fourier transform given the codeword c. Thus the number of information symbols equals n-2t and there are two parity symbols for every error to be corrected.

This is not the only way to encode the n-2t information symbols into codewords - others may yield a simpler implementation - but this method is the most convenient to deal with here.

The proof that this construction does indeed give a code that corrects t-errors consists of describing the decoder. The codeword c is transmitted and the channel makes errors described in the vector e which is nonzero in not more than t places. The received word v is written componentwise as

$$v_i = c_i + e_i \qquad i = 0, \ldots, n-1$$

The decoder must process the received word so as to remove the error word $\underline{e}$; the information is then recovered from $\underline{c}$. The syndromes of this noisy codeword $\underline{v}$ are defined by the following set of equations

$$S_j = \sum_{i=0}^{n-1} \omega^{ij} v_i \qquad j = 1, \ldots, 2t$$

Obviously, the syndromes are computed as 2t components of a Fourier transform. The received noisy codeword has Fourier transform given by $V_j = C_j + E_j$ for $j=0, \ldots, n-1$, and the syndromes are the 2t components of this spectrum from 1 to 2t. But, by construction of a Reed–Solomon code, the parity frequencies are spectral components equal to zero.

$$C_j = 0 \qquad j = 1, \ldots, 2t$$

Hence

$$S_j = V_j = E_j \qquad j = 1, \ldots, 2t$$

The block of syndromes gives us a window through which we can look at 2t of the n components of the spectrum of the error pattern. The decoding is seen as a problem in spectral estimation. Find the spectrum of the error pattern given a segment of length 2t of the spectrum and the side information that at most t components of the time-domain error pattern are nonzero.

Suppose there are $\nu \leq t$ errors at locations with index i_k for $k = 1, \ldots \nu$. Define the polynomial

$$\Lambda(x) = \prod_{k=1}^{\nu} (1-x \, \omega^{i_k})$$

which is known as the error-locator polynomial. The vector Λ whose components Λ_j are coefficients of the polynomial $\Lambda(x)$ has inverse transform

$$\lambda_i = \frac{1}{n} \sum_{j=0}^{n-1} \Lambda_j \, \omega^{-ij}$$

which is just $\Lambda(x)$ evaluated at $x = \omega^{-i}$. But then

$$\lambda_i = \frac{1}{n} \prod_{k=1}^{\nu} (1-\omega^{-i} \, \omega^{i_k})$$

Hence $\lambda_i = 0$ if and only if $e_i \neq 0$. That is $\lambda_i e_i = 0$, so the convolution in the frequency domain is zero.

$$\underline{\Lambda} * \underline{E} = 0$$

but $\Lambda_j = 0$ for $j > t$ and $\Lambda_0 = 1$, so this can be written

$$\sum_{j=1}^{t} \Lambda_j E_{k-j} = -E_k \qquad k = 0, \ldots, n-1$$

This convolution is a set of n equations in n-t unknowns; t unknown components of $\underline{\Lambda}$ and n-2t unknown components of $\underline{E}$, and in 2t known values of $\underline{E}$ given by the syndromes. This computation can be described as the operation of a linear feedback shift register with tap weights given by the coefficients of $\Lambda(x)$. It is an autoregressive filter. Of the n equations, the t equations

$$\sum_{j=1}^{t} \Lambda_j S_{k-j} = -S_k \qquad k = 1 + t, \ldots, 2t$$

involve only the known syndromes and the t unknown components of Λ. These t equations are always solvable for the t unknown components of Λ. Massey viewed this computation as the task of designing an autoregressive filter of length t (or less) that would recursively produce a specified sequence of 2t outputs.

After the shift register $\underline{\Lambda}$ is computed, the remaining components of $\underline{E}$ can be obtained by recursive extension. That is, sequentially computed from $\underline{\Lambda}$ using the above convolution equation written in the form

$$E_k = -\sum_{j=1}^{t} \Lambda_j E_{k-j} \qquad k = 0, \ldots, n-1$$

In this way all components of the vector $\underline{E}$ are computed. Then

$$C_j = V_j - E_j$$

An inverse Fourier transform recovers the initial codeword with all errors corrected. The information symbols may then be read out in accord with the method of encoding.

FOURIER TRANSFORMS IN EXTENSION FIELDS

The construction of an extension of a Galois field using a polynomial representation is familiar. The same construction also can be used to extend the rationals. A polynomial extension of the rationals is not often seen in engineering applications since one usually works in the real number system, and when necessary one extends the real field immediately to the complex field. However, all numbers in practice are rationals, and there are many extension fields of the rationals. We will introduce some reasons for using these other extension fields, but first we look at more familiar problems in the Galois fields.

It is well-known that the real field does not contain an nth root of unity (except when n equals two). This is why Fourier transforms of real-valued signals have complex-valued components. In a Galois field $GF(q)$ an nth root of unity may or may not exist. If it does not exist in $GF(q)$ then it may exist in some extension field $GF(q^m)$. For some n, an nth root of unity does not exist in any extension of $GF(q)$.

In the language of digital signal processing, a Fourier transform of blocklength n over $GF(q)$ may lie in $GF(q)$, may lie in the extension $GF(q^m)$, or may not exist. If the Fourier transform of blocklength n over $GF(q)$ lies in $GF(q)$, then there is a Reed-Solomon code of blocklength n over $GF(q)$ as was described in the previous section. If there is a Fourier transform only in the extension field $GF(q^m)$, then a code over $GF(q)$ known as a BCH code can be constructed.

The construction of the previous section still can be used, but some modifications are necessary. As before the spectrum $\underline{C}$ of the codeword $\underline{c}$ satisfies the constraint $C_i = 0$ for $j = 1, \ldots, 2t$ and this is all that is needed to make the decoding procedure work. However, the spectrum consists of elements of $GF(q^m)$ while the codeword can only have components in $GF(q)$. Other constraints on the spectrum are needed to ensure that the codewords are in $GF(q)$. In particular, the spectral components must satisfy

$$C_j^q = C_{((qj))}$$

where the double parentheses denote modulo n on the indices. In the complex field, this becomes the familiar relationship on the spectrum

$$C_j^* = C_{n-j}$$

if the inverse Fourier transform is to be real.

Hence to obtain a BCH code over $GF(q)$, we need the two constraints on the spectrum

$$C_j = 0 \qquad\qquad j = 1, \ldots, 2t$$

$$C_j^q = C_{((qj))} \qquad\qquad j = 0, \ldots, n-1$$

Except for the second constraint, everything about Reed-Solomon codes also applies to BCH codes even when the BCH codes are defined over the rational field.

We will define appropriate extensions of the rational field, then define Reed-Solomon codes as burst-correcting codes in the rational field.

Let v_i, $i=0, \ldots, n-1$ be a vector of rational numbers. The discrete Fourier transform of $\underline{v}$ is a vector of complex numbers called the spectrum. However, an arbitrary complex number cannot occur in the spectrum. Complex numbers can occur only in the subfield known as $Q(\omega)$ or, more simply, as Q^m. This is the smallest subfield of the complex field containing ω. It also contains all rationals since every subfield of the complex field contains all rationals.

Let x^n-1 be factored into its prime factors over the rationals.

$$x^n-1 = p_0(x)p_1(x) \ldots p_S(x)$$

The factors are known as cyclotomic polynomials, and when n is small, have coefficients only equal to -1, 0, or +1. Since ω is a zero of x^n-1, it is a zero of one of the cyclotomic polynomials, say $p(x)$ a polynomial of degree m with leading coefficient equal to one.

$$p(x) = x^m + p_{m-1}x^{m-1} + \ldots = p_1x + p_0$$

Since $p(\omega) = 0$, this gives

$$\omega^m = -p_{m-1}\omega^{m-1} - \ldots - p_1\omega-p_0$$

Hence ω^m can be expressed in terms of lesser powers of ω. However if i is less than n, then ω^i cannot be so expressed because if it could, ω would be a zero of another polynomial of degree smaller than the degree of $p(x)$.

The field $Q(\omega)$ can be represented as the set of polynomials in ω of degree $m-1$ or less with rational coefficients. The polynomials are represented by a list of m coefficients. Addition is polynomial addition, and multiplication is polynomial multiplication modulo $p(\omega)$. In a practical implementation, it takes m words of memory to store one element of $Q(\omega)$ instead of taking two words of memory which suffices for the usual complex numbers.

In order to emphasize that the number representation consists of the polynomials themselves, and not the complex value the polynomial takes at ω, we will use the variable x in place of ω. Then the numbers are

$$a = a_{m-1}x^{m-1} + a_{m-2}x^{m-2} + \ldots + a_1 x + a_0$$

as represented by the list of coefficients. Of course, if we wanted to know the "true" complex value of a, we can just substitute ω for x and carry out the indicated calculations. However, our aim is to use the polynomial representations as intermediate variables to derive codes and algorithms. For example, if n is a power of two, say 2^m, then

$$x^n - 1 = (x^{n/2} + 1)(x^{n-4} + 1) \ldots (x+1)(x-1)$$

The cyclotomic polynomial $x^{n/2} + 1$ leads to an extension field whose elements are all the rational-valued polynomials of degree less than n/2; addition is polynomial addition; multiplication is polynomial multiplication modulo $x^{n/2} + 1$.

The discrete Fourier transform takes values only in the extension field $Q(\omega)$ is $\underline{v}$ is a vector over the rationals (or if $\underline{v}$ only takes values in $Q(\overline{\omega})$). In the polynomial representation this becomes

$$V_k = \sum_{i=0}^{n-1} x^{ik} v_i \qquad (\text{mod } p(x)) \qquad i = 0, \ldots, n-1$$

Nussbaumer[4] introduced this and called it a polynomial transform. Our development shows that Nussbaumer's polynomial transform is a notational variation of a Fourier transform . The polynomial transform is simple to evaluate because multiplication by x is an indexing operation, and the modulo p(x) reduction consists of at most m additions. Any FFT algorithm, such as the Cooley-Tukey FFT can be used for the computation, but a multiplication by x is replaced by at most m real additions and a polynomial addition is replaced also by m real additions.

Let m be the degree of the cyclotomic polynomial having ω as a zero. Then $Q(\omega)$ is represented by the set of polynomials of degree less than m. The Fourier transform maps n-vectors of (m-1)-degree polynomials into n-vectors of (m-1)-degree polynomials. The components of the vectors are rationals.

To construct a burst-correcting code of real numbers, take an (n,k) Reed-Solomon code in Q^m where k = n−2t. Each symbol of the codeword is a polynomial over Q which can be represented by m rationals in sequence. There are a total of n such symbols in a codeword, requiring a sequence of nm rationals. The code can correct any t symbols in error, so it will correct any single burst of m(t−1) + 1 errors and many patterns of multiple burst errors.

Each of the rationals may actually be a fixed-point number of limited precision. If the information consists of b-bit fixed-point numbers that are systematically encoded, then the codeword will consist mostly of these same sixteen bit numbers. Parity symbols will require wordlengths greater than b-bits.

THE BERLEKAMP-MASSEY ALGORITHM

We will now turn back to the study of algorithms for decoding. We interrupted this discussion after writing the equation

$$\sum_{j=1}^{t} \wedge_j S_{k-j} = -S_k \qquad k = 1 + t, \ldots, 2t$$

which must be solved for a vector $\underline{\wedge}$. This set of linear equations can be solved by computing a matrix inverse. This is the task of designing an autoregressive filter to have a given output sequence; a task well-known in digital signal processing and arising in problems such as maximum-entropy spectral estimation or LPC speech compression.

If there are exactly t errors, then it is well known that there is exactly one solution to this system of linear equations. If there are less than t errors, then the determinant of the matrix M with $m_{ij} = S_{i-j}$ will equal zero and there will be more than one solution for $\underline{\wedge}$. Normally one solves for that $\underline{\wedge}$ corresponding to a polynomial $\wedge(\overline{x})$ of smallest degree.

The problem of solving for $\underline{\wedge}$ is the problem of inverting a system of Toeplitz equations. There are many ways of dealing with a Toeplitz system of equations. This instance has an added property that the vector on the right side of the equation is related to elements of the Toeplitz matrix in a special way. The most popular algorithm for solving this system of equations for error-control decoders is the Berlekamp-Massey algorithm stated as follows:

Let $S_1, \ldots, S_{2t}$ be given. Let the following set of recursive equations be used to compute $\wedge^{(2t)}(x)$:

$$\Delta_r = \sum_{j=0}^{n-1} \Lambda_j^{(r-1)} S_{r-j}$$

$$L_r = \delta_r(r - L_{r-1}) + (1 - \delta_r) L_{r-1}$$

$$\begin{bmatrix} \Lambda^{(r)}(x) \\ \\ B^{(r)}(x) \end{bmatrix} = \begin{bmatrix} 1 & -\Delta_r x \\ \\ \Delta_r^{-1} \delta_r & (1-\delta_r)x \end{bmatrix} \begin{bmatrix} \Lambda^{(r-1)}(x) \\ \\ B^{(r-1)}(x) \end{bmatrix}$$

for $r = 1, \ldots, 2t$. The initial conditions are $\Lambda^{(0)}(x) = 1$, $B^{(0)}(x) = 1$, and $\delta_r = 1$ if both $\Delta_r \neq 0$ and $2L_{r-1} \le r-1$ and otherwise $\delta_r = 0$. Then $\Lambda^{(2t)}(x)$ is the smallest-degree polynomial with the properties that $\Lambda_0^{(2t)} = 1$ and

$$S_k + \sum_{j=1}^{n-1} \Lambda_j^{(2t)} S_{k-j} = 0 \qquad k = L_{2t} + 1, \ldots, 2t$$

The Berlekamp-Massey algorithm can be used in any field but it seems not to have attracted much attention for problems in the real or complex field - perhaps because of the division by Δ which can be a small number.

The Berlekamp-Massey algorithm has $2t$ iterations and each iteration can have on the order of t operations, so the complexity is on the order of t^2. There are also several Fourier transforms to support it and these can have on the order of n^2 operations. Later we shall discuss ways to reduce this complexity. First we discuss a time-domain decoder that may be simpler for small blocklength.

A TIME-DOMAIN DECODER

By recognizing the problem of decoding Reed-Solomon codes as a computation in the Fourier transform domain, we have opened other possibilities for the processing. The Berlekamp-Massey algorithm processes the spectrum of the received word. It is preceded by a Fourier transform and is followed by a Fourier transform in some way. However, instead of pushing the received word into the frequency domain, it is possible to push the Berlekamp-Massey algorithm into the time-domain. This makes the Fourier transforms simply vanish. On the other hand, the frequency-domain vectors of length t are replaced by time-domain vectors of length n; algorithms that in the frequency-domain have complexity t^2 or nt become algorithms in the time-domain that have complexity nt or n^2. The time-domain decoder is structurally simple and is useful in applications where structural simplicity is important and the number of iterations is not important.

To pass the Berlekamp-Massey equations into the time-domain, simply replace Λ_i and B_i by the time-domain variables λ_i and b_i, replace the delay operator x by ω^{-1}, and replace product terms by convolution terms. One then obtains the following.

Let $\underline{v}$ be the received noisy BCH codeword. Let the following set of recursive equations be used to compute $\lambda_i^{(2t)}$ for i = 0, ..., n-1.

$$\Delta_r = \sum_{i=0}^{n-1} \omega^{ir} (\lambda_i^{(r-1)} v_i)$$

$$L_r = \delta_r (r-L_{r-1}) + (1-\delta_r) L_{r-1}$$

$$\begin{bmatrix} \lambda_i^{(r)} \\ \\ b_i^{(r)} \end{bmatrix} = \begin{bmatrix} 1 & -\Delta_r \omega^i \\ \\ \Delta_r^{-1} \delta_r & (1-\delta_r) \omega^i \end{bmatrix} \begin{bmatrix} \lambda_i^{(r-1)} \\ \\ b_i^{(r-1)} \end{bmatrix}$$

r=1, ..., 2t. The initial conditions are $\lambda_i^{(0)} = 1$ for all i, $b_i^{(0)} = 1$ for all i, and $\delta_r = 1$ if both $\Delta_r \neq 0$ and $2L_{r-1} \leq r-1$, and otherwise $\delta_r = 0$. Then $\lambda_i^{(2t)} = 0$ if and only if $e_i \neq 0$.

For nonbinary codes, it does no good to compute only the error locations in the time-domain, we must also compute the error magnitudes. In the frequency domain the error magnitudes are computed by the following recursion.

$$E_k = -\sum_{j=1}^{t} \Lambda_j E_{k-j} \qquad k = 2t+1, ..., n-1$$

It is not possible to just write the Fourier transform of this equation; some restructuring is necessary. The following set of recursive equations for r = 2t+1, ..., n

$$\Delta_r = \sum_{i=0}^{n-1} \omega^{ir} v_i^{(r-1)} \lambda_i$$

$$v_i^{(r)} = v_i^{(r-1)} - \Delta_r \omega^{-ri}$$

is suitably restructured. Starting with $v_i^{(2t)} = v_i$, and $\lambda_i = \lambda_i^{(2t)}$, the last iteration results in

$$v_i^{(n)} = e_i \qquad i = 0, ..., n-1$$

The time-domain decoder has no Fourier transforms; it has only one major computational block which is easily designed into digital logic. It does, however, always deal with vectors of length n rather than vectors of length t as in the frequency domain. Hence there will be hardware/speed trades.

FAST CONVOLUTION ALGORITHMS

Algorithms for fast convolution can be constructed by using the Chinese remainder theorem as was suggested by Winograd. Let

$$s(x) = g(x) d(x) \qquad (\text{mod } m(x))$$

This formula includes a linear convolution simply by choosing $m(x)$ as any polynomial whose degree is larger than that of $s(x)$. The formula includes cyclic convolutions where $m(x) = x^n - 1$. Let $m(x)$ be written in terms of its prime factors:

$$m(x) = m^{(0)}(x) \, m^{(1)}(x) \, \ldots \, m^{(K)}(x)$$

Then by taking residues, the original polynomial product becomes a set of polynomial products

$$s^{(k)}(x) = g^{(k)}(x) \, d^{(k)}(x) \qquad (\text{mod } m^{(k)}(x))$$

for $k=0, \ldots, K$. Then $s(x)$ can be recovered using

$$s(x) = \sum_{k=0}^{K} a^{(k)}(x) \, s^{(k)}(x) \qquad (\text{mod } m(x))$$

for some set of polynomial $a^{(k)}(x)$ given by the Chinese remainder theorem.

THE WINOGRAD FAST FOURIER TRANSFORM

We have seen that a Reed-Solomon code can be decoded by a Fourier transform, followed by inverting a Toeplitz system, followed by an inverse Fourier transform. Fast Fourier transform algorithms can be applied. The best-known FFT is the radix-two Cooley-Tukey FFT. Because the blocklength of Reed-Solomon codes is usually not a power of two, the radix-two FFT algorithms cannot be used. However mixed-radix Cooley-Tukey FFT algorithms can be used to advantage.

There are other FFT algorithms that will do better than the Cooley-Tukey FFT. The Winograd FFT is intrinsically a mixed-radix FFT. The Winograd FFT can be best described as a Winograd small FFT and a Winograd large FFT. The first is a way of constructing highly optimized routines for small blocklengths – blocklengths of 2, 3, 4, 5, 7, 9, and 16 are popular. The Winograd large FFT is a method of binding these small pieces together to get a large FFT. For a Reed–Solomon code of blocklength 255, one must use Winograd small FFT algorithms of blocklength 3, 5, and 17, and bind these with the Winograd large FFT into an algorithm of blocklength 255.

The Winograd small FFT uses the Rader prime algorithm (or a generalization) to change a Fourier transform into a convolution; then an efficient Winograd convolution algorithm is used for the convolution.

Let us construct a binary five-point Winograd FFT in the field GF(16) that will compute

$$V_j = \sum_{i=0}^{4} \omega^{ij} v_i \qquad j = 0, \ldots, 4$$

First use the Rader prime algorithm. The Rader prime algorithm can be used to compute the Fourier transform in any field GF(q) whenever the blocklength n is a prime. Because n is a prime, we can make use of the structure of GF(n). This prime field is not to be confused with GF(q) nor with any subfield of GF(q). A generalized form of the Rader algorithm can be used if the blocklength is a prime power.

Choose a primitive element π in the field GF(n). Then each nonzero value of the indices i and j can be expressed as a power of π. (The zero frequency index and the zero time index must be treated specially since they cannot be expressed as powers of π.) The reason for doing this is that that product ij in the exponent becomes recast in the form $\pi^{\ell - r}$ to make the computation a convolution. - All that this entails is a scrambling of the components of the input and output vectors.

In GF(5), the element 2 is primitive, and so in GF(5) we have: $2^0 = 1$, $2^1 = 2$, $2^2 = 4$, $2^3 = 3$ which defines new indices on v_i. To define new indices on V_j, use negative exponents: $2^0 = 1$, $2^{-1} = 3$, $2^{-2} = 4$, $2^{-3} = 2$.

The five-point Fourier transform is

$$
\begin{bmatrix} V_0 \\ V_1 \\ V_2 \\ V_3 \\ V_4 \end{bmatrix} = \begin{bmatrix} 1 & 1 & 1 & 1 & 1 \\ 1 & \omega^1 & \omega^2 & \omega^3 & \omega^4 \\ 1 & \omega^2 & \omega^4 & \omega & \omega^3 \\ 1 & \omega^3 & \omega^1 & \omega^4 & \omega^2 \\ 1 & \omega^4 & \omega^3 & \omega^2 & \omega^1 \end{bmatrix} \begin{bmatrix} v_0 \\ v_1 \\ v_2 \\ v_3 \\ v_4 \end{bmatrix}
$$

From this we obtain

$$V_0 = v_0 + v_1 + v_2 + v_3 + v_4$$

and

$$
\begin{bmatrix} V_1 - V_0 \\ V_2 - V_0 \\ V_3 - V_0 \\ V_4 - V_0 \end{bmatrix} = \begin{bmatrix} \omega^1 - 1 & \omega^2 - 1 & \omega^3 - 1 & \omega^4 - 1 \\ \omega^2 - 1 & \omega^4 - 1 & \omega^1 - 1 & \omega^3 - 1 \\ \omega^3 - 1 & \omega^1 - 1 & \omega^4 - 1 & \omega^2 - 1 \\ \omega^4 - 1 & \omega^3 - 1 & \omega^2 - 1 & \omega^1 - 1 \end{bmatrix} \begin{bmatrix} v_1 \\ v_2 \\ v_3 \\ v_4 \end{bmatrix}
$$

By the scrambling rules of the Rader algorithm, this becomes

$$
\begin{bmatrix} V_1 - V_0 \\ V_2 - V_0 \\ V_4 - V_0 \\ V_3 - V_0 \end{bmatrix} = \begin{bmatrix} \omega^1 - 1 & \omega^3 - 1 & \omega^4 - 1 & \omega^2 - 1 \\ \omega^2 - 1 & \omega^1 - 1 & \omega^3 - 1 & \omega^4 - 1 \\ \omega^4 - 1 & \omega^2 - 1 & \omega^1 - 1 & \omega^3 - 1 \\ \omega^3 - 1 & \omega^4 - 1 & \omega^2 - 1 & \omega^1 - 1 \end{bmatrix} \begin{bmatrix} v_1 \\ v_3 \\ v_4 \\ v_2 \end{bmatrix}
$$

which can be recognized as the matrix representation of a cyclic convolution

$$
\begin{bmatrix} V_1 - V_0 \\ V_2 - V_0 \\ V_4 - V_0 \\ V_3 - V_0 \end{bmatrix} = \begin{bmatrix} g_0 & g_3 & g_2 & g_1 \\ g_1 & g_0 & g_3 & g_2 \\ g_2 & g_1 & g_0 & g_3 \\ g_3 & g_2 & g_1 & g_0 \end{bmatrix} \begin{bmatrix} v_1 \\ v_3 \\ v_4 \\ v_2 \end{bmatrix}
$$

where $g_0 = (\omega - 1)$, $g_1 = \omega^2 - 1$, $g_2 = \omega^4 - 1$, and $g_3 = \omega^3 - 1$. This convolution is performed by an efficient algorithm for cyclic convolution. There is a four-point cyclic convolution algorithm in fields of

characteristic 2 with nine multiplications. We can rewrite this to
do the Fourier transform. Incorporate the scrambling and unscram-
bling operations into the convolution by scrambling the appropriate
rows and columns. Also the coefficients of $g(x)$ are fixed constants in $GF(16)$, so it is possible to precompute terms involving
only $g(x)$. When these changes are made to the four-point convolution algorithm, and the terms V_0 and v_0 are included, it becomes
the five-point Winograd small FFT. It can be written in the form

$$V = BDAv$$

as

$$
\begin{bmatrix} V_0 \\ V_1 \\ V_2 \\ V_3 \\ V_4 \end{bmatrix}
=
\begin{bmatrix}
1 & 0 & 0 & 0 & 0 & 0 & 0 & 0 & 0 & 0 \\
0 & 1 & 0 & 1 & 1 & 0 & 0 & 0 & 0 & 1 \\
0 & 1 & 1 & 0 & 0 & 1 & 1 & 0 & 0 & 0 \\
0 & 1 & 0 & 1 & 0 & 1 & 0 & 0 & 1 & 0 \\
0 & 1 & 1 & 0 & 1 & 0 & 0 & 1 & 0 & 0
\end{bmatrix}
\begin{bmatrix}
1 \\
& \alpha^{13} \\
& & \alpha^{9} \\
& & & \alpha^{10} \\
& & & & \alpha^{6} \\
& & & & & 1 \\
& & & & & & \alpha^{4} \\
& & & & & & & \alpha^{14} \\
& & & & & & & & \alpha \\
& & & & & & & & & \alpha^{8}
\end{bmatrix}
\begin{bmatrix}
1 & 1 & 1 & 1 & 1 \\
0 & 1 & 0 & 0 & 0 \\
0 & 1 & 0 & 0 & 1 \\
0 & 1 & 0 & 0 & 1 \\
0 & 1 & 0 & 1 & 0 \\
0 & 1 & 1 & 0 & 0 \\
0 & 1 & 1 & 1 & 1 \\
0 & 1 & 1 & 1 & 1 \\
0 & 1 & 1 & 1 & 1 \\
0 & 1 & 1 & 1 & 1
\end{bmatrix}
\begin{bmatrix} v_0 \\ v_1 \\ v_2 \\ v_3 \\ v_4 \end{bmatrix}
$$

Notice that the matrix of preadditions A and the matrix of post-
additions B are not square. The five-point input vector is expanded
to a ten-point vector, and this is where the multiplications occur
as represented by the diagonal matrix D. The top row inside the
braces has to do with V_0; and has no multiplications. The other
nine rows come from the four-point cyclic convolution algorithm.
One of the multiplying constants turns out to be a one, so there
are really only eight multiplications in the algorithm.

The Winograd small FFT of length n can be derived in this way
whenever n is a prime. A Winograd small FFT also can be derived
whenever n is a prime power. Generally, one constructs the
Winograd small FFT only for fairly small blocklengths. For large
blocklengths, one prefers to use something with a little more
structure even if there is a slight penalty in the number of
multiplications. The Winograd large FFT satisfies this need.

The large Winograd FFT has a blocklength n which is a product
of small primes or small prime powers. We will discuss the case

with two factors. Then $n = n'n''$. The first step is to change the one-dimensional Fourier transform into a two-dimensional Fourier transform using the Good–Thomas indexing scheme.

The Good–Thomas indexing scheme is based on the Chinese remainder theorem for integers. The input index is described by its residues as follows

$$i' = i \pmod{n'}$$
$$i'' = i \pmod{n''}$$

This is the map of the input index i down the extended diagonal of a two-dimensional array indexed by (i', i''). By the Chinese remainder theorem there exist integers N' and N'' such that

$$N'n' + N''n'' = 1 \qquad \pmod{n}$$

Then the input index can be recovered as follows:

$$i = i'N''n'' + i''N'n' \qquad \pmod{n}$$

The output index is described somewhat differently. Define

$$j' = N''j \qquad \pmod{n'}$$
$$j'' = N'j \qquad \pmod{n''}$$

The output index j can be recovered as follows:

$$j = n''j' + n'j'' \qquad \pmod{n}$$

To verify this, write it out.

$$j = n''(N''j + Q_1 n') + n'(N'j + Q_2 n'') \qquad \pmod{n'n''}$$

$$= j(n''N'' + n'N') \qquad \pmod{n}$$

$$= j$$

Now, with these new indices, we convert the formula

$$V_j = \sum_{i=0}^{n-1} \omega^{ij} v_i$$

into

$$\omega^{ij} = \omega^{(i'N''n''+i''N'n')(n''j'+n'j'')}$$

$$= \omega^{N''(n'')^2 i'j'} \omega^{N'(n')^2 i''j''}$$

$$= \beta^{i'j'} \gamma^{i''j''}$$

where $\beta = \omega^{N''(n'')^2}$ is an element of order n' and $\gamma = \omega^{N'(n')^2}$ is an element of order n''. This is now in the form of a two-dimensional n' by n'' point Fourier transform.

$$V_{j,j''} = \sum_{i'=0}^{n'-1} \sum_{i''=0}^{n''-1} \beta^{i'j'} \gamma^{i''j''} v_{i',i''}$$

The individual components of this two-dimensional Fourier transform can be computed by an n'-point Winograd FFT and an n''-point Winograd FFT respectively. This consists of taking an n'-point Fourier transform of each row followed by an n''-point Fourier transform of each column.

There is yet one more step before we have the Winograd large FFT. Since it does not matter whether the rows or the columns of the two-dimensional Fourier transform are transformed first, it seems that it may be possible somehow to do them together. This is what the Winograd FFT does. It binds together the row computations and the column computations in a way that reduces the total number of multiplications.

Let W' and W'' be matrix representations of Fourier transform of size n' and n'' respectively. That is:

$$V' = W'v'$$

$$V'' = W''v$$

are matrix representations of the Fourier transforms

$$V'_j = \sum_{i=0}^{n'-1} \beta^{ij} v'_i$$

$$V''_j = \sum_{i=0}^{n''-1} \gamma^{ij} v''_i$$

An n' by n'' two-dimensional Fourier transform of the two-dimensional signal $v_{i',i''}$ is obtained by applying W' to each row, and then applying W'' to each column.

We can write the two-dimensional Fourier transform as

$$V = W'W''v$$

140

The notation is unconventional because v is a two-dimensional array. The matrix W" multiplies every row of v and the matrix W' multiplies every column of v. We are free to interchange the order of these operations

$$V = W''W'v$$

because this amounts simply to an interchange in the order of summation.

Whenever we have Winograd FFT algorithms of length n' and n", then we have the matrix factorizations

$$W' = B'D'A'$$

$$W'' = B''D''A''$$

where A', A", B' and B" are matrices of integers of the field; and D' and D" are diagonal matrices with elements from GF(q). The multiplication by matrix D' or D" is where the Winograd algorithm collects all of its multiplications. Then, because we can interchange the order of prime terms with double prime terms,

$$V = (B'D'A') (B''D''A'')v$$

$$V = (B'B'') (D'D'') (A'A'')v$$

Hence, we have an n'n" Fourier transform algorithm, again in the form of the Winograd FFT. In this way Winograd large FFT algorithms can be built up from small ones.

A FAST BERLEKAMP-MASSEY ALGORITHM

The Berlekamp-Massey algorithm has an asymptotic computational complexity that is proportional to t^2. In this section, we will show a way to reduce the computational complexity for long codes.

The fast algorithm can be derived as a doubling algorithm. Split the iterations into two halves and modify the equations so that each half can be solved separately. The two half solutions are merged into the desired solution. Because the form of the equations is unchanged, the same idea can be applied to each of the two halves. Thus, a recursive algorithm is obtained.

The development begins with a more compact organization of the Berlekamp-Massey algorithm. Recall that the computations of the Berlekamp-Massey algorithm reside primarily in the two equations

$$\Delta_r = \sum_{j=0}^{n-1} \Lambda_j^{(r-1)} s_{r-j}$$

$$
\begin{bmatrix} \Lambda^{(r)}(x) \\ B^{(r)}(x) \end{bmatrix} = \begin{bmatrix} 1 & -\Delta_r x \\ \Delta_r^{-1}\delta_r & (1-\delta_r)x \end{bmatrix} \begin{bmatrix} \Lambda^{(r-1)}(x) \\ B^{(r-1)}(x) \end{bmatrix}
$$

$$
= \begin{bmatrix} 1 & -\Delta_r x \\ \Delta_r^{-1}\delta_r & (1-\delta_r)x \end{bmatrix} \dots \begin{bmatrix} 1 & -\Delta_1 x \\ \Delta_1^{-1}\delta_1 & (1-\delta_1)x \end{bmatrix} \begin{bmatrix} 1 \\ 1 \end{bmatrix}
$$

Define the matrix $\underset{\sim}{\Lambda}^{(r)}(x)$ by

$$
\underset{\sim}{\Lambda}^{(r)}(x) = \begin{bmatrix} 1 & -\Delta_r x \\ \Delta_r^{-1}\delta_r & (1-\delta_r)x \end{bmatrix} \dots \begin{bmatrix} 1 & -\Delta_1 x \\ \Delta_1^{-1}\delta_1 & (1-\delta_1)x \end{bmatrix}
$$

From this matrix $\Lambda^{(r)}(x)$ and $B^{(r)}(x)$ can be obtained by the expression

$$
\begin{bmatrix} \Lambda^{(r)}(x) \\ B^{(r)}(x) \end{bmatrix} = \underset{\sim}{\Lambda}^{(r)}(x) \begin{bmatrix} 1 \\ 1 \end{bmatrix}
$$

It serves just as well to update $\underset{\sim}{\Lambda}^{(r)}(x)$ as to update $\Lambda^{(r)}(x)$ and $B^{(r)}(x)$, although updating $\underset{\sim}{\Lambda}^{(r)}(x)$ directly can involve about twice as many multiplications because it has four elements rather than two. We will replace the iterates $\Lambda^{(r)}(x)$ and $B^{(r)}(x)$ by the iterate $\underset{\sim}{\Lambda}^{(r)}(x)$ and accept the doubling of the number of multiplications. This penalty will be overcome by reorganizing the computations.

The recursive form of the Berlekamp-Massey algorithm is built around the two equations

$$
\Delta_r = \sum_{j=0}^{n-1} \Lambda_{11,j}^{(r-1)} S_{r-j} + \sum_{j=0}^{n-1} \Lambda_{12,j}^{(r-1)} S_{r-j}
$$

$$
\underset{\sim}{\Lambda}^{(r)}(x) = \begin{bmatrix} 1 & -\Delta_r x \\ \Delta_r^{-1}\delta_r & (1-\delta_r)x \end{bmatrix} \underset{\sim}{\Lambda}^{(r-1)}(x)
$$

To split the algorithm into halves, let

$$
M^{(2t)}(x) = M'^{(t)}(x) \, M''^{(t)}(x)
$$

where

$$M'^{(t)}(x) = \prod_{r=2t}^{t+1} \begin{bmatrix} 1 & -\Delta_r x \\ \Delta_r^{-1}\delta_r & (1-\delta_r)x \end{bmatrix}$$

$$M''^{(t)}(x) = \prod_{r=t}^{1} \begin{bmatrix} 1 & -\Delta_r x \\ \Delta_r^{-1}\delta_r & (1-\delta_r)x \end{bmatrix}$$

We will compute the two halves separately and then multiply them together. This will entail less work than the original organization.

We will also need to reorganize the equations for Δ_r. Think of Δ_r as the rth coefficient of the first component of the two-vector of polynomials.

$$\begin{bmatrix} \Delta(x) \\ \Delta'(x) \end{bmatrix} = \begin{bmatrix} \Lambda_{11}^{(r-1)}(x) & \Lambda_{12}^{(r-1)}(x) \\ \Lambda_{21}^{(r-1)}(x) & \Lambda_{22}^{(r-1)}(x) \end{bmatrix} \begin{bmatrix} S(x) \\ S(x) \end{bmatrix}$$

Hence for r larger than t

$$\begin{bmatrix} \Delta(x) \\ \Delta'(x) \end{bmatrix} = M'^{(r-1)}(x)M''^{(t)}(x) \begin{bmatrix} S(x) \\ S(x) \end{bmatrix}$$

$$= M'^{(r-1)}(x)\, S^{(t)}(x)$$

where

$$S^{(t)}(x) = M''^{(t)}(x) \begin{bmatrix} S(x) \\ S(x) \end{bmatrix}$$

This completes the splitting of the Berlekamp-Massey algorithm. The basic algorithm now is written

$$\Delta_r = \sum_{j=0}^{n-1} M_{11,j}^{(r-1)} S_{1,r-j} + \sum_{j=0}^{n-1} M_{12,j}^{(r-1)} S_{2,r-j}$$

$$M^{(r)}(x) = \begin{bmatrix} 1 & -\Delta_r x \\ \Delta_r^{-1}\delta_r & (1-\delta_r)x \end{bmatrix} M^{(r-1)}(x)$$

where $M^{(r)}(x)$ may represent either $M'^{(r)}(x)$ or $M''^{(r)}(x)$, and $(S_1(x), S_2(x))$ represent $(S(x), \tilde{S}(x))$ in the first half of the computation, and is updated to $S^{(t)}(x)$ in the second half. After both halves are complete, $M^{(2t)}(x)$ is obtained by multiplying its two halves.

The Berlekamp-Massey algorithm is now split into two halves. Notice that each of the halves itself is a Berlekamp-Massey algorithm and can in turn be split. If t is a power of two, the splitting can continue until pieces that are only one iteration long are reached. These are executed, but are quite trivial. (If t is not a power of two the computation can be padded with dummy iterations.)

The multiplication load of the recursive procedure is almost entirely in the convolutions. This is because when an iteration of the Berlekamp-Massey algorithm is finally executed, it occurs as a single iteration and has just one multiplication (by Δ_m^{-1}), which can be neglected when counting multiplications. The computational complexity of the recursive procedure depends directly on the computational complexity of the convolution algorithms. Its successful use presumes the availability of a good set of convolution algorithms.

Convolution algorithms of complexity proportional to n log n can be used. Then because log n recursions will completely split the algorithm, the complexity is proportional at most to n log² n. One can do even better by using better convolution algorithms.

144

References

1.   Blahut, R. E., Theory and Practice of Error Control Codes,
      Addison-Wesley, 1983.

2.   Wolf, J. K., "Redundancy, the Discrete Fourier Transform, and
      Impulse Noise Cancellation," Proc Princeton Conf Inf Sci Syst,
      Princeton, NJ, 1982.

3.   Marshall, T. G., Jr., "Real Number Transform and Convolutional
      Codes," Proceedings of the 24th Midwest Symposium on Circuits
      and Systems, Ed S. Karne, Albuquerque, New Mexico,
      June 29-30, 1981.

4.   Nussbaumer, H. J. Fast Fourier Transform and Convolution
      Algorithms, Springer Verlag, Berlin, 1981.

5.   Winograd, S., "On Computing the Discrete Fourier Transform,"
      Math Comp Vol 32, pp 175-199, 1978.

6.   Miller, R. L., T. K. Truong, and I. S. Reed, "Efficient
      Program for Decoding the (255, 223) Reed-Solomon Code Over
      $GF(2^8)$ with Both Errors and Erasures, Using Transform
      Decoding," IEEE Proceedings, Volume 127, pp 136-142, 1980.

7.   Blahut, R. E., "Efficient Decoder Algorithms Based on Spectral
      Techniques," IEEE Abstracts of Papers - IEEE International
      Symposium on Information Theory, Santa Monica, California,
      1981.

AVERAGE-CASE OPTIMIZED BUFFERED DECODERS.

Elwyn R. Berlekamp   Robert J. McEliece

Cyclotomics, Inc.   California Inst. of Technology
Berkeley, CA   94704   Pasadena, CA   91125

1.   INTRODUCTION

In many error-correction coding systems, the required decoding time is a random variable which depends on the noise severity. When there are few errors, the decoder has a relatively easy time, but when there are many errors, the decoder must work much harder. If the "many errors" situation is relatively rare, the average decoding time may be much less than the maximum decoding time. If this is so, it will be possible to increase the decoder's effective speed dramatically, through the use of a buffered decoder architecture.

In this paper we shall describe in detail how the addition of a buffer (RAM-simulated delay line) can allow a dramatic increase in decoder speed. Our general ideas are of wide applicability, but we shall begin by considering a specific example, the Cyclotomics Model 120 GF1^{TM} decoder, which uses an average-case optimized, buffered architecture.

The underlying code is a (255, 243) Reed-Solomon code, which is capable of correcting up to 6 character errors in each block. In the Cyclotomics implementation, it takes 271 microseconds to correct 6 errors. This suggests that the data rate should be (243x8)/271 μs = 7.2 megabits per second. However, in one application, this decoder actually runs at over 120 M bps! The secret is that the decoding time is not a constant 271 microseconds, but depends on the noise severity, as shown in Table 1.

<——————— decoding time ———————>

No. errors (k)	Machine cycles (T_k)	Microseconds
0	0	0
1	700	42
2	850	51
3	3280	197
4	3680	221
5	4100	246
6	4510	271
$\geq$ 7	4080	245

Table 1. Decoding time as a function of noise severity for Cyclotomics' Model 120 GF1^{TM} decoder (one machine cycle = 60 nanoseconds).

(In the Cyclotomics decoder there is a special device which allows the decoder to be bypassed entirely if the syndrome is zero. If only one or two errors have occurred, special algebraic tricks can be used to speed up the decoding significantly. Thus the k = 0, 1, 2 decoding times in Table 1 are much smaller than the others. In fact the k $\geq$ 3 decoding times are somewhat longer than necessary so that the presence of k $\leq$ 2 errors can be quickly detected. The average decoding time has been lowered at the expense of increasing the maximum decoding time.)

If we model the number of errors in a codeword as a Poisson random variable with mean λ, then according to Table 1 the average decoding time is (in machine cycles)

$$\overline{T}(\lambda) = \sum_{k=0}^{6} T_k e^{-\lambda} \frac{\lambda^k}{k!} + T_{\geq 7} \sum_{k>7} e^{-\lambda} \frac{\lambda^k}{k!}$$

If a decoded bit error probability of 10^{-9} is required (not exceptionally stringent for computer-memory applications), the required channel symbol error probability ([1], Table II) is 1.33 E-3, an average of .3392 characters per block. For λ = .3392, the above formula gives

$\overline{T}$ = 221 clock cycles

= 12 microseconds,

corresponding to a data rate of 147 Megabits per second. Thus the average decoder speed is more than 20 times faster than the worst-case decoder speed. If the actual channel speed were

120 Mbps* the decoder could easily keep up, on the average. And yet 12 µs isn't enough time to decode even <u>one</u> channel error! How can the decoder keep up when it gets several noisy codewords in a row? The answer is that it employs a <u>buffer</u>, as illustrated in Figure 1.

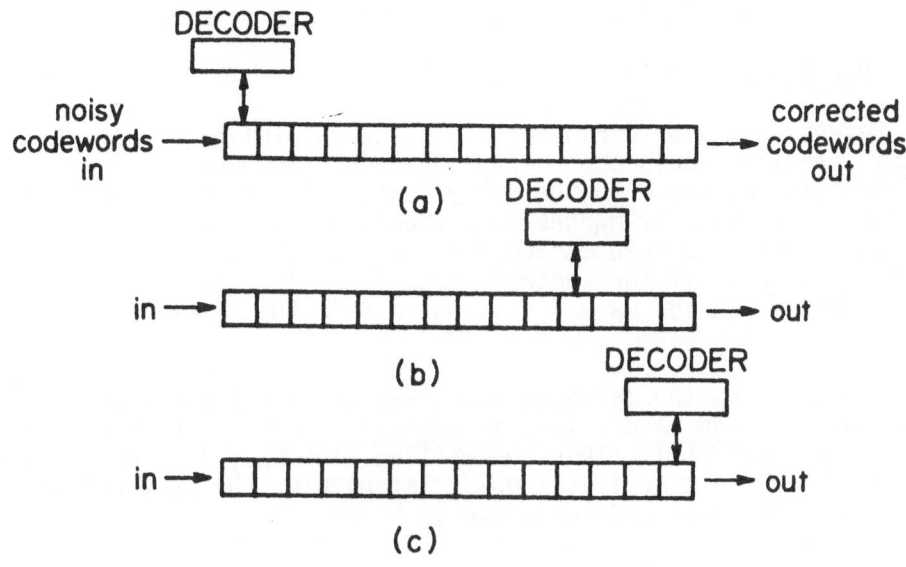

(a)

(b)

(c)

<u>Figure 1</u>.   A buffered decoder.

In Figure 1 the "buffer" is actually just a <u>delay-line</u> simulated by VLSI RAMs. The delay line operates completely independently from the rest of the machine; it accepts one byte of input every clock cycle and emits the same byte a fixed time later. Internally this RAM-based delay line is organized so that each read is followed by a write onto the same address. However, in looking at Figure 1 it is best to imagine that the codewords move sequentially from left to right through the delay line. From this conceptual viewpoint, the location of the decoder's pointer indicates its current "backlog". During a quiet period, after several error-free codewords in a row, the decoder will be "caught up", as shown in Fig. 1(a). When several noisy codewords in a row have been received, the backlog will increase, as shown in Figure 1(b). Finally, in an "overflow

---

*In practice this particular decoder is limited to 120 Mbps because it can only read/write a byte once per 60 ns clock cycle.

148

situation", the backlog will exceed the total delay in the delay line, and "buffer overflow" will occur. In practice, this rare occurrence will be handled gracefully by having the decoder refuse to decode a codeword when time is too tight, and go on to the next customer. These undecoded codewords will naturally contribute to the overall bit error probability. In the next section we give an analysis of the probability of B.O., using techniques from <u>queueing theory</u>. We show that B.O. occurs with a probability that is an exponentially decreasing function of the buffer length.

2.   BUFFER OVERFLOW AND D/G/1 QUEUES.

Referring again to Figure 1, we take the following viewpoint. A semi-infinite sequence of noisy codewords, numbered 1, 2, 3, ..., arrive at the entrance to the delay line, which we henceforth call the queue. We note by $\overline{W}_n$ the time (measured in machine cycles) from the moment the n-th codeword arrives at the queue until its decoding <u>begins</u>. If the codeword must leave the queue before de-coding begins, we set $\overline{W}_n = B + 1$, where B is the total length of the queue.

Denote by D_n the decoding time (service time) required by the n-th codeword, and assume that <u>interarrival interval</u> between code-words is a constant T. Then the waiting time $\overline{W}_{n+1}$ is T units less than $\overline{W}_n$ (because it arrives T units later) plus the service time required by the n-th codeword, i.e.,

$$\overset{?}{\overline{W}_{n+1}} = \overline{W}_n - T + D_n \tag{1}$$

However Eq. (1) is not quite correct, since $\overline{W}_{n+1}$ cannot be negative, and cannot (by definition) exceed B + 1. Hence if we define, for any integer k

$$(k)_+^L = \begin{cases} 0 \text{ if } k \leq 0 \\ k \text{ if } 0 \leq k \leq 1 \\ L \text{ if } k \geq L, \end{cases} \tag{2}$$

the correct recursion is not (1) but

$$\overline{W}_1 = 0$$
$$\overline{W}_{n+1} = (\overline{W}_n - T + D_n)_+^{B+1} \tag{3}$$

With this notation, the probability that the n-th codeword will not be completely decoded before it must leave the queue of Figure 1 is given by

$$Pr\{B.O.(n)\} = Pr\{\overline{W}_n + D_n \geq B + 1\}. \tag{4}$$

We assume throughout that the service times are independent, identically distributed random variables. In practice it will usually be necessary to interleave the codewords in order to safely make this assumption. The design of interleavers is not part of this discussion, but it too is greatly facilitated by the availability of fast cheap RAMs. See [2] for much more about interleavers.

The computation of the probability in (4) proves to be too difficult for us, and so we adopt the following approximative approach, which leads to very tight upper bonds on the overflow probability (4).

We define, in place of $\{\overline{W}_n\}$, a new random process $\{W_n\}$ as follows:

$$W_1 = 0 \tag{5}$$
$$W_{n+1} = (W_n - T + D_n)_+,$$

where $(x)_+$ denotes the positive part of x, i.e.,

$$(x)_+ = \begin{cases} 0, & \text{if } x \leq 0 \\ \\ x, & \text{if } x \geq 0. \end{cases} \tag{6}$$

Then (as is easily proved by induction)

$$\overline{W}_n \leq W_n, \text{ for } n = 0, 1, 2, \ldots . \tag{7}$$

The numbers W_n can be interpreted as the waiting times for a sequence of customers, arriving at a constant rate (one every T cycles) with independent service times D_1, D_2, D_3,..., where there is no limit to the number of customers who may be in the queue at a given time. It is as if the buffer in Figure 1 were infinite, and each codeword were released as soon as it was decoded (Figure 2).

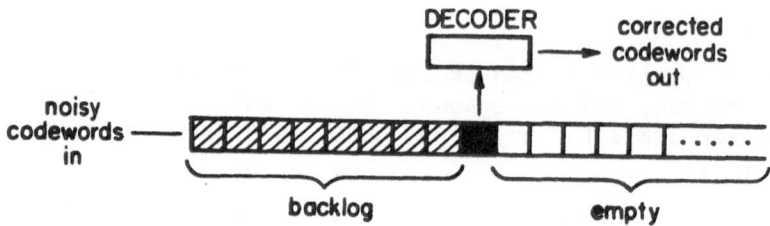

<u>Figure 2</u>. An infinite delay line.

Although this model does not correspond to a physically realizable system, in view of (4) and (7) we may write

$$\Pr\{\text{B.O.}(n)\} \leq \Pr\{W_n + D_n \geq B + 1\}. \tag{8}$$

We shall now consider the problem of computing the right side of (8).

In the standard notation of queueing theory, a system like the one depicted in Fig. 2 is called a <u>D/G/1 queue</u> [5]. "D" means that the interarrival time between consecutive customers is <u>deterministic</u>, i.e., constant; "G" means that the customer service times are <u>general</u>, i.e. arbitrary; and "1" means that there is only one server (decoder). Although the general tools developed e.g. in [5, vol. 1, chapter 8] for G/G/1 queues can be applied to this situation, we feel that it is worthwhile to present a detailed analysis for this particular discrete D/G/1 queue in the next section.

3. A CLOSED-FORM SOLUTION FOR THE D/G/1 QUEUE.

Motivated by the discussions above, we restate our problem as follows. Let $D_1, D_2, \ldots$ be i.i.d. discrete random variables, with common distribution function

$$d(j) = \Pr\{D_n = j\}, \quad j = 0, 1, 2, \ldots, M, \tag{9}$$

where M is the maximum possible service time. If T denotes the interarrival time between customers, we assume that

$$E(D_n) = \sum_j jd(j) < T. \tag{10}$$

Equation (10) simply means that the server can keep ahead of the traffic, on the average. We define the <u>waiting times</u> $\{W_n\}$ by (cf(5))

$$W_1 = 0$$
$$W_{n+1} = (W_n - T + D_n)_+. \tag{11}$$

Our first problem is to find the distribution functions $w_n(k)$ for the random variables W_n:

$$w_n(k) = \Pr\{W_n = k\}. \tag{12}$$

The recursive definition (11) implies that

$$w_{n+1}(k) = \sum_j d(j) \, w_n (k + T - j), \quad k \geq 1 \tag{13a}$$

$$w_{n+1}(0) = \sum_j d(j) \sum_{k \leq T-j} w_n(k), \tag{13b}$$

$$w_{n+1}(k) = 0. \qquad\qquad k \leq -1. \tag{13c}$$

We will return to the detailed recursion (13) in the next section, but for now let us observe that if we define the <u>stationary waiting times</u> $w(k)$ by

$$w(k) = \lim_{n \to \infty} w_n(k) \tag{14*}$$

then (13a) becomes

$$w(k) = \sum_j d(j)w(k + T - j), \qquad k \geq 1. \tag{15}$$

---

*The limit in (14) is known to exist [5, chapter 15], provided that the possible service times and the interarrival time T do not lie in an arithmetic progression of span ≥ 2. See Remark 2 after the proof of Lemma 3, below.

We now employ the method of generating functions, and define, for a complex variable z,

$$D^*(z) = \sum_{j=0}^{M} d(j)z^j, \tag{16}$$

$$W^*(z) = \sum_{k \geq 0} w(k)z^k. \tag{17}$$

Now if we multiply both sides of (15) by z^{k+T} and sum for $k \geq 1$, we obtain

$$\sum_{k \geq 1} w(k)z^{k+T} = \sum_j d(j)z^j \sum_{k \geq 1} w(k + T - j)z^{k+T-j}. \tag{18}$$

If the m-th truncated generating function for w(k) is defined for any integer m by

$$W_m^*(z) = \sum_{k \leq m} w(k)z^k, \tag{19}$$

then (18) becomes

$$z^T\{W^*(z) - W_0^*(z)\} = \sum_j d(j)z^j\{W^*(z) - W_{T-j}^*(z)\}, \tag{20}$$

which can be rearranged to give

$$W^*(z) = \frac{E^*(z)}{D^*(z) - z^T}, \text{ where} \tag{21}$$

$$E^*(z) = \sum_j d(j)z^j W_{T-j}^*(z) - z^T W_0^*(z) \tag{22}$$

$$= \text{a polynomial of degree} \leq T.$$

As we shall see, the location of the complex roots of the denominator $D^*(z) - z^T$ plays a crucial role. We need three easy technical lemmas.

<u>Lemma 1.</u> $W^*(z)$ has no poles in the unit disk $|z| \leq 1$.

<u>Proof:</u> if $|z| \leq 1$ then $|W^*(z)| \leq \sum_k w(k)|z|^k \leq \sum w(k) = 1$. Hence ([3, Sec. 3.33]) the radius of convergence of $W^*(z)$ is at least 1. This rules out any poles with $|z| \leq 1$.   QED

<u>Lemma 2.</u> $D^*(z) - z^T$ has exactly T zeroes inside the unit disk $|z| \leq 1$.

<u>Proof:</u> For any z we clearly have

$$|D^*(z)| \leq \sum_j d(j)|z|^j = D^*(|z|).$$ (23)

On the other hand by (10) we have, for all sufficiently small positive ε,

$$D^*(1+\varepsilon) < (1+\varepsilon)^T.$$ (24)

It follows from (23) and (24) that on the circle $|z| = 1+\varepsilon$,

$$|D^*(z)| \leq D^*(1+\varepsilon) < (1+\varepsilon)^T = |z|^T,$$

and so by Rouche's Theorem [3, sec. 6.21], z^T and $D^*(z)-z^T$ have the same number of zeroes in the region $|z| < 1+\varepsilon$. Since z^T clearly has T zeroes in this region, for all $\varepsilon \geq 0$, it follows that $D^*(z)-z^T$ has T zeroes in $|z| \leq 1$, as asserted.   QED.

Lemma 3. If $D^*(z)-z^T$ has a zero with $|z| > 1$, let z_0 be one of the smallest modulus. Then $r_0 = |z_0|$ is also a zero of $D^*(z)-z^T$.

Proof: Let $G(z) = D^*(z)-z^T$. Clearly $G(1) = 0$, and by (10), $G'(1) < 0$. If $G(r_0) > 0$, there would be a real root of $G(z)$ in the interval $(1, r_0)$, which contradicts the minimality of $|z_0|$. If $G(r_0) < 0$ we would have by (23) $|D^*(z_0)| \leq D^*(r_0) < r_0^T = |z_0|^T$, another contradiction, since $D^*(z_0) = z_0^T$. Thus $G(r_0) = D^*(r_0) - r_0^T = 0$.   QED.

Remarks:

1) $D^*(z)-z^T$ need not have any zeroes with $|z| > 1$, but this only happens when M, the longest possible service time, is $< T$, the interarrival time. This is a trivial case for which $W_n = 0$ for all n.

2) It is easy to show that there can be nonpositive zeroes of $D^*(z)-z^T$ on the circle $|z| = r_0$ only if the possible service times and T all lie in an arithmetic progression of span ≥ 2. If this happens, the problem can and should be simplified by renormalization. For example (cf. Example 1 below) if T = 5 and $D^*(z) = \frac{1}{2}z + \frac{1}{2}z^7$, the times {1, 5, 7} lie in the a.p. {1, 3, 5, 7, 9, 11, ...} with span 2. The waiting times for this queue are however the same as those for the queue with T = 4 and $D^*(z) = \frac{1}{2} + \frac{1}{2}z^6$, and those in turn are exactly twice those for the queue with T = 2 and $D^*(z) = \frac{1}{2} + \frac{1}{2}z^3$. We shall return to this (renormalized) example below.

With the help of these lemmas, we can simplify the expression (21) for $W^*(z)$ considerably. By Lemma 1, every root of the denominator $D^*(z)-z^T$ with absolute value ≤ 1 must also be a root of the numerator $E^*(z)$. By Lemma 2 $D^*(z)-z^T$ has T such zeroes, and by direct computation $E^*(z)$ has degree $< T$. Hence $E^*(z)$ must be a scalar multiple of $D^*(z)-zT$, and so in fact,

$$W^*(z) = \frac{f(1)}{f(z)}, \tag{25}$$

where $f(z)$ is that factor of $D^*(z)-z^T$ containing all its roots with $|z| > 1$. (The constant $f(1)$ is forced by the condition $W^*(1) = 1$.). If we are interested in the steady-state system time (wait plus service) $S = W + D$, the appropriate generating function is clearly

$$S^*(z) = D^*(z)W^*(z). \tag{26}$$

We can use (25) or (26) to estimate the rate of decay of the steady-state probabilities $w(j)$ and the steady-state overflow probabilities. For by Lemma 3, the radius of convergence of the power series $W^*(z)$ and $S^*(z)$ (by (25) and (26)) are both r_0. But it is well known [3, Sec. 3.33] that if $A(z) = \Sigma a_k z^k$, its radius of convergence is

$$r_0 = \lim_{k\to\infty} \inf a_k^{-1/k}. \tag{27}^*$$

It follows that we have <u>approximately</u>

$$w(j) = r_0^{-j+a_1}(j) \tag{28}$$

$$\Pr\{W + D \geq L\} = r_0^{-j+a_2}(j) \tag{29}$$

We illustrate with two examples.

<u>Example 1.</u> Let $T = 2$ and $D^*(z) = \frac{1}{2} + \frac{1}{2}z^3$. Here customers arrive every two cycles, and the average service time is $1\frac{1}{2}$ cycles. We have

$$D^*(z)-z^T = \frac{1}{2} + \frac{1}{2}z^3 - z^2 = \frac{1}{2}(z-1)(z^2-z-1). \tag{30}$$

The roots of z^2-z-1 are $(1 \pm \sqrt{5})/2$. Thus, in agreement with Lemma 2 $D^*(z)-z^T$ has two roots in $|z| \leq 1$, viz. 1 and $(1 - \sqrt{5})/2 = -0.62$, and one root with $|z| > 1$, viz. $(1 + \sqrt{5})/2 = +1.62$. We therefore have by (25)

$$W^*(z) = \frac{1-r_0^{-1}}{1-r_0^{-1}z} = \frac{3-\sqrt{5}}{z} \sum_{k=0}^{\infty} (\frac{\sqrt{5}-1}{2})^k z^k, \text{ and} \tag{31}$$

---

*Since by (25) and (26) $W^*(z)$ and $S^*(z)$ are rational functions of z, the "lim inf" in (27) can be replaced by "lim".

$$S^*(z) = D^*(z)W^*(z) = \frac{3-\sqrt{5}}{2} \{1+r_0^{-1}z+r_0^{-2}z^2+(r_0^{-3}+1)z^3 +$$

$$(r_0^{-4}+r_0^{-1})z^4 + \ldots \}. \tag{32}$$

Thus the steady state probabilities are given explicitly by

$$w(j) = \frac{3-\sqrt{5}}{2} r_0^{-j} \tag{33}$$

$$Pr\{W + D \geq L\} = r_0^2 r_0^{-L} \tag{34}$$

which should be compared to the approximations (28) and (29). Let us try a more serious example.

Example 2. Returning to the data in Table 1, if we assume T = 255 cycles (one RS character enters the delay line each clock cycle), we have

$$D^*(z) = e^{-\lambda} + \lambda e^{-\lambda} z^{700} + \frac{\lambda^2}{2} e^{-\lambda^2} z^{850} +\ldots+$$

$$\sum_{k \geq 7} \frac{\lambda^k}{k!} e^{-\lambda} z^{4080} - z^{255}, \tag{35}$$

which is a polynomial of degree 4510! Lemma 1 tells us that this polynomial has 255 roots inside the unit circle and 4255 outside, but this isn't much help. For any fixed λ it would be possible to locate the crucial real zero r_0 described in Lemma 3, but even this (cf. (28) and (29)) would not allow us to give a very confident estimate for the overflow probability. In a complex example like this, it is clear that another tool is needed. We give this in the next section.

4. THE KINGMAN BOUND FOR D/G/1 QUEUES.

The formulas (25) and (26) are satisfying from a theoretical viewpoint, and informative in at least a qualitative way (cf. (28) and (29)), but there are shortcomings. First, as we have seen in Example 2, if $D^*(z)-z^T$ is of high degree, it may be impractical to compute $W^*(z)$ or $S^*(z)$ from (25) or (26). Second, even if (as in Example 1) these generating functions can be computed, they will only give information about the steady-state probabilities, and not the actual waiting times W_n or $W_n + D_n$. In this section we present a very simple and useful technique due to Kingman [5, vol. 2, Sec. 2.4] which largely overcomes both of these difficulties.

Theorem (The Kingman Bound):

Using the previously-introduced notation,

$$\Pr\{W_n \geq L\} \leq r_0^{-L} \qquad\qquad (36)$$

$$\Pr\{W_n + D_n \geq L\} \leq r_0^{-L+T}, \text{ for all } L, \text{ all } n \geq 1. \qquad (37)$$

<u>Proof</u>: We first prove (36), which is true by definition for $n = 1$ (see (11)). We proceed by induction, assuming (36) is true for n. Since (36) is trivial for $L \leq 0$, we assume $L \geq 1$. Then

$$\Pr\{W_{n+1} \geq L\} = \sum_{k \geq L} w_{n+1}(k)$$

$$= \sum_j d(j) \sum_{k \geq L} w_n(k+T-j) \quad \text{(by (13a))}$$

$$\leq \sum_j d(j) r_0^{-(L+T-j)} \qquad \text{(by induction)}$$

$$= r_0^{-(L+T)} D^*(r_0)$$

$$= r_0^{-L} \qquad\qquad \text{(since } D^*(r_0) = r_0^T\text{).}$$

This proves (36). To prove (37) note that

$$\Pr\{W_n + D_n \geq L\}$$

$$= \sum_j d(j) \Pr\{W_n \geq L - j\}$$

$$\leq \sum_j d(j) r_0^{-L+j} \qquad \text{(by (36))}$$

$$= r_0^{-L} D^*(r_0)$$

$$= r_0^{-L+T} \qquad\qquad \text{(since } D^*(r_0) = r_0^T\text{).}$$

We conclude with a return to our two examples. QED.

<u>Example 1 (bis)</u>.

We saw above that $r_0^{-1} = (\sqrt{5}-1)/2$, and so by (36) and (37)

$$\Pr\{W_n \geq L\} \leq \left(\frac{\sqrt{5}-1}{2}\right)^L \qquad\qquad (38)$$

$$\Pr\{W_n + D_n \geq L\} \leq \left(\frac{\sqrt{5}-1}{2}\right)^{L-2} \qquad\qquad (39)$$

For a fixed L, (33) and (34) show that these bounds are exact, as n → ∞.

Example 2 (bis).

Here we have $D^*(z)-z^T$ given by (35). If we chose $\lambda = .3392$ (corresponding to a decoded bit error probability of 10^{-9}), a simple computer program gives $r_0 = 1.000375$. Hence for a delay-line B cycles long, from (39)

$$Pr\{B.O.(n)\} \le r_0^{-(B+1-255)}$$

For this probability to be 10^{-10} or less, this means that B should be chosen around 60,000. Thus a 64 K byte RAM would easily suffice to make the B.O. contribution to the decoded bit error probability negligible. In 1983 this is of course a very inexpensive and reliable device. With a 60 ns clock, the total (constant) delay for the delay line would be about 4 milliseconds. This is a small price to pay for a tenfold increase in decoder speed!

In Table 2 we give some more numerical data for this particular decoder, as a function of P_{bit}, the desired decoded bit error probability. In Table 2 λ denotes the average number of character errors which correspond to the given P_{bit} (obtained by inverting the formula given in Table II of [1]), $\overline{T}$ is the corresponding average decoding time, measured in clock cycles, and $\overline{S}$ is the corresponding average decoder speed, measured in bits per second. The next column, T gives a supposed value for the interarrival time measured in cycles between codewords, and S is the corresponding data rate. (These numbers are chosen so that the ratio $S/\overline{S}$ has the constant value .87.) Finally the column labelled "r_0-1" gives the value of the crucial real zero r_0 of $D^*(z)-z^T$, and B gives the required buffer size (estimated from (37)) so that the probability of B.O. is less than 1/10 the value of P_{bit}.

5.   CONCLUSION.

We conclude that with a delay-line architecture like the one in Figure 1, many decoders can be made to run at their average speed instead of their slowest speed. The probability of "buffer overflow", which might better be called "late output" is an exponentially decreasing function of the buffer size and can usually be made negligibly small with (by modern standards) a very small buffer and very short (constant) delay. The appropriate analytical tool for estimating the probability of B.O. is the Kingman bound (37), which requires only the computation of the least positive root r_0 of the key equation $D^*(z)-z^T = 0$ with $|z| > 1$.

158

6. ACKNOWLEDGEMENTS.

We wish to thank Bruce Hajek for showing us the Kingman bound, and Eugene Rodemich for help with the crucial Lemma 2.

P_{bit}	λ	$\bar{T}$	$\bar{S}$, Mbps	T	S	r_0-1	B
10^{-9}	.339	221	146	255	127	3.75E-4	61,000
10^{-8}	.479	311	104	360	90	3.32E-4	62,000
10^{-7}	.683	450	72	520	62	2.81E-4	65,000
10^{-6}	.984	674	48	780	42	2.39E-4	67,000
10^{-5}	1.44	1060	31	1225	26	2.18E-4	63,000
10^{-4}	2.19	1732	19	2000	16	2.43E-4	47,000
10^{-3}	3.57	2836	12	3250	10	4.76E-4	19,000

Table 2. Some parameter values for the Cyclotomics Model 120 GF1^{TM} Decoder (see text for key).

7. REFERENCES

1. Berlekamp, E., "The Technology of Error-Correcting Codes," Proc. IEEE 68 (1980), pp. 564-593.
2. Berlekamp, E., and Po Tong, "Interleaving," IEEE Trans. Inform. Theory, in press.
3. Copson, E. T., An Introduction to the Theory of Functions of a Complex Variable. London: Oxford University Press, 1960.
4. Feller, W., An Introduction to Probability Theory and Its Applications, Volume I (3rd Ed.), New York: Wiley, 1968.
5. Kleinrock, L., Queuing Systems, Vols. 1 and 2, New York: Wiley, 1975.

CODE STRUCTURE AND DECODING COMPLEXITY

P. G. Farrell

Electrical Engineering Department, The University, Manchester.
M13 9PL, U.K..

Presented at the NATO ASI on "The Impact of Processing Techniques
on Communications", Chateau de Bonas (Gers), France, 11-22 July,
1983.

1.   INTRODUCTION

The decoder for an error-correcting code is, in almost all
cases of interest, several orders of magnitude more complex than
the corresponding encoder. So the selection of a code is in
practice usually determined by the choice of decoding method,
rather than the other way round. Possible decoding methods are
determined by the constraints of the information transmission,
storage or processing system which requires protection against
errors; the code selected will be the one which offers the highest
error-control capability from among those codes which can be
decoded cost-effectively within the system constraints. A simpler
decoding method will, other things being equal, permit the use of
a more powerful error-control code; thus there is a continuing
requirement to devise simpler and more effective decoding algorithms.
One way to achieve this is to exploit as much as possible any
sturcture which exists in the code; several ways of doing this are
reviewed and presented in this contribution.

1.1 Motivation

The contribution is concerned with decoding methods for
linear systematic block codes (1,2,3,4), and in particular, those
for the two main classes of useful linear block code: cyclic codes
and array codes. Array codes are extended and generalised forms of

product and iterated codes (5). Both these classes of codes exist
for a wide range of code parameters (block length n, number of
information digits k, number of check digits c = n − k), and are
capable of protecting against random, burst or diffuse error
patterns. They can be decoded in a number of different ways, many
of which are effective in practice.

The study reported here was motivated by several considerations:-

(i) As noted above, simpler decoding algorithms lead to better
error-control system performance, because more powerful codes can
be used. This is of particular interest for codes in the short to
medium range of block length ($15 \leqslant n \leqslant 128$) where there are many
potential trade-offs and applications.

(ii) There is a need for very simple serial and parallel decoders
for certain applications, such as regenerative satellite trans-
ponders, where power dissipation and circuit complexity are
severely restricted.

(iii) There is also a need for fast decoders with short decoding
delay (latency), capable of operating with channel transmission
rates greater than 8 Mbit/sec. This means that the decoding delay
must be as small as possible, which gives simple parallel decoding
algorithms an advantage.

(iv) In order to make error-control coding really cost-effective,
it is necessary to be able to use soft-decision decoding techniques;
which, as is well established, can double the error-control
capability of a code (6). There is, therefore, a very great need
to devise relatively simple soft-decision decoding algorithms.

(v) As noted above, an effective way of achieving decoder
simplicity is to exploit code structure and symmetry as fully as
possible. This is one reason why cyclic and array codes are widely
used; they possess an extensive mathematical and conceptual basis
which gives them well defined structural and symmetrical properties.
Many well established decoding methods, however, do not fully
exploit this structure. For example, in some parts of the decoding
process for a cyclic code, use is made of the linear and cyclic
nature of the code (e.g., syndrome calculation and error search),
but in other parts (e.g., error location and error magnitude
calculation) only the linearity properties are used. This is
particularly apparent in soft-decision decoding algorithms; until
recently it was not at all clear how to devise an effective and
simple soft-decision decoder for a cyclic code.

1.2 Complexity of Decoding

Decoding complexity is notoriously difficult to define and

determine since, following Savage (7), it is a function of at least three measures: circuit (hardware) complexity, decoding delay (decoder processing time), and computational complexity (number of operations or computations performed by the decoding process). Circuit complexity on its own is not a very good measure of decoding complexity, except in the case of a comparison between decoders with approximately equal decoding delays. In other cases the apparent decoding simplicity of less complex hardware circuits operating repeatedly in a serial or hybrid serial-parallel mode is misleading because of the increased decoding delay. The circuit complexity measure itself is difficult to define precisely, because it can be done either in terms of the number of basic functional circuit elements (e.g., 2-input gates), or in terms of the number of circuit components (e.g., IC, MSI, LSI or VLSI packages), although .the development of uncommitted logic array (ULA) and microprocessor implementation techniques has tended to blur this distinction. In practice, it is also important to take account of decoder input and output buffers, and timing and synchronisation circuits. Apparently simple decoding processors often require quite complex buffering and timing arrangements.

Decoding delay (or latency) is also difficult to define, because of the many different decoder input-output arrangements that can be envisaged. Conceptually, decoding can not begin until a complete block has been input to the decoder. For this reason, and in order to facilitate comparisons between serial and parallel types of decoder, it is most convenient to define the decoding delay as being the time that elapses during the decoding process, beginning after arrival of the complete block, and continuing until the corrected information digits start to become available. In practice, though, with certain types of serial decoder it is possible to start delivery of corrected digits before arrival of the whole block; these decoders appear to exhibit negative decoding delay (decoding advance!), but a re-definition of the measure would unfairly penalise serial decoders over parallel decoders in most practical situations. Such "look-ahead" decoders are therefore, for comparison purposes, normally deemed to have zero (or unit) decoding delay. Of course, they are advantageous when either the block length of the code is long, or there is a need for rapid estimation of information digits (e.g., "quick-look" decoding) followed by full decoding later (in real or non-real time). The situation is further complicated by the fact that in many decoders the decoding delay changes from block to block, depending on the error pattern affecting each word.

Because circuit complexity on its own is not a good measure of decoding complexity, since it does not take into account the time factor, it is necessary to use a computational complexity measure. Again, it is not realistic just to take the number of computations required to decode a block (which is often fairly easy to calculate), since this ignores advantages that may be gained by using parallel processing and by ordering the computations efficiently in time. Thus a more practical computational complexity measure is the product of the circuit complexity and the decoding delay (Savage (7) calls this computational work). It allows the various trade-offs between circuit complexity and decoding delay to be assessed and quantified. However, the various difficulties associated with the definition and determination of circuit complexity and decoding delay carry over into the decoding complexity measure, so that no reliance can be placed on it in an absolute sense, but only in a relative sense when applied to decoders in comparable situations. The reality is that no one decoding complexity measure can be defined which will permit assessment and comparison of decoder implementations with many different constraints and for widely differing applications; just as there is no one measure (e.g., coding gain, quality factor, etc.) which accurately assesses decoder performance. This is analogous to the status of device, circuit and system reliability measures (e.g., mean time between failure calculations, etc.), which are similarly limited to rather restricted comparative assessments. If an absolute complexity measure is required, as a fundamental quantity to compare with practical values, then it seems more fruitful to pursue the information theoretic approach of Chaitin (8) and others (including the contributions of Cover, Massey and Ziv in this volume).

1.3 Summary of the Contribution

In section 2 hard and soft-decision decoding methods for block codes are briefly reviewed. Four main decoding methods are mentioned: full minimum distance decoding, syndrome decoding, majority logic (threshold) decoding, and trellis decoding. Syndrome decoding techniques, including list, Meggitt, algebraic, systematic search, step-by-step, error trapping, permutation, and trial-and-error algorithms are covered in section 3. The emphasis is on methods for cyclic codes. In section 4, error trapping decoding is re-examined, and related to other decoding methods. Minimum weight decoding for cyclic codes is introduced and assessed, for both hard and soft-decision decoding of either random or burst errors. Decoding algorithms for array codes are described in section 5; serial and parallel algorithms for decoding bursts of errors or erasures, for decoding diffuse erasures; and for soft-decision decoding of random, burst or diffuse errors, are presented and assessed. The contribution finishes with a concluding discussion.

2. DECODING METHODS FOR BLOCK CODES

The various decoding methods which are mentioned in this section and the next are comprehensively described in references

(1,2,3,4 and 9). The emphasis will be on binary, linear, cyclic and array codes, with exceptions where noted. Decoding computational complexity values, Δ, will be calculated using the following assumptions, again with exceptions where noted:-

(a) circuit complexity, ν, computed as the equivalent number of 2-input gates (reckoning a shift-register stage as 6 2-input gates), and ignoring input and output buffers and timing circuits;

(b) decoding delay, τ, calculated as the number of clock pulses required to carry out the decoding process, with the clock pulse repetition rate being equal to the bit rate of the input to the decoder;

(c) regardless of input and output buffer arrangements, a decoder will be defined as parallel when it delivers all the corrected digits together after one ($\tau = 1$) or more ($\tau > 1$) clock pulses, and as serial when corrected digits are delivered one at a time after one ($\tau = 1$) or more ($\tau > 1$) clock pulses. A hybrid decoder delivers sub-sets of the corrected digits, one sub-set at a time, after one or more clock pulses. Note that both serial and parallel decoders can be implemented a combination of serial and parallel processing, with consequent affect on the values of ν, τ and $\Delta = \nu.\tau$.

The four main types of decoding method for block codes will now be reviewed.

2.1 Full Minimum Distance Decoding

Algorithm: (i) Compute the distance between the received block and all the code words in the code.

(ii) Select the word with minimum distance to the received block; this word is the best estimate of the corrected digits.

The distance measure can be Hamming, burst (10), or Euclidean distance, chosen appropriately for hard or soft-decision decoding of random, burst, or diffuse error patterns. Because the algorithm does not presuppose any structure in a code, it can be applied to any type of code, including non-linear codes. In the non-linear case, storage (or generation) of the code words for comparison with the received block has a complexity proportional to N, the number of code words in the codes. A serial decoder implementation (hard or soft-decision) has decoding delay proportional to N, as does the circuit complexity of a parallel decoder. Thus the overall decoding computational complexity is proportional to N^2. If the code is linear, then generation of the code words has a complexity proportional to k (the number of information digits in a code word) or less if the code is cyclic, so that Δ is dominated by the factor $N = 2^k$. This factor increases exponentially with k, so the algorithm is still only effective with short or low rate codes. If the code is transparent (the logical complement of every code word is also in the code) then Δ is halved.

2.2 Syndrome Decoding

Minimum distance decoding algorithms are complex because they do not make much use of any structure that the code may have. If a code is linear, then it has considerable structure, and hard decision minimum distance decoding can be considerably simplified, with no loss in performance (i.e., still maximum likelihood). This arises from the well-known fact that in a linear code there is a one-to-one correspondence between correctable error patterns and syndromes (1-4, 9):

$$[y].[H]^t = \{[x] + [e]\}.[H]^t$$

$$= [e].[H]^t \quad \text{as} \quad [x].[H]^t = [0]$$

$$= [S]$$

where [y] is the received block, $[H]^t$ is the transpose of the parity check matrix of the code, [x] is the transmitted code word, [e] is the error pattern, and [S] is the syndrome. For binary linear codes, and particularly for cyclic and array codes, computation of the syndrome (syndrome calculation) is trivial compared to the complexity of determining the error pattern which corresponds to the syndrome (error correction). This is not always true for q-ary (multi-level) codes. So the computational complexity of syndrome decoding is dominated by the error correction process; Δ is proportional to a factor not greater than 2^c, and in many cases (particularly for burst errors) the factor is much smaller. This makes syndrome decoding feasible for long, high rate codes. The various syndrome decoding algorithms are reviewed and assessed in section 3 below.

Syndrome decoding algorithms, when applied to the hard-decision decoding of linear block codes, very effectively reduce the complexity of minimum distance decoding; they are not nearly so successful when used for soft-decision decoding. The most promising general algorithms (11,12,13) operate by generating possible error patterns in order of decreasing likelihood as determined by the soft-decision confidence information of the received block, stopping when the corrected block is a word in the code. Each candidate error pattern is tested by effectively computing a syndrome; this is the extent to which the linear properties of the code are used. These algorithms appear to have complexities roughly similar to those exhibited by sequential or Viterbi decoding algorithms for convolutional codes with comparably error-control capability (17). Results presented in (12) indicate that the average number of

candidate error patterns that need to be generated increases
asymptotically towards $2^c = 2^{n-k}$ as the signal-to-noise ratio
(assuming additive white Gaussian noise) decreases. This makes the
algorithm attractive for high rate codes. These algorithms are
examples of a number of algorithms (14-17) which select a small set
of candidate code words (or error and erasure patterns) for testing
to determine the most likely code word to have been transmitted.
The algorithms make use of "masking" (14) and "projecting" (15)
techniques which to some extent invoke the structure of the code in
order to reduce the complexity. All these algorithms suffer from
the disadvantage that the actual number of computations required
for maximum likelihood decoding is a random variable; in most cases,
however, the algorithms can be truncated without much loss in
performance.

An earlier, alternative approach which does not suffer from the
variable computation problem, called weighted erasure decoding
(18-20), implements soft-decision decoding by computing, for each
decoder output digit, a weighted sum of the outputs of $r = \lfloor (Q+1/2) \rfloor$
hard-decision syndrome decoders operating in parallel (or with r
times faster clock rate, serially) on the received soft-decision
demodulated digits; where Q is the number of soft-decision levels,
and $\lfloor x \rfloor$ is the largest integer $\leq$ x. This algorithm makes better
use of any simplifying advantage that syndrome decoding has to offer
(see section 3), but at the cost of replicating the circuit
complexity (or decoding delay). The technique has been extended to
burst-error correction (10), and to array codes (18). There also
are some relatively efficient soft-decision syndrome decoding
algorithms which make use of the structure and symmetries of
particular codes, such as the Hamming codes (21), the Golay code
(22), the BCH (128,106) code (23) and certain Reed-Muller codes (24).

2.3 Majority Logic Decoding

This is a method of decoding (1-4, 9) for any linear code,
provided that the parity check equations of the code are self-
orthogonal (or can be orthogonalised); in the special sense of
orthogonal that guarantees one information digit appearing in each
of a set of parity check equations, with all the other (information
and check) digits appearing once only somewhere in the set. Thus
k sets of orthogonal equations are required for a linear code, but
only one set if the code is cyclic. There are many variants of
majority logic decoding (9); certain types are forms of syndrome
decoding, and others are forms of simplified minimum distance
decoding. Very low complexity hard-decision majority logic decoders
can be devised for cyclic and array codes, for random and burst
error correction:-

Algorithm: (i) Using the received block, re-compute the set of
orthogonal parity check equations for the first information digit

in the block.

(ii) If a majority of the re-computed checks fail, then an error is indicated; there is no error otherwise.

(iii)Repeat (i) and (ii) for the remaining information digits.

The disadvantage of majority logic decoding are that relatively few codes are orthogonal or can be orthogonalised, and in general it is not possible to correct up to the full capacity of the code. Weighted majority, or threshold, decoding can be used to decode any linear code, however, at the cost of a considerable increase in complexity (15,26).

Soft-decision decoding has been applied to majority logic decodable codes with rather more success (in terms of complexity) than is the case with syndrome decoding. Perhaps the most funda-mental approach is that of reference (27), which gives maximum likelihood (per symbol) performance. It can be viewed as an extension of threshold decoding, which uses every word in the dual (1-4) code (dual-code-domain decoding) so that the complexity is proportional to 2^C, favouring high rate codes. Simpler decoders, with asymptotically optimum performance, may be derived from the fundamental approach, and are called algebraic analog decoders (28). Useful soft-decision decoding extensions of threshold decoding and of generalised minimum distance decoding (29) have been reported (30-32). These and similar schemes are closely related, and in fact references (30) and (31) independently describe almost the same algorithm. A practical soft-decision scheme with moderate complexity, based on noise digit estimation, is reported in (33). The above algorithms have been developed mainly for cyclic codes, but can also be applied to array (product) codes (31).

2.4 Trellis Decoding

Trellis decoding was originally devised as a method for decoding convolutional codes (2-4, 9), but it can also be used to decode block codes. The code words of a linear block code are all the paths in a trellis with 2^C nodes (states) and n+1 segments. If the code is cyclic, then the trellis is periodic; for an array code (product code) the number of nodes can be reduced to much less than 2^C. The code can be maximum likelihood (hard or soft-decision) decoded by means of the Viterbi algorithm (2-4, 9) applied to the trellis, with complexity proportional to 2^C in the general case. This makes the technique very suitable for high rate codes. Reduced-search decoding algorithms which trade off complexity against performance have also been reported (34). The trellis algorithm is related to the fundamental soft-decision dual-code-domain algorithm (27) mentioned in section 2.3 above; reference (35) brings together these and related ideas, forming a unified approach to soft-decision decoding by means of a trellis, based on the concept of replication decoding.

3.   SYNDROME DECODING METHODS

The principal syndrome decoding methods are reviewed below; they are described in detail in one or more of references (1-4) and (9). As noted in section 2.2, syndrome decoding consists of two basic steps: syndrome calculation, and error correction by finding the error pattern corresponding to the syndrome: the interest lies in the second basic step. Hard decision decoding is implied, unless otherwise noted.

3.1 List Decoding

Conceptually, the easiest way to relate a syndrome to its corresponding error pattern (random or burst) is to have a stored list of syndromes and error patterns. In practice, the syndrome can be used as the address for a ROM in which the error pattern is stored. Thus the circuit complexity is proportional to $k.2^c$; if the code is cyclic, this reduces to 2^c, since the syndrome need only determine the correctness of one information digit at a time. If the code structure and symmetry is exploited then the syndrome list can be considerably shortened (23,38). In all cases, the decoding delay can be unity (one clock pulse), with a serial form of decoder. The computational complexity is thus proportional to 2^c, which limits the straightforward table-lookup method to codes with $c \leq 18$ (a 256K memory).

3.2 Meggitt Decoding

Meggitt decoding (3,4) operates by transforming the syndrome into the error pattern by means of a logic circuit. Again, the circuit complexity is much reduced if the code is cyclic, being proportional to approximately n^t, the number of error patterns that the logic circuit must recognise. This makes the method clumsy for $t > 3$, but in practice list decoding has superseded Meggitt decoding because of the advent of large cheap memories.

3.3 Algebraic Decoding

The method described so far have only made use of the cyclic shift property of a cyclic code. The considerable mathematical structure of cyclic codes can be used, however, to convert the syndrome into its error pattern, as follows in outline (1-4):-

Algorithm: (i) Compute the syndrome polynomial.

(ii) Use the syndrome to find the error location polynomial and the error value polynomial.

(iii) Locate the errors by finding the roots of the location polynomial; each position in code word represents a possible root, and all are searched in succession (Chien search).

(iv) Calculate the error values (not required if the code is binary).

The algorithm can also be applied to alternant codes (1), a very large class of linear codes which includes Goppa, Shrivastava and generalised BCH codes, as well as many classes of cyclic codes. In the binary case, the complexity of the decoding algorithm is dominated by step (ii), and is proportional to $n \log^2 n$. For q-ary (multi-level) codes, steps (i) and (iv) are non-trivial, and approximately double the complexity. The non-exponential growth of Δ with n makes this algorithm the simplest available for hard decision decoding of long codes (n > 127). Its only disadvantages are that it is unnecessarily complex for medium and short length codes, and it is not directly compatible with soft-decision decoding. The nature of the algorithm is such that it is best implemented with hybrid processing in a serial decoder. The decoding delay, depending on the clock rate, may have to be one or more block lengths ($\tau \geq n$), to give time for computation of the error location and error value polynomials. It can be speeded up for low weight errors (34). It is essentially a random-error-correcting algorithm. Transform techniques can be used to implement a variety of alternative algebraic decoders; see the contribution by Blahut in this volume.

3.4 Systematic Search Decoding

Another conceptually very simple decoding method for random or burst errors is as follows:-

Algorithm: (i) Compute the syndrome of the received block.

(ii) Generate a correctable error pattern, and compute its syndrome.

(iii) Compare the syndromes in (ii) and (iii), and stop is they are the same. If not, then go to (iv).

(iv) Repeat (ii) and (iii) until all the correctable error patterns have been exhausted.

This form of systematic search decoding is equivalent to adding correctable error patterns to the received block until one is found which forces the syndrome of the sum to zero; the errors have then been corrected. The decoding delay of this essentially serial processing parallel decoder is bounded by the number of error patterns; i.e., $\tau \leq 2^c$. The hardware required is a syndrome decoder and an error pattern generator, so the circuit complexity can be quite low, but it is dominated by the complexity of error generation unless only a few errors are to be corrected (t $\leqq$ 3). The decoding delay can be reduced by using a hybrid form of implementation. The apparent simplicity of this algorithm is deceptive, for it is not a practical decoding algorithm except in very limited situations (for burst correction, perhaps). The main reason for this is that, once again, very little account is being taken of the structure of the code.

3.5 Step-by-Step Decoding

Algorithm: (i) Compute the syndrome of the received block, and determine the weight of the corresponding error pattern.

(ii) Invert the first information digit of the block, re-compute the syndrome, and re-determine the error pattern weight.

(iii) If the re-determined weight is higher than in (i), then re-invert the digit back to its original value; if lower, then leave inverted.

(iv) Repeat (ii) and (iii) for successive information digits until the error weight falls to zero.

The crucial step in this random-error-correcting algorithm is the determination of the error pattern weight from the syndrome. If this can be done with a complexity less than that of list decoding, then the algorithm is potentially useful. Meggitt decoding is a form of step-by-step decoding.

3.6 Error Trapping Decoding

This decoding method relies on the fact that if an error pattern lies wholly within the parity check section of a received block, then the syndrome is identical to the error pattern; that is, the ONES in the syndrome are the errors. If the code is cyclic, and the error pattern spans c or fewer digits, then after sufficient cyclic shifts of the received block, the errors will all lie within the check section. The syndrome of the shifted block may be obtained by cyclically shifting, in the syndrome calculating register, the syndrome of the original (unshifted) received block. If the error pattern is correctable, and has weight e (i.e., contains e errors), then the shifted syndrome has weight e (and is identical to the error pattern) only when the error pattern lies wholly within the check section; it has weight greater than e whenever any part of the pattern lies outside the check section. Hence, for a code capable of correcting t random errors:-

Algorithm: (i) Calculate the syndrome of the received block.

(ii) Compute the weight of the syndrome. If the weight is t or less, then the errors are trapped and the syndrome is the error pattern. If the weight is greater than t, then continue.

(iii) Shift the syndrome once in the syndrome calculating register, repeat (ii), and continue until the errors are trapped.

For a code capable of correcting all bursts of length $\leq$ b:-

Algorithm: (i) Calculate the syndrome of the received block.

(ii) Test for c-b consecutive ZEROS in the left-most stages of the syndrome calculating register. If found, then the burst is trapped in the right-must be stages of the register; if not, then continue.

(iii) Shift the syndrome in the register until the burst is trapped.

This is the simplest of all the syndrome decoding methods for cyclic codes. The circuit complexity of a serial decoder is dominated by the syndrome register plus a buffer to store the received block during step (iii) of the algorithm. Thus $\nu \cong 6(n+c)$. For the (24,15) b = 3 shortened cyclic burst-error-correcting code, $\nu \cong 220$. The decoding delay is between 1 and n; in practice it is convenient to set $\tau = n$. Thus Δ is proportional to n^2. The decoding delay can be reduced to unity by using a parallel form of decoder, each shift in the syndrome being produced by a combinatorial logic circuit. In this case, Δ is proportional to kn^2. The disadvantage of error trapping for random-error-correcting cyclic codes is that it can only deliver maximum likelihood performance (i.e., cope with all correctable patterns) if n/k > t. Thus it is most effective with low rate and small t codes. An interesting type of serial decoder for single-error-correcting (t=1) cyclic codes, devised to take advantage of magnetic bubble devices, is described in reference (36).

3.7 Permutation Decoding

Error trapping succeeds because cyclic shifting is a code-preserving permutation of the received digits, so that the same decoder can be used after shifting. If other, non-cyclic, code-preserving permutations of the received digits can be found, then these can also be used to trap errors, thus extending the range of error patterns for which trapping is effective (1). If transform domain techniques are used, then non-code-preserving permutations are also possible (38).

3.8 Trial-and-Error Decoding

This is an extension (4) of error trapping. Suppose that a t-error-correcting cyclic code is such that only up to t-1 errors can be trapped. To correct a t-error pattern invert, one at a time, the information digits of the received block, attempting to trap at each stage. At some point, one of the t errors will be inverted, and the remaining t-1 errors will then be trapped. Clearly, this is also a modified form of step-by-step decoding (section 3.5). Lin and Costello (3) call this algorithm systematic search decoding; it is in fact a hybrid combination of systematic search (section 3.4) and step-by-step decoding.

4.   SYNDROME DECODING RE-EXAMINED

Syndrome decoding methods, reviewed in the previous section (see Table 1) offer the simplest hard-decision decoding algorithms for cyclic and array codes. The only exceptions are certain majority logic decoding algorithms, which are applicable to a restricted set of codes. There remains a major problem, however, with syndrome decoding techniques: the fact that they apparently are not by their nature easily extendable to soft-decision decoding. In addition, syndrome decoding technqiues as applied to cyclic codes, though conceptually and computationally simple, do not make full use of cyclic code structure and symmetry. Cyclic codes derive their structure from the fact that they may be represented by, and generated from, a single polynomial. This fact is not explicitly used to any extent in the decoding process, except during syndrome calculation (and even here is not fully exploited). Intuitively, it can be conjectured that there must be a better way of decoding cyclic codes; a way which goes beyond treating the code as just a rather special linear code, probing much more deeply into its cyclic nature, but hopefully without losing the basic simplicity of syndrome decoding. This section sets out such a decoding method, and assesses its potential application. The method is a combination of error trapping, systematic search, and step-by-step decoding, which turns out to be a form of reduced search minimum distance decoding; it can be called minimum weight decoding.

In addition, efficient parallel and serial forms of decoder for array codes will be described, which exploit the specific two-or-more-dimensional structure of array codes.

4.1 Minimum Weight Decoding

The error trapping decoding algorithm for cyclic codes is given in section 3.6 above. It consists of calculating the syndrome, and shifting it in the syndrome register until the weight is $\leq$ t. Calculating the syndrome is done by dividing the received block (polynomial) by the generator sequence (polynomial); the syndrome is the remainder of the division process. Shifting the syndrome in the syndrome register is the same as continuing to divide by the generator polynomial. Division is repeated subtraction, which is the same as addition in the binary case (i.e., in GF(2)). As an example, let the code be the Hamming (7,4) single-error-correcting code with t = 1 (d = 3), and generator polynomial $G(x) = x^3 + x + 1 \equiv 1011$. Also let the received block, $Y(x)$, be the all-ZERO code word with an error in the second position. Then a "simulation" of the error trapping decoding process is:-

```
0 1 0 0 0 0 0         Y(x)
  1 0 1 1             G(x)
0 0 0 1 1 0 0
    1 0 1 1
0 0 0 0 1 1 1         Syndrome, S(x), Weight w = 3
      1 0 1 1
0 0 0 0 1 0 1         "Shifted Syndrome", w = 2
        1 0 1 1
0 0 0 0 0 0 1         "Shifted Syndrome", w = 1
                      Error in Second Position
```

which can conveniently be re-arranged as follows.

Example 1

```
0 1 0 0 0 0 0         Y(x)
  1 0 1 1             G(x)
0 0 0 1 1 0 0
      1 0 1 1
0 0 0 0 1 1 1         S(x), w = 3
1       1 0 1
1 0 0 0 0 1 0         w = 2
1 1       1 0
0 1 0 0 0 0 0         w = 1, Error in Second Position
```

As all single errors can be trapped, this "simulation" will work for an error in any position of any possible code word.

As another example, consider the BCH (15,5) t = 3 (d = 7) code.

Example 2

```
1 1 1 1 1 1 1 1 1 1 1 1 1 1 1     Code word, X(x)
0 0 1 0 0 1 0 0 0 0 0 0 0 0 0     Error Pattern, E(x), w = 2
1 1 0 1 1 0 1 1 1 1 1 1 1 1 1     Y(x)
1 0 1 0 0 1 1 0 1 1 1             G(x)
0 1 1 1 1 1 0 1 0 0 0 1 1 1 1
  1 0 1 0 0 1 1 0 1 1 1
0 0 1 0 1 1 1 0 0 1 1 0 1 1 1
    1 0 1 0 0 1 1 0 1 1 1
0 0 0 0 0 1 1 1 1 1 0 1 0 1 1     S(x), w = 8
1           1 0 1 0 0 1 1 0 1 1
```

```
1               1 0 1 0 0 1 1 0 1 1
1 0 0 0 0 0 1 0 1 1 1 0 0 0 0      w = 5
1 1             1 0 1 0 0 1 1 0 1
0 1 0 0 0 0 0 0 0 1 1 1 1 0 1      w = 6
1 0 1 1 1         1 0 1 0 0 1
1 1 1 1 1 0 0 0 0 0 1 0 1 0 0      w = 7
1 1 0 1 1 1         1 0 1 0 0
0 0 1 0 0 1 0 0 0 0 0 0 0 0 0      w = 2, Errors (w ⩽ 3) .
```

So this pair of errors is trapped, as expected, since its span is less than c = 10. With a different code word, the result is still the same.

Example 3

```
1 1 1 1 0 1 0 1 1 0 0 1 0 0 0      X(x)
0 0 1 0 0 1 0 0 0 0 0 0 0 0 0      E(x)
1 1 0 1 0 0 0 1 1 0 0 1 0 0 0      Y(x)
1 0 1 0 0 1 1 0 1 1 1              G(x)
0 1 1 1 0 1 1 1 0 1 1 1 0 0 0
  1 0 1 0 0 1 1 0 1 1 1
0 0 1 0 0 1 0 0 0 0 0 0 0 0 0
    1 0 1 0 0 1 1 0 1 1 1
0 0 0 0 1 1 0 1 1 0 1 1 1 0 0
      1 0 1 0 0 1 1 0 1 1 1
0 0 0 0 0 1 1 1 1 0 1 0 1 1      S(x), w = 8
1               1 0 1 0 0 1 1 0 1 1
1 0 0 0 0 0 1 0 1 1 1 0 0 0 0      w = 5
1 1             1 0 1 0 0 1 1 0 1
0 1 0 0 0 0 0 0 0 1 1 1 1 0 1      w = 6
1 0 1 1 1         1 0 1 0 0 1
1 1 1 1 1 0 0 0 0 0 1 0 1 0 0      w = 7
1 1 0 1 1 1         1 0 1 0 0
0 0 1 0 0 1 0 0 0 0 0 0 0 0 0      w = 2, Errors (w ⩽ 3)
```

Consider now the all-ONES code word, again, as in Example 2, but with a different error pattern.

Example 4

```
1 1 1 1 1 1 1 1 1 1 1 1 1 1 1      X(x)
1 0 0 0 0 1 0 0 0 0 1 0 0 0 0      E(x)
0 1 1 1 1 0 1 1 1 1 0 1 1 1 1      Y(x)
  1 0 1 0 0 1 1 0 1 1 1            G(x)
0 0 1 0 1 0 0 0 1 0 1 0 1 1 1
    1 0 1 0 0 1 1 0 1 1 1
0 0 0 0 0 0 0 1 0 0 0 1 0 1 1      S(x), w = 4
1 1 1             1 0 1 0 0 1 1 0
1 1 1 0 0 0 0 0 0 1 0 1 1 0 1      w = 7
1 0 1 1 1         1 0 1 0 0 1
```

```
1 0 1 1 1             1 0 1 0 0 1
0 1 0 1 1 0 0 0 0 0 0 0 1 0 0   4
0 0 1 1 0 1 1 1             1 0 1
0 1 1 0 1 1 1 1 0 0 0 0 0 0 1   7
0 1 0 0 1 1 0 1 1 1           1
0 0 1 0 0 0 1 0 1 1 0 0 0 0 0   4
    1 0 1 0 0 1 1 0 1 1 1
0 0 0 0 1 0 1 1 0 1 1 1 1 0 0   7
      1 0 1 0 0 1 1 0 1 1 1
0 0 0 0 0 0 0 1 0 0 0 1 0 1 1   4, same as S(x)
- - - - - - - - - - - - - - -
```

From here on, the division process repeats itself. No "shifted syndrome" (error pattern) of weight ≤ 3 has emerged. This is because the error pattern is untrappable. To find it, use the trial-and-error algorithm set out in section 3.8.

Example 5

↓INVERTED

```
1 1 1 1 1 0 1 1 1 1 0 1 1 1 1   Y(x) First Digit Inverted
1 0 1 0 0 1 1 0 1 1 1           G(x)
0 1 0 1 1 1 0 1 0 0 1 1 1 1 1
  1 0 1 0 0 1 1 0 1 1 1
0 0 0 0 1 1 1 0 0 1 0 0 1 1 1
          1 0 1 0 0 1 1 0 1 1 1
0 0 0 0 0 1 0 0 0 1 0 0 0 0     S(x), w = 2, Errors
```

A pair of errors is trapped, so the first digit inverted must also have been an error, and the untrappable three-error pattern is corrected. This method works for the (15,5) code, and for several other codes like the Golay code and the BCH (31,21) t = 2 (d = 5) code.

The really interesting thing about these examples of error trapping and trial-and-error decoding is that they are really forms of bounded minimum distance decoding (section 2.1) in disguise. This is because the code words of a cyclic code are made up of sums of cyclic shifts of the generator sequence (polynomial), G(x); and it is sums of shifts of G(x) that are being compared with Y(x) during the division (repeated subtraction $\equiv$ repeated addition) process that generates the syndrome and traps the errors. In Example 1, the four code words compared with Y(x) are

1.   0 1 0 1 1 0 0

```
     0 1 0 1 1 0 0
  ⊕  0 0 0 1 0 1 1
2.   0 1 0 0 1 1 1
```

2. $\overline{0\ 1\ 0\ 0\ 1\ 1\ 1}$

      0 1 0 1 1 0 0
$\oplus$ 0 0 0 1 0 1 1   0 1 0 0 1 1 1
    $\underline{1\ 0\ 0\ 0\ 1\ 0\ 1}$ $\equiv$ $\underline{1\ 0\ 0\ 0\ 1\ 0\ 1}$ $\oplus$
3.   1 1 0 0 0 1 0   1 1 0 0 0 1 0

and

      0 1 0 1 1 0 0
      0 0 0 1 0 1 1
      1 0 0 0 1 0 1 $\equiv$ 1 1 0 0 0 1 0
    $\underline{1\ 1\ 0\ 0\ 0\ 1\ 0}$ $\underline{1\ 1\ 0\ 0\ 0\ 1\ 0}$ $\oplus$
4. 0 0 0 0 0 0 0   0 0 0 0 0 0 0

The weights of the syndrome and "shifted syndromes" are the Hamming distances between the four code words and the received block, $Y(x)$. Since this code is a single-error-correcting code, then if a "shifted syndrome" of weight 1 appears, it must be the error pattern, since the syndrome itself was non-zero, indicating the presence of at least one error. The same point applies to the (15,5) code of Examples 2-5. In Example 1, only 4 code words needed to be compared with $Y(x)$, out of a total of 16. In Example 2, it is 7 out of 32; Example 3, 8 out of 32; Examples 4 and 5, 11 out of 32.

Thus error trapping and trial-and-error decoding (i.e., step-by-step decoding) is a form of reduced-search bounded minimum distance decoding, where the code words which are compared with $Y(x)$ are generated during the division by $G(x)$ process. A considerable reduction in the number of code words that need to be compared with $Y(x)$ is thereby achieved. In addition, a number of other interesting results emerge.

(i) Since at each step of the division, a code word comparison is taking place, the weight of the result is always relevant; that is, there is no need to wait until $S(x)$ is generated. Thus, in Example 1, the weight of $Y(x)$ itself is 1, so this must be the error! The transmitted code word is the all-ZEROS word. In Example 3, the weight at the third comparison (arrowed) is 2; therefore this must be the errors, and there is no need to continue the process.

(ii) Some untrappable error patterns can be decoded.

Example 6

  1 1 1 1 0 1 0 1 1 0 0 1 0 0 0 $X(x)$
  $\underline{0\ 0\ 0\ 0\ 1\ 0\ 0\ 0\ 0\ 1\ 0\ 0\ 0\ 0\ 1}$ $E(x)$
  1 1 1 1 1 1 0 1 1 1 0 1 0 0 1 $Y(x)$, w = 11
  $\underline{1\ 0\ 1\ 0\ 0\ 1\ 1\ 0\ 1\ 1}$ $G(x)$
  0 1 0 1 1 0 1 1 0 0 1 1 0 0 1   w = 8

```
0 1 0 1 1 0 1 1 0 0 1 1 0 0 1    w = 8
  1 0 1 0 0 1 1 0 1 1 1
0 0 0 0 1 0 0 0 0 1 0 0 0 0 1    w = 3, Errors
```

Note that a standard error trapping decoder would not have corrected this pattern, except by using trial-and-error.

(iii) The weight structure of the code can be used to further reduce the number of steps required to decode an error pattern. To illustrate this, consider the (15,5) code of Examples 2-6. This code has the following weight structure:-

w	A_w
0	1
7	15
8	15
15	1

where A_w is the number of code words of weight w. Therefore if Y(x) has weight 13, as in Example 2, or weight 12, as in Example 4, then the ZERO digits in Y(x) must be the error positions; that is, the ONES cannot be errors. So these two examples can be decoded by means of a single weight calculation. Similarly, if a weight t + 1 emerges during the division process, this can only occur as the result of a pattern of t errors being added to a code word of weight 2t + 1 = d. Hence the t + 1 ONES occurring at the step of the division process cannot be errors. In this way, the three distinct sequences of weight 4 that result from the division process in Example 4 combine to leave only three possible error positions (positions 1, 6 and 11), confirming the previous deduction. Another example is given next.

Example 7

```
1 1 1 1 0 1 0 1 1 0 0 1 0 0 0    X(x)
1 0 0 0 0 1 0 0 0 0 1 0 0 0 0    E(x)
0 1 1 1 0 0 0 1 1 0 1 1 0 0 0    Y(x)
  1 0 1 0 0 1 1 0 1 1 1          G(x)
0 0 1 0 0 0 1 0 1 1 0 0 0 0 0    w = 4, Not Errors
    1 0 1 0 0 1 1 0 1 1 1
0 0 0 0 1 0 1 1 0 1 1 1 1 0 0    7
      1 0 1 0 0 1 1 0 1 1 1
0 0 0 0 0 0 0 1 0 0 0 1 0 1 1    4, Not Errors
1 1 1         1 0 1 0 0 1 1 0
1 1 1 0 0 0 0 0 0 1 0 1 1 0 1    7
1 0 1 1 1         1 0 1 0 0 1
0 1 0 1 1 0 0 0 0 0 0 0 1 0 0    4, Not Errors
```

Once again, the only positions where errors can be is 1, 6 and 11.

This is a particularly fortuitous property of the (15,5) code; in general, however, the number of positions where errors can be located is much reduced, and the trial-and-error search significantly shortened, if the weight structure of the code is used. By this means, trial-and-error decoding can be made to succeed even when many error patterns of weight t - 1 or less are still untrappable.

(iv) The method extends naturally to q-ary (multi-level) codes. Consider the Reed-Solomon (7,3) code, with $q = 2^3 = 8$, t = 2 (d = 5) and $G(x) \equiv 1 \; \alpha^4 \alpha^2 \alpha^4 \; 1$.

Example 8

1 1 1 1 1 1 1	X(x)
1 0 0 1	E(x)
0 1 1 0 1 1 1	Y(x), w = 5
1 $\alpha^4 \alpha^2 \alpha^4$ 1	G(x)
0 0 $\alpha^5 \alpha^2 \alpha^5$ 0 1	w = 4
$\alpha^5 \alpha^2$ 1 $\alpha^2 \alpha^5$	
0 0 0 0 $\alpha^4 \alpha^2 \alpha^4$	w = 3
$\alpha \alpha^4 \quad \alpha^4 \alpha \alpha^6$	
$\alpha \alpha^4$ 0 0 0 $\alpha^4 \alpha^3$	w = 4
$\alpha^6 \alpha \alpha^4 \quad \alpha^4 \alpha$	
$\alpha^5 \alpha^2 \alpha^4$ 0 0 0 1	w = 4
$\alpha^4 \alpha^2 \alpha^4$ 1 $\quad$ 1	
1 0 0 1 0 0 0	w = 2, Errors

(v) The weight metric used during the decoding process can be soft (Euclidean) distance, instead of hard (Hamming) distance. This means that soft-decision minimum weight decoding is feasible, with very little additional complexity over the hard decision case. The actual processing can be done using the received decision digits (i.e., with binary sequences), the confidence digits being used in the weight computations, but it is conceptually easier and more convenient to set out an example in terms of positive integers and an integer Euclidean metric. Consider the Hamming (7,4), t = 1 (d = 3), code with Q = 4 soft-decision detection levels. The soft distance of the code is $d_s = 3(Q-1) = 9$, digits can take the values x_i (or y_i) = 0, 1, 2 or 3, and the soft distance between a pair of digits, d_{sij}, or the soft weight of a digit, w_{si}, also takes the values 0, 1, 2 or 3. Note that $d_{ij} = |x_i - x_j| = |w_{si} - w_{sj}|$.

Example 9

3 3 3 0 3 0 0	X(x)
0 2 0 0 2 0 0	E(x)

```
0 2 0 0 2 0 0          E(x)
3 1 3 0 1 0 0          Y(x)   w_S = 8
3 0 3 3                G(x)
0 1 0 3 1 0 0          w_S = 5
        3 0 3 3
0 1 0 0 1 3 3           8
3 3         3 0
3 2 0 0 1 0 3           9
0 3 3       3
3 1 3 0 1 0 0          8, same as Y(x)
```

The process now repeats itself (the double-error pattern is untrappable), so apply the trial-and-error algorithm:-

↓INVERTED

```
0 1 3 0 1 0 0          Y(x) With 1 Digit Inverted, w_s = 5
      3 0 3 3
0 1 0 0 2 3 0          6
3         3 0 3
3 1 0 0 1 3 3          11
3 3         3 0
0 2 0 0 1 0 3          6
0 3 3       3
0 1 3 0 1 0 0          5, Repeating
```

↓INVERTED

```
3 2 3 0 1 0 0          Y(x) With 2 Digit Inverted, w_s = 9
3 0 3 3
0 2 0 3 1 0 0          6
  3 0 3 3
0 1 0 0 2 0 0          3, Errors (w_s ≤ 4)
```

which successfully completes the decoding. Soft-decision decoding of codes with medium and long block lengths, and with rates $R \geq \frac{1}{2}$ is quite feasible and practical by this method.

(vi) Error trapping decoding is, of course, the simplest method for burst-error-correction with cyclic codes. Using the technique illustrated in (v) above, it can easily be extended to soft-decision burst-error-correction; again, with very little additional complexity. Consider the (7,3) b = 2 code, for which $G(x) \equiv 1\ 1\ 1\ 0\ 1$, again with Q = 4 soft-decision levels. The burst-2-distance, d_{b2}, of this code (10) is 3 (since it is single-burst-correcting), so the soft burst-2-distance is

$$d_{bs2} = d_{b2} (Q - 1) \quad = 3.3 \quad = \quad 9$$

Thus with soft decision decoding the code is capable of correcting any set of bursts, each of length 2, and aggregate soft burst weight ≤ 4. Note that b_2 is the number of bursts of length 2 (computed cyclically).

Example 10

```
3 0 3 0 0 3 3      X(x)
0 2 2 0 2          E(x)
3 2 1 0 2 3 3      Y(x), b₂ = 3, w_bs = 8
3 3 3 0 3          G(x)
0 1 2 0 1 3 3      b₂ = 3, w_bs = 6
    3 3 3 0 3
0 1 1 3 2 3 0      3, 7
3     3 3 3 0
3 1 1 0 1 0 0      3, 5
3 3 3 0 3
0 2 2 0 2 0 0      2, 4, Errors
```

It may be necessary to use trial-and-error (note that there is no need to return to the received block when inverting digits):-

Example 11

```
3 0 3 0 0 3 3      X(x)
0 2 2 0 2 2 0      E(x)
3 2 1 0 2 1 3      Y(x), b₂ = 3, w_bs = 7
3 3 3 0 3          G(x)
0 1 2 0 1 1 3      b₂ = 3, w_bs = 6
    3 3 3 0 3
0 1 1 3 2 1 0      3, 5
3     3 3 3 0
3 1 1 0 1 2 0      3, 6
3 0 3       3 3
0 1 2 0 1 1 3      3, 8, Repeating
3 1 2 0 1 1 3      1st Digit Inverted, 3, 9
3 3 0 3       3
0 2 2 3 1 1 0      3, 6
  3 3 3 0 3
0 1 1 0 1 2 0      2, 3, But not Errors because the
                       Inverted Digit Lies Outside
                       the Two Bursts

3 0 3       3 3
3 1 2 0 1 1 3      3, 9  Repeating
0 2 2 0 1 1 3      2nd Digit Inverted, 3, 6
3 3 0 3       3
3 1 2 3 1 1 0      3, 7
3 3 3 0 3
0 2 1 3 2 1 0      3, 6
  3 3 3 0 3
0 1 2 0 2 2 0      2, 4, Errors
```

In the running prose above, the subscripted variables are: b_2, w_{bs}.

and to continue until the least weight pair of length 2 bursts emerges as the error pattern:-

Example 12

3 0 3 0 0 3 3	X(x)
0 0 2 3 0 0 0	E(x)
3 0 1 3 0 3 3	Y(x), $b_2 = 3$, $w_{bs} = 9$
3 3 3 0 3	G(x)
0 3 2 3 3 3 3	3, 9
3 3 3 0 3	
0 0 1 0 3 0 3	3, 7
0 3 3 3 3	
0 3 1 0 0 3 0	2, 6
3 0 3 3 3	
3 3 2 0 0 0 3	2, 6
3 3 0 3 3	
0 0 2 3 0 0 0	1, 3, Errors

A special case of two bursts each of length 2 is a single burst of length 4:-

Example 13

3 0 3 0 0 3 3	X(x)
0 2 2 2 2 0 0	E(x)
3 2 1 2 2 3 3	Y(x), $b_2 = 4$, $w_{bs} = 9$
3 3 3 0 3	G(x)
0 1 2 2 1 3 3	3, 6
3 3 3 0 3	
0 1 1 1 2 3 0	3, 5
0 3 3 3 3	
0 2 1 1 1 0 3	3, 6
3 3 0 3 3	
3 1 1 2 1 0 0	3, 5
3 3 3 0 3	
0 2 2 2 2 0 0	2, 4, Errors

so that the burst length of a code is virtually doubled by using soft decision decoding.

   All the above examples are forms of serial decoder with decoding delays in the approximate range n to n^2, and circuit complexities about the same as for error trapping with trial-and-error decoding. A decoding delay of n^2, however, is very much the worst case (trial-and-error of all n block digits); on average it will be much less, and in practice it can be reduced by using parallel processing (see section 3.6). Thus the overall computational complexity will normally be much less than n^3, and probably less than n^2.

4.2 Decoding for Array Codes

Serial and parallel hard-decision decoders for burst-correcting array codes have been described previously, in references (5) and (39), respectively. The serial decoder is a form of error trapping decoder, modified to take account of the structure of the array code (two-dimensional in the case of reference (5)). An example of this type of decoder is shown in Figure 2, for the (5x3, 4x2) b = 2 array code given, together with its encoder, in Figure 1. Note that the code word digits are lettered in the order in which they are transmitted and received. The complexity of this type of decoder is similar to that of an error trapping decoder for a cyclic code.

The parallel decoder is a modified form of Meggitt (logic) syndrome decoder, again modified to take account of the array code structure. From another point of view, the parallel decoder is a modified form of majority logic decoder (section 2.3). An example of this type of decoder, for the same code as in Figure 1, is given in Figure 3. The basic principle of the decoder is that a particular information digit is in error if:

(i) the vertical and horizontal checks corresponding to the digit both fail (i.e., the corresponding syndrome digits are ONES); and

(ii) no errors have occurred in other digits corresponding to correctable error patterns which involve the same parity checks. For the code in Figure 1, for example, digit "b" will be in error if checks 2 and 4 both fail, and error patterns "de" and "no" have not occurred; i.e., there is not an error in digit "d" or digit "n". Hence the logic circuit shown in Figure 3. The parallel decoder for a (4x4, 8) b = 3 array code (40) is shown in Figure 5, by way of another example; the code and its encoder are given in Figure 4. These parallel decoders have decoding delay $\tau = 1$, and circuit complexities of about 5n.

The parallel type of array decoder can be modified to carry out erasure correction, or erasure and error correction; the circuit complexity is approximately doubled in the case of erasure correction ($\cong$ 10n), and trebled for erasure and error correction ($\cong 15n \equiv n^2$). The code of Figure 1, for which b = 2, is capable of correcting all erasure-only bursts of length up to 5, or all double-adjacent error bursts with one more adjacent erasure, or various combinations of single or dobule-adjacent errors or erasures with random single or double-adjacent erasures (e.g., diffuse patterns such as random pairs of double-adjacent erasures).

A serial soft-decision decoding algorithm for array codes was described in (39). In outline it consists of the following steps.

Algorithm: (i) Compute the overall soft-decision confidence sum of each row and column in the received array code block.

(ii) In decreasing order of overall confidence sum, soft-decision minimum distance decode rows and columns in the array, alternately.

(iii) Any digits corrected during (ii) have their confidence value set to zero, and any digits not corrected have their confidence values set to the maximum; these amended values are used during the rest of the decoding process.

The minimum weight decoding method described in the previous section can be used to speed up and simplify step (ii) of the algorithm.

This soft-decision algorithm for array codes cannot cope with certain extreme (highest confidence) error patterns (e.g., for the code of Figure 1, a single extreme error, or a burst of two adjacent extreme errors). The problem can be overcome by implementing the soft-decision algorithm and one of the hard-decision algorithms (serial or parallel) both together in the decoder, but there are more effective possibilities. In particular, certain array codes can be transformed into cyclic codes by a simple re-arrangement of the digit transmission order (41). Then the minimum weight decoding method can be used to efficiently decode the cyclically transformed array code, thus correcting both hard and soft errors simultaneously.

5.   CONCLUSIONS

This contribution has reviewed and assessed decoding methods for block codes. Full minimum distance decoding is conceptually the simplest method: it applies to any type of code, it can be used to combat random, burst or diffuse error patterns, with hard or soft-decision metrics. It offers maximum likelihood performance, but it suffers from the serious disadvantage that its complexity grows exponentially with the number of information digits in the code word. This means that full minimum distance decoding is only practical for short and low rate codes.

Syndrome decoding methods apply to linear codes, and are particularly effective for the two main practical classes of linear codes: cyclic codes, and array (including product and iterated) codes. The computational complexity of syndrome decoding algorithms at worst grows exponentially with the number of check digits in the code word, and in certain cases (error trapping decoding for burst-error-correction, and algebraic decoding for random-error-correction) the complexity falls substantially below the upper bound. Some syndrome decoding methods offer maximum likelihood performance, others not; but any loss in performance can usually be traded for a compensating reduction in decoder complexity. Thus syndrome methods offer cost-effective decoding for a wide range of error control requirements and code parameters; and they are particularly useful for high rate codes. The main problem with standard syndrome techniques is that they can not be adapted easily for soft-decision decoding, which restricts or prohibits their use in many cases of practical interest.

Majority logic decoding methods are related to syndrome methods, and they offer particularly simple decoders for certain codes. This is their main disadvantage: that they are only applicable to a very restricted, and often relatively poor performance, set of codes. Majority logic algorithms, on the other hand, have successfully been adapated for soft-decision decoding.

Finally, trellis decoding methods, originally developed for convolutional codes, can be applied to the decoding of linear block codes. These methods offer maximum likelihood performance and complexities which grow exponentially with the number of check digits in a code word. Their use is restricted, therefore, to high rate codes.

This study of the various decoding methods reveals that the simplest decoding algorithms exploit code structure and symmetry to the fullest possible extent. The less that a decoding method relies on code structure, the more complex the decoding process is likely to be (full minimum distance decoding is an extreme example). Syndrome (and majority logic) methods are effective because they take advantage of cyclic and array code structure. Two major questions remain:

(i) how to further exploit the structure of cyclic codes, to find even more efficient syndrome decoding algorithms; and

(ii) how to adapt syndrome algorithms for soft decision decoding, without incurring severe complexity penalties?

A deeper re-examination of syndrome decoding for cyclic codes reveals, however, that the simplest technique, error trapping, can be combined with trial-and-error (systematic search) and step-by-step methods to yield a very effective reduced-search bounded-minimum-distance decoding algorithm, with near-maximum-likelihood performance. The method, for which the name minimum weight decoding is suggested, makes full use of the nature of a cyclic code; this answers question (i) above. In addition, becuase minimum weight decoding is essentially minimum distance decoding, its adaption for soft-decision decoding is trivial, merely requiring the use of a soft (Euclidean) distance instead of Hamming distance. Thus minimum weight decoding offers high performance hard or soft-decision decoding of cyclic codes, with low complexity, proportional to the square of the block length or less. This makes it of great interest for decoding short and medium length codes, and it may also be

useful for long codes. It can be easily modified for shortened
cyclic codes, and for q-ary (multi-level) cyclic codes. It permits
effective use of metrics developed for burst-correcting-codes
(burst distance, reference (10)), with hard or soft decision
decoding. Array decoding techniques, also presented in this
contribution, which make considerable use of array code structure,
become even more effective with the use of minimum weight decoding,
particularly in the soft-decision case. This advantage also applies
to the recursive coding techniques developed by Tanner (42). The
parallel hard-decision array decoding technqiue can be used with
any burst-correcting code.

Just as in the case of replication decoding (35) among the
trellis decoding methods, minimum weight decoding is an interpre-
tative and unifying concept, as well as being a simple and effective
decoding algorithm. It explains the fundamental nature of syndrome
decoding methods, and relates apparently distinct algorithms within
a coherent framework. It makes explicit the intuitive concept of a
soft syndrome, and it provides the first really practical example
of an implementation of step-by-step decoding. Like algebraic
decoding, minimum weight decoding appears to be essentially a serial
algorithm; it will be very interesting to devise parallel processors
for this decoding method.

REFERENCES

1. Macwilliams, F. J. and N. J. A. Sloane: The Theory of Error
    Correcting Codes, North-Holland, 1977.
2. McEliece, R. J.: The Theory of Information and Coding,
    Addison-Wesley, 1977.
3. Shu Lin and D. J. Costello: Error-Control Coding-Fundamentals
    and Applications, Prentice-Hall, 1983.
4. Peterson, W. W. and E. J. Weldon: Error-Correcting Codes,
    2nd Edition, MIT Press, 1972.
5. Farrell, P. G.: Array Codes, Chap. 4 in "Algebraic Coding Theory
    and Applications", Ed. G. Longo, Springer-Verlag, 1979.
6. Farrell, P. G., E. Munday and N. Kalligers: Digital
    Communications Using Soft-Decision-Detection Techniques, AGARD
    Symposium on Dig. Comms. in Avionics, Munich, June 1978.
7. Savage, J. E.: Three Measures of Decoder Complexity, IBM Jour.
    Res. and Dev., July 1970, pp. 417-425.
8. Chaitin, G. J.: Information-Theoretic Computational Complexity,
    IEEE Trans, Vol. IT-20, No. 1, Jan. 1974, pp. 10-15.
9. Farrell, P. G.: A Survey of Error-Control Codes, Chap. 1 in
    reference 5 above.
10. Wainberg, S. and J. K. Wolf: Burst Decoding of Binary Block
    Codes on Q-ary Output Channels, IEEE Trans, Vol. IT-18, No. 5,
    Sept. 1972, pp. 684-686.
11. Dorsch, B. G.: A Decoding Algorithm for Binary Block Codes and
    J-ary Output Channels, IEEE Trans, Vol. IT-20, No. 3, May 1974,
    pp. 391-4.
12. Tait, D. J.: Soft-Decision Decoding of Block Codes, IEE
    Colloquium on Practical Applications of Channel Coding
    Techniques, London 17 Feb. 1983 (IEE Colloq. Digest No. 1983/15,
    pp. 6/1-6/4).
13. Greene, E. P.: Minimum Decoding of Cyclic Block Codes, NTC'77
    Conference Record, Vol. 2, pp. 26:8-1/3.
14. Greenberger, H. J.: Approximate Maximum Likelihood Decoding of
    Block Codes, JPL Publication No. 78-107, Feb. 1979.
15. Hwang, T.-Y.: Decoding Linear Block Codes for Minimizing Word
    Error Rate, IEEE Trans, Vol. IT-25, No. 6, Nov. 1979, pp. 733-6.
16. Hwang, T.-Y.: Efficient Optimum Decoding of Linear Block Codes,
    IEEE Trans, Vol. IT-26, No. 5, Sept. 1980, pp. 603-6.
17. Tanaka, H. and K. Kakigahara: Simplified Conclation Decoding
    by Selecting Possible Code Words Using Erasure Information,
    IEEE Int. Symp. on Information Theory, Les Arcs, France,
    June 1982.
18. Weldon, E. J.: Decoding Binary Block Codes on Q-ary Output
    Channels, IEEE Trans, Vol. IT-17, No. 6, Nov. 1971, pp. 713-718.
19. Reddy, S. M.: Further Results on Decoders for Q-ary Output
    Channels, IEEE Trans, Vol. IT-20, No. 4, July 1974, pp. 552-4.
20. Hong, Y.K. and S. Tsai: Simulation and Analysis of Weighted
    Erasure Decoding, IEEE Trans, Vol. COM-27, No. 2, Feb. 1979,
    pp. 483-488.

21. Sundberg, C.-E.: Asymptotically Optimum Soft-Decision Decoding Algorithms for Hamming Codes, Elec. Letters, Vol. 13, No. 2, 20 Jan. 1977, pp. 38-40.

22. Hackett, C. M.: An Efficient Algorithm for Soft-Decision Decoding of the (24,12) Extended Golay Code, IEEE Trans, Vol. COM-29, No. 6, June 1981, pp. 909-911 (and correction in Vol. COM-30, No. 3, March 1982, p. 554.

23. Berlekamp, E. R.: The Construction of Fast, High-Rate, Soft-Decision Block Decoders, IEEE Trans, Vol. IT-29, No. 3, May 1983, pp. 372-377.

24. Seroussi, G. and A. Lewpel: Maximum Likelihood Decoding of Certain Reed-Muller Codes, IEEE Trans, Vol. IT-29, No. 3, May 1983, pp. 448-50.

25. Massey, J. L.: Threshold Decoding, HIT Press, 1963.

26. Rudolph, L. D.: Threshold Decoding of Cyclic Codes, IEEE Trans, Vol. IT-15, No. 3, May 1969, pp. 414-8.

27. Hartmann, C. R. P. and L. D. Rudolph: An Optimum Symbol-by-Symbol Decoding Rule for Linear Codes, IEEE Trans, Vol. IT-25, No. 4, July 1979, pp. 514-7.

28. Rudolph, L. D., C. R. P. Hartmann, T.-Y. Hwang and N. Q. Duc: Algebraic Analog Decoding of Linear Binary Codes, IEEE Trans, Vol. IT-25, No. 4, July 1979, pp. 430-40.

29. Forney, G. D.: Generalised Minimum Distance Decoding, IEEE Trans, Vol. IT-12, April 1966, pp. 125-31.

30. Tanaka, H., K. Furusawa and S. Kaneku: A Novel Approach to Soft Decision Decoding of Threshold Decodable Codes, IEEE Trans, Vol. IT-26, No. 2, March 1980, pp. 244-246.

31. Yu, C. C. H. and D. J. Costello: Generalised Minimum Distance Decoding Algorithms for Q-ary Output Channels, IEEE Trans, Vol. IT-26, No. 2, March 1980, pp. 238-243.

32. Hwang, T.-Y. and S.-U. Guan: Generalised Minimum Distance Decoding on Majority Logic Decodable Codes, IEEE Trans, Vol. IT-28, No. 5, Sept. 1982, pp. 790-2.

33. Goodman, R. M. F.: Soft-Decision Threshold Decoders, Chap. 11 in reference 5 above.

34. Matis, K. R. and J. W. Modestino: Reduced-Search Soft-Decision Trellis Decoding of Linear Block Codes, IEEE Trans, Vol. IT-28, No. 2, March 1982, pp. 349-355.

35. Battail, G., M. C. Decouvelaere, and P. Godlewski: Replication Decoding, IEEE Trans, Vol. IT-25, No. 3, May 1979, pp. 332-345.

36. Ahamed, S. V.: Serial Coding for Cyclic Block Codes, BSTJ, Vol. 59, No. 2, Feb. 1980, pp. 269-276.

37. Chen, C. L.: High-Speed Decoding of BCH Codes, IEEE Trans, Vol. IT-27, No. 2, March 1981, pp. 254-256.

38. Campello de Souza, R. M. and P. G. Farrell: Finite Field Transforms and Symmetry Groups, paper accepted for presentation at the Int. Symp. on Algebra and Error-Correcting Codes, Toulouse, France, June 18-30, 1983.

39. Farrell, P. G. and S. J. Hopkins: Decoding Algorithms for a Class of Burst-Error-Correcting Array Codes, IEEE Int. Symp. on Information Theory, Les Arcs, France, June 21-25, 1982.
40. Daniel, J. S.: Array Codes for Error Control, M.Sc. Thesis, University of Manchester, June 1983.
41. Smith, R. J. G.: Private Communication.
42. Tanner, R. M.: A Recursive Approach to Low Complexity Codes, IEEE Trans, Vol. IT-27, No. 5, Sept. 1981, pp. 533-547.

TYPE	$\sim\!\Delta$	COMMENTS
LIST	2^c	LIST CAN BE SHORTENED
MEGGITT	n^t	SUPERSEDED BY LIST
ALGEBRAIC	$n^2 \log^2 n$	VERY GOOD FOR LONG CODES
SYST. SEARCH	$2^c.(?)$	DECEPTIVELY SIMPLE!
STEP-BY-STEP	$\leq 2^c$	HOW CALC. WEIGHT?
ERROR TRAP.	n^2	VERY SIMPLE CONCEPT
PERMUTATION	$n^2.(?)$	HOW FIND PERMS?
TRIAL-AND-ERROR	$\leq k.n^2$	HYBRID S.B.S/SYST. SEARCH

TABLE 1 : COMPARISON OF SYNDROME DECODING METHODS

Figure 1 : Structure and Encoder for (15,8) b=2 Array Code

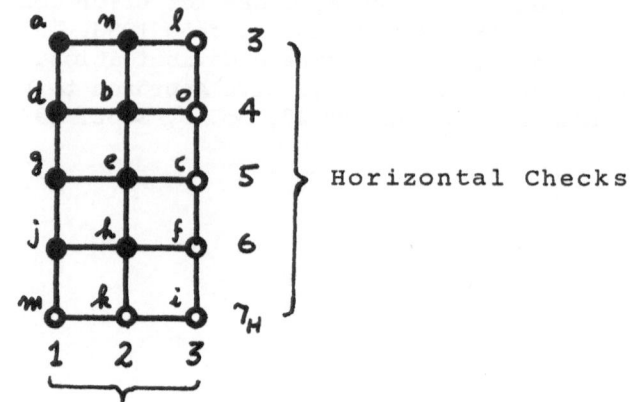

Horizontal Checks

Vertical Checks

●   Information Digit

○   Check Digit

⊕   EX-OR Gate

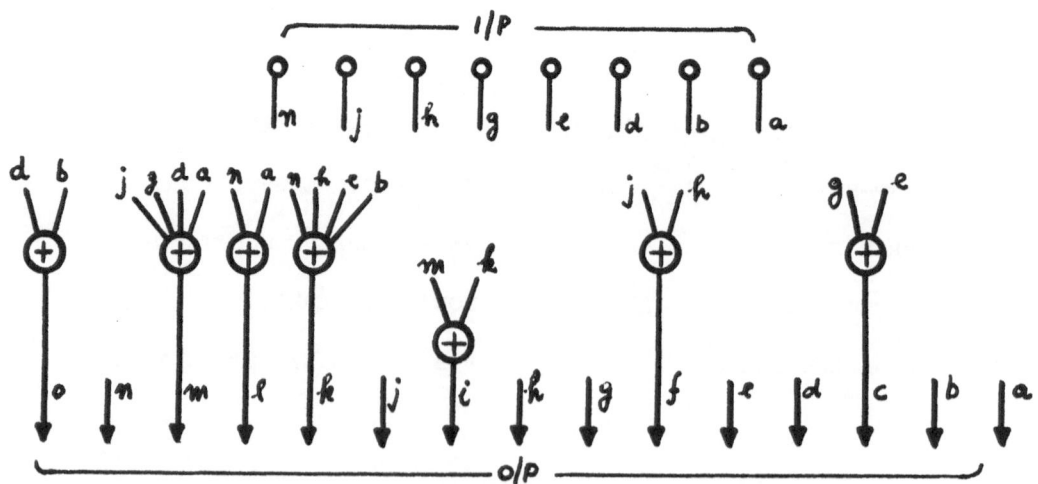

189

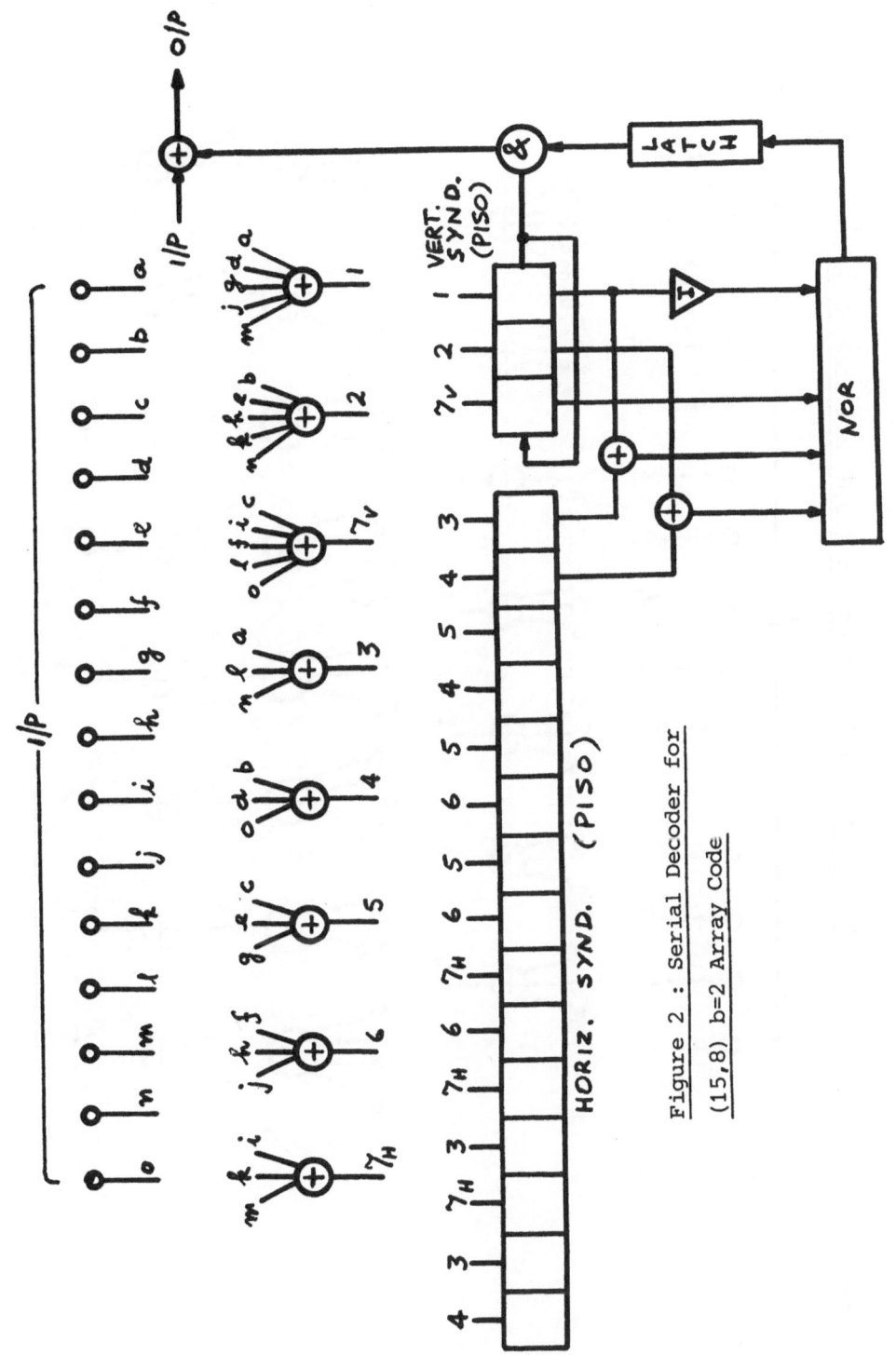

Figure 2 : Serial Decoder for
(15,8) b=2 Array Code

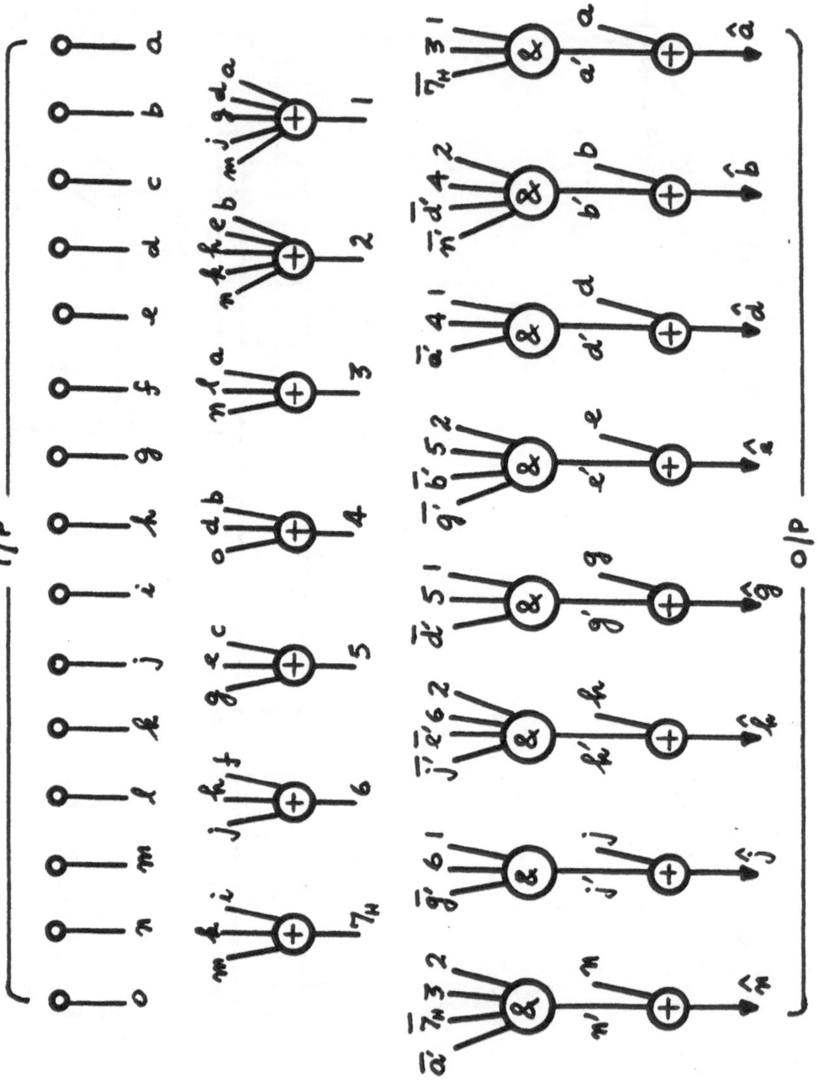

190

Figure 3 : Parallel Decoder for (15,8) b=2 Array Code

Figure 4 : Structure and Encoder for (16,8) b=3 Array Code

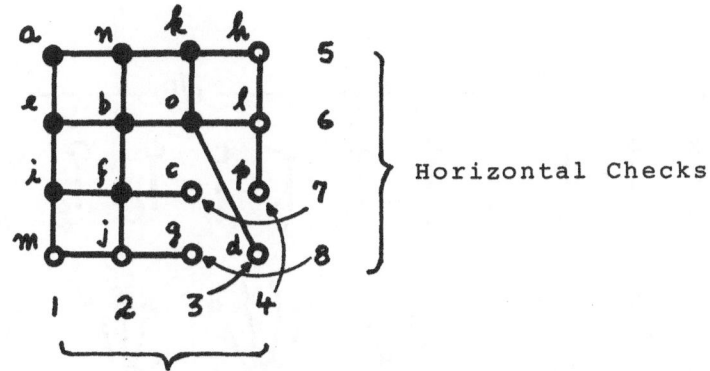

Horizontal Checks

Vertical Checks

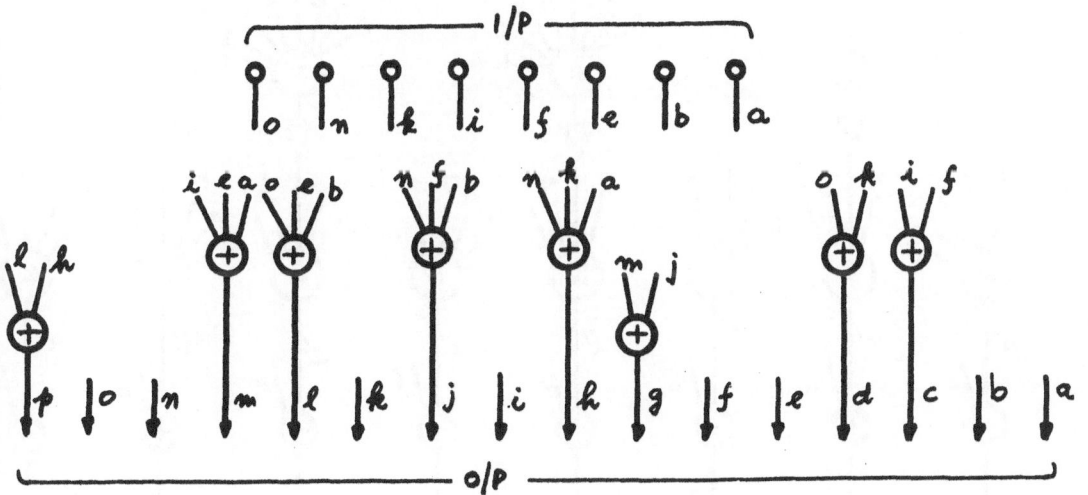

Figure 5 : Parallel Decoder for (16,8) b=3 Array Code

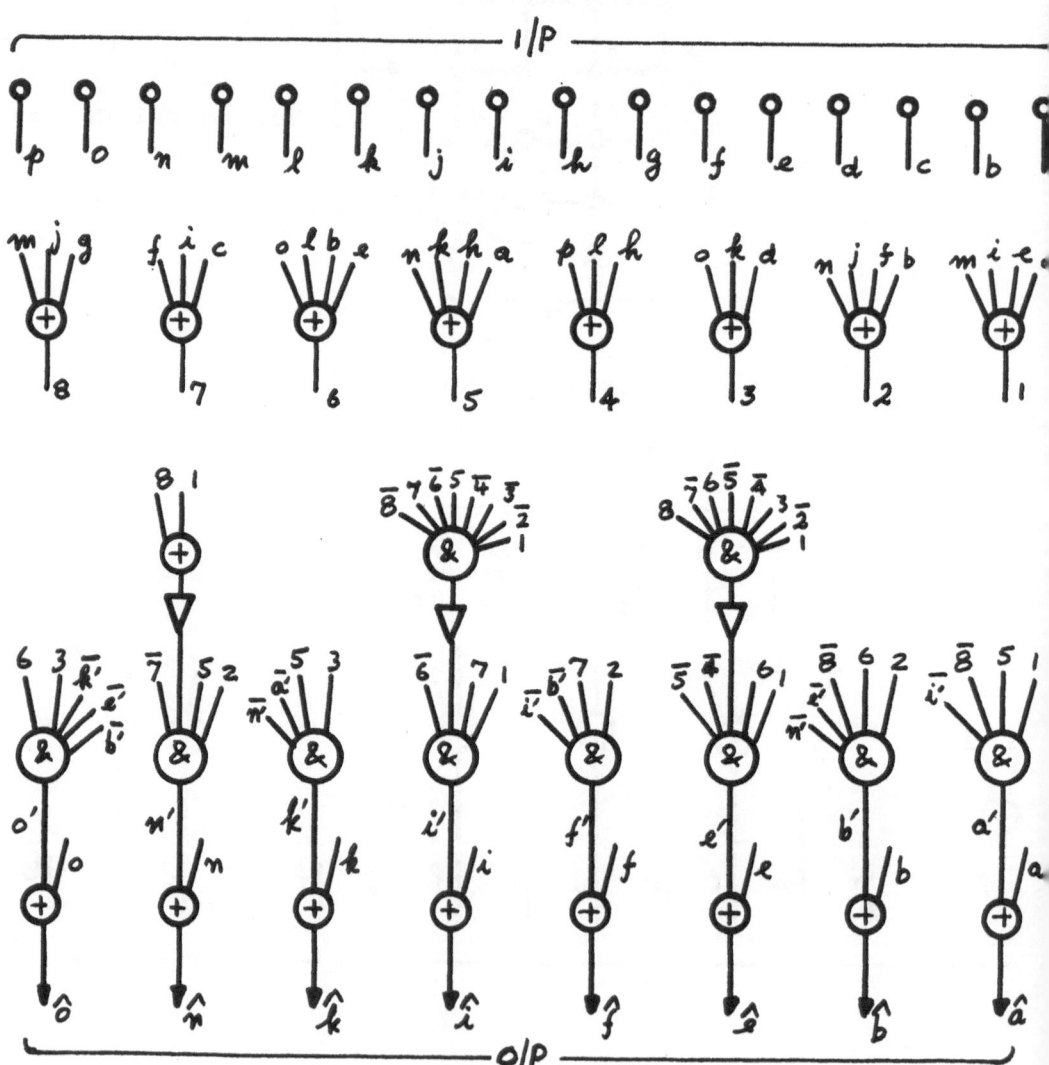

PANEL DISCUSSION

ON

TWO DIMENSIONAL SIGNALS AND CODES

Tuesday 12th July 1983 at 20.30 hours

Chairman and Organiser:	Professor S.W. Golomb	University of Southern California, Los Angeles, CA, U.S.A.
Panel Members:	Professor R. McEliece	CIT, Pasadena, CA, U.S.A.
	Professor P.G. Farrell	University of Manchester, U.K.
Contributors:	Mr. F.P. Coakley	University of Essex, Colchester. U.K.
	Dr. E.T. Cohen	Cyclotomics Inc., Berkeley, CA, U.S.A.
	Professor A. Ephremides	University of Maryland, MD, U.S.A.
	Dr. R.M.F. Goodman	University of Hull, U.K.
	Professor Ingemarsson	Linköping University, Sweden.
	Professor J.L. Massey	ETH-Zentrum, Zürich, Switzerland.
	Dr. S. Matic	Institute 'B. Kidric', Beograd, Yugoslavia.
	Professor D.G. Messerschmitt	University of California, Berkeley, CA, U.S.A.
	Dr. J. O'Reilly	University of Essex, Colchester, U.K.
	Professor J.K. Wolf	University of Massachusetts, Amherst, MASS, U.S.A.

<u>Professor Golomb</u> - It is hard to imagine any problem that does not
have some two-dimensional (2D) aspects. When you draw a graph with
a horizontal axis and a vertical axis, that is 2D. If you write
things out on a page, that page is 2D. Let me first quote the
abstract which I prepared for this panel discussion. You will see
there what came to my mind in the communications areas which have
clear 2D features:

'Finding good frequency-hop patterns is an example of 2D
signal design in time and frequency. Applications of these signals
include: frequency hopping for spread-spectrum communications to
achieve either anti-jamming (A-J) or low probability of intercept
(LPI), or for frequency diversity, or for multiple-access, and
frequency-hop signals for radar, sonar, and 2D synchronisation.
The transmission of pictorial data provides another natural context
for 2D signal design. A variety of transforms (e.g., 2D Fast
Fourier transforms, Hadamard transforms, etc.) are used to encode
such data. Inverse problems of creating accurate 2D and even 3D
reconstructions of surfaces from electromagnetic wavefronts arise
in contexts as diverse as radar mapping of planetary surfaces and
computer-aided X-ray tomography ('CAT-scans')'.

The first theme I referred to above is closest to me since I
have been working on this. If you try to design signal patterns in
both time and frequency then you are designing some kind of 2D
pattern. In particular, if you are working with frequency-hop
signals then you basically have the problem of designing some kind
of a matrix. There are intervals along the time axis and there are
positions and intervals on the frequency axis. When you select
where you want to put a one in this matrix, then at that particular
instant of time, or in that particular interval of time you will
send the corresponding tone or corresponding frequency. Frequency
hop patterns are used in quite a number of applications. One of
them is the spread spectrum communication imaginatively termed
"frequency hop spread spectrum". Frequency hop patterns are also
now being used in a number of cases in both radar and sonar systems
and in the closely related application of certain types of 2D
synchronisation. Frequency hop spread spectrum is used for a
variety of reasons, such as establishing anti-jamming features, or
spreading spectrum so thinly that you achieve a low probability of
intercept. There are some environments where you do not know which
frequencies are likely to be the good ones and then frequency
hopping gives you frequency diversity to help the signal to get
through. This is also true in certain multi-access systems where
different users have different patterns of the frequency hopping in
a sort of mesh hopefully constructed on a non-interfering basis.

Something that I think many people would consider in terms of
2D signals are codes which involve various forms of transmission of
pictorial data, which is a very big topic these days. First of all
there is the whole series of 2D transforms, such as the 2D Fourier

transforms, Hadamard transforms and many other varieties, which are really generalisations of Walsh transforms. These are used to encode 2D pictorial data. Today someone was talking to me about the problem of a fascimile transmission standard, and I first heard a talk about it at the WESCON meeting in San Francisco in 1957, which is now quite some time ago [1]. There were some people from the Sylvania Company in Mount View, California, who were proposing a fascimile transmission scheme which basically consisted of using a Huffman code to encode the spaces between black cells in a picture, and 26 years later this method is now standard. The facsimile business has had many attempted bursts of interest as a widespread form of communication, and it keeps being proposed for wide adoption, but never quite making it. People keep hoping that perhaps we can reduce the volume of physical mail that gets shipped around by express, and that has been the business of such companies as DHL and others. In all cases what they are trying to do is to transmit text or diagrams that can be represented perfectly and adequately on a 2D surface. This would be a quick and reliable way to transmit everything, even physical objects from one place to another, and it seems to me that it could also be used for package delivery and for special express mail services, though it would put them out of business. Yet for some reason this has not happened yet. It even looks as if the chief candidate for making this happen might well be the old standby telephone, which has many other capabilities, but it does not seem to be optimised for efficient communication of facsimile information. It is a system that everybody has and the network exists. Someone in the facsimile business said to me many years ago that he really wanted to set up a seance with the ghost of Alexander Bell to find out how to create a network for this business. At the start, there are not many people out there who are already plugged in to it, so how can we persuade people to sign up when there is no one else to talk to? This is the major hurdle that one must to overcome.

One of the things that people have particularly looked at in connection with facsimile picture transmission is television data compression. For this and for similar applications the main problem is how to make really efficient use of the fact that the data has redundancy in two directions and not just in one. Most of the experience has been that you have to expand a great deal of extra effort to get extremely small benefits in terms of bandwidth compression to get any effective use out of the second dimension of redundancy. Considering the theme of this Institute, it occurs to me that at least in the limit this processing problem depends on the VLSI technology. We are reaching the point where we can anticipate an era not far in the future where everything might be viewed in this way: No matter how complicated the algorithm, the implementation of it in VLSI hardware or whatever the technology might be, will make it so cheap that we will not even think of the cost. If we can save communication costs with more complicated processing techniques, then the main challenge could be how to design an efficient 2D method of compressing data contained in pictures. One caveat might be added to this, and that is to give some thought to ensembles of pictures having different statistical properties. I strongly suspect that wiring diagrams form very different ensembles from television pictures associated with soap operas, and that an efficient compression for one might be rather inefficient for the other.

The other class of 2D signals and codes which I wish to mention at the outset of this discussion are used in several different fields, although the basic techniques are the same. One can also say that these get into the 3D class. Thus, some of my friends at the Jet Propulsion Laboratory have been constructing the surface of the planet Venus from radar data. Very much the same technique is used to reconstruct the inside of a spinal cavity, or the inside of a brain, with the help of the so called 'Computer Aided Tomography' ('CAT-scans'). What you get is basically an electromagnetic wave front, and you try to process the data in order to decode it. The encoding is in some sense natural, and the decoding is aimed to re-interpret this as a 2D picture or a 3D surface. Great strides have been made in this technology, and in fact new techniques are coming along such as the nuclear magnetic resonance method (NMR) which stirs up great excitement among neuro-technologists. They can get better resolution and actually image the soft tissue with no ionising radiation at all. The techniques for re-interpretation of wave fronts as pictorial surfaces are already well worked out and we do not have to re-invent that sort of process.

At some point, especially if we are running short of material, I can show you some of my own pictures to describe the area of work in which I have been engaged, but I will save this till after we have had a chance to introduce you to the panel members who have volunteered to tell us something. I see that Professor McEliece is among us and he will be the first of these. Each of the panel members will talk for perhaps eight or ten minutes. After that we shall have an opportunity for repartee between and among the panel members. Then we shall have a free-for-all with the entire audience. I have a question from the floor.

Mr. Coakley - May I have a repartee with the chairman? The Japanese in their telex network, working since 1976, are already developing a facsimile transmission system which will be operational in 1990. That is a pretty large growth. I am working in 2D coding and when you look at the complexity of such methods, some nasty things can happen. Thus, when you lose a character in 2D decoding you tend to lose complete lines. If you go to higher dimensions and still higher definitions you may get into still bigger problems.

Professor Golomb - It is certainly true that in Japan they have made much progress in this technology. As with other products, they have a large domestic market to refine their techniques. Then they hope to sell it outside.

Professor Messerschmitt - I think that probably the major reason the Japanese have made this advance is that their character set cannot be transcribed by ordinary means. I should also mention another thing in your introduction, namely the coding of television signals, whether for telecom or for ordinary TV. That really is also a 3D problem because the time aspect has to be used in the so-called interframe coding. It is a fact that there is not much change in pictures between one frame and the next and this fact can be exploited.

Professor Golomb - I might mention that the Japanese can use fairly comfortably and simply their phonetic system called "Kana" for such things as telegraphy transmission. This has been their practice for a hundred years or so. Thus it is not compelling for them to use Chinese characters. They may choose to do so and then have a market in countries which use Chinese ideographs.

Mr. Coakley - Returning to Professor Messerschmitt's comment, it is not only the time dimension that you use in TV, since you have voice as well; the same applies to TV satellite systems. This is one reason why the time dimension is not much used in TV encoding.

Professor Messerschmitt - It is used for encoding of digital TV.

Professor Golomb - I would like now to invite Professor McEliece to make the first contribution to this discussion.

Professor McEliece - I am going to talk about something that is not controversial. Our director has asked me whether I am doing anything on 2D coding and I said 'NO', but that did not help me, for he then asked me whether I know of any 2D work and I had to say 'YES'. So he sent me to our chairman. I will also say a bit more about this subset in my lecture [2]. It concerns a paper by E. Berlekamp and P. Tong which is going to appear next year in the IEEE Transactions on Information Theory [3], so most of you have not heard about it. It is a new view of interleaving which, as some of you know, is a technique for scrambling encoded data prior to transmission, so that 'burstiness' in the transmission media would be, after un-scrambling, distributed independently among various code words. It is a very old idea and it is to me quite amazing that anything essentially new can be said about it.

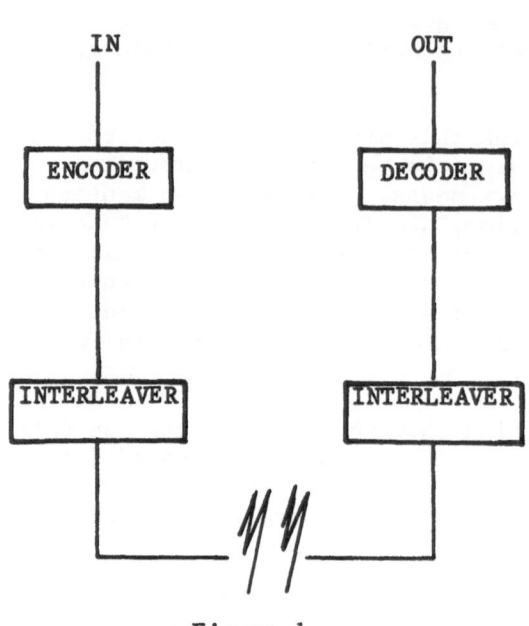

IN   OUT

ENCODER   DECODER

INTERLEAVER   INTERLEAVER

Figure 1.

Block Diagram of the Interleaver

System

The block diagram looks like this (Figure 1). The data goes in and is then encoded. The channel is represented by 'lighting bolts' and is supposed to have some burst characteristics, so that say a Hamming code is not able to deal with it directly. One way around that is to scramble or to interleave the data before transmission. Then the data, having been garbled by all these bursts, has to be unscrambled, though I call both these devices 'interleavers', since the function of one is the inverse of the other. Both these devices permute the codes; most of you are familiar this idea. The most common method is to use a block interleaver and

I will give you an example. Here is a 2D array (Figure 2), a 4 x 3 block dealing in this instance with a code of length 4. I want to interleave this to depth 3. The codewords (that is the way the encoder sees them) are written vertically — three code words of length 4, namely ABCD, EFGH and IJKL. This is how they come out of the encoder. However they are interleaved to depth 3; that is, the channel receives them in another direction in Figure 2, and reads the code symbols in a different order, namely AEI, BFJ, etc. This

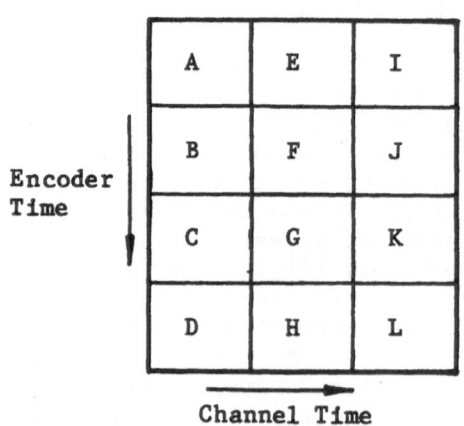

A	E	I
B	F	J
C	G	K
D	H	L

Encoder Time

Channel Time

Figure 2.

Example of a 4 x 3 Block Interleaver

is the action of a depth 3 block interleaver for code words of length 4.

The question is, how do we actually build these things? We can implement them simply with 'Random Access Memory' (RAM). In this particular example we can build a 4 x 3 interleaver like the one in Figure 2 by considering a 12 x 1 array (see Figure 3), so that in each of the little boxes we put one of the twelve symbols we had in Figure 2. At the first unit of time, assuming that the RAM is filled with some 'letters' from the code alphabet in its boxes, we read the letter and send to the channel the contents of location '0'. Then we write into the location '0' the current input to the RAM. We do the same at the second unit of time with location '1', then with location '2' and so forth. This sequence of addresses should be no mystery, for instead of A to L in Figure 2 we have written 1 to 11. Here it is:

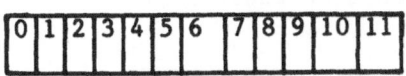

Figure 3
RAM Address Sequence

0 1 2 3 4 5 6 7 8 9 10 11 0 8 1 5 9 2 6 10 3 7 11

and so it repeats itself periodically. It is clear that you can implement such a block interleaver with this RAM, and this sequence of addresses.

Suppose now that we start thinking before each more generally. We have some data coming into a RAM and before each 'read/write' cycle we get a new addresses from a device which I call the 'address sequencer' (Figure 4). At every unit of time we get a new address. We read out the contents of that particular address and then we write into that address the current input bits. By using an address sequence with nice combinatorial properties, you can also get interleavers with every interesting properties. We can get interleavers with many other interesting properties in this way.

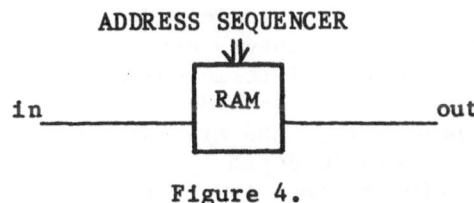

Figure 4.

'One-RAM Interleaver'

Here is a simple example. Suppose the RAM has four cells labelled as follows:

0	1	2	3

Suppose next that we have this sequence of addresses:

0 0 1 2 0 0 3 1 0 0 2 3 ...

which is periodic of period 12. What kind of scrambling does this cause? The answer is that it gives exactly the same functional interleaving as does the 4 x 3 interleaver in Figure 2. Let me explain this with the help of Figure 5.

Here we have a 2D array of codewords. As the codewords (in this example of length 4) arrive at the interleaver, they are written vertically in Figure 5. For example one 4-bit codeword would be written in the underlined positions "0012" in the first column. The next codeword is written in the second column, one position lower: "0031"; the next, in the third column, "0023"; the next, back in the first column, "0012". The pattern is helical. Notice, however, that the pattern 003100230012 also occurs if you read the helix horizontally.

0	0	3
1	0	0
2	3	0
0	1	2
0	0	3
1	0	0
2	3	0
0	1	2
0	0	3
1	0	0
2	3	0
0	1	2
0	0	3

What this means (you can convince yourself of this in a minute or two) is that a 4-cell RAM controlled by the periodic address sequence 003100230012 behaves exactly the same as the conceptual interleaver of Figure 5 in which the codewords are read in vertically (helically), and are read out horizontally. This is depth 4 interleaving just as effective as that in Figure 2, which, however, needs only 4 cells of storage instead of 12. Notice also that because of the nice symmetry of Figure 5, the corresponding descrambler needs only syndromization mod 4, instead of mod 12, which is needed by the block interleaver.

Figure 5.

Conceptual Interleaver for the Sequence 001200310023

All of this generalizes nicely to bigger "helical" interleavers. I refer you again to Berlekamp and Tong's paper. Dr. Cohen will now explain a further advantage of hetrical interleaving, namely "burst forecasting".

Dr. Cohen - I wish to show you some work done by Dr. Berlekamp on
burst forecasting. A frequent problem with block coding schemes is
that you do not always obtain erasure information from whatever
channel you have, and it is more difficult to decode everything on
a hard decision basis. Let us then look again at the original
helical interleaver system as introduced here by Professor McEliece
(Figure 6). I have drawn staggered codewords, and there are a
number of these that wrap around the helix (compare with Figure
5).

Now, in order for the decoder to determine synchronisation, it
is a common practice that instead of a normal code character, you
insert a synch character periodically in the sequence. What you
actually do for symmetry's sake (and in order to keep all the
codewords the same length) is to use one character per codeword as
a 'synch' character. This character will have a known value in
each codeword. It is then pretty easy to design some autonomous
digital hardware that determines 'synch' from this character in
each codeword. You get the obvious benefit of finding block
'synch' for the decoder, and you also get the feature of burst
forecasting, whereby the decoder, while looking at each codeword
that it is decoding, can see whether or not the 'synch' character

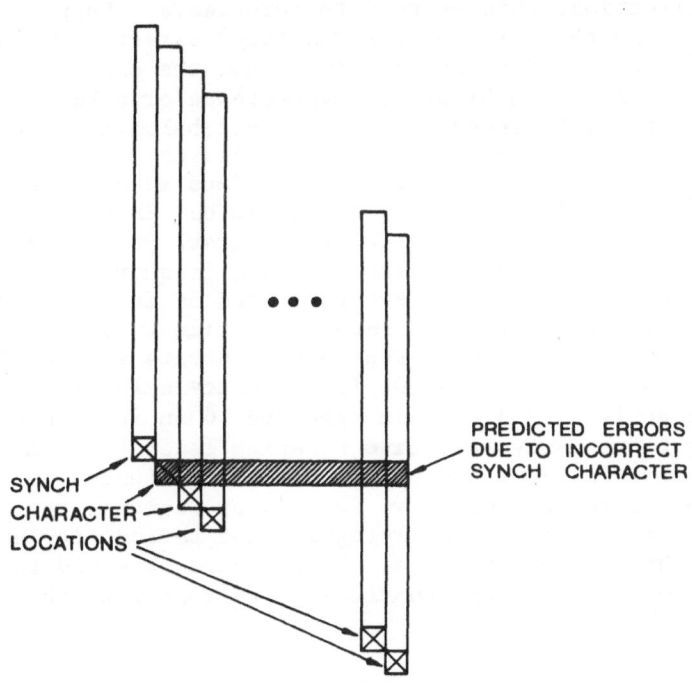

Figure 6.

Helically Interleaved Codewords

has the correct value. Of course, we assume here that the hardware 'synch' system is robust enough and does not easily lose 'synch' if one, or even several 'synch' characters in a row, are garbled. The decoder checks this 'synch' character and determines whether it has the proper value. If it does not have the proper value, then the decoder can use this information and decide that the channel is in the middle of a burst, starting at that particular location in the interleaver block. In other words, in a burst environment, an incorrect 'synch' character is used to forecast the start of a burst, thus obtaining erasure information for the decoder. Any comments?

Dr. Goodman - There is absolutely no collusion between me and Professor McEliece. I just want to make a comment on 2D interleaving for we also have a problem there. Consider a low speed multi-carrier modem. Take a 3 kHz bandwidth (see Figure 7a) and transmit low speed 75 baud data on say 48 subcarriers. Then the subcarriers will be separated by 75 Hz. Here we see one time frame horizontally and the next frame comes below etc. along the time direction. So we have a 2D pattern in frequency and time. That is what the modem typically does and we have a problem. For error correction we have a serial bit stream that maps on to the structure in Figure 7. If we code it with a code that is not burst error correcting, then we need to interleave. Suppose we interleave in the horizontal (frequency) direction. Then if we get a flat fade, when for say one time frame the entire 3 kHz band dies, we would get a blast of interference or a burst. So interleaving in frequency alone does not help in the case of flat fades. So let us try a 'normal' interleaver, in the vertical (time) direction to a certain depth. Thus adjacent bits go out on the same subcarrier. Then say we get in our HF radio a frequency selective fade. This again causes a burst. So what do we do? We have to interleave in both the time and frequency directions, or in other words, we have to select some kind of interleaving in a diagonal direction on the frequency time 2D array. We have to optimise this. Suppose we separate by 6 bits in frequency and by 8 bits in time (6 x 8 = 48). Or if frequency selective fading is bad we could do it with 3 bits in time and 16 in frequency. However, we can get frequency selective sweeping fades (see Figure 7b) and that leads to a burst again! Within a few minutes we would be prone to stationary selective fading, or we could get minutes of sweeping fades. You could recognise the periodicity of these fades running through the spectrum. In any case it would be nice to change the interleaving dynamically, to cope with these varying situations.

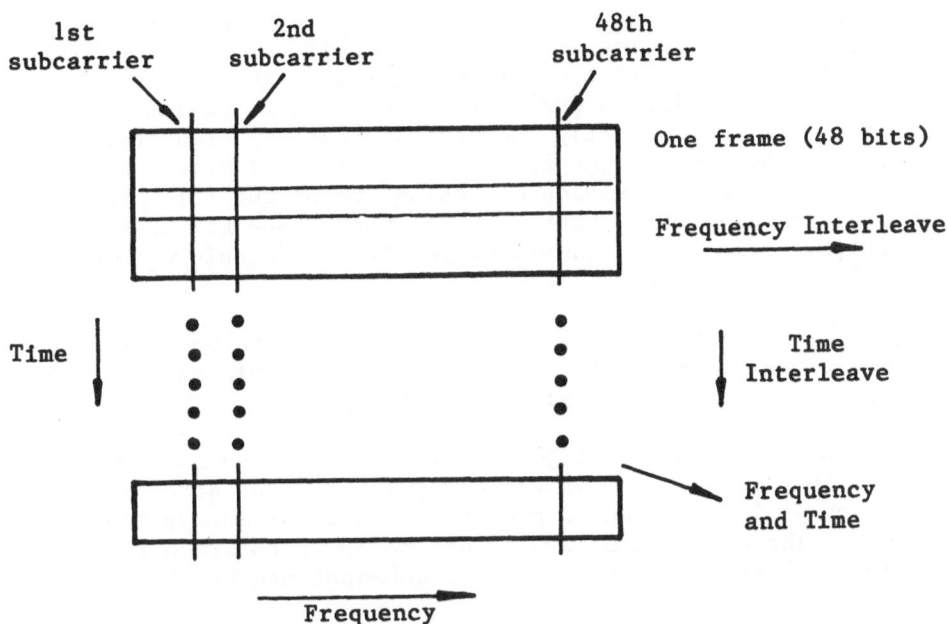

Figure 7a. Multi-subcarrier low-speed parallel modem

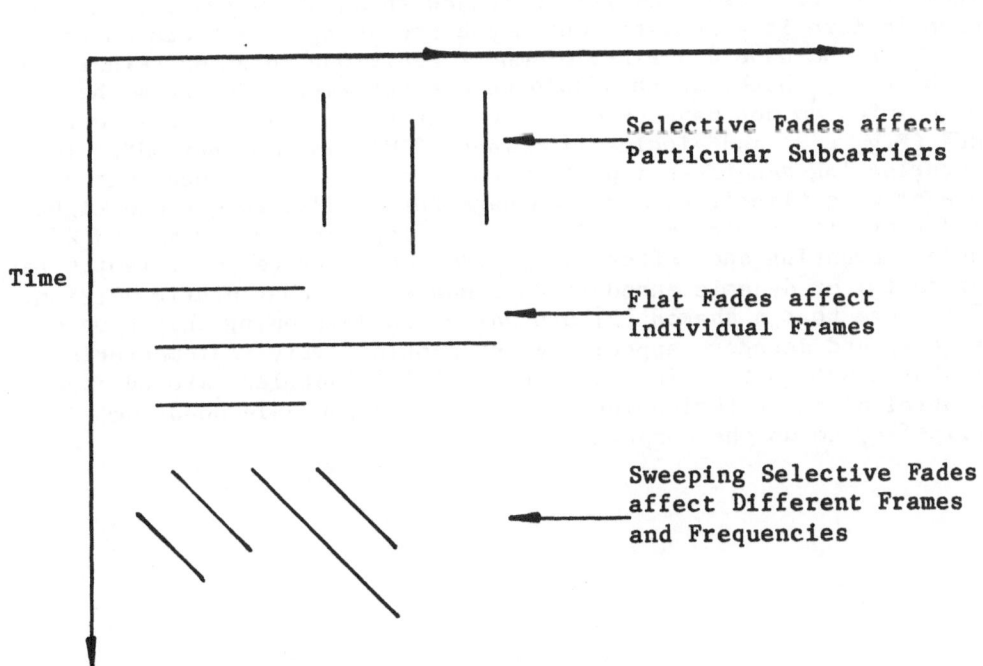

Figure 7b. Types of Fades

Professor Messerschmitt – Just one comment on the interleaving theory. There is a lot of material on this, based around digital switching. We have time slots coming into a switch, as well as a number of those signals and then you do what is called 'time slot exchange' which is exactly the same processing. You write in a certain order and you get it out in a different order; all this is based on the 'time division switching' technique [4]. There is a lot of very elaborate combinatorial theory used for this for about 50 years. This may be something to offer to people working on this interleaving problem.

Mr. Cockley – Yes, there is some work being done on this in the International Tele-Traffic Company, based on this time switching idea, to get certain swopping facilities.

Professor Ingemarsson – I just want to remind you that the same device that Professor McEliece was talking about here, has been used, or at least is being proposed to use for encryption of speech. There you have samples of speech which are permuted in time. I think the same device was commissioned by the Brown Boveri Company in Switzerland.

Professor Golomb – Now it is the turn of Professor Farrell, who has promised to tell us something new.

Professor Farrell – What I want to do first is to generalise the idea which Professor McEliece proposed in such a nice way. As an example here is a register which you are using as a block encoder and here you have a variety of mod 2 additions going on (Figure 8), with parity checks which adjoin to the information to form the code block. You do not have to do it in that way. You can do it with a MUX (Figure 9) and single flip-flops (F/F), and another MUX. The interleaving generates a pattern which switches the input through one or more flip-flops and then back to the MUX, to get the right order of bits coming out. These flip-flops are acting as mod 2 adders counting ones effectively. You can also get different types of serial or dynamic encoders to compare with that static version. I believe that a Ahamed [5] has worked on developing this type of encoder and decoder, especially for bubble-memory implementation. It turns out that it is very easy to shift 'bubbles' around under control of an 'interleaving' sequence, and you only need one flip-flop to do the computation.

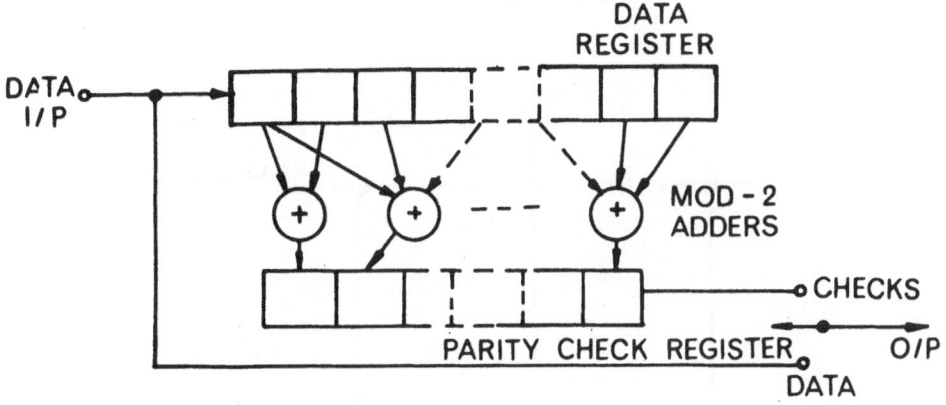

Figure 8   Static Encoder

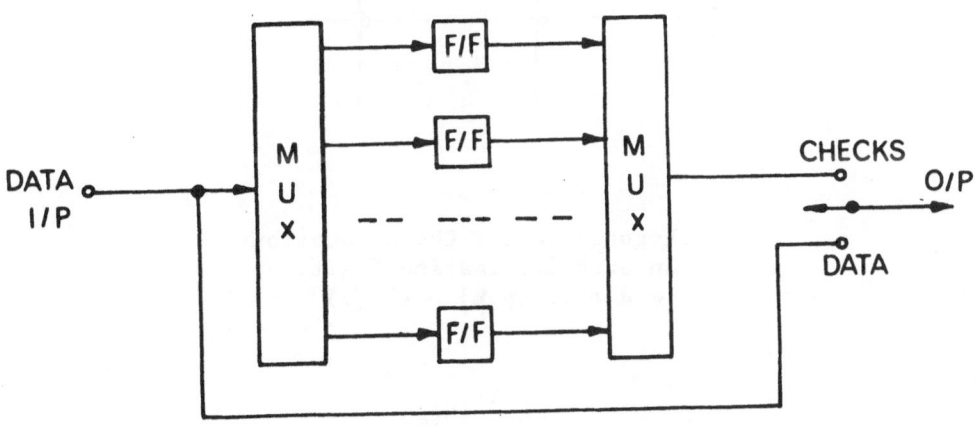

Figure 9   Dynamic Encoder

206

COLUMN CHECKS

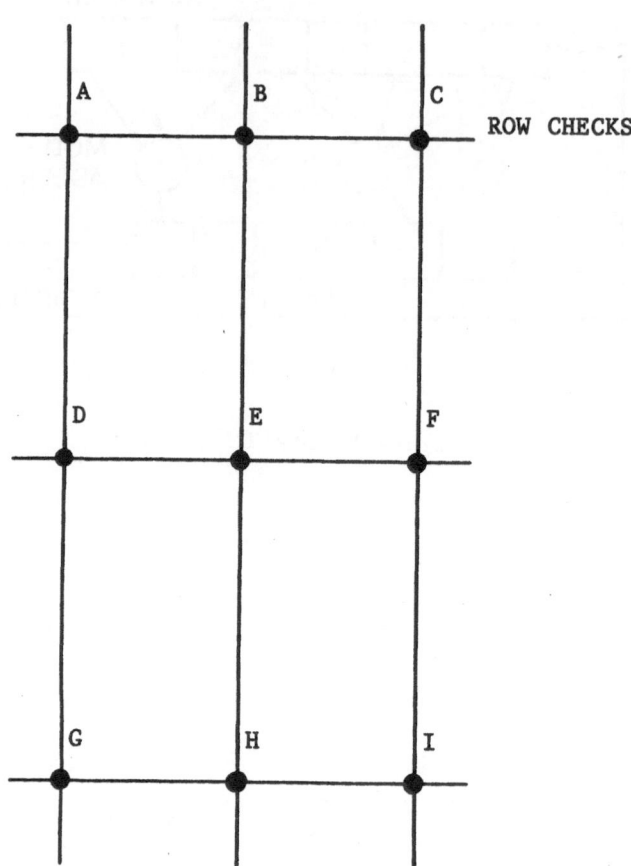

ROW CHECKS

2 Orthogonal Parity Check Equations
on each Information Digit
→ d = 3; (n,k) = (15,9)

Figure 10(a)

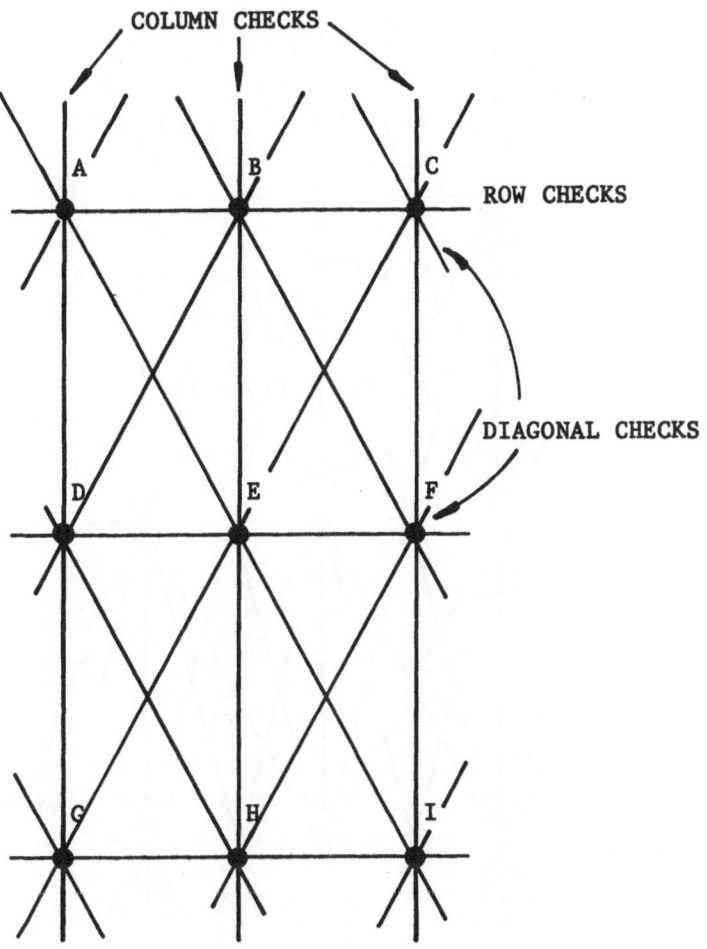

4 Orthogonal Parity Check Equations
on each Information Digit
→ d = 5; (n,k) = (25,9)

Figure 10(b)

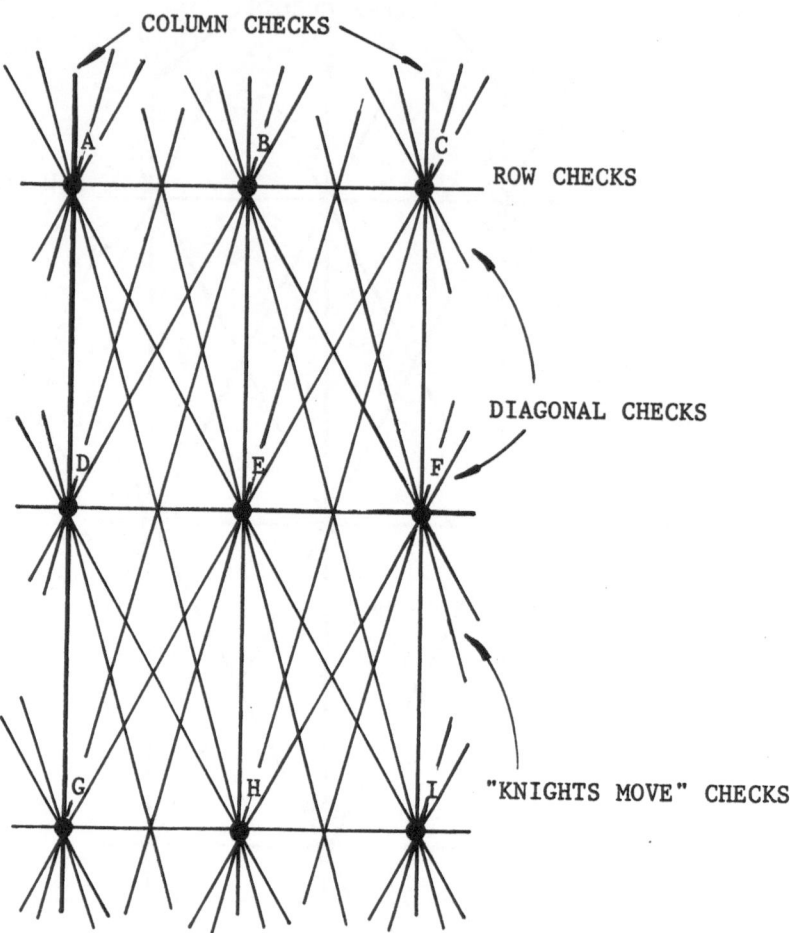

6 Orthogonal checks → d = 5; (n,k) = (39,9)

Q Orthogonal checks → d = Q + 1

Figure 10(c)

The thing I was really going to say was prompted by our director not understanding me during my lecture [6]; that is, to give you a little commercial on array codes, and then to say only one thing on convolutional codes.

I want to go back to a simple 2D array code. So we have information bits (Figure 10a) A,B,C...H,I, and for the moment assume that we have no check bits. First I will introduce checks along rows and columns (e.g. ABC, DEF, GHI and ADG, BEH, CFI respectively). These are single error correcting checks, or orthogonal parity checks on each information digit, with Hamming distance $d = 3$. The reason that these are 'single-error correcting' is because each information has two orthogonal parity checks in the standard majority logic decoding way. If you want to get something better than the single-error correction, then you put on diagonal checks (i.e. AEI or CEG etc, as in Figure 10b). Now you have four orthogonal parity checks for each information digit and that gives you the Hamming distance $d = 5$. Of course, you pay for this because there are a lot more check bits now. If you want to go further, you put on 'knight's move' checks (e.g. AH, BI, BG and CH etc, Figure 10c), i.e diagonal lines going through points two down and one across. Now you have six orthogonal parity checks with Hamming distance $d = 7$. You can go on that way with Q directions, to get a Hamming distance of $d = Q+1$.

However, you soon lose interest in this for random error correction, because you have so many check bits that the rate goes down very fast.

Some time ago we have investigated how to get round this problem. The obvious thing to do is here shown by a simple example (Figure 11). The way is not to have straight lines for your check equations, but to make them bend round the place. It so happens, for this example, that you get the horizontal (ABC, DEF, GHI). You also get the dashed set which is diagonal one way and is interleaved (AHF, DBI, GEC), and the wavy set which is the other way round (AEI, DHC, GBF). For example, you have 9 information bits, 9 checks and you get the Hamming distance of 4. It turns out that we have another reason for producing this. Here (Figure 11) p is prime. The number of information bits k is p^2. The Hamming distance is less then or equal to $p+1$. The number of checks is up to p^2 or less. These turn out to have very comparable parameters to simple cyclic codes which are also majority logic type codes (SOQC codes). They do the same job, but they are derived in a different way. This is a nice little confirmation that you can do this for random error correction. I talked a bit in my lecture [6] on error burst correction and now I want to go back to the basics again.

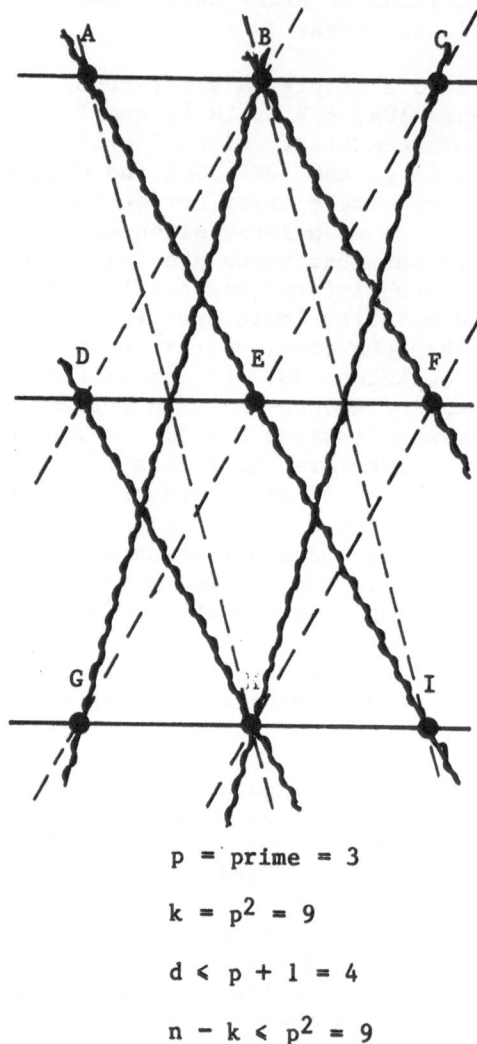

$$p = \text{prime} = 3$$

$$k = p^2 = 9$$

$$d \leqslant p + 1 = 4$$

$$n - k \leqslant p^2 = 9$$

Optimum Random-Error Correcting Array Codes

Figure 11

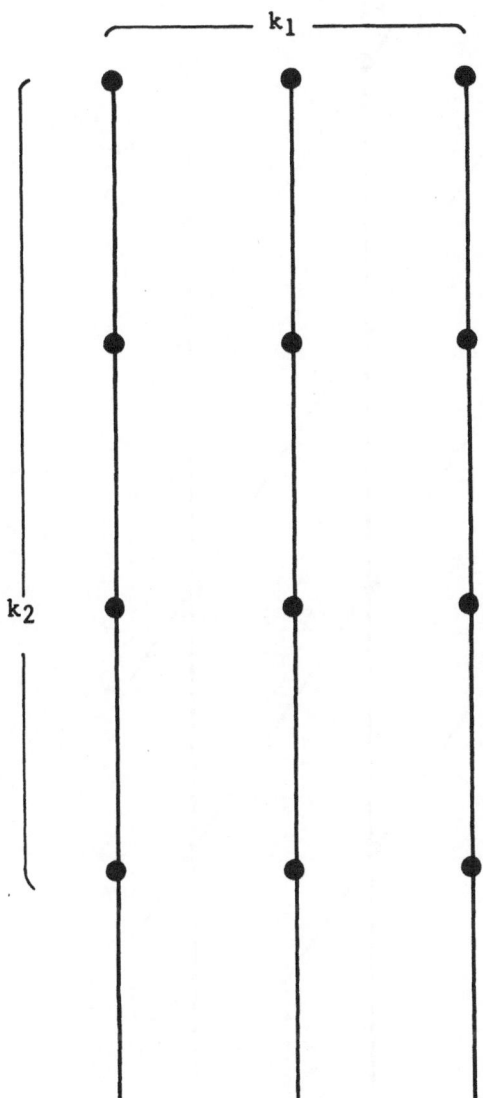

Vertical Checks

$$c = k_1$$
$$b = k_1$$

Burst-Error Detection
or
Burst-Erasure Correction

Figure 12(a)

212

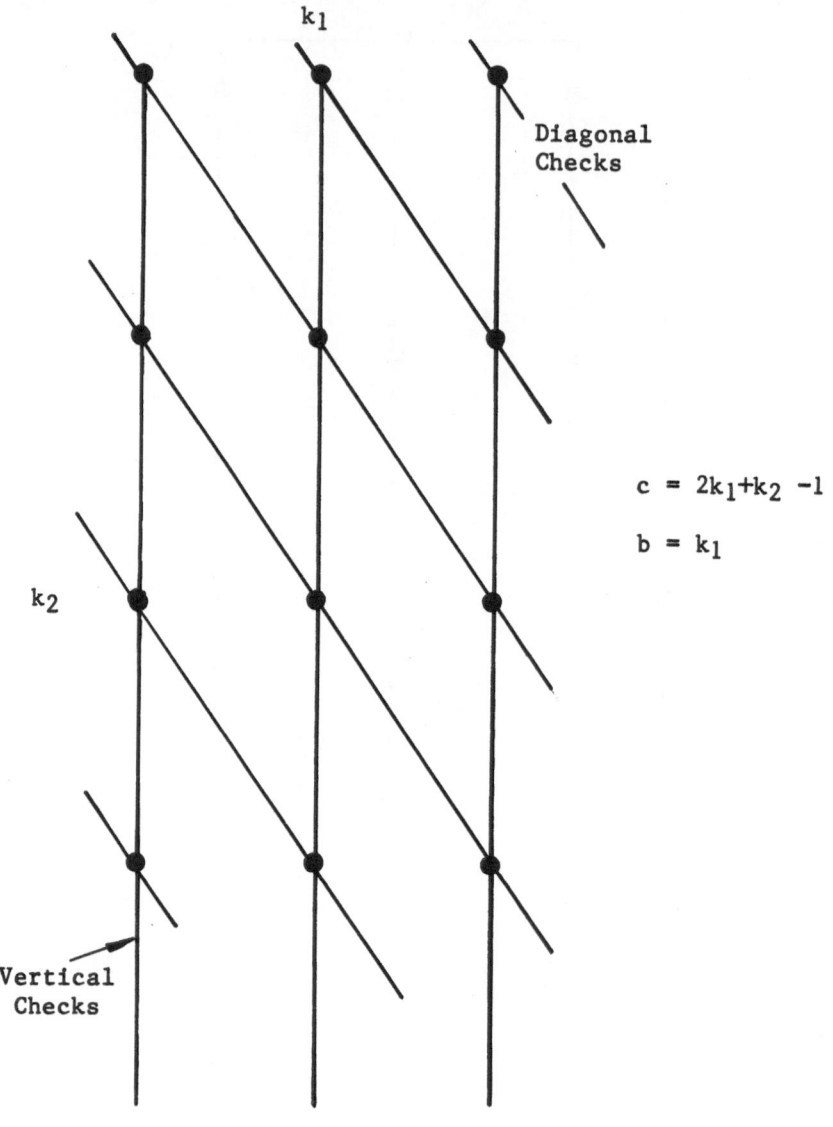

$$c = 2k_1 + k_2 - 1$$

$$b = k_1$$

$$c = k1$$
$$b = k_1$$

Burst-Error Correction

Figure 12(b)

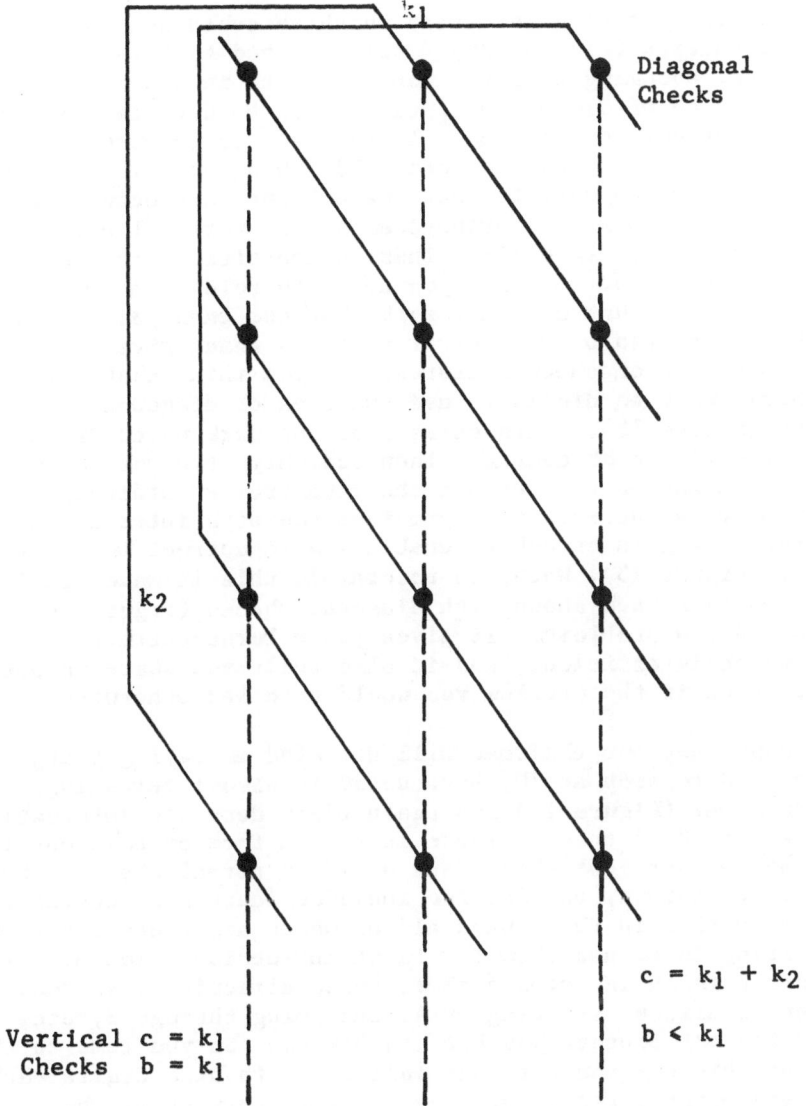

$c = k_1 + k_2$

$b < k_1$

Vertical $c = k_1$
Checks $b = k_1$

Row-burst Error Correction

Figure 12(c)

Here we have a rectangular array which we could use for burst-error detection (Figure 12a). All you need to have are vertical checks, assuming that you can read data in binary words. Any burst of length of dimension k_1 can either be detected or you can have burst-erasure correction. If you want to do burst correction with an array code (Figure 12b), then you have to find another direction orthogonal to that where bursts are occurring, which is across the rows, and orthogonal to the vertical parity checks. So you put on diagonals. This is inefficient at the corners, as you can see. So what you do is to fold them round (Figure 12(c)). But that does not work, because then you can find that you could have two or more error patterns which give you exactly the same set of check failures. So you think what can one do about this? What we did is to use the idea of diagonal interleaving (Figure 13). This comes from the work which Dr. E. Munday has done [7] on HF coding. Then suddenly it occurred to us that this is the way of reading out the bits from an ordinary row-and-column code, because this provides you with interleaving, and the interleaving is enough to enable you to correct burst of dimension k_1 (Figure 13). When you unscramble this it makes it look like a code which I have shown with diagonal checks (Figure 14). This has solved two problems: it gives you a burst-correction capability which is efficient, and it also tells you where to put the checks, which is the problem you would have had otherwise.

Now I hope that our chairman will not mind me saying that there is no need to stop at 2D, because 3D is also interesting. Here is a 3D array (Figure 15) and again black dots are information bits. There are 12 of these. There is also a face of horizontal checks, a face of vertical checks and a set of normal checks. So you get a lot of checks, but it also includes 'checks on checks'. You can compute them in three ways all of which are useful, but the most interesting is to use them for burst correction. You have to find a read-out which is orthogonal to three directions, so you read it down in slices, starting at A, and going through B, etc, to F. Then you run out because you hit the bottom. So you come to the top at G; and this way you get four 'slices' with nine digits each. That gives you burst correction. You can also correct patches of errors in 2D arrays which is quite nice.

Now something completely different. This is really derived from the recent work of Ungerboeck [8]. I want to remind everybody of the power of this technique of coding linked to a type of modulation (Figure 16). On the transmitter side of a QASK modem for transmitting data you start with a series/parallel converter, go through an encoder and then to the modulator, which is based on a 16-point constellation. Thus you have a 6-bit address selecting one of the points for transmission along the channel. This is better than putting the encoder outside, before the conversion, because it turns out to be more practical for a number of

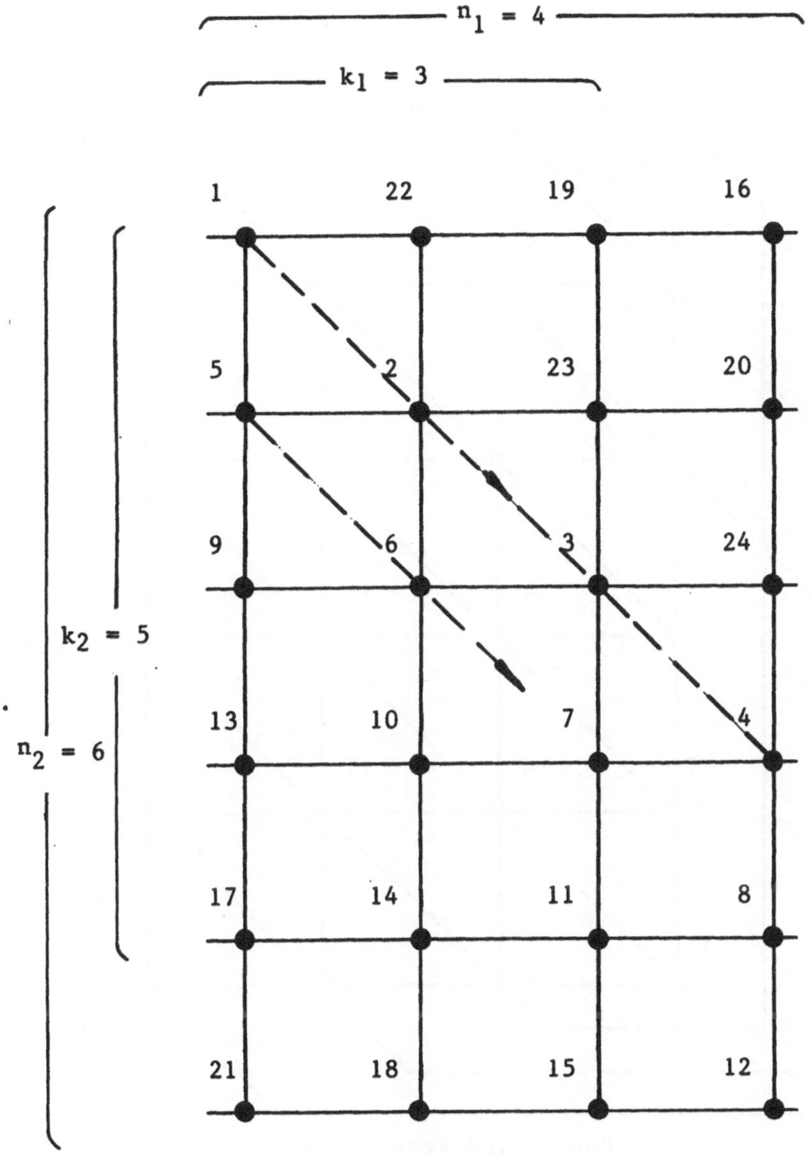

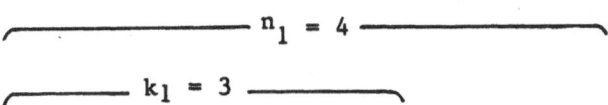

1   22   19   16

5   2   23   20

9   6   3   24

13   10   7   4

17   14   11   8

21   18   15   12

$k_2 = 5$

$n_2 = 6$

$n_1 = 4$

$k_1 = 3$

(24,15) b = 3

In general b = k_1; $k_2 > 2(k_1-1)$

Burst-Error Correction
With
Diagonal Read-Out

Figure 13

216

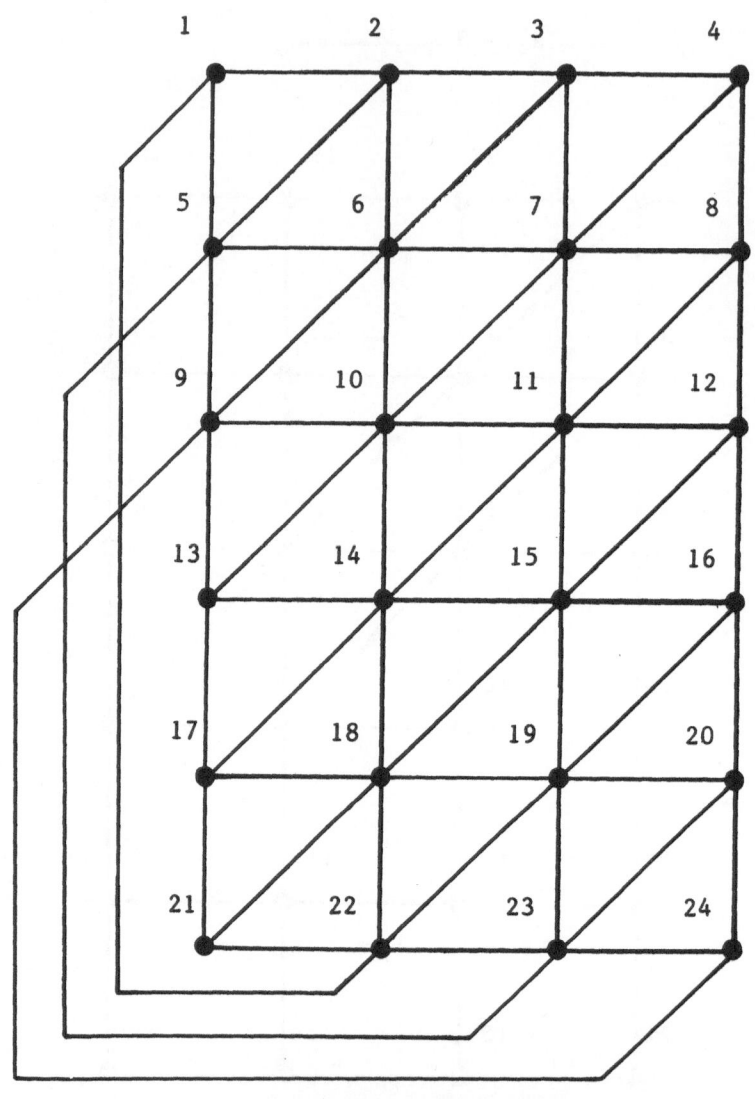

<u>Unscrabled Array Code</u>

Figure 14

217

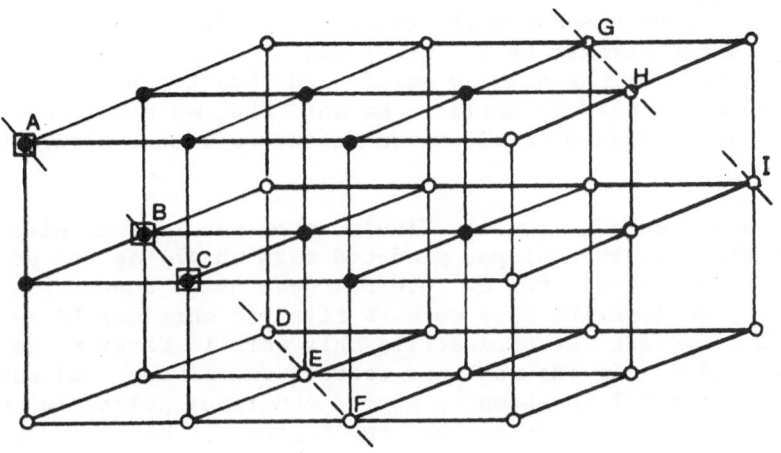

● Info

○ Check

$k = 3.2.2 = 12$

$n = 4.3.3 = 36$

◉ OR ◉ Diagonal Slices 4 @ 9
$b = 8$

Three-Dimensional Array Code

Figure 15

options - and it does work extremely well. For instance, using a simple 5/6 rate coder (Figure 17) you can get the performance shown in Figure 18. Here, with soft decision Viterbi algorithm decoding, you can get about 3.5 dB gain at 10^{-4} error rate. Curiously enough, you get -0.5 dB gain if you use hard decision decoding. Thus, it is not good to use hard decision in this case. This leaves the intriguing thought this 4 dB from hard to soft may be due to the fact that we are working in a 2D situation (2 dB for each dimension).

Professor Wolf - I believe what Professor Farrell has said is very important, for he has asked the right questions. We have been trying to apply hammers to the wrong nail (or vice versa), using codes designed for the Hamming metric and then trying to decode them using the Euclidean metric. He said that we should design codes in the Euclidean metric if we intend to decode using this metric.

The other subject I would like to show you is something that is quite old. In 1962 Elspas produced work which, as far as I know is unpublished, except for an internal government report [9]. The easiest way to describe this work is first to consider 1D cyclic codes. The easiest way to describe this work is first to consider 1D cyclic codes. If one wants to correct one dimensional bursts, one must be careful in choosing a good generator polynomial since two generator polynomials of the same degree can have widely different burst error correction capabilities. Every one dimensional cyclic code, however, with r parity bits will detect all bursts of length r or less and will fill in the same length for a burst of erasures. If you take a 2D cyclic code (Figure 19) with a block of k_1 and k_2 information digits and use r_1 parity digits in each row and r_2 parity digits in each column, then no matter what generators you choose for the row and column code that code will correct every 2D burst of dimensions $r_1 \times r_2$. That is, any kind of errors fitting into a rectangular $r_1 \times r_2$ is correctable, no matter what generators you use. The proof is very easy. Since every 1D cyclic code will detect every burst of length r or less one can first detect all rows and columns containing errors and then erase the digits in a row and column with detected errors. You now fill in the erasures using the erasure burst capability of cyclic codes.

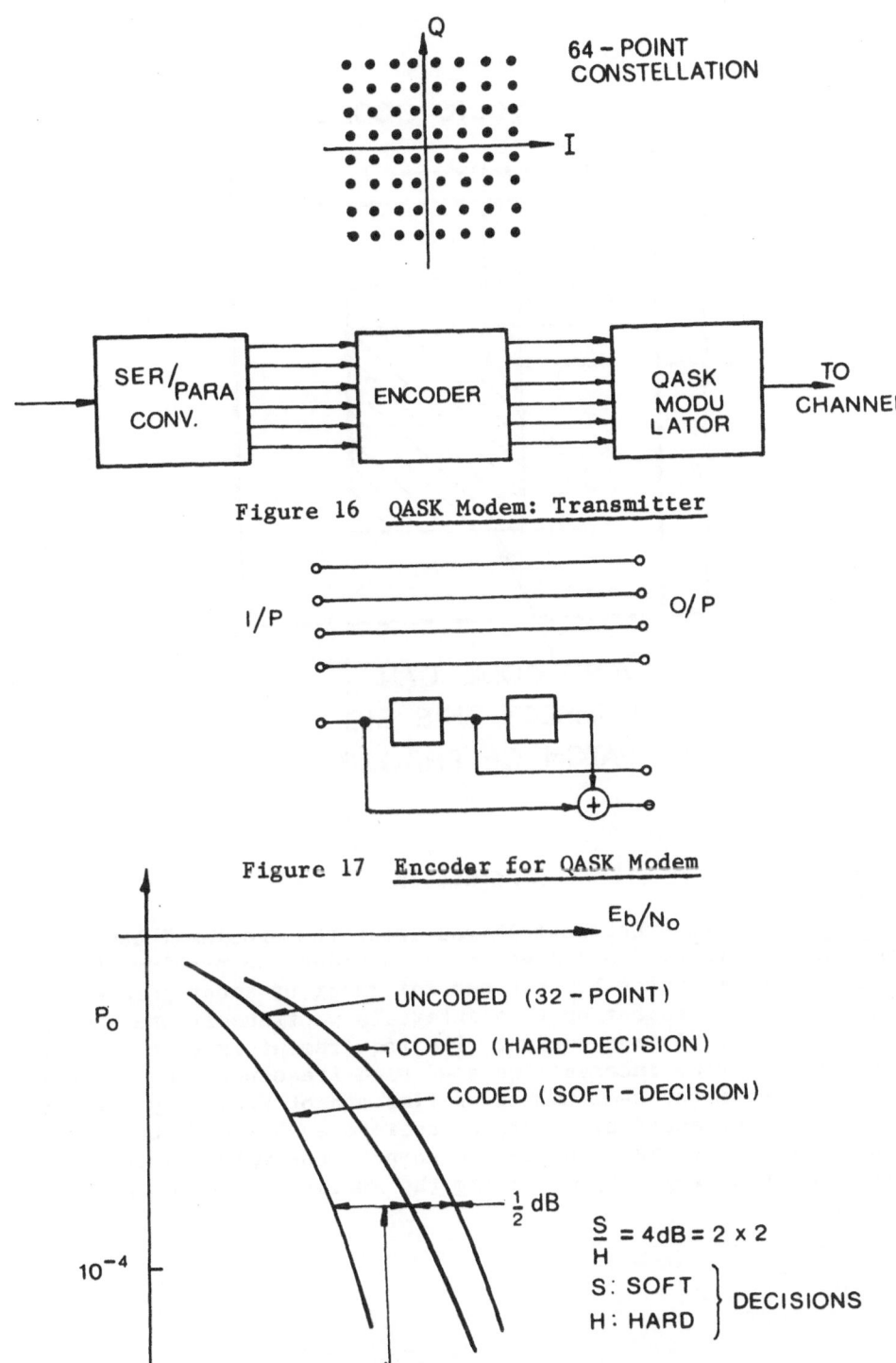

Figure 16   QASK Modem: Transmitter

Figure 17   Encoder for QASK Modem

Figure 18   Performance of QASK Scheme

CYCLIC CODES

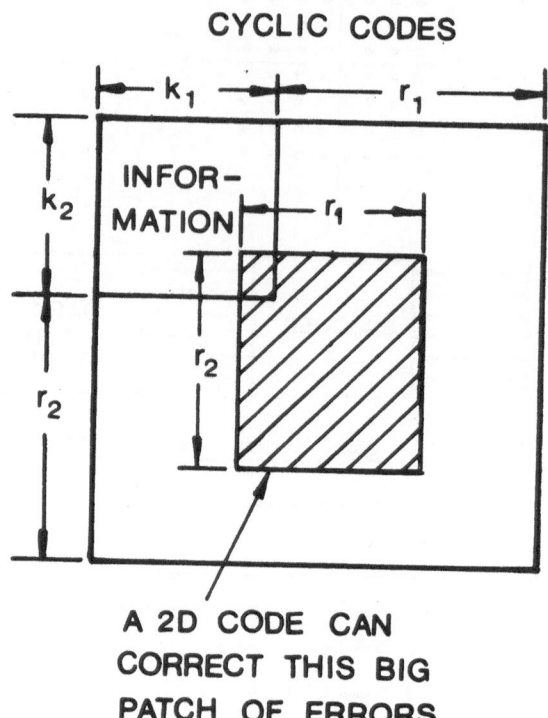

A 2D CODE CAN
CORRECT THIS BIG
PATCH OF ERRORS

Figure 19

<u>Professor Golomb</u> — Let me tell you about the problem I have been
working on lately. We are going to consider the problem of
designing a frequency-hop pattern for radar or sonar purposes,
where at each a number of time intervals we transmit one and only
one from a set of frequencies. The requirement is that over the
whole set of time intervals we send each frequency once and only
once. We have the same number of time and of frequency intervals.
The signal reflected by a target, therefore shifted both in time
and in frequency, is then used to compute the velocity and the
range of this target by measuring the amounts of these shifts.

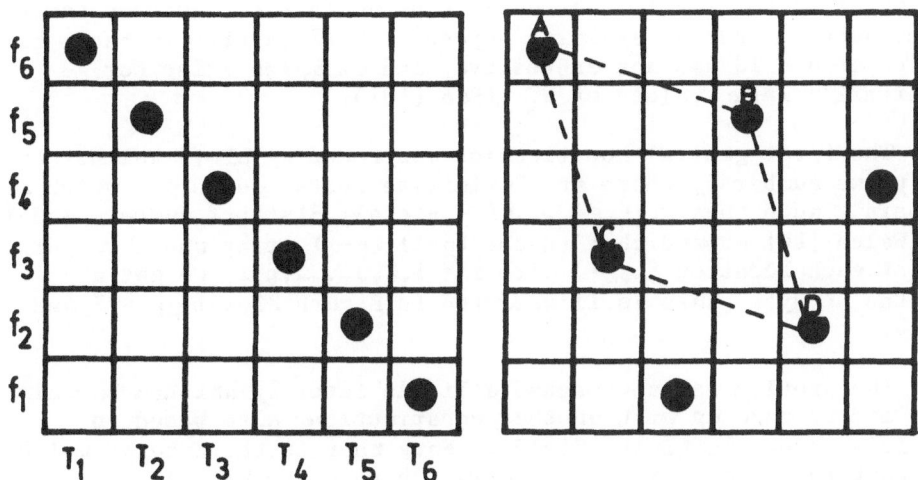

Figure 20   Figure 21

I deliberately picked a terrible way to do this (Figure 20), because if a shift is one unit in both time and frequency, then the position of the reflected signal will agree with this model in five out of six possible places. Then in case of noise I would have ambiguities in the estimates of both time and frequency shifts. In fact a 2D auto-correlation function of this type of signal has been known in the radar literature as the 'ambiguity function'. This pattern has a very bad ambiguity function compared to a spike of height 6, but generally that spike is constructed by a combination of time and frequency factors. I have here a pattern that is a little better, but is still not optimal (Figure 21). Here if you look carefully, you will find that the same shift which moves the dot A to position B, also moves the dot C to D. You can see this easily by constructing suitable parallelograms on this pattern. So this is not quite an optimal pattern. An optimal pattern would be one where any shift other than the identity shift gives you at most one coincidence, that is a dot on a dot. Obviously we cannot avoid having any coincidences because you can deliberately shift any dot from its position onto any other dot; but in doing this you can also avoid having 2nd or further multiple coincidences occurring.

The problem was first brought to my attention by John P. Costas, of 'Costas Loop' fame. For a sonar signal design problem he was interested in, he wanted to find examples of these 2D synchronization patterns, which in his honour I call 'Costas Arrays'. With the aid of a computer search, he found examples of all sizes up to 12 x 12, but could find no 13 x 13, and was tempted to conjecture that no examples beyond 12 x 12 exist. In fact, his search at n = 13 was not exhaustive, and examples exist for arbitrarily large values of n. (See [10]).

The first general construction was found by Lloyd Welch. For any prime number p, there are "primitive roots modulo p", which are numbers g such that $g, g^2, g^3, \ldots g^{p-1}$ are all distinct numbers modulo p. Welch [12] showed that in the $(p-1) \times (p-1)$ array where we put a dot at each location (i, g^i) with $i = 1, 2, 3, \ldots, p-1$, we get a 'Costas Array'. This is illustrated in Figure 22 with p = 7 and g = 3.

The problem was approached a little later by Abraham Lempel [11] and he came up with another construction, also based on primitive roots in finite fields. Note that in the case of Welch the pattern has no axis of symmetry, whereas the Lempel construction has a symmetry along the main diagonal (see Figure 23). I might also mention that the Welch construction always gives a dot in the lower left-hand corner, since $g^{p-1} \pmod{p} = 1$. Thus, I can always remove the left column and the bottom row to get the Costas pattern which is of (p-2) size. If further, I have a prime which has 2 as a primitive root (i.e. g = 2 is primitive mod p), then the top row has a dot at (1,2). p=7 does not have 2 as a primitive root, but 5 does, 11 does, 13 does etc. The conjecture is that there are infinitely many primes which have 2 as a primitive root. So you can remove the row and column for primes having 2 as a primitive root. The Lempel construction works whenever you have a finite field with q elements, with a primitive element α. You look at all the solutions of the equation:

$$\alpha^i + \alpha^j = 1$$

and this gives you dot coordinates as the index pair (i,j). Obviously the equation shows symmetry in i and j. This accounts for symmetry in dot distribution along the main diagonal. It is relatively easy to prove that if you have q elements in a field (in Figure 23 we have a field with 9 elements), you have i and j running from 1 to (q-2). The Lempel construction shown has no dots in the corners, but if you work with a case where q is prime, and a primitive root in GF(q) is 2, then it works out that you get a dot in the lower right-hand corner (because $2^{-1} + 2^{-1} = 1$, and since 2 is primitive, so is 2^{-1}). This is a very special case, for then the Welch construction has no symmetry, and the Lempel one has, though the parameters are the same.

Figure 22   Figure 23

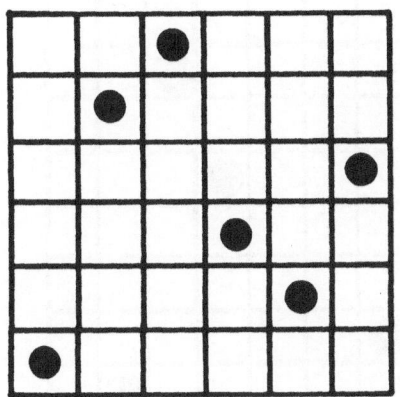 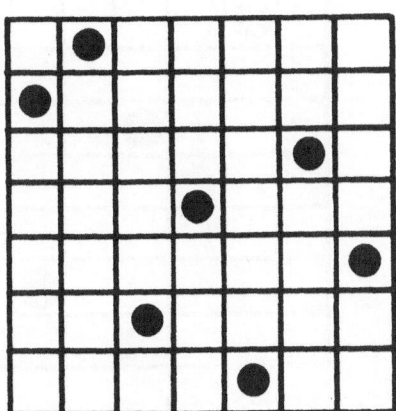

WELCH CONSTRUCTION

(i, g^i) modulo p
Illustrated case for
p = 7 and g = 3.
Other examples are with
n = p-1, n = p-2
and n = p-3 (if g = 2
is primitive mod p).

LEMPEL CONSTRUCTION

$\alpha^i = + \alpha^j = 1$, $\{(i,j)\}$
where α is primitive in
GF(q). Illustrated
case for j = 1 + i.
Other examples are for
n = q - 2 (for all q)
and for n = q - 3 if q
is prime and α = 2 is
primitive in GF(q).

Figures 22 and 23

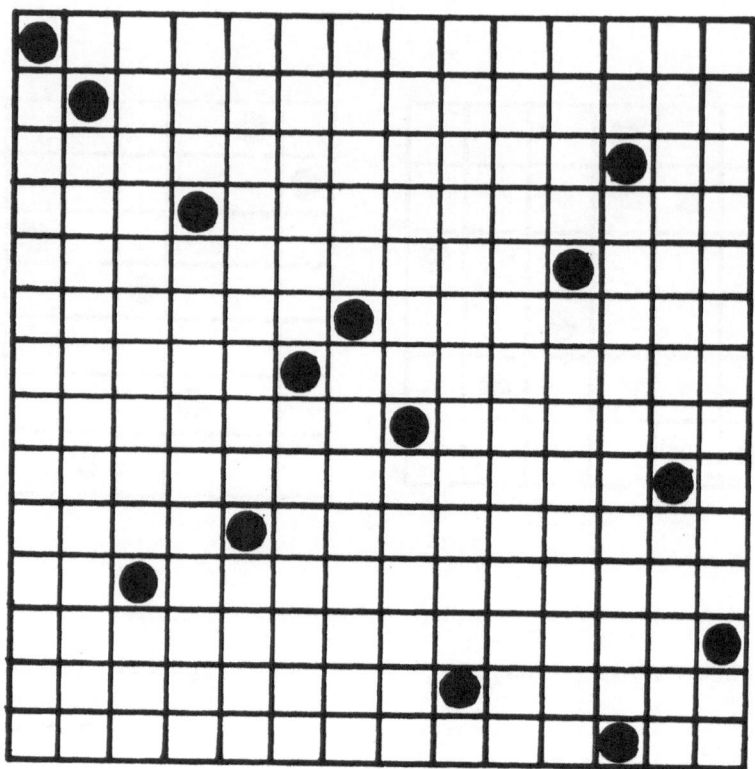

GOLOMB CONSTRUCTION

CF(16), $\alpha^4 = \alpha + 1$, $\beta = \alpha^4$,

$\alpha^i + \beta^j = 1$ for all $\{(i,j)\}$.

Also gives examples for n = 14, n = 13 and n = 12.

Figure 24

Looking through this, I discovered a rather remarkable generalisation of Lempel's construction. I take for instance GF(16) and have the generating equation

$$\alpha^i + \beta^j = 1,$$

where α and β are different primitive roots in the same field. You do not have to use the same primitive root. I can get something particularly nice if I can get α and β in such a way that their sum is 1, because then I have a dot at $(1,1)$, since $\alpha + \beta = 1$. Figure 24 gives such an example based on the field GF(16), which is generated by α which satisfies the irreducible polynomial

$$\alpha^4 = \alpha + 1.$$

For β we take α^4 $(\beta = \alpha^4)$, since the period of α is 15 and since 4 is relatively prime to 15. Then it is rather obvious that

$$\beta + \alpha = \alpha^4 + \alpha = 1.$$

I get this particular 14 x 14 pattern in which there is a dot in the top left corner ($\alpha + \beta = 1$) and since we are working mod 2, this guarantees that $\alpha^2 + \beta^2 = 1$ mod 2.

Then, going from 14 x 14 we can do the same thing for 13 x 13 (where Costas never got to) and then to 12 x 12, where we also get symmetry properties. This works out because in each case α and β are conjugate roots of the same quadratic over GF(4). That leads us to some technical computations in finite field theory, and these will show you that if $\alpha^i + \beta^j = 1$, then also $\alpha^j + \beta^i = 1$, in this case. However, my generalisation of Lempel's construction frequently leads to patterns with no symmetry. In general, by means of all these constructions, I can get a number of other special cases that enable me to remove extra rows and columns. We have systematically investigated this for all GF(g) and have discovered relatively few holes, certainly less than 14% for integers n up to 300. We do not have systematic algebraic constructions for all n, for our constructions are obviously related to the occurrence of primes and of prime powers. We know however that these patterns exist for arbitrarily large values of n. Costas had originally asked me whether I could produce a pattern for 50 x 50, or for 100 x 100. The answer is yes, since 53 is prime, and is near enough to 50. Also 101 and 103 are primes, and are near enough to the 100 x 100 case. Thus we can get examples which are near any desired size, though how near depends on the distribution of the primes. We have tried to find how many patterns there altogether for the cases upto 10 x 10, and by an exhaustive search we have found that at least up to 10 x 10 the number of inequivalent patterns increases monotonically with n. Thus you can presume that they exist for every n, and the number of them gets larger and larger with n. The constraints are such that

the construction does not lead up from one size to the next larger
size. For this reason we know of no easy way to get an estimate
for the number of patterns, given the size of n.

I just wanted to mention, as an example of 2D signal design,
that you can, subject to practical constraints, use the same sort
of technique for any 2D alignment that you may want to have for a
constallation of given pattern characteristics. The objective of
our research has been that we do not want any shifts in both time
and frequency that would produce more than one coincidence. Thus,
if I shift 2 units in time and one in frequency, we still will not
shift one dot on the top of another. Therefore, if I am in error
about the amount of time or frequency shift, I will not get much
misdirection from the pattern itself. It does not generate
self-noise that leads me into confusion over whether a target is in
the right position and has the right velocity.

We are now at a point in this discussion, where we can turn
over to the audience, to invite them to say anything they want to
add or to question.

<u>Professor Massey</u> - I have a question to Professor McEliece. Why
does it pay to do this sliding down, and then when restricting
arises, going up the helix again? Why does it pay to use block
interleaving, rather than convolutional interleaving? I recall a
paper by Ramsey in about 1960 [12] where he did prove that the
absolute minimim amount of storage that you can get for a given
interleaving depth was with the convolutional interleaving. So why
to have this predilection for block interleaving?

<u>Professor McEliece</u> - I just reported a case and I cannot answer
your question, and yet I am in agreement with the questioner.

<u>Dr. Goodman</u> - About six years ago I started to play with
multi-dimensional convolutional codes. Unfortunately this did not
work. The basic idea was this. Let us have a convolutional code
with bit sequences stretched horizontally as in Figure 25, and with
bit streams shown dotted. Then I tried to make a product
convolutional code in the way showed by the arrowed curve. The
arrowed curve shows the 'product' code in a similar way to row and
column block parity checks, except that the product code wraps
round from the bottom of one column to the top of the next. In fact
this whole convolutional code is equivalent to another
convolutional code with another generator. So that does not work.
What we have done then is to use a 3D coding (the name is a bit of
a cheek!). Again we have some horizontal bit streams (Figure 26)
which require us to have a very fast decoding situation. Assume we

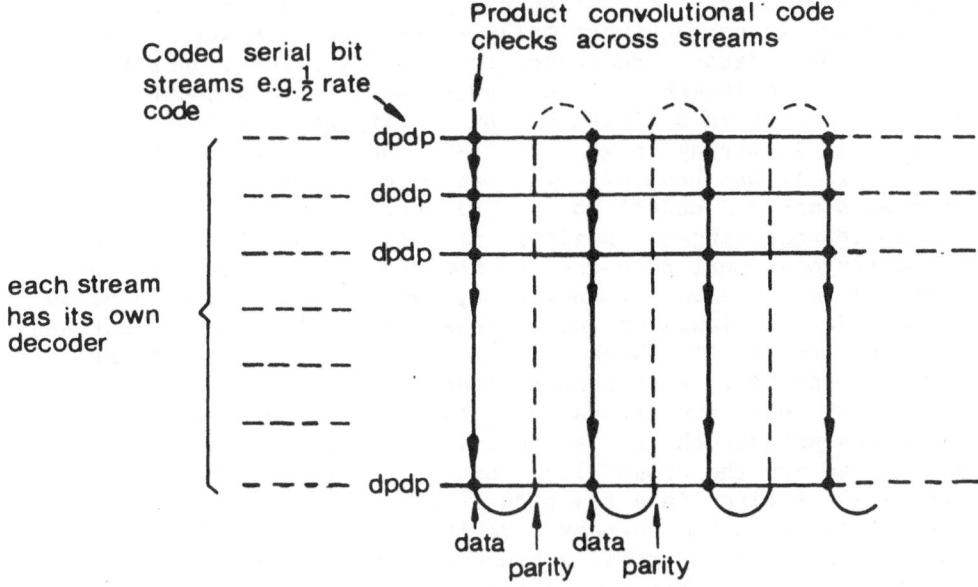

Figure 25

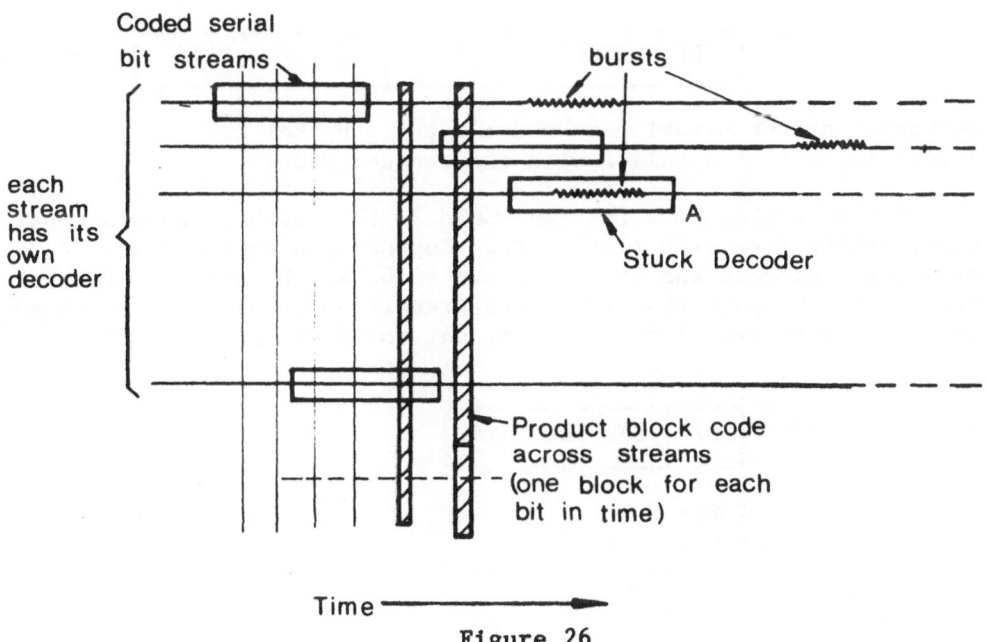

Figure 26

also have a 'bursty' situation along each bit stream, but, for no
particular reason, not vertically across the streams. Let us have a
convolutional decoder dedicated to each bit stream, operating on a
fixed decoding length. You can think of it as a sequential
decoder, but it is a little bit of a dirty sequential decoder
because it is trying to go very fast. Quite often it refuses to
decode, if it has been hit by a bad burst of errors. Then this
decoder stops and cannot go on. So what the diagram shows is a
window representing the position of each convolutional decoder.
These convolutional decoders are now in different places, because
some are having a rather easy time, but some are stuck (e.g. A).
So down in the figure we put a cross-product block code (hatched),
where we take parity checks across the bit streams in the standard
product code way. What happens then is that the bit streams that
are racing ahead are already decoded, while others have to go via
the cross-product check. Because in this case, if we cannot
decode, we use the cross-block code to try to help the stuck
decoder by filling in a few bits in the segment and then we
submitting it to the decoder. So this may be a 3D case, because we
have a convolutional code, a product code and progress in another
direction making three dimensions.

<u>Professor Massey</u> - I just want to show that the convolutional code
is much easier to frame synchronise than is the block code.
Generally we set up a sequence to initiate the communication, and
we put a 'synch' pattern, say usually the 01 pattern, as shown:

... 0101010 | FRAME SYCHRONISATION | ENCODED DATA |

For frame synchronisation we usually put a Barker sequence of
length 13, and then follow with the encoded data.

Let me explain now how you would do this with a convolutional
code, rather than with block code. Suppose you take a rate 1/2
convolutional code and this happened to be at the all-one state of
the coder. Suppose that the state transition is such that when you
put in a "one" bit of information, out comes 01 (see below):

... 101010 | 01 | ENCODED DATA |

2 bit
frame sequence

In fact, that will always be the case, because if both generators have even parity in a two-generator convolution code, the code would be catastrophic. You can see now what you are doing, for whenever you find the bit 'synch' sequence, you just tell the decoder that OK, you are now in the all-one state. The decoder

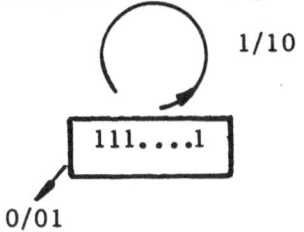

1/10

0/01

moves ahead and the rest of the bit synch sequence looks like the encoded bit stream for input zeroes. When you want a signal frame 'synch', all do you is to put 01, which will be a transition from the 'all 1' state. You use two bits of transmitted data to give frame 'synch'. You then have the same reliability in frame 'synch' as you have when you are decoding all information bits.

You can either use a Viterbi decoder, or whatever other scheme you want to use with the convolutional code. The only other thing you cannot do is to employ as many people as at present to build frame synchronizers, as we need only those that are building encoders. Thus, anything you can do with a block code, you can readily do with a convolutional code, and these codes do it better. So we should try that.

Professor McEliece - I can respond to that. There exist things called orchard codes which I believe are array convolutional codes. Dr. Matic would know something about it.

Dr. Matic - I can give you some information about it, but I have no results to show. The basic idea is very simple with this scheme. You are passing by some orchard (Figure 27A), then a specific tree can be spotted from a number of places, say from three places, as shown by the arrows. The first proposed orchard scheme is shown in Figure 27B [13]. You see there data streams and one parity check stream. This pattern appears in every bit shift if you are computing appropriate data. Up to now I have found only one paper relating to this problem. It was published some years ago [14] and it introduced the specific orchard pattern shown in Figure 27C. This led to a 2D error-correcting code with rate $(k-1)/k$. In the example you have $k = 6$, so you have a 5/6 rate code. To compute a specific check bit you have to take into account bits marked by circles. It has been proven in that paper that this code can correct 2-bit errors and that it can detect 3-bit errors; also, that it is equivalent to a kind of convolutional code. Computation of this parity bit (the cross in Figure 27C) is based on the five bits in the column above it and also on the remaining bits covered by the orchard pattern.

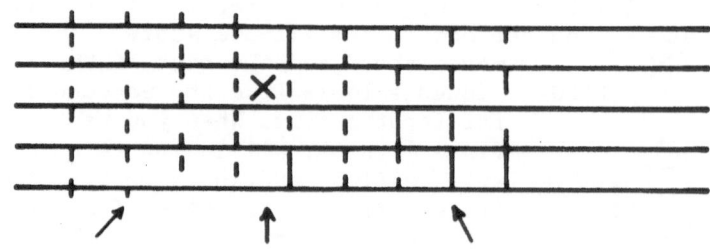

Figure 27A. The idea which has led to 'orchard' coding, scheme: any particular tree in a regular orchard can be seen from a number of places.

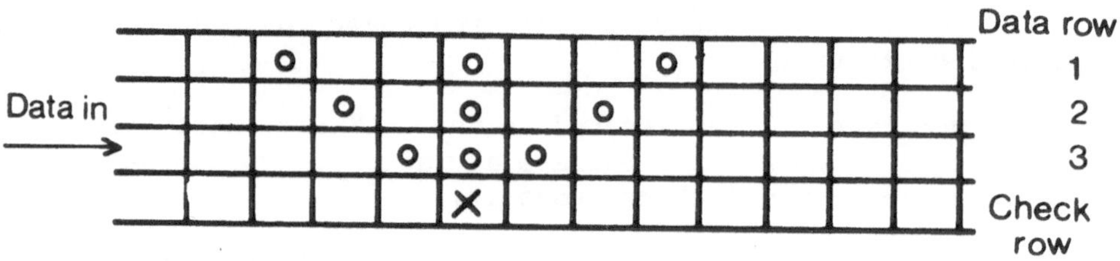

Figure 27B. One of the first proposed orchard coding patterns.

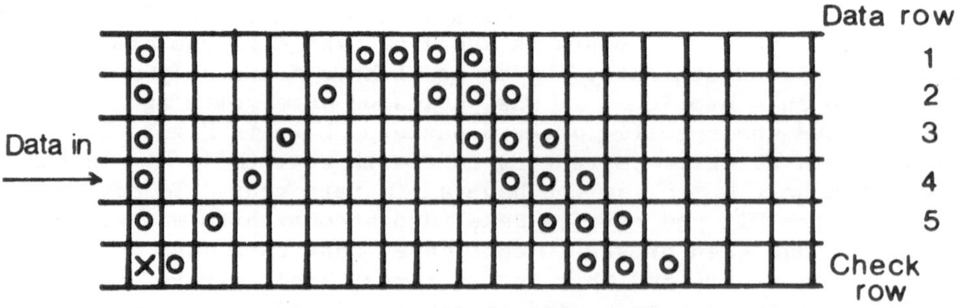

Figure 27C. Coding pattern for an orchard code rate R = 5/6, which corrects all two-bit errors and detects three-bit errors.

Professor Ephremides – You cannot see a single tree in an orchard from all directions.

Professor McEliece – There are many more 2D regular arrays than trees in orchards. If you drive past an orchard, you see many patterns – as trees are lined up in various straight lines. The places where you cannot see a tree are when it is on a diagonal as you go by.

Professor Golomb – We still have time for one more comment.

Dr. O'Reilly – I just want to raise a question which has been prompted by our chairman's initial remarks. This is another aspect of 2D imagery and this is not my field, but I have come across it recently, when a colleague of mine was working on his thesis. Suppose you try to eliminate the blocking structure that you get by using some transform coding of pictures. Say you take a TV picture and you want to do some transform coding. You do not want to do the whole lot at once, so you break up the picture into appropriate blocks (Figure 28). Then you take some sort of transform of each of these cells, and may be with this you throw away some of the boundary pixels. One of the consequences of this is that the block structure is visible in the resultant decoded picture. So you want somehow to fiddle around with this to get rid of the block structure. There are a number of things that you can do. They have all relative advantages and disadvantages. You may try to do some averaging over the edges of blocks and in this way try to remove the block structure. Otherwise, you could re-order the cells, so instead of actually using rectangular blocks, you can employ some interlocking to get this pattern out; for instance, you could employ a tasselation pattern. This gives you some visual filtering and it works quite well.

When I was examining this graduate student, being myself basically a 1D person, my immediate approach to this problem was to think in terms of vector space and inner products. What then frustrated me about this was that when I threw at him a suggestion that I knew would work in 1D, his answer was that he did not think it would work in 2D. In 1D, apart from quantisation, by throwing away higher order terms, I can get rid of the blocking structure, I could choose an inner product that would give me zero error. However, this is of no value in 2D. This is both a frustration and annoyance that there is a nice and elegant technique in 1D, that does not apply in 2D.

Professor Golomb – This is the end, and I would like to thank all of you for attending this late night panel (Applause).

References

[1] WESCON Convention Record, Western Electronics Conference, San
     Francisco, 1957.

[2] R.J. McEliece, "Some Applications on VLSI RAMs to Coded
     Communication Systems" (in this volume).

[3] E. Berlekamp and P. Tong, "Interleaving". To appear in the
     IEEE Transactions on Information Theory.

[4] V. Benes, "Mathematical Theory of Connecting Networks and
     Telephone Traffic", New York, Academic Press, 1965.

[5] S.V. Ahamed, "Serial Coding for Cyclic Block Codes", BSTJ,
     Vol. 59, No. 2, February 1980, pp269-276.

[6] P.G. Farrell, "Code Structure and Decoding Complexity" (in
     this volume).

[7] E. Munday, "Soft-Decision Decoding for HF Data
     Communications", Ph.D. Thesis, University of Kent at
     Canterbury, 1978.

[8] G. Ungerboeck, "Channel Coding with Multi-Level/Phase
     Signals", IEEE Transactions, Vol. IT-28, No. 1, January 1982,
     pp55-67.

[9] B. Elspas, "Design and Instrumentation of Error-Correcting
     Codes". Final Report under Contract AF30(602)-2327,
     RADC-TDR-62-511, October 1962.

[10] S.W. Golomb and H. Taylor, "Two Dimensional Synchronization
     Patterns for Minimum Ambiguity", IEEE Transactions on
     Information Theory, Vol. IT-28, No. 4, July, 1982,
     pp.600-604.

[11] S.W. Golomb, "Algebraic Constructions for Costas Arrays",
Journal of Combinatorial Theory (A). To appear in 1984.

[12] J.L. Ramsey, "Realisation of Optimum Interleavers", IEEE
Transactions, Vol. IT-16, pp338-345, May 1970.

[13] E. Scott and D. Goetschel: "One Check Bit Per Word can Correct
Multibit Errors", Elctronics, Vol. 54, No. 9, 1981,
pp 130-134.

[14] A. Shiozaki: "Proposed for a New Coding Pattern in Orchard
Scheme", Information and Control, Vol. 51, No. 3, December
1981, pp 209-216.

Part 3.

PROCESSING IN COMMUNICATION NETWORKS

Organised by:

K.W. Cattermole
Department of Electrical Engineering Science
University of Essex, U.K.

and

J.K. Wolf
Department of Electrical and Computer Engineering
University of Massachusetts, U.S.A.

PRINCIPLES OF GROUP TESTING AND AN APPLICATION TO THE DESIGN AND ANALYSIS OF MULTI-ACCESS PROTOCOLS

Jack Keil Wolf

Department of Electrical and Computer Engineering, University of Massachusetts, Amherst, Massachusetts 01003

1. INTRODUCTION

In group testing we are concerned with determining the state of items, each one of which can be either in a good state (denoted "g") or a defective state (denoted "d"). We assume that we have a universal test which can be applied simultaneously to any number of the items. In classical group testing there can be two outcomes of the universal test. One of the outcomes is that all items tested are good (denoted "G") and the other outcome is that there was at least one defective in the group tested (denoted "D"). The choice of the items to be tested in any given test can depend upon the entire history of the testing to that point. That is, the items to be tested in the i-th test can depend upon which items were tested in all preceding tests and the outcomes of these tests. A test plan is a set of tests and the corresponding rules for choosing which items are to be tested in each test. If the number of items is finite, the test plan is concluded when the states of all items are determined. The problem of group testing is to specify an efficient test plan. Our criterion for efficiency will be discussed later.

2. EXAMPLE

At this point an example is worthwhile. Assume that we have four items (x_1, x_2, x_3, x_4) to be tested. Assume, a priori, that the probability of the items being defective is governed by a binomial distribution with paramater p. That is, the

$x_3 x_4$ / $x_1 x_2$	dd	dg	gg	gd
dd	p^4	$p^3 q$	$p^2 q^2$	$p^3 q$
dg	$p^3 q$	$p^2 q^2$	pq^3	$p^2 q^2$
gg	$p^2 q^2$	pq^3	q^4	pq^3
gd	$p^3 q$	$p^2 q^2$	pq^3	$p^2 q^2$

Figure 1   Probabilities for States of Items

probability that each item is defective is equal to p and the probability of having i defects out of the 4 is equal to $\binom{4}{i} p^i (1-p)^{4-i}$, i=0,1,2,3,4. Figure 1 gives in tabular form the probability of each of the 2^4=16 possible states for the items. If we tested each of the items individually 4 tests would be required to determine the correct state of the items. A test on item x_1 would correspond to determining whether or not the true state was in the last two rows of the table. Similarly a test on x_4 alone tests whether or not the true state is in the middle two columns. This information is also shown in Figure 1 by labeling appropriate pairs of rows of columns with x_i,i=1,2, 3,4. Testing the items individually seems reasonable if p≃.5 since in that case all 16 states would be roughly equally probable. Suppose, however, that p is much smaller than 1. Then some of the states are much more likely than others. The most likely state for very small p is that all items are good. A group test that would test for this possibility directly would be to test all 4 items. If they are all good (that is, the result of the test is "G") then the testing is completed. If they are not all good, further tests are required. Figure 2 shows some of the tests and results of these tests for a specific test plan appropriate for moderate values of p. For each test, the items tested are identified by a rectangle. If the outcome of the test is "G" then the true state corresponds to a square in that rectangle. First x_1, x_2, and x_3 are tested and results in a defective (i.e., a "D") test outcome. Then x_1 and x_2 are tested and results in a good (i.e., "G") test

TEST x_1, x_2, x_3; OUTCOME D

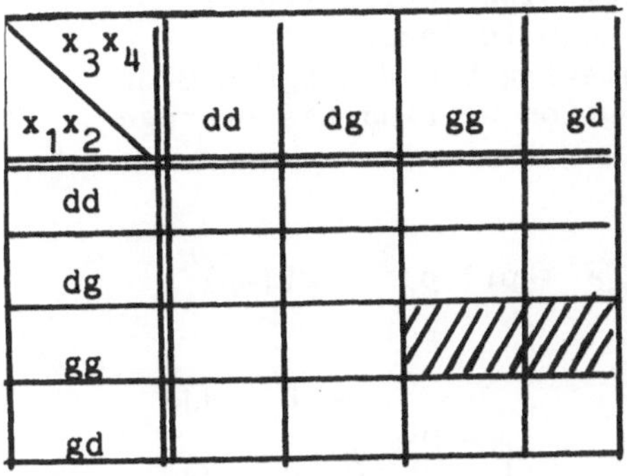

$x_3 x_4$ / $x_1 x_2$	dd	dg	gg	gd
dd				
dg				
gg			▨	▨
gd				

Not in last two squares of
3-rd row.

(a)

TEST x_1, x_2; OUTCOME G

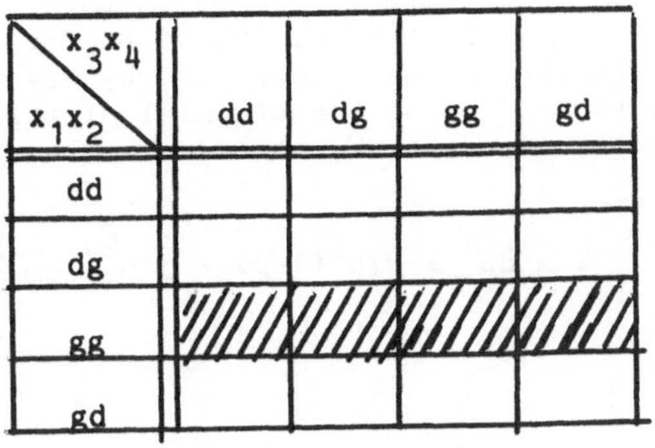

$x_3 x_4$ / $x_1 x_2$	dd	dg	gg	gd
dd				
dg				
gg	▨	▨	▨	▨
gd				

In first two squares of
3-rd row.
$(x_1 \to g, x_2 \to g, x_3 \to d)$

(b)

TEST x_4; OUTCOME D

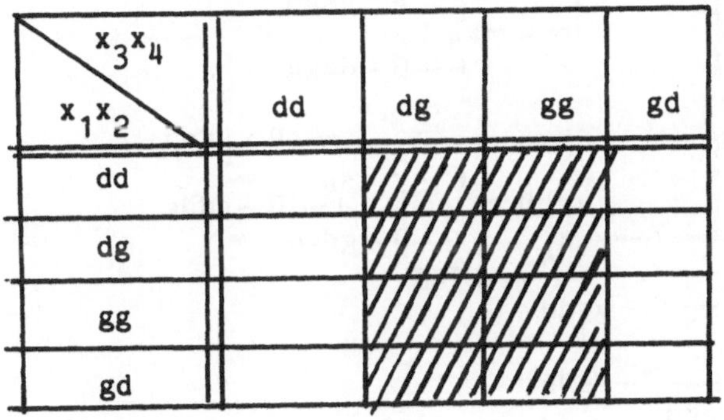

$x_3 x_4$ / $x_1 x_2$	dd	dg	gg	gd
dd		▨	▨	
dg		▨	▨	
gg		▨	▨	
gd		▨	▨	

In first square of 3-rd row.
$(x_1 \to g, x_2 \to g, x_3 \to d, x_4 \to d)$

(c)

Figure 2. Part of a Test Plan

outcome. At this point we know that x_1 and x_2 are good and x_3 is defective. Finally we test x_4 to finish our test plan. In this case we required three tests instead of the four that would be necessary if we tested the items individually. A complete test plan is given in Figure 3 in the form of a binary tree. The number of tests in this test plan varies from a minimum of 2 to a maximum of 8. We chose to measure the efficiency of this test plan by calculating the average number of tests. Using the binomial distribution we find the average number of tests, $\overline{T}$, to be:

$$\overline{T} = 8p^4 + 27p^3(1-p) + 31p^2(1-p)^2 + 14p(1-p)^3 + 2(1-p)^4.$$

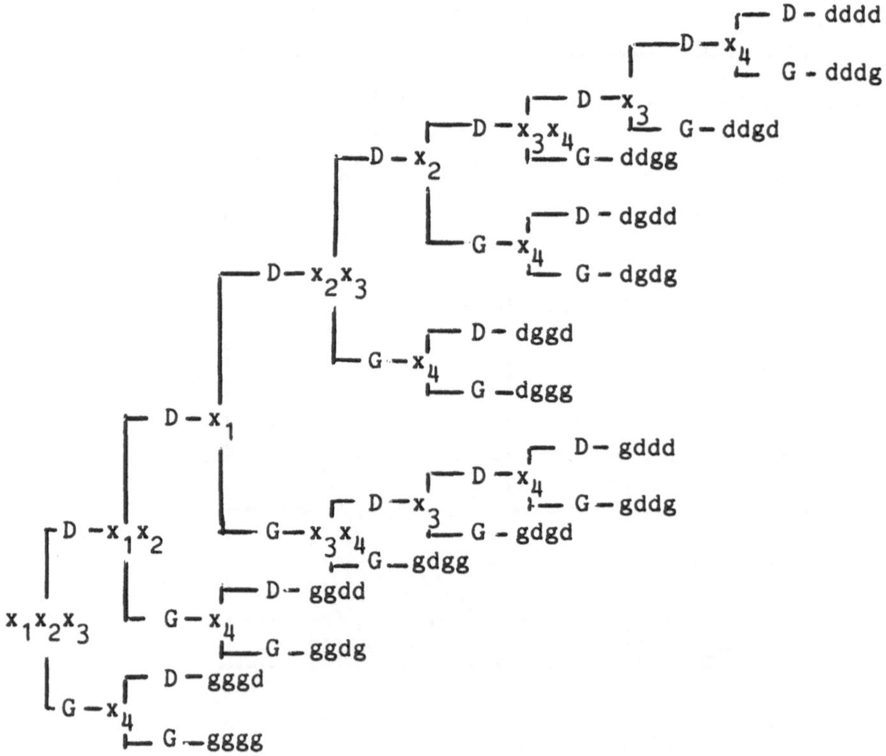

Figure 3. A Complete Test Plan

Table 1 gives numerical results for $\overline{T}$ for various values of p. It should be noted that for p larger than approximately .32, $\overline{T}$ exceeds 4, the number of tests which result from testing these items individually.

TABLE 1 Average Number of Tests, $\overline{T}$, vs. p.

p	$\overline{T}$	p	$\overline{T}$
.01	2.060	.26	3.610
.02	2.120	.27	3.673
.03	2.181	.28	3.736
.04	2.242	.29	3.800
.05	2.302	.30	3.863
.06	2.363	.31	3.926
.07	2.425	.32	3.990
.08	2.486	.33	4.053
.09	2.547	.34	4.116
.10	2.609	.35	4.180
.11	2.671	.36	4.243
.12	2.733	.37	4.306
.13	2.795	.38	4.370
.14	2.857	.39	4.433
.15	2.919	.40	4.496
.16	2.982	.41	4.559
.17	3.044	.42	4.622
.18	3.107	.43	4.685
.19	3.169	.44	4.748
.20	3.232	.45	4.811
.21	3.295	.46	4.874
.22	3.358	.47	4.937
.23	3.421	.48	5.000
.24	3.484	.49	5.062
.25	3.547		

3. SOME HISTORY

The early history of group testing is interesting in its own right. Dorfman [1] noted that a particular blood test (the Wasserman test) was so sensitive that it could detect the presence of syphilis in a mixture of blood samples from many persons if one or more persons (whose blood was in the mixture) had the disease. The Dorfman procedure consisted of separating a blood sample taken from an individual into two samples. One sample was pooled along with samples from (K-1) other individuals and the pool was tested. If the test was negative, all persons whose blood was in the tested mixture were assumed free of the disease. If, however, the test was positive, then the Dorfman procedure dictated that the other half of each blood sample from the persons whose blood was in the tested mixture (called the current defective group) be tested individually. Dorfman calculated the optimum value of K for a binomial population. Sterrett [2] suggested an improvement to the

Dorfman procedure whereby when a positive result was obtained in an individual test, the remaining samples in the defective group were returned to the untested population. Sterrett was on firm ground in his suggestion since in the binomial model, the remaining samples in the pool again are described by a binomial distribution. Sobel and Groll [3] noted that when there was a group of size M>1 in which there was known to be at least one positive sample there could be an advantage in testing a subset of size greater than 1 of this defective group. Sobel and Groll also foresaw many applications other than blood testing for group testing. These included checking devices which possibly have leaks, checking circuits, etc. A well known (seasonal) demonstration of this idea is the series circuit used in inexpensive Christmas lights where all lights are extinguished when one or more of the bulbs becomes defective! Recently, group testing has been applied to screening a large number of experimental variables [4], pattern recognition [5], questionnaire theory [6], and binary identification [7].

After the work of Sobel and Groll, a vast literature emerged on this subject. The subsequent work can be subdivided into two areas: a probabilistic one and a deterministic one. In the probabilistic model, the criterion for choosing the best testing plan has been called the "optimal criterion" which is to minimize the _average_ number of tests to find all defectives. In the case where the population size is infinite this "optimal criterion" chooses that testing plan with the minimum average number of tests per item classified. In the deterministic model the criterion for choosing a testing plan is called the "minimax criterion." Here we minimize the _maximum_ number of tests to find all defectives.

The optimal criterion has been applied to the situation where the population's statistics are binomial [3,8,9,10,11]. In this case, the statistical behavior of the items are governed by the two parameters "p" and "N" where p is the constant probability of each item being "defective" and "N" is the total number of items to be tested. A generalization of the binomial problem is where the probability of an item being "defective" can vary from item to item [12].

The minimax criterion has been applied to the situation where we know a priori the number of defective items. Here, the statistics of the set is hypergeometric [13,14] and again two parameters are needed: d, the number of defectives and N, the total number of items to be tested. In a generalization of this case, d is taken to be an upper bound to the number of defectives [14].

An important subclass of testing strategies is called "nested" [3,8,9,11]. Here, no restrictions are placed on which items are to be tested until one group, A, is found to contain one or more defective items. We call the group A, the current defective group. The next test must be a proper subset, X, of A. If this test has a defective result, then X replaces A as the current defective group. Otherwise every item in X is classified "good" and A − X replaces A as the current defective group. This process is continued until A has cardinality 1 in which case we classify the one item in A as defective and return to our unrestricted testing. Of course the procedure terminates when all items are classified. Nested strategies, in general, may not be as efficient as non-nested strategies but usually are easier to analyze (and perhaps to carry out). "Mixing" strategies can be employed in non-nested test plans where we test items both from a defective group and from an untested group in the same test. A portion of a non-nested test plan employing mixing is shown in Figure 4 for the case of four items.

Two problems in group testing which have received a great deal of attention are:

(1) What is the best strategy to follow if one is satis-
    fied with finding only one defective item [10]?
(2) When is it better to test the items individually
    rather than in groups [15]?

TEST x_1, x_2, x_3; OUTCOME D   TEST x_1, x_2, x_4; OUTCOME G

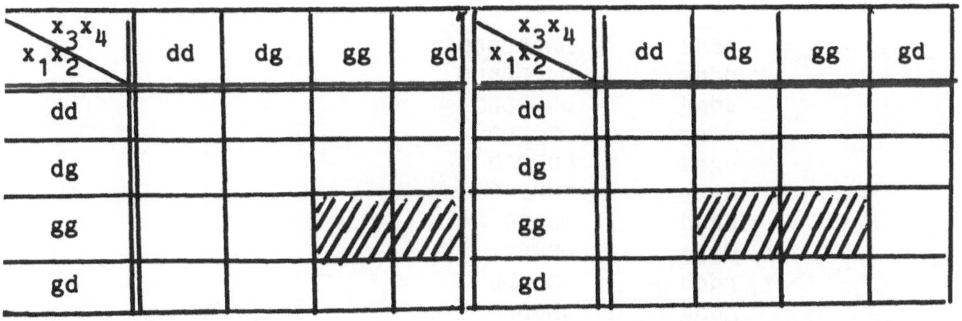

$x_1 x_2$ \ $x_3 x_4$	dd	dg	gg	gd	$x_1 x_2$ \ $x_3 x_4$	dd	dg	gg	gd
dd					dd				
dg					dg				
gg			////	////	gg			////	////
gd					gd				

Not in last two squares of   In second square of 3rd row
3rd row $(x_1 \rightarrow g, \ x_2 \rightarrow g, \ x_3 \rightarrow d, \ x_4 \rightarrow g)$

Figure 4. Part of a Test Plan with Mixing

Regarding the latter question, Ungar [15] showed that for $p \geq (3-\sqrt{5})/2$ no group testing plan has a smaller average number of tests than the number obtained by testing each item individually.

An interesting relationship between group testing and information theory was explored by several researchers [9,16]. Since each test in a test plan has a binary outcome, the result of the tests in a test plan can be thought of as a variable length binary encoding of the 2^N states of the N items. This is shown in Table 2 for the test plan of Figure 3. The average number of tests for any test plan can be no smaller than the average length of a Huffman code [17] for a source with 2^N symbols having the same probabilities as the states. Any lower bound to the average length of a Huffman code is thus a lower bound to the average number of tests in the best possible test plan. Thus N h(p) is such a lower bound [18], where h(p) is the binary entropy function.

$$h(p) = -p \log_2 p - (1-p) \log_2 (1-p).$$

The optimal test plan may require an average number of tests strictly larger than the average length of the corresponding Huffman code. This is because the tree corresponding to the optimal Huffman code may not correspond to a realizable test plan.

Table 2   Encoding of States

State	Result of Tests
dddd	DDDDDDDD
dddg	DDDDDDDG
ddgd	DDDDDDG
ddgg	DDDDDG
dgdd	DDDDGD
dgdg	DDDDGG
dggd	DDDGD
dggg	DDDGG
gddd	DDGDDD
gddg	DDGDDG
gdgd	DDGDG
gdgg	DDGG
ggdd	DGD
ggdg	DGG
gggd	GD
gggg	GG

4.   SOME RESULTS FROM CLASSICAL GROUP TESTING
      (Probabilistic Model)

      Sobel and Groll presented an algorithm for finding the
nested test plan that satisfied the optimal criterion (the
least average number of tests) for a binomial population of
size N and parameter p. In the statement of the algorithm to
follow, Q denotes the set of N items to be tested, U is the set
of items for which nothing more is known than that they contain
items from a binomial population, A is a set of items for which
it is known that at least one item in this set is defective, K
is the set of items whose states are known and $\emptyset$ denotes the
empty set.

<u>Algorithm</u>

0.   U:=Q; K:=$\emptyset$; A:=$\emptyset$;

1.   if U=$\emptyset$ then goto 3
                else test a subset X $\subseteq$ U
                    if response = G
                        then K:=K+X;
                             U:=U-X;
                             goto 1;
                    if response = D
                        then A:=X;
                             U:=U-X;
                             goto 2;

2.   if A contains one item
          then K:=K+A;
               A:=$\emptyset$;
               goto 1;
          else test a subset X$\subseteq$A
                if response = G
                    then K:=K+X;
                         A:=A-X;
                         goto 2;
                if response = D
                    then U:=U+A-X
                         A:=X;
                         goto 2;

3. end

The sizes of the subsets X are chosen to minimize the expected
number of queries as described by the recurrence equations to
follow.

Let H(m,n) denote the expected number of tests remaining in a nested algorithmn which satisfies the optimal criterion when the set A contains m users and the set U contains (n-m) users, for all m and n such that $0 \leq m \leq n \leq N$. Then it is easy to show that:

$$H(0,0)=0, \quad H(1,n) = H(0,n-1), \quad n=1,2,\ldots,N,$$

$$H(0,n)=1+\min_{1 \leq x \leq n} \{(1-p)^x H(0,n-x) + (1-(1-p)^x) H(x,n)\},$$

$$2 \leq n \leq N,$$

$$H(m,n) = 1 + \min_{1 \leq x \leq m} \left\{ \left[\frac{(1-p)^x - (1-p)^m}{(1-(1-p)^m)} \right] H(m-x,n-x) \right.$$

$$\left. + \left[1 - \frac{(1-p)^x - (1-p)^m}{(1 - (1-p)^m)} \right] H(x,n) \right\},$$

$$2 \leq m \leq n \leq N.$$

The order in which these quantities are determined is:

$$H(0,0), \ H(1,1), \ H(0,1), \ H(1,2), \ H(2,2), \ H(0,2), \ \text{etc.}$$

Numerical results obtained from these equations are given by Sobel and Groll, [3] and Towsley and Wolf [19]. Some sample values for H(0,N) for $p\epsilon(.1,.2,.3,.4,.5)$ and $N\epsilon(2,4,8,16)$ are given in Table 3.

Towsley and Wolf [19] also give the performance of a simpler nested algorithm where whenever the set A is not empty, a test is performed on one half of the subset of A. (More precisely the test is performed on a number of items equal to the largest integer less than or equal to half the number in the subset.) The optimum size for testing a subset of U is still chosen by optimization of the recurrence relation. The difference in numerical values between the performance of the original algorithm and the "halving" algorithm is insignificant.

The nested algorithm which satisfies the optimal criterion for the case of an infinite number of items can be completely specified. Furthermore a closed form expression for its performance can be given. In this case, the nested algorithm which satisfies the optimal criterion has the smallest average number of tests per item classified over all nested algorithms. These results follow from the following observations [20]:

Table 3 - Average Number of Tests for Optimal
Nested Algorithm - Classical Group Testing

N\p	.1	.2	.3	.4	.5
2	1.29	1.56	1.81	2.00	2.00
4	2.05	3.01	3.60	4.00	4.00
8	3.76	5.91	7.17	8.00	8.00
16	7.22	11.73	14.28	16.00	16.00

1. Assume we have a test plan (called TP-1) that identifies the <u>first</u> defective item in the smallest average number of tests. Then the nested procedure that classifies all items and for which the ratio of the average number of tests per item classified is a minimum is obtained by repeatedly applying procedure TP-1.
2. The average number of tests for procedure TP-1 is greater than or equal to the average length of a Huffman code for code words having a geometric distribution with parameter p [10].
3. The code words of the Huffman code for the geometric distribution can be put in one-one correspondence with outcomes of test plan TP-1.

The complete test plan for TP-1 follows: Choose the integer "s" such that

$$(1-p)^s + (1-p)^{s+1} \leq 1 \leq (1-p)^{s-1} + (1-p)^s$$

For any 0<p<1 there is a unique "s" satisfying this equation, namely

$$s = -\lceil \log(2-p)/\log(1-p) \rceil$$

where $\lceil x \rceil$ is the smallest integer greater than or equal to x. Test groups of items of size "s" until a defective group is found. Define t as $t = \lfloor \log_2 s \rfloor$ where $\lfloor x \rfloor$ is the largest integer smaller than or equal to x. The defective group of size "s" is split into two groups as follows

(a) If $s \leq 3 \cdot 2^{t-1}$ then group 1 contains 2^{t-1} items and group 2 contains $s - 2^{t-1}$ items.

(b) If $s > 3 \cdot 2^{t-1}$ then group 1 contains $s-2^t$ items and group 2 contains 2^t items.

Then test group 1 uses (t-1) tests and group 2 uses "t" tests. The average number of tests to find the first defect can be shown to be given by the equation [10,21]

$$\overline{T} = \lfloor \log_2 s \rfloor + 1 \frac{(1-p)^k}{1-(1-p)^s}$$

where $k = 2^{\lfloor \log_2 t \rfloor + 1} - t$. Table 4 gives some typical values of the parameters which pertain to this algorithm. It also gives an entropy bound which is a lower bound to the average number of tests per item for any group testing procedure (nested or non-nested).

It should be noted that the minimum number of tests per item for the best nested algorithm for N=∞ is a lower bound to the minimum number of tests per item for the best nested algorithm for any finite value of N. This follows from noting that any nested algorithm for finite N, could be used as a nested algorithm for the case of infinite N.

Table 4 Performance of Optimal Nested Algorithms (N=∞)

p	Best Group Size s	Average No. of Tests to Find First Defective $\overline{T}$	Average No. of Tests per Item $\overline{T}_p$	Entropy Bound h(p)
.01	69	8.105	.0811	.0808
.05	14	5.762	.2881	.2864
.1	7	4.725	.4725	.4690
.5	4	4.092	.6138	.6098
.2	3	3.639	.7278	.7219
.25	2	3.286	.8215	.8113
.3	2	2.961	.8882	.8813
.35	2	2.732	.9561	.9341
.4	1	2.500	1.0000	.9710
.45	1	2.222	1.0000	.9928
.5	1	2.000	1.0000	1.0000

5. GENERALIZED GROUP TESTING AND APPLICATIONS TO MULTI-ACCESS PROTOCOLS

Many generalizations of group testing could be considered. Here we consider the specific generalization of a test with three outcomes:

(1) no defective items in tested group, (denoted I),
(2) exactly one defective item in the tested group, (denoted S),
(3) two or more defective items in the defective group, (denoted C).

The reason for considering this generalization is that now the items to be tested are transmitters which are in one of the two following states:

State 0 – No message to be transmitted (occuring with probability (1-p)),
State 1 – Message to be transmitted (occuring with probability p).

Again, we use a binomial model to describe the states of N transmitters. A test consists of giving permission to (or "enabling") specific transmitters to transmit. If two or more transmitters transmit messages, a collision between these messages occurs requiring them to be retransmitted. This corresponds to the test outcome C. If, however, exactly one transmitter transmitted a message (resulting in test outcome S), that message is assumed to be correctly received. Finally, if no transmitters transmitted messages, the test outcome is I. The letters I, S, and C were chosen to denote an "idle" outcome, a "success", and a "collision" respectively. A test plan is concluded when all messages have been successfully transmitted. Note that at the end of the test plan the states of all of the transmitters may not be known since if two or more transmitters were enabled and the test result was S, we do not know which of the enabled transmitters actually transmitted the message.

The problem described above is one version of a problem which has been termed <u>multi-access communications with ternary feedback</u> [22-24]. Much of the previous literature concerned with this problem [25-30] has concentrated on the case where the number of transmitters N is infinite and p—>0 in such a way that Np $\longrightarrow \lambda$ resulting in Poisson statistics. One exception is the work of Molle [30] who also treats the case of non-zero p.

As discussed by Berger [31], Pippenger [32] suggested that group testing principles could be employed to design efficient multiple access protocols and Berger incorporated this idea in a research proposal [33]. Towsley and Wolf [34], independently pursued this approach and first published results on the use of group testing in the design of multiple access protocols.

A group testing algorithm for a test with three outcomes is described below. It is not strictly nested but seems to be the natural generalization of nested algorithms to this type of test. If we assume the transmitters are ordered and always enable the lowest numbered transmitters consistent with the algorithm specification, then successful messages are trans-mitted in order--i.e., transmitter i successfully transmits a message before j if and only if i < j.

We describe this algorithm in terms of four sets of transmitters, U, A(1), A(2), and K. Here U contains trans-mitters of which we know nothing. The set K contains all of the transmitters that are known to have no messages to transmit or have successfully transmitted their messages. For i=1,2, A(i) is a set which is known to contain at least i active transmitters. If both A(1) and A(2) are nonempty, then A(1) is a proper subset of A(2). Again Q is the total population of transmitters.

Algorithm

0. U:=Q; K:=∅; A(1):=∅; A(2):=∅;

1. if U=∅ then goto 4
            else enable a subset $X \subseteq U$;
                if test result is $\bar{I}$ or S
                    then K:=K+X;
                         U:=U−X;
                         goto 1;
                if test result is C
                    then A(2):=X;
                         U:=U−X;
                         goto 2;

2. enable a subset $X \subseteq A(1)$ or X=A(1)+Y where $Y \subseteq (A(2)−A(1))$
            if test result is I
                then K:=K+X;
                A(2):=A(2)−X;
                if $X \subseteq A(1)$ then A(1):=A(1)−X;
                goto $\bar{2}$;
            if test result is S
                then K:=K+X;
                A(1):=A(2)−X;
                A(2):=∅;
                goto 3;

```
        if test result is C
           then U:=U+A(2)-X;
                A(2):=X;
                if X ⊆ A(1) then A(1):=∅;
                goto 2;

3.  enable subset X ⊆ A(1) or X=A(1)+Y where Y ⊆ U
           if test result is I
              then K:=K+X;
                   A(1):=A(1)-X;
                   goto 3;
           if test result is S
              then K:K+X;
                   U:=U+A(1)-X;
                   A(1):=∅;
                   goto 1;
           if test result is C
              then A(2):=X;
                   U:=U+A(1)-X;
                   if A(2) ⊆ A(1) then A(1):=∅;
                   goto 2;

4.  end.
```

Recursion relations can be derived for this algorithm. The optimum sizes of the subsets in the above algorithm which lead to the minimum average number of tests required to successfully transmit all messages from N transmitters are found by solving these recurrence equations. Some numerical results obtained from these equations are given below in Table 5.

Table 5

Average Number of Tests for Towsley-Wolf Algorithm
– Generalized Group Testing

N \ p	.1	.2	.3	.4	.5
2	1.02	1.08	1.18	1.32	1.5
4	1.13	1.48	2.01	2.63	3.0
8	1.59	2.88	4.02	5.18	6.0
16	3.05	5.71	8.02	10.34	12.0

Towsley and Wolf [19] also give results for a simplified version of this algorithm where if A(1) = A(2) = Ø then the optimum choice for size of the subset of U is chosen but if A(2) ≠ Ø and A(1) = Ø then half of the group A(2) is chosen while if A(1) ≠ Ø and A(2) = Ø then the entire group A(1) is enabled. In this simplified algorithm the situation never occurs where both A (1) and A (2) are not empty. The performance of this modified algorithm differs only slightly from the performance of the full algorithm (where all group sizes are optimized).

6. BOUNDS ON PERFORMANCE OF MULTI-ACCESS PROTOCOLS

The usual performance measure for a multi-access protocol is its throughput, η, which is taken to equal the percentage of tests resulting in successful outcomes (i.e. an "S" outcome). For the case of a finite number of transmitters described by a binomial distribution with parameters N and p, the throughput is given as $\eta = Np/\overline{T}$, where $\overline{T}$ is the average number of tests.

Very clever strategies have been employed to find upper bounds to the throughput over all possible test plans. These strategies are based upon (a) information theory, (b) clever calculations of the expectation of the number of tests conditional on various events and (c) bounds based upon the presence of "genies." We will only briefly describe the first two of these methods and give a more complete description of a special case of the third.

Information Theory Bounds [27]

Here we argue that the difference between the uncertainty in our knowledge of the states of the transmitters before testing and our uncertainty in this knowledge after testing must be equal to the average information provided by the test outcomes. Using this approach, Pippenger found an upper bound to the throughput for the case of Poisson statistics to be .744.

Conditional Expectation Bounds [35]

The basic approach here is to find a set of mutually exclusive and exhaustive events $A_1, A_2, \ldots, A_z$ such that one can find a lower bound to the expected number of tests conditional on events A_i which is independent of i. This lower bound to the conditional expectation then serves as a lower bound to the unconditional expectation. Using this approach Tsybakov and

Mikhailov [26] have upper bounded the throughput for the
Poisson model as .5875.

Genie-Aided Testing [30,36]

The basic idea here is to assume the presence of a genie
that supplies additional information regarding the test
results. Certainly the maximum throughput of a test plan which
utilizes the genie's information can be no smaller than the
maximum throughput of test plans which do not have this added
information. Thus any upper bound to the throughput of a
"genie-aided" test plan serves as an upper bound to the
throughput of any test plan which must function without this
information. In order to derive a tight upper bound one must
create a genie which does not give too much information but for
which an upper bound on the throughput can be calculated.

Molle [30] suggested a genie, which after a test which
resulted in a collision (test result "C"), identified the two
transmitters of lowest order (assuming all transmitters are
numbered) that transmitted packets during that test. Molle
then showed that if $N=\infty$, the strategy which yields the largest
throughput with this added information is as follows:

Genie-Aided Algorithm

1. Enable "X" transmitters
            if test result is S or I
                then goto 1
            if test result is C
                then enable each of two transmitters identified
                by genie individually;
                        goto 1.

The only free parameter in this algorithm is "X" and it is
chosen to optimize the throughput for a given value of p. It
is found that for .707 $\leq p \leq 1$, the best value of "X" is 1 while
for .568 $\leq p \leq$ .707 the best value of "X" is 2. Since for X=1 or
2 the genie does not furnish any information we find that:

    (a)   For .707 $\leq p \leq 1$, no multiple access algorithm with
        ternary feedback outperforms enabling the trans-
        mitters individually (called time-division-multiple-
        access or TDMA).
    (b)   For .568 $\leq p \leq$ .707, the best multiple access algorithm
        with ternary feedback enables two users at a time and
        if a collision results, enables each of the two
        individually.

254

    (c)   For $0 \leq p \leq .568$ the maximum throughput of the genie algorithm (maximum over choice of "X") serves as an upper bound to the throughput achievable for any multiple-access scheme (with ternary feedback).

Using the result of (c) above, Molle showed that for the Poisson case, the throughput of any multiple access scheme with ternary feedback cannot exceed .673. Cruz and Hajek [36], using a less informative genie, improved the upper bound for the Poisson case to .6126. The best lower bound to the throughput for the Poisson case stands at .488 using the constructive algorithm of Gallager [26] as improved by Mosely. [29].

Recently Rueppel [37] has found that a generalization of the technique utilized by Tsybakov and Mikhailov [35] allows for a unified derivation of all previously mentioned upper bounds. He also extended the region of p for which pairwise enabling (i.e., X = 2) is optimal in the case of an infinite number of users to $.433 \leq p \leq .707$

ACKNOWLEDGMENT

This research was supported by the National Science Foundation under grant NSF ECS 7921140. Most of this work was done in collaboration with my colleague Don Towsley at the University of Massachusetts. The work reported in section 4 regarding the optimum nested algorithm for the infinite population case was done in collaboration with Professor James Massey at the E.T.H., Zurich, Switzerland.

REFERENCES

1. R. Dorfman, "The Detection of Defective Members of Large Populations," Annals of Mathematical Statistics, Vol. 14, pp. 436-440, 1943.

2. A. Sterrett, "On the Detection of Defective Members of Large Populations," Annals of Mathematical Statistics, Vol. 28, pp. 1033-1036, 1957.

3. M. Sobel and P.A. Groll, "Group Testing to Eliminate Efficiently All Defectives in a Binomial Sample," Bell Systems Technical Journal, Vol. 38, pp. 1178-1252, 1959.

4. C.H. Li, "A Sequential Method for Screening Experimental Variables," Journal American Statistical Association, Vol. 57, pp. 455-477, 1962.

5. A. Gill and D. Gottleib, "The Identification of a Set of Successive Interactions," Information and Control, Vol. 24, pp. 20-35, 1974.

6. G.T. Duncan, "Heterogeneous Questionnaire Theory," SIAM Journal of Applied Mathematics, Vol. 27, pp. 59-71, 1974.

7. M.R. Garey, "Optimal Binary Identification Procedures," SIAM Journal of Applied Mathematics, Vol. 23, pp. 173-186, 1972.

8. M. Sobel, "Group Testing to Classify Efficiently All Units in a Binomial Sample," Information and Decision Processes, edited by Robert Machol, McGraw-Hill, New York, pp. 127-161, 1960.

9. M. Sobel, "Optimal Group Testing," Colloquium on Information Theory, Bolyai Mathematics Society, Debrecen, Hungary, 1967.

10. F.K. Hwang, "On Finding a Single Defective in Binomial Group Testing," Journal of the American Statistical Association Theory and Methods Section, Vol. 69, pp. 146-150, 1974.

11. F.K. Hwang, "On Optimum Nested Procedures in Binomial Group Testing," Biometrics, Vol. 32, pp. 939-943.

12. F.K. Hwang, "A Generalized Binomial Group Testing Problem, Journal of the American Statistical Association Theory and Methods Section, Vol. 70, pp. 923-926, 1975.

13. F.K. Hwang, "Hypergeometric Group Testing Procedures and Merging Procedures," Bulletin of the Institute of Mathematics, Academy Senica, Vol. 5, pp. 335-343, 1977.

14. F.K. Hwang, "A Note on Hypergeometric Group Testing Procedures," SIAM Journal of Applied Mathematics, Vol. 34, pp. 371-375, 1978.

15. P. Ungar, "The Cutoff Point for Group Testing," Communications on Pure and Applied Mathematics, Vol. XIII, pp. 49-54, 1960.

16. F.K. Hwang, S. Lin and C.L. Mallows, "Some Realizability Theorems in Group Testing," SIAM Journal on Applied Mathematics, Vol. 37, pp. 396-400, 1979.

17. D.A. Huffman, "A Method for the Construction of Minimum Redundancy Codes," Proceedings of the Institute of Radio Engineers, Vol. 40, p. 1098, 1952.

18. C.E. Shannon, "A Mathematical Theory of Communication," Bell System Technical Journal, Vol. 27, pp. 379-423, 623-656, 1948.

19. D. Towsley and J.K. Wolf, Group Testing, Polling, and Multi-Access Communications," submitted for publication to IEEE Transactions on Communications.

20. J. Massey and J.K. Wolf, unpublished research conducted in August 1982 at the E.T.H., Zurich, Switzerland.

21. R.G. Gallager and D.C. Van Voorhis, "Optimal Source Codes for Geometrically Distributed Integer Alphabets," IEEE Trans. on Information Theory, Vol. 21, pp. 228-229, 1975.

22. M. Schwartz, Computer-Communication Network Design and Analysis, Prentice-Hall, Inc., Englewood Cliffs, N.J., 1977: Chapters 1, 12, and 13.

23. L. Kleinrock, Queueing Systems, Vol. II: Computer Applications, John Wiley & Sons, New York, 1976: Chapters 4, 5 and 6.

24. A. Tanenbaum, Computer Networks, Prentice-Hall, Inc., Englewood Cliffs, N.J., 1981: Chapters 6 and 7.

25. N. Abramson, "Packet-Switching with Satellites," AFIPS Conference Proceedings, National Computer Conference, Vol. 42, pp. 695-702, 1973.

26. R.G. Gallager, "Conflict Resolution in Random Access Broadcast Networks," Proc. AFOSR Workshop on Communication Theory and Applications, pp. 74-76, 1978.

27. N. Pippenger, "Bounds on the Performance of Protocols for a Multiple-Access Broadcast Channel," IEEE Transactions on Information Theory, Vol. IT-27, pp. 145-151, 1981.

28. J.I. Capetanakis, "Tree Algorithms for Packet Broadcast Channels," IEEE Transactions on Information Theory, Vol. IT-25, pp. 505-515, 1979.

29. J. Mosely, An Efficient Contention Resolution Algorithm for Multiple Access Channels, LIDS-TH-918, Laboratory for Information and Decision Systems, MIT, Cambridge, MA, 1979.

30. M.L. Molle, "On the Capacity of Infinite Population Multiple Access Protocols," IEEE Transactions on Information Theory, Vol. IT-28, pp. 396-401, 1982.

31. T. Berger, letter to J. Hayes, dated February 1, 1983.

32. N. Pippenger, private conversation with Toby Berger, February 1977.

33. T. Berger, proposal to National Science Foundation, 1979.

34. D. Towsley and J.K. Wolf, "An Application of Group Testing to the Design of Multi-User Access Protocols," Proceedings of Nineteenth Annual Allerton Conference on Communications, Control and Computing, Urbana, Illinois, pp. 397-403, 1981.

35. B.S. Tsybakov and V. Mikhailov, "An Upper Bound to Capacity of Random Multiple Access Systems," Probl. Peredach. Inform., Vol. 17, January, 1981.

36. R. Cruz and B. Hajek, "A New Upper Bound to the Throughput of a Multi-Access Broadcast Channel," IEEE Trans. on Information Theory, Vol. IT-28, pp. 402-405, 1982.

37. R. Rueppel, "A Unification of All Existing Upper Bounds for the Random-Access Problem," IEEE International Symposium on Information Theory, St. Jovite, Quebec, Canada, September 26-30, 1983.

NETWORK ORGANISATION, CONTROL SIGNALLING AND COMPLEXITY

Kenneth W. Cattermole

Department of Electrical Engineering Science, University of Essex, Colchester, CO4 3SQ, Essex, England.

1. INTRODUCTION

A switched communication network is a distributed processing system, in that its control organs - be they people, relay sets, computers or whatever - are numerous and are spread over several locations between which the only links are provided by telecommunication channels. The primary purpose of the network is the conveyance of information (user messages), but the primary channels must be paralleled by ancillary channels for the conveyance of control signals (operating messages). The ancillary channels may be associated with or multiplexed onto the primary channels in some way; or they may be distinct in routing, in technology or both. It is important that the ancillary channels be properly balanced in relation to the capacity and usage of the primary network. They must not dominate it, but they must be adequate to serve its needs.

The control signalling problem has two aspects, which are distinguishable but not independent: the properties of the set of signalling channels, and the procedures for which they provide a material basis. (Each of these sets can, of course, be viewed as the conjunction of several levels of protocol in a layered architecture, but the binary distinction will serve most of the needs of this paper.) In the design of communication networks, the control signalling technique has always had a strong influence, and has at times been the dominant limitation, on the procedures available and thence on the network architecture and on the services provided.

The evolution of networks over the last two decades has exhibited several trends which are largely interdependent, namely:

(a) More complex network topologies and routing procedures

(b) Increasing variety of services and facilities

(c) More powerful and versatile control processors

(d) More versatile signalling systems.

If one looks at the specification of a modern integrated-services digital network (ISDN), with its powerful processing complexes at switching centres, high-capacity common-channel signalling links between the centres, and signalling links of unprecedented versatility between terminals and switching centres, it is clear that many of the signalling limitations of the older telephone systems have been overcome. Packet-switched networks, widely used for computer communications, were designed from the beginning with flexibility in mind. It is tempting to suppose that control signalling is no longer a problem: at least in respect of the material basis, the way should be smooth, though of course the development and validation of procedures may still call for large endeavours. Indeed, much current work seems to be directed towards the development and validation of procedures, abstracted from the material basis which is assumed to be adequate.

The main purpose of this paper is to suggest that the limitations of the material basis for control signalling, though much alleviated, have not been removed: and to identify some potential problem areas.

2. TYPES OF SIGNALLING LIMITATION

There would appear to be five types of limitation, for each of which some real or potential examples can be identified. In general terms, they are as follows.

(a) Inadequate signalling repertoire. A channel designed for a specific code, such as telephonic on-hook/off-hook and decadic signalling, or 5-baud teletype characters, may be difficult or impossible to adapt for anything else.

(b) Limited disposition of signalling paths, as a permanent property. For example, a signalling channel intended primarily for operation in one direction may lack a return path: or a through channel may not be accessible at an intermediate point.

(c) Limited disposition of paths, temporarily as the result of a procedure. For example, if a control organ has a temporary function during call set-up, its signalling paths may be released after the requisite operation and - more importantly - may be impossible to re-establish.

(d) Incompatibility between two different signalling systems.
     This arises most readily if two networks operating on
     different principles are interconnected (some examples
     are given in section 5) but may also occur as the result
     of technical evolution in a nominally unified network.

(e) Limited capacity. Even a signalling system free from the
     foregoing deficiencies may suffer from overload. This
     may arise from heavy traffic of a conventional kind.
     Another problem is that certain strategies or procedures
     may demand a signalling capacity so far beyond the norm
     as to be uneconomic or infeasible.

3. SIGNALLING TECHNIQUES

     It is not the purpose of this section to give an extensive
review; control signalling is the subject of at least one book (1)
and a fairly large scattered literature. We merely list the major
techniques, in fairly broad terms.

3.1 Channel-Associated Signalling

     In telephone practice, a signalling channel may be associated
physically with a speech channel. This is natural, and probably
inevitable, in subscriber loops: it also occurs on trunk and
junction circuits in the older systems which are still numerically
predominant in most territories.

(a) D.c. signalling. Loop/disconnect signals for seize/hold/
     clear, decadic dialling etc.: also various other battery
     signalling arrangements. This has severe limitations on
     signalling rate and on transmission properties of the path:
     it can, however, convey simple indications such as 'hold'
     without mutual interference with the speech channel using
     the same circuit.

(b) A.c. outband signalling. In telephonic parlance, the
     apparently generic terms 'a.c.' and 'outband' have specific
     connotations, namely tone signals respectively below and
     above the speech frequency band, used to convey substan-
     tially the same repertoire as d.c. signalling. Transmission
     is better adapted to certain kinds of link: for our present
     purposes, the essential property is a limited signalling
     capacity, without mutual interference with the speech
     channel, and utilised on a link basis rather than end-to-end.

(c) Voice frequency signalling. More rapid signalling, with a
     wider repertoire, can be conveyed by multi-frequency signals
     in the voice band, with the obvious limitation that the
     signalling channel can be used for call set-up and clear-down

but not during conversation. This technique is commonly
used for inter-register signalling over trunks and junctions,
and (with a different format) over local loops with key
phones. It can be used link by link, or over an extended
speech path comprising transmission channels of any kind
with intermediate switches. It can be bi-directional, though
certain systems are in practice uni-directional.

(d) Radio-telephony channels. There are various inband, outband
and carrier on/off techniques whose functions and capacity
limitations are broadly similar to those listed above.

(e) Digital signalling associated with p.c.m. channels. All
pulse code modulation formats currently used for telephony
allocate some bits in the word or frame to signalling.
The standard 30-channel p.c.m. group allots a 64 kbit/s
time slot, which can be used in the common-channel mode
(see below) but can be, and quite often is, used in a channel-
associated mode. In the latter case there is a capacity
of 2 kbit/s per channel, which even with some constraints
on its utilisation is ample for all normal purposes.
Digital subscriber access to the ISDN is not yet fully
standardised: the emerging practices allow an 8 kbit/s
control channel in addition to one or more 64 kbit/s
channels. The general implication is that in digital
systems there is an adequate outband signalling capacity,
but it is to be used link-by-link rather than end-to-end.

3.2 Message-Associated Signalling

In message-switching and packet-switching systems, control
signals relating to address, routing, class of service, message
sequencing etc. are sent as headers associated with a message or
packet.

3.3 Common-Channel Signalling

Many modern telephone networks utilise common-channel signalling over trunk and junction routes, the standard systems being the
CCITT No. 6 and No. 7 (2) and the Bell CCIS. The principle common
to all these systems is that messages known as 'signalling units'
are sent by data-transmission techniques over a shared link dedicated
to control signalling. Each signalling unit comprises signalling
information denoting an action or a condition, and an indication of
the identity of the traffic circuit to which this information relates.
The signalling links must be connected to control equipment which can
handle messages in this form: this will normally be a common control
processor in a switching centre, or a message switch (signal transfer
point) which can forward the signalling unit to its destination.

There are several modes of usage for common-channel signalling systems, as follows:

(a) Associated mode: associated, that is, with a route rather
    than a channel. Signalling units are transferred over a
    direct channel between two switching centres, and they
    relate to traffic circuits on the same route: the only
    identity required is a circuit number. Similarly, in the
    quasi-associated mode, signalling units proceed over an
    indirect but fixed route via one or more signal transfer
    points.

(b) Disassociated mode. Signalling units proceed via a message
    switching network comprising common-channel links and sig-
    nal transfer points, the routing being flexible: 'identity'
    now implies full addressing information.

(c) The systems cited were originally intended to carry out
    telephonic signalling functions on trunk routes. However,
    it is clear that the principles can be applied more widely:
    compatible techniques can be used on trunks and junctions,
    between main exchanges and distributed concentrators, bet-
    ween exchanges and maintenance centres, and for internal
    message transfer within main exchanges. This is one of the
    principles of System X (3).

(d) In the most general mode of usage, a network of common-
    channel links and signal transfer points is considered
    as a general-purpose data network linking all the pro-
    cessors of its principal network and associated services.
    It carries telephonic control signals in the disassociated
    mode, but can simultaneously be used for any other purpose,
    for example, network management information having no
    specific association with calls or channels (4).

It will be clear that common-channel signalling is potentially
very versatile, and even its earliest implementations had both an
extensive repertoire and some provision for extension (on the
latter point see section 3.5 below). Total capacity limitations
may, however, be significant, especially in early systems using a
2.4 kb/s data link to support a large trunk group. The latest
systems (CCITT No. 7 and similar) are based on a 64 kbit/s link,
normally a time slot in a multiplex p.c.m. frame: these are clearly
adequate for traditional telephone signalling but could be stretched
by some potential new applications.

3.4 Broadcast Signalling

In all the foregoing techniques, the signal is typically
directed to a selected destination only, either by association with
a channel or a message or by selective addressing. There are other

systems in which signals are broadcast to a large number of stations, each of which must recognise a coded address indicating a signal for which it is the destination. Examples are as follows:

(a) Mobile radio communications, in several different forms

(b) Multiple-access satellite systems, again in several different forms

(c) Local area networks in the form of rings or busses

(d) In addition to the foregoing, which are necessarily in a broadcast mode, it will be clear that common-channel mode (d) and packet-switching techniques could be used in broadcast or multi-address modes, even though single-address is the norm.

The dominant hazards of broadcast techniques are (i) the problem of contention, for which there are diverse solutions, (ii) the sheer volume of signalling messages in a large system, which volume may well be augmented by procedures for avoiding or recovering from contention, (iii) the risk of complete failure as a response to overload.

3.5 Signalling Codes and Repertoires

Any signalling technique provides a set of code words which may be allocated to signalling conditions or messages in one of several ways.

(a) The set may be fully allocated, for example, ten signal codes may be allocated to the ten values of a decimal digit. There is then no room for extension to new applications. (In the decimal-digit example, it may be possible to use numbers with more digits: but even this may be forbidden by other features of a system, e.g. limited register capacity.)

(b) There may be some allocated code words, and some unallocated which can be either standardised at a later date, or used with ad hoc senses for any specific purpose.

(c) The basic set may be allocated, with one or more allocations using an 'escape character' indicating that subsequent code words have an altered significance. Familiar examples are the "letter" and "figure" characters of the teletype alphabet, which permit a 5-bit word structure to convey more than 32 characters.

(d) A combination of escape characters and unallocated words, for example by having a code word for escape from the standard repertoire to a user-defined repertoire, is obviously versatile. This is equivalent to variable word length with a substantial number of unallocated words.

Modern common-channel signalling systems do have this
degree of flexibility. To make full use of it clearly
requires (i) appropriate capabilities in the control
processors, (ii) the acceptance of a reduced signalling
rate, as compared with a system restricted to the basic
repertoire.

4.   SIGNALLING REQUIREMENTS

4.1 The Well-Established Cases

It is a tribute to the art and science of communication engineering that certain types of network, potentially full of problems,
are in fact well established. These are:

(a)   Public telephone networks, national and international, provi-
      ding regular telephone service by means of determinate
      routing over a basically hierarchical network. The signalling
      techniques involved are those of 3.1 and 3.3 above, and are
      described very fully by Welch (1). The principles of cer-
      tain other networks, such as public telex and various private
      networks, are very similar.

(b)   Computer networks based on packet switching and using a
      limited amount of alternate routing. Their basic control
      protocols, for which a measure of standardisation has
      emerged, are the subject of a large literature to which a
      good introduction is the book by Davies *et al* (5).

There are, however, numerous further developments in networks and
services which make greater demands on signalling systems. For the
remainder of this section, we outline the categories of problem:
Section 5 then cites some specific instances.

4.2 New Facilities on Two-Party Calls

A large category of new modes and facilities in a modern telephone system or ISDN require a basic station-to-station connection
through the network, together with additional operations marked by
class-of-service indications, priority or transfer conditions,
special numerical codes, etc. From the signalling point of view,
the prime requirement is for a repertoire of signals from terminal
to exchange, and between exchanges, which shall be significantly
beyond the capabilities of classical d.c. or a.c. telephone signalling but well within those of modern v.f. or digital signalling
systems. Thus, on new and future systems there is no major problem.

4.3 Multi-Party Calls

This category includes conference calls, and also those forms
of transfer which involve the simultaneous holding of three parties.

They require more radical changes in both connection and control architecture than the two-party calls discussed above. In modern practice, however, they are fairly readily implemented so long as all parties are on the same (new style) exchange. The difficulties lie in achieving the same result with terminals spread over several exchanges. Concentrating on signalling once more, it is necessary for signalling channels to be available after achievement of a connection (when in the usual two-party case they would be dropped). Common-channel signalling overcomes this problem in principle.

4.4 Network Topology and Routing Practices

The classical telephone trunk network is hierarchical, in the form of a tree with a branching factor averaging about 8 at each node. This has two useful properties (i) many thousands of exchanges can be connected with a maximum path length of about 7 links, (ii) there is a unique route from any exchange to any other, which simplifies path search, signalling and control functions generally. Two developments have led to a departure from this principle, in each case for good technical and economic reasons, but at the expense of complexity.

(a) Within a basically hierarchical network, an overlay of direct routes is now very common. There may then be two or more alternative routes. In present practice these are chosen sequentially at each switching centre, so that the basic control algorithm is fairly simple. However, as the alternative-routing principle spreads, the implications for network management and congestion control become more complex. Current proposals for digital trunk networks include mutual overflow, which is likely to demand more active real-time network management than has been the norm; which will require control signals associated with routes and sub-networks rather than with channels.

(b) Networks can also be planned on an entirely non-hierarchical basis. This is the case with many computer networks, military networks for tactical communications, and other special purpose networks of moderate size. It is now being seriously considered by several national PTT's as the pattern for telephone trunk networks linking perhaps 100-200 trunk exchanges (local exchanges then being dependent on one or perhaps more of the trunk exchanges). We can identify three interconnection principles:

(i) Fully-connected networks with direct paths between any pair of nodes (feasible in the small scale cases but probably not for large numbers).

(ii) Partially connected networks of irregular structure, with the choice of linkages based on geography and traffic pattern (this seems to be usual in computer networks).

(iii) Partially-connected networks with regular structure, for example, product networks (6). No large-scale network has been implemented on this principle, though the regularity properties would appear to simplify both design and operation.

Path search in large non-hierarchical networks can demand many exploratory signals per completed path, especially under congested conditions where numerous trials are made before a free path is found. The widely-ramifying effects of mutual overflow would seem to call for network management algorithms of some complexity, implying in turn a substantial management signalling load.

4.5 User-Addressed Connections

Almost all the networks considered so far are basically station-addressed: to call up a person, you have to know his telephone or other station number, and if he is not there he will not receive the message. There is now a good deal of interest in the possibilities of addressing a user by personal number, wherever he may be (with the obvious restriction that he has a terminal accessible somewhere in the network). The only systems of this type so far implemented are of the broadcast-signalling type using radio-communication, and this is the obvious approach if the user population is of moderate size and located in a fairly compact geographical area. The inte-resting question is, how far can we go in extending the scope to larger user populations and larger geographical areas?

One can envisage several different principles for the operation of such a user-addressed network. The following sequence (probably not exhaustive) places progressively fewer limitations on the mobi-lity of the addressed user, but implies progressively more explora-tory signalling around the network in order to track him down.

(a) User stipulates in advance* precisely where he will be: the current address being stored in a location addressible by his regular number. This could be rather similar in principle to "transfer calls", with the complication that

* "In advance" does not necessarily mean the day before: it implies that the location is known before a call is initiated. A travelling subscriber may, for example, report his arrival at a destination, against the possibility of future calls being directed to him.

transfers may frequently be out of area. There could also be other types of implementation.

(b)   User stipulates in advance that he will be in a certain area within which broadcast signalling is possible. This does not necessarily imply radio communication, though obviously the principle fits in well with several kinds of mobile radio-telephony system. One can also imagine a set of interlinked local area networks operated in this way. As in (a), the current area address is stored in a location addressible by the user's regular number.

(c)   User's presence in an area is reported to a control centre for that area, perhaps by action on the part of the user or maybe by some more passive means. A call directed to him initiates action to locate his current area, perhaps by broadcast signalling over an ancillary network linking area processors: his precise location within his current area being then established so that connection can be made.

(d)   User's location is known only at that location, until such time as he responds to some sort of scan or broadcast signal. It is difficult to believe that a really large network could function on this principle: probably the nearest practical approach is a version of (c) in which the presence report is accomplished without the user's initiative.

4.6   Network Interconnection

Some of the most acute signalling problems arise when it is necessary to interconnect two networks which use different techniques or principles. The following examples have in common the historical fact that at least one of the inter-connected components was designed without any thought for the interconnection problem, usually because of historical priority.

(a)   Circuit-switched telephone network and packet-switched network (the former, of course, having historical priority by about 90 years). Many similar problems carry over into the interconnection of the circuit-switched ISDN and packet-switched networks, despite their contemporaneous evolution.

(b)   National telephone networks and new competitive or complementary services established under the anti-monopoly legislation in the U.S. and the U.K.

(c)   Static networks and mobile radio-telephony.

(d)   Telephone networks and paging systems.

(e)   Local area networks and trunk communication facilities based on radically different technology and protocols (e.g. multiple-access satellite).

(f)   New services (e.g. electronic funds transfer) which might well be designed around the public telephone network or the public packet-switched network, but which for various practical reasons require access to both.

(g)   The variable-bit-rate digital network has been suggested as a generalisation of the ISDN capable of carrying a wider variety of services by reason of its ability to provide a range of channel capacities (7). If such a technique is introduced, then over a very long transitional period it will have to interconnect with just about everything we can think of, and probably several things we haven't thought of yet.

Some of these examples are discussed further in section 5. In every case, one of the components requires procedures and control signals having no counterpart in the normal operation of the other component; reliance is placed on each component having enough signalling transparency for adaptation to be possible.

5.   SOME PRACTICAL PROBLEMS

The following rather random selection shows some practical problems associated with the qualitative signalling limitations (a)-(d) of section 2.

5.1 Liberalisation of Telephone Networks

There are current problems of this kind both in the U.S. and the U.K.: we will express one of them in fairly general terms. A network operator is obliged by legislation to give a telephone subscriber A access through the existing local network B to a new competitive trunk network C, whence a call may proceed to one of many selected destinations, probably via another local network D. On initiating a call, A dials a prefix α (a few decimal digits) indicating that he wants a connection to C: followed by an area code β (several decimal digits) which will determine the routing within C, and a local number γ (several more decimal digits) denoting the final destination within D. Digits α go to the local end office in whatever form is customary: how do digits β and γ reach the appropriate control organs? Maybe B uses a d.c. signalling path which is not normally extended to the trunk network, hence cannot be connected to C. A voice circuit is extended to C, so v.f. signalling could be used - if A and B are equipped for keyphones. Probably the basic network uses v.f. inter-register signalling, so perhaps the first register normally encountered could store β and γ, and repeat

them over a v.f. path: but this new mode of operation implies
longer numbers, and there may be a limitation on the length of
number the register can store. There are as many permutations of
this problem as can be generated by the variety of signalling techniques in use; and there seems to be no completely general solution
which does not require new apparatus in several places.

5.2 Interconnections of ISDN and PSS

The foregoing example may suggest that problems are due to old-
fashioned telephone equipment, in which limitations 2(a)-(c) are
quite severe. However, there are problems with two modern network
developments - ISDN and PSS. Full interconnection of these is quite
difficult, there have been various proposals for packet assembler/
disassembler and other interface arrangements, which appears to be
useful and are now in use (on a fairly small scale). A favoured
arrangement for further development is to provide transparent access
to PSS for ISDN subscribers with full X25 protocol operation by using
the ISDN switch to make a level 1 connection and act transparently
for levels 2 and 3. Fine for an out-going call from the ISDN subscriber: but what about an incoming call? It can be accepted only
if a level 1 connection has already been established (7).

5.3 Cellular Mobile Radio

This type of service has from its inception required some quite
complex control and signalling arrangements to hand-off mobile subscribers in the course of travelling across call boundaries within
the catchment area of one switching office (8). However, the problems of hand-off when travelling across boundaries between catchment
areas are more complex again, and any solution will clearly require
a good deal of trunk network signalling which is feasible only in
the best modern trunk networks.

A further problem arises in a radio-telephony system integrated
with a fixed network. Scarce radio channels may be tied up unpro-
ductively while signalling operations are going on elsewhere: and
there has been some controversy about how far "off-air" set-up is
feasible and economic. Rival systems differ in this factor (9),
though at present without any very clear balance of advantage.

5.4 Distributed Mobile Radio Networks

The foregoing example assumes a central base controlling connection to mobile users. Interconnection of a mobile population by
means of distributed control is a different, and very difficult,
problem. Ephrimedes (10) describes a constructive approach but
remarks that the problem is so complex as to make it hard to specify
what we are trying to do.

5.5 Nationwide Extension of Telephonic Facilities

We have pointed out previously (section 4.3) that multi-party facilities such as transfer and conference calls are (a) readily available on current PABXs, (b) partially available on some modern public local exchanges, (c) still very far from being available over a multi-exchange area or a full national network. The same is true of certain 2-party facilities such as calling line identification. However, these are areas of current development where the problems do appear to be soluble.

6.   THE PROBLEM OF SIZE

6.1 Introduction

A fairly general aphorism for the communication engineer seems to be: you can do almost anything on a small scale, it is the extension to large scale which raises the worst problems. In a different but not unrelated field, the central question of algorithmic art and science is: how rapidly does the computational requirement grow with the numerical parameters of the problem? We may ask a similar question concerning control signalling within distributed networks: how fast does the signalling burden grow as the network size increases? Can we keep it in reasonable proportion to the useful message traffic?

Clearly the answer depends on the type of procedure being implemented. The major categories appear to be as follows. As we go down the list, the rate of growth with network size increases.

(a)   Procedures implemented by means of connections or virtual connections between two parties having fixed locations over a fixed routing.

(b)   As (a), for three or more parties.

(c)   Procedures implemented by means of connections or virtual connections between fixed locations but with variable routings chosen from a predefined subset of the network.

(d)   As (c) but with a wider class of variable routings, limited only by the avoidance of deadlocks and crashes.

(e)   Procedures requiring the establishment of calls or virtual calls involving user-addressing, when the user's location is incompletely defined.

(f)   As (e), with the user's location undefined.

A full analysis based on precisely defined techniques and protocols would need a very long treatment. The remainder of this section is a sketch, based on simple approximate models, to show that the asymptotic variation of signalling demand with network size does, indeed, differ for several of the categories listed above. We shall show that the number of signal occupancies of a link, for one user demand, contain in various cases factors of the order

    (i) $\log N$

    (ii) $(\log N)^2$

   (iii) $N^{\frac{1}{2}}$

    (iv) N

for N terminals.

We look first at path length in networks, then at signalling requirements for paths of given length: for each of these some substantial analysis is possible. The problem of user-addressing is then discussed more briefly.

6.2 Hierarchical Networks

The hierarchical principle is taken as a basis (sometimes much modified) in all large communication networks, such as the world-wide telephone network and all the major national or regional networks. We illustrate a pure form in Figure 1(a). The branching factor at each vertex is irregular in practical networks, but there is commonly an upper bound set by switching technology (such as the classical decadic signalling in telephone networks). Supposing this upper bound to be m, then for a hierarchy of n levels the number of nodes is bounded by

$$N \leq \sum_{i=o}^{n-1} m^i = \frac{m^n - 1}{m - 1} \tag{1a}$$

while the path length between two nodes expressed as a number of links, is bounded by

$$L \leq 2 (n - 1) \tag{1b}$$

A common alternative form, shown in Figure 1 (b), has a fully-connected set of nodes as the topmost level in the hierarchy, and a tree depending from each. In this case

$$N \leq \sum_{i=1}^{n} m^i = \frac{m (m^n - 1)}{m - 1} \tag{2a}$$

and

$$L \leq 2n - 1 \tag{2b}$$

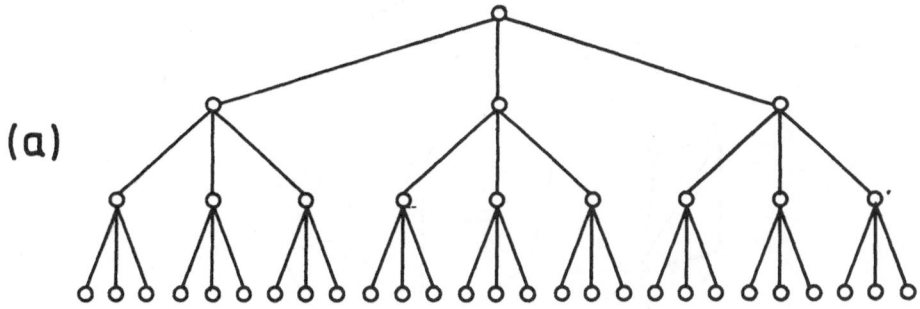

(a)

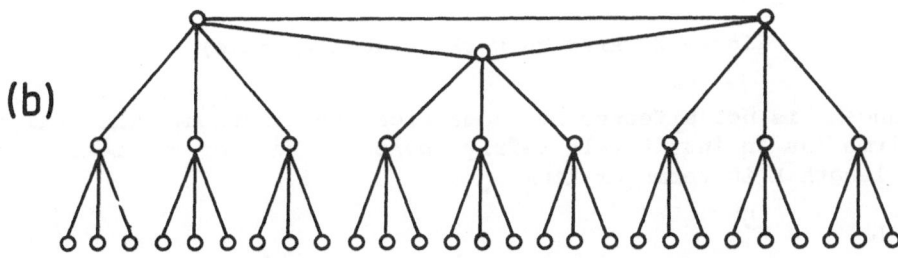

(b)

Figure 1   Hierarchical networks

It is clear in each case that in the limit L varies with log N. Specifically, from our second case

$$L > 2 \left\{ \frac{\log N - \log(\frac{m}{m-1})}{\log m} \right\} - 1 \qquad (3)$$

The hierarchical network is so widely used in telephony precisely because it is necessary, for both signalling and transmission purposes, to restrict the number of tandem links while being able to read a very large number of exchanges.

6.3 Area Networks

At the other extreme from the hierarchy is the area network. This is not well defined, but has been widely used to signify a non-hierarchical network each of whose nodes is surrounded by a catchment area of approximately similar size, each node being connected by a link to its immediate neighbours. In practice, it is likely to be somewhat irregular (as in our randomly-chosen example, Figure 2), but we can estimate some of its properties from a regular model, namely a hexagonal close packing: the topological properties of interest, such as mean path length expressed a number of links

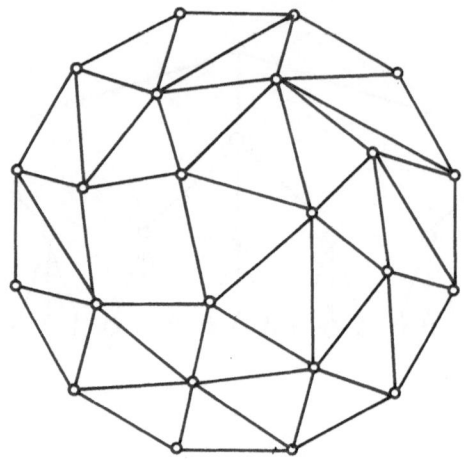

Figure 2   Area Network (random example)

in tandem, is not affected by geometrical deformation. The analysis
confirms (as in intuitively fairly obvious) that for any measure of
path length L in terms of links,

$$L \simeq k N^{\frac{1}{2}}$$

(4)

On the hexagonal analysis, for mean path length $k \simeq 0.524$ for large
N. Various trial networks (including Figure 2) have exhibited values
between 0.524 and 0.540. Maximum path length has $k \simeq 1.155$ for large
N. Various trial networks have exhibited values between 1.043 and
1.170.

6.4   A Bound on Regular Networks

Non-hierarchical, multiply-connected networks are sometimes
thought to be limited to small sizes, perhaps by induction from
the properties of area networks and other irregular structures.
This is not necessarily the case. In this section, we derive a
limiting length of the order log N for networks whose regularity is
purely one of degree. In the following section we discuss a class
of networks with a further regularity of structure.

We suppose a regular graph of degree k, and nominate one vertex
as a root. Consider all the vertices accessible from the root in
1, 2 ... L steps. (Figure 3a). There are k edges from the root to
the first level: (k - 1) edges from each vertex at the first level
to the second level: and so on. Applying this construction to an
arbitrary graph, some of these vertices might be coincident, but
clearly the number is maximised if all are distinct. Continuing to
level L on the assumption of distinct vertices, we generate a tree
whose vertices number

$$N_L = 1 + \sum_{i=1}^{L} k (k - 1)^{i - 1} = \frac{k (k - 1)^L - 2}{k - 2} \tag{5}$$

The mean length of the shortest path from the root to all other vertices is

$$M_L = \frac{L (k - 1)^L}{(k - 1)^L - 1} - \frac{1}{k - 2} \tag{6a}$$

which is bounded by

$$L - \frac{1}{k - 2} < M_L \leq L \tag{6b}$$

By hypothesis, the vertices of the tree are the vertices of a regular graph, while the edges of the tree are a subset of its edges. The remaining edges must join vertices at the topmost level. Moreover, for regularity they must join vertices whose paths to the root are disjoint. It follows that each vertex has k disjoint paths from the root: a vertex at level i has one path of length i, and k - 1 paths of length 2L + 1 - i. The mean length of all disjoint paths from the root to all other vertices is

$$P_L = (1 - \frac{1}{k}) (2L + 1) - (1 - \frac{2}{k}) M_L \tag{7a}$$

which is bounded by

$$L + 1 - \frac{1}{k} \leq P_L < L + 1 \tag{7b}$$

This analysis is based on the nomination of one vertex as a root: but clearly in a regular network any vertex can be taken as a root, and we can envisage a network in which the path length measures derived are true for all roots. Such regular graphs do not exist for all integer pairs (k, L): Cerf et al (11) give a necessary condition for what they call 'Moore graphs'. Existence or otherwise does not affect the validity of the bound, but of course it is interesting to know that the bound is sometimes attained and so is not too loose to be of value.

By inversion of equation (5) it is easy to show that

$$L \gtrsim \frac{\log N_L - \log (\frac{k}{k - 2})}{\log (k - 1)} \tag{8}$$

and the other measures are not very different, as can be seen from (6b) and (7b).

Now let us consider any regular network with N vertices, connectivity λ and at least one path of length $\leq L$ between any vertex pair. N cannot exceed the value N_L for the appropriate k, and $k \geq \lambda$. So it follows that

$$L > \frac{\log N - \log (\frac{\lambda}{\lambda - 2})}{\log (\lambda - 1)} \tag{9}$$

Further, if the mean shortest path length is M, then

$$M > \frac{\log N - \log (\frac{\lambda}{\lambda - 2})}{\log (\lambda - 1)} - \frac{1}{\lambda - 2} \tag{10}$$

It is clear that the bound cannot be attained at values of N intermediate between the N_L of equation (5). Only integer values of L have any meaning, but mean path length may be fractional; so we derive a tighter bound to M based on a generalisation of the Moore graph. Figure 3(b) shows a tree similar to Figure 3(a) save that the topmost level is incomplete. Suppose the number of vertices N satisfies $N_L \leq N \leq N_{L+1}$. The $N - 1$ vertices distinct from the roo comprise $N_L - 1$ vertices with mean shortest path M_L, and another $N - N_L$ vertices with shortest path $L + 1$. The mean shortest path fo the incomplete tree is therefore

$$M = \frac{(N_L - 1) M_L + (N - N_L) (L + 1)}{N - 1} \tag{11}$$

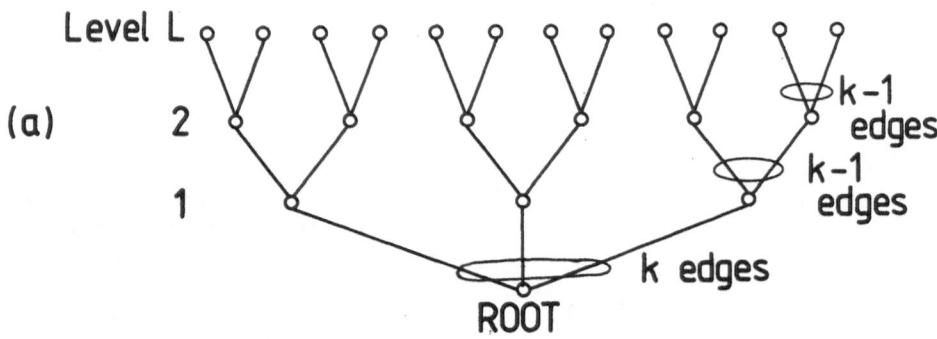

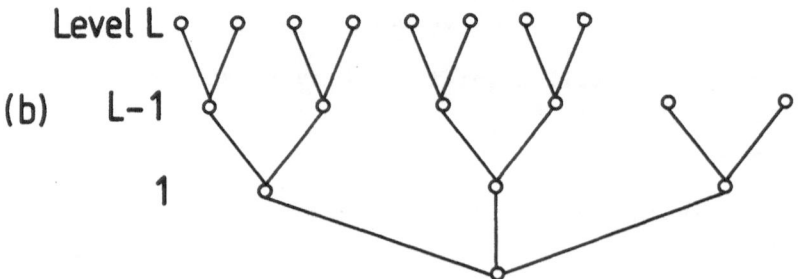

Figure 3   Tree for Network Bound

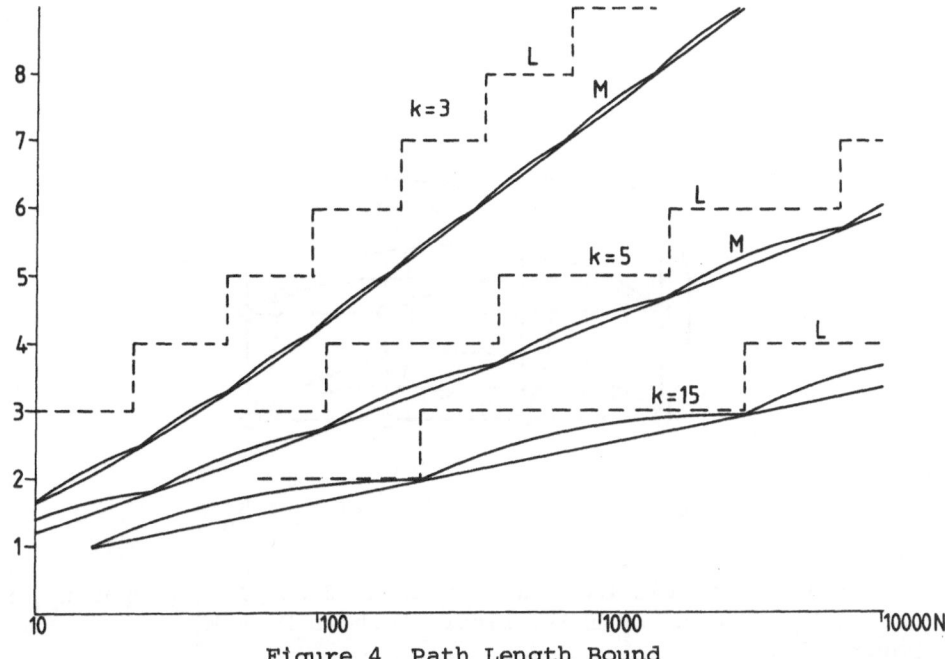

Figure 4   Path Length Bound

The bounds (10) and (11) are plotted in Figure 4, for the cases λ = 3, 5 and 15. The curves (11) are a series of hyperbolic segments falling between envelopes whose spacing depends on k but is asymptotically independent of N. The lower envelope is the bound (10): the upper envelope is separated from it by a quantity

$$\frac{1}{k-2} + \frac{1}{\log (k-1)} \left\{ \log \left[\frac{k-2}{\log (k-1)} \right] - 1 \right\} \qquad (12)$$

which is small and increases slowly with k, being 0.086 for k = 3 and 0.258 for k = 10. Our general statement, that bounds to mean shortest path length increase logarithmically with N, is confirmed by this analysis.

6.5 Product Network

The analysis of the last section gave us a bound for path length in relation to network size and connectivity: but in the absence of a construction, we cannot be sure what the routing principles and signalling requirements of a near-limiting network might be. We proceed to show that a logarithmic path length increase is in fact exhibited by a class of networks for which a general construction is available and some possible routing strategies are known.

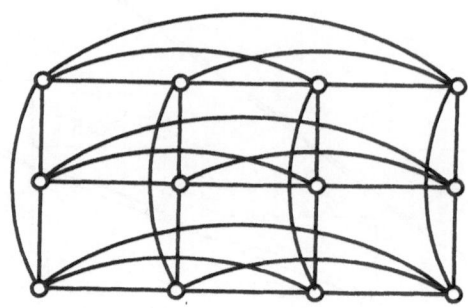

Figure 5   Product Network K_3 x K_4

Consider an m-fold Cartesian product of complete graphs K_{n1} x K_{n2} x ... K_{nm}. As an example, Figure 5 shows K_3 x K_4. The number of vertices is

$$N = \prod_{i=1}^{m} n_i \tag{13}$$

The connectivity is

$$\lambda = \sum_{i=1}^{m} (n_i - 1) \tag{14}$$

The distribution of path lengths between two vertices depends on the dimension of the smallest subgraph containing the vertex-pair, but the following statements are always true (6).

(i)   There is at least one path of length $\leq m$

(ii) There are λ disjoint paths of length $\leq m + 1$.

Several examples have shown that under random traffic the mean path length is not very far removed from m. Given these facts, we may reasonably take L = m as a measure of path length.

From (13), N is a product of factors $n_i \geq 2$ which from (14) are an L-fold partition of λ. Consequently

$$2^L \leq N \leq (1 + \frac{\lambda}{L})^L \tag{15}$$

The upper bound is attained only if λ is divisible by L. In other cases, the nearest integer solutions are given by

$$n_i = 1 + \left[\frac{\lambda + i - 1}{L}\right], \quad i = 1, 2 \dots L \tag{16}$$

where the brackets $[\ ]$ denote the integer part. Trial calculations show that these solutions correspond to a path length not far from the limit which we shall derive from (15).

The bounds in (15) may be inverted - the lower one rather obviously, the upper with a good deal of manipulation - to yield

$$\log_2 N \geq L \geq \log_2 N - 0.335 \lambda \tag{17}$$

(the numerical value 0.335 is an approximate solution to a transcendental equation).

Product networks, though they can be fairly near the fundamental bounds for small values of N, depart further as N increases: this is to be expected, since the coefficient of log N in (17) is greater than in (9) if λ > 3 (which it is for all save trivial products). They are included here, not with any suggestion of optimality, but as evidence for the logarithmic relationship in a feasible class of networks.

6.6 Path Search

The last four sections have established some properties of the path length distribution in networks. We now turn to the question of the signalling requirements imposed by a given length of path. A fundamental operation is the setting up of such a path through a switched network, and we take this as the primary problem.

The search for a free multi-link path through a network will require control signals to be passed backwards and forwards between the originating point and the various nodes at which searches for free links have to be made. We wish to estimate the number of such signals, weighted by the number of links over which they have to pass, in order to assess their contribution to the signalling traffic of the network. This number will depend on the technology and protocols: here we attempt to estimate reasonable bounds by considering minimal and maximal signalling strategies in a network where signalling is associated with links (either channel-associated, or common-channel in the associated mode).

The simplest network model is that of a pre-determined path over n tandem links, each with an independent probability that there is no free circuit. Numbering the links 1, 2 ... n from origin to destination, it is clear that a test of link i or a

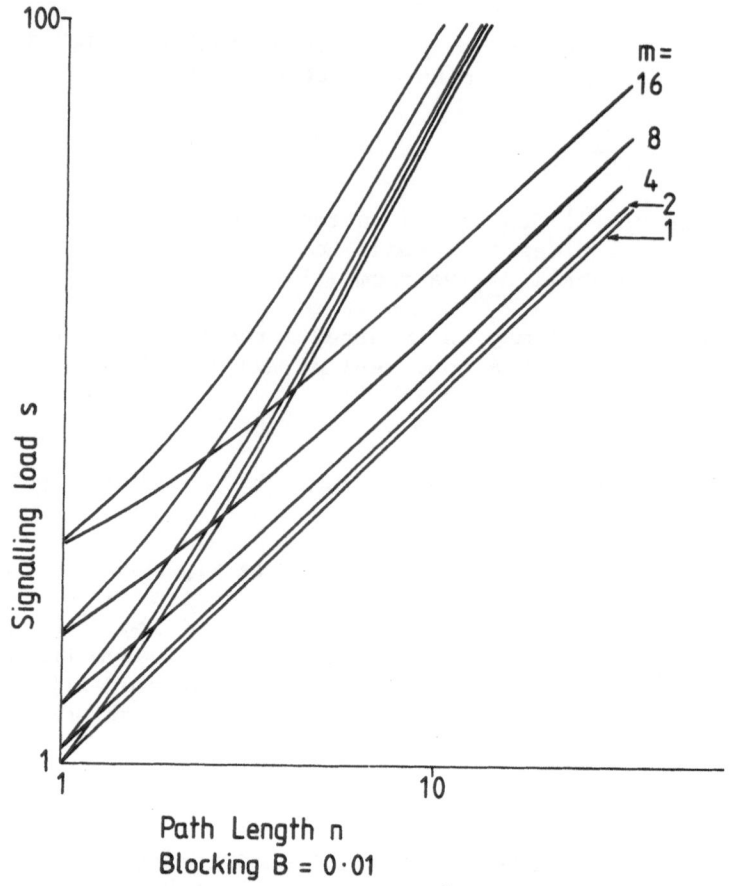

Path Length n
Blocking B = 0·01

Figure 6   Signalling Load Bounds

response from it implies a signal over links of numbers < i, for
i > 1: we will generalise this to include link 1 by postulating a
signalling path (link O), possibly though not necessarily internal
to a node, from the control processor to the access switches for
link 1.

A minimal signalling strategy will require:

(a) A forward signal over each free link in turn

(b) A return signal from the end (if a path is found) or
     from the first busy link encountered.

This assumes that forward signals can be repeated link-by-link
without intervention of the originating control processor. Some
systems make much more use of compel modes, handshakes or other
forms of acknowledgement at all stages. Allowing for this, we
define as a maximal signalling strategy

(a) for each test of a link i, a forward signal over links 0,1 ...
    (i - 1)

(b) for each such test, a return signal denoting the status of link i.

An attempt at making a connection may be successful, or it may be blocked at the ith stage (i = 1, 2 ... n). Let the links have independent blocking probabilities $p_i = 1 - q_i$, and let the weighted signalling load in each direction be s_i for a call which proceeds as far as testing link i. The overall probability of blocking on the route is

$$B = 1 - \prod_{i=1}^{n} q_i \tag{18}$$

The mean signalling load S may be calculated as follows. If the route is free up to and including the (i - 1)th link, then an additional signalling load $(s_i - s_{i-1})$ is incurred as a result of testing the ith link. Summing all such contributions with appropriate probabilities gives

$$S = \sum_{i=1}^{n} (s_i - s_{i-1}) \prod_{j=1}^{i} q_{j-1} \tag{19}$$

with the conventions $s_o = 0$, $q_o = 1$.

As an example, consider the case of equiprobable link occupancy $p_i = p = 1 - q$, with minimal and maximal signalling strategies. With the minimal strategy, $s_i = i$, and

$$S \equiv \overset{v}{S} = \sum_{i=1}^{n} q^{i-1} = \frac{1 - q^n}{1 - q} \tag{20a}$$

$$= \sum_{i=1}^{n} \binom{n}{i} (-p)^{i-1} \tag{20b}$$

$$\simeq n - \tfrac{1}{2}n (n-1) p + O (p^2) \tag{20c}$$

$$\leq n \tag{20d}$$

With the maximal strategy, $s_i = \tfrac{1}{2}i (i + 1)$ and

$$S \equiv \hat{S} = \sum_{i=1}^{n} i q^{i-1} = \frac{1 - q^n}{(1 - q)^2} - \frac{n q^n}{1 - q} \tag{21a}$$

$$= \sum_{i=1}^{n} i \binom{n+1}{i+1} (-p)^{i-1} \tag{21b}$$

$$\approx \tfrac{1}{2} n (n + 1) - \tfrac{1}{3} n (n + 1) (n - 1) p + O (p^2) \tag{21c}$$

$$\leq \tfrac{1}{2} n (n + 1) \tag{21d}$$

In each case the inequality is obvious since $S \leq s_n$. With practical values of blocking, the upper bound is a good approximation: we have developed the general expression for its use in a wider class of network models.

The next network model has an alternative routes, tested in sequential order. Let the jth route have blocking probability B_j and signalling load estimator S_j. The overall probability of blocking is (assuming independence)

$$B = \prod_{j=1}^{m} B_j \tag{22}$$

All call attempts are offered to route 1 and educe a mean contribution S_1 to the mean signalling load. Calls are offered to route $j > 1$ if and only if routes 1, 2 ... $j - 1$ are blocked, and in this event a mean contribution S_j is added. Summing all such contributions with appropriate weight gives

$$S = S_1 + \sum_{j=2}^{m} S_j \prod_{i=1}^{j-1} B_i \tag{23a}$$

$$\leq \sum_{j=1}^{m} S_j \tag{23b}$$

The upper bound is obvious, but is not a good approximation for $m > 1$.

As an example consider the case of similar alternative routes with $B_i = B_1$, $S_i = S_1$. Then

$$S = S_1 \sum_{j=1}^{m} B_1^{j-1} = S_1 \left(\frac{1 - B_1^m}{1 - B_1} \right) \tag{24a}$$

$$= S_1 \left(\frac{1 - B}{1 - B^{1/m}} \right) \tag{24b}$$

For a given overall blocking B, S increases only slowly with m. Numerical calculations show that S is doubled by increasing m from 1 to somewhere between 6 and 20, depending on the values of B and n chosen (within a realistic range $B \leq 0.05$ and $n \geq 2$). Figure 6 shows $\hat{S}$ and $\check{S}$ (the signalling loads for maximal and minimal strategies) as a function of n and m for the blocking probability 0.01.

The main aim of this analysis is to find the dependence of signalling load on path length, as a means of estimating its growth with network size.   Approximately, we can say that for path length L

$$S \geq \overset{\vee}{S} \simeq k^L \qquad\qquad (25a)$$

$$\leq \hat{S} \simeq \tfrac{1}{2} kL^2$$

where k is a factor, depending on blocking and path set,   unlikely to fall outside the range $1 \leq k \leq 2$ in practice.

6.7 User-addressed connections

The basic problem is that to locate a called party who is known only to be at one of n locations requires either

(a)   sequential polling of all locations until he is identified, which requires a maximum of n test signals each way and a mean of $\tfrac{1}{2}$n: or

(b)   some form of block testing, whereby test signals are sent to a large number of locations simultaneously.   For example, if $n = q^m$ the sequence could comprise up to $(q - 1)$ tests of blocks of q^{m-1} locations, then up to $(q - 1)$ tests of blocks of q^{m-2} locations, and so on (effectively, encoding the location address as a number to base q).   This speeds up the search, but it does not reduce the number of tests required if each is counted with its multiplicity.

It appears therefore that if a network linking N locations has user-addressing then the number of tests is of the order of N, unless some additional organising principle be introduced.

The obvious organising principles require the N locations to be partitioned into n_1 areas each with an average of n_2 locations $(N = n_1 n_2)$.   Then, as discussed in section 4.5, we have the possibilities that

(a)   A user is known to be located in a specific area, in which case $O(n_2)$ tests are needed to locate him: or

(b)   His presence in an area is known to the control centre of that area, but not to the calling party or his agent: in which case $O(n_1)$ tests are needed to locate the area, possibly followed by $O(n_2)$ tests to locate the called party.

Both the number and size of areas would probably increase with N: so the number of tests might increase as N^α where α is a fractio maybe about $\frac{1}{2}$, almost certainly in the range 1/3 - 2/3.

6.8 Size Limitations

We have shown that factors of log N or $N^{\frac{1}{2}}$ can arise from the path length of networks of differing topology: there are also intermediate orders which we have not considered, arising for example from semi-hierarchical arrangements. Such factors may enter into signalling load either linearly or squared, according to signalling protocols. User-addressing may introduce factors of the order N^α, where α is in the range 1/3 to 1.

Asymptotic behaviour might be unimportant, or even misleading, if we were only ever concerned with small values of N. In tele-communications the general trend has been towards large integrated networks. To indicate the scale of the magnitudes concerned, we tabulate a few factors

$\log_2 N$	$\dfrac{N^{\frac{1}{2}}}{\log_2 N}$	$\dfrac{N}{\log_2 N}$	N	Set of size ~ N
4	1	4	16	
6	1.3	10.7	64	
8	2	32	256	TE in Switzerland
10	3.2	102	~ 1k	
12	5.3	341	~ 4k	TE in UK
14	9.1	1170	~ 16k	TE in US
....				
20	51	~ 54k	~ 1M	TS in Switzerland
24	171	~700k	~ 17M	TS in UK
....				
27	429	~ 5M	~134M	TS in US

TE = Telephone exchanges

TS = Telephone subscribers

It is absolutely clear that some practices which are quite feasible in small networks become impossible in large networks. A regular structural principle, routing principles, a large measure of station-addressing or at least area-addressing, seem inescapable.

It is an important task to combine with these some measure of flexibility and innovation. For example, there is no doubt that alternative routing could be much more widely used than at present,
without coming up against any barriers from network size, so long
as an appropriate routing strategy is used. There is much less
certainty about the scope for wide-area user-addressing, though
this is certainly a topic which will arise as more mobile users
become linked to the main telephone networks.

7.   CONCLUDING COMMENT

There is a great deal of interest in network control algorithms,
but these depend on a material basis in the form of a set of control
signalling channels. The limitations of the latter are dominant in
some older network technologies, and by no means trivial in the
newer technologies. We must consider

(a) qualitative limitations, as discussed in sections 2 - 5
     of this paper

(b) quantitative limitations, for procedures whose demands
     increase rapidly with network size, as discussed in
     section 6

(c) the hierarchical nature of the problem: procedures piled
     upon procedures.

We illustrate this last point with one example, which suggests that
the topic is still a very open one. Hemrin and Rasmussen (4) advo-
cate a fairly general form of packet-switched network, based on the
CCITT No. 7 signalling system, as an ancillary to a telephone network or ISDN and dealing with its problems of routing, congestion
control, etc. They point out, however, that there is a meta-
problem of congestion control in the ancillary network, for which
present proposals are quite crude and rudimentary. How do we
solve this one? *Quis custodiet ipsos custodes?*

REFERENCES

(1) S. Welch: *Signalling in Telecommunications Networks*
     (Peter Peregrinus, London, 1979).

(2) CCITT. *Common Channel Signalling System No. 7: Recommenda-*
     tions Q701-707, Q721-725. (Yellow Book Vol. VI)

(3) T.W. Pritchard: Common Channel Signalling and the Message
     Transmission Sub-System. *P.O.E.E.J.* $\underline{73}$, 165-170 (1980).

(4) P. Hemrin and G. Rasmussen: An Approach to the Planning of
     Signalling Networks using CCITT Signalling System No. 7.
     I.S.S. 81, Paper 23B3.

(5) D.W. Davies, D.L.A. Barker, W.C. Price and C.M. Solomonides:
     Computer Networks and their Protocols. (Wiley, 1979).

(6) K.W. Cattermole and J.P. Sumner: Communication Networks
     Based on the Product Graph. *Proc. IEE, 124, 38-48* (1977)

(7) C.J. Hughes: Evolution of Switched Telecommunication Networks.
     ICL Tech. J, May 1983, pp. 313-329.

(8) Z.C. Fluhr and P.T. Porter: Advanced Mobile Phone Service:
     Control Architecture. *BSTJ 58, 43-69* (1979).

(9) H. Pfannschmidt: MATS E, an Advanced Cellular Radio Telephone
     System. *Philips Telecom Review, 41, 17-27* (1983).

(10) A. Ephrimedes: Distributed Protocols for Mobile Radio Net-
     works. NATO Advanced Study Institute on *The Impact of
     processing techniques on communications,* 1983.

(11) V.G. Cerf, D.D. Cowan, R.C. Mullin and R.G. Stanton:
     Computer networks and generalised Moore graphs. *Proc. 3rd
     Manitoba Conference on Numerical Math., 1973, pp. 379-398.*

DISTRIBUTED PROTOCOLS FOR MOBILE RADIO NETWORKS

Anthony Ephremides

Department of Electrical Engineering, University of Maryland, College Park, MD 20742

ABSTRACT

We examine ways of organizing mobile radio users into robust, connected networks by means of distributed protocols. We review the basic self-organization algorithms, the link activation algorithms, the inherent jamming protection that the resulting architecture provides, some routing alternatives, and, finally, an analytical technique for performance evaluation.

I. INTRODUCTION

The complexity of the design of a mobile-user (generally multi-hop) radio network makes the task of designing such a system formidable. Without sufficient theoretical support for modeling such a design problem we are often forced to make hard choices based on qualitative and uncertain guidelines. In this paper we would like to explore one case study of such a design, that is general enough to be useful and specific enough to put to a test the capabilities of the different theories and approaches that have been proposed in connection with mobile radio networks.

Suppose that we want to specify the rules of transmission and retransmission, choice of waveform, rules of error and flow control, relaying, etc. for a finite number of users operating under power and equipment constraints and moving over a large geographical area without central control or guidance and in the presence of other, possibly hostile, users. The first question to ask is what performance criteria or measures are to be used. The answer to this question shows how complex the problem is right from the beginning. There are several performance measures. We may classify them in the following way.

1) Measures of Effectiveness

a) average delay per message; this is defined as the average time from generation of a message to reception by the intended receiver. Obviously this time includes several "hop" cycles of queueing, processing, propagation, and transmission times if there is relaying in the network. Theoretical evaluation of delay is very difficult in realistic environments due to the highly complex nature of interacting queueing systems.

b) throughput; this is related to the delay but not in a well understood way. It is defined as the average rate of successfully delivered messages in the network. It is not clear that there are protocols that minimize delay and at the same time maximize throughput, as we would like them to do.

c) stability; this is difficult even to define. Contention based protocols, such as ALOHA or even CSMA, sometimes display a dynamic behavior that results in total degradation of performance, namely a reduction of the throughput to the lowest levels and a simultaneous increase of the delay to intolerable levels. Such behavior is called unstable and is obviously unacceptable.

2) Measures of Efficiency

This relates vaguely to the essence of the entire field of networking, namely the shared use of resources. How efficiently is the bandwidth utilized? How often are there wasted time (or frequency) slots due to collisions or due to idleness of users to whom they are dedicated? This is a subtle concept which is related to the preceding criteria.

3) Measures of Survivability

a) robustness; this requires the maintenance of satisfactory performance over a wide range of values of the critical parameters of the network such as traffic statistics, number of users, topological layout etc. Often robustness cannot be achieved without adaptability.

b) adaptability; when the critical parameters of the network are time varying it is desirable or, sometimes, imperative to adapt the protocol to the changes in order to maintain satisfactory performance. Robustness and adaptability together constitute ingredients of survivability and graceful degradation. The latter also requires fail-safety.

c) <u>fail-safety</u>; when some nodes or links fail due to various reasons, the ability of the network to operate must not be totally impaired. This is perhaps the most fundamental requirement for certain networks. It is also an overriding criterion, since it may be incompatible with minimizing delay, maximizing throughput, and maintaining high efficiency. It requires that the protocol rules are such that no deadlocks due to errors or data base inconsistencies occur and that no node or link is indispensable. The latter requirement often dictates a distributed control architecture (ruling out, for example, polling schemes) and at the same time it implies the possibility of inconsistencies, instabilities, and deadlocks.

II. THE BASIC ORGANIZATION ALGORITHMS

The awesome complexity of the total problem suggests a blend of rigor and "wisdom" in the approach or a mixture of science and "art" or, plainly, an engineering solution. Details of the total problem can be found in a number of publications [1-5]. Here we present a summary of the problem. Let us start with an envisioned architecture as in Figure 1. As a first step, we would like to invent a <u>distributed</u> algorithm that will allow the nodes to connect themselves as shown in the figure. We consider two procedures (very similar to each other). We describe them first in a fictitious centralized version and then in the implementable, distributed one. Note that three "types" of nodes are shown in the figure. The first type, denoted by a square and situated in the center of the circle, we call <u>clusterhead</u>. The second type, denoted by a triangle, we call <u>gateway</u>. The third type denoted by a dot, we call <u>ordinary</u>. The meaning of these identities and

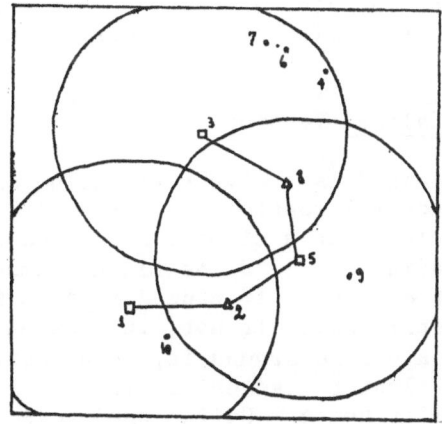

Figure 1

their role in the operation of the network is not important for the moment and will be discussed later.

At this point it is important to realize that any node is equipped to acquire any one of these three identities (unlike the case of SRI's packet radio where some nodes were designated as stations a priori and possessed special equipment). Here, the very essence of the approach is to develop a distributed algorithm that will enable the nodes to decide on their status unambiguously and with consistency.

1st Method (Centralized Version)

This method produces the node clusters shown in Figure 2. The nodes are assumed numbered from 1 to N. The fictitious central controller ("genie") starts with the highest numbered node, say node N, and declares it a cluster head. Then it draws a circle around that node N with radius equal to the range of communication. If propagation range is not constant in all directions an appropriate contour is drawn instead. The nodes inside the circle form the first cluster. It then considers whether there are nodes outside this circle. If there are, it tentatively considers drawing a circle about node N-1. Should any nodes lie within this circle that were not already within the first circle, node N-1 becomes a cluster head and a circle is drawn about it. Then consideration of tentative cluster head status for nodes N-2, N-3, etc. follows, until all nodes lie within at least one circle. The resulting arrangement provides every node with a cluster head, provided the group of nodes is not inherently disconnected. Any pair of clusters may be directly linked, they may even cover one another, they may simply overlap, or they may be disconnected. In the last two cases, selected nodes must serve as gateways for the interconnection of the cluster heads. This issue will be addressed in the discussion of the distributed version of the algorithm.

2nd Method (Centralized Version)

The alternative method is a slight variation of the one just described. The central controller starts with the lowest numbered node, node 1, declares it a cluster head, and draws a circle around it with radius equal to the fixed communication range, thus forming the first cluster. If node 2 lies in this circle it does not become a cluster head. If not, it does become a head and the controller draws a circle around it. Continuing in this manner; node i becomes a cluster head unless it lies in one of the circles drawn around earlier nodes. Unlike the previously desribed case, here no cluster can cover another nor can two clusterheads be directly linked. To facilitate comparisons the nodes are numbered

in reverse order from that shown for the 1st method. Thus, nodes
1,2,3, etc. of the 1st method are nodes N,N-1,N-2, etc. of the
corresponding 2nd method in Figure 3.

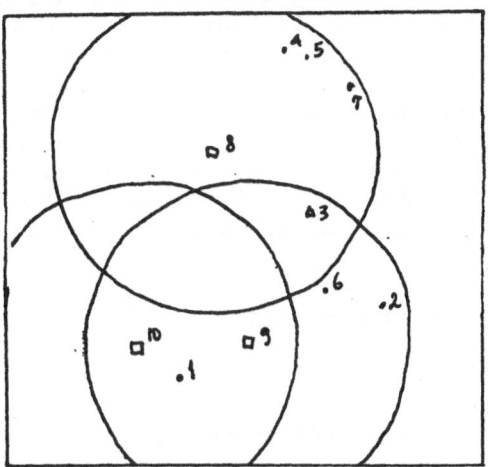

Figure 2

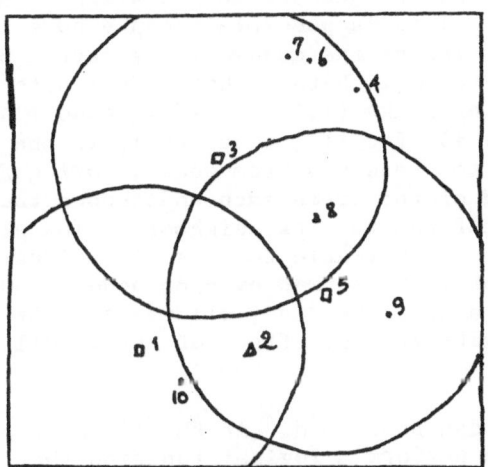

Figure 3

We now describe the distributed versions of these algorithms, the
details of which can be found in [2].

1st Method (Distributed Version)

The algorithm has two logical stages: first, the formation of
clusters and second, the linking of the clusters. It also has a
communication stage in which probing messages are exchanged among

neighbors, and a computation stage in which the logical stages are performed based on the local information acquired during the communication stage. Assuming initially some level of synchronization among the N participating nodes we consider two TDMA frames each consisting of N slots. In frame 1, in its assigned ith slot, node i broadcasts the identitites of the nodes it has heard from during the earlier slots of this frame. (Thus node i also receives partial connectivity information of the nodes that it can hear). So by the end of this frame, node i has filled in some of the entries of its connectivity matrix*. In particular, it can fill in the elements above the main diagonal, i.e., the elements (i,j) of the ith row that satisfy $j>i$. The element (i,j) is set equal to 1 if:

a)   node i heard from node j

and

b)   node i appears in the connectivity list broadcast by node j.

In frame 2, each node broadcasts in its assigned slot its full connectivity row. This is possible because node i has completely filled in the ith row of the connectivity matrix by the time of the ith slot of Frame 2. Here is how node i determines the bidirectionality of links (i,j) for $j<i$. Node i sets connectivity matrix element (i,j), for $j<i$, equal to 1, if the ith elements of the connectivity row received from node j during Frame 2 is equal to 1. By the end of the frame each node knows the two-way connectivities for itself and for its neighbors. The global connectivity matrix is not available to every individual node – only a partial version of it is formed by each node. However, for the case of error-free transmissions, all versions are consistent with the global true matrix. The effect of errors will be considered later.

Now the clusters can be formed. At the ith slot of the second frame, node i can perform a logical function that permits it to determine whether it is a head. It can then transmit one extra bit along with its row connectivity, if it decides to become a clusterhead. The node with the highest identy number among a group of nodes is the first candidate to claim clusterhead status. Thus node i first checks its own connectivity row. If there is no neighbor with higher identity number, node i becomes a cluster head. If another neighbor exists with higher identity number, that neighbor will become a cluster head, so i doesn't have to.

_____

* A connectivity matrix is an NxN matrix with ones in the (i,j) entries for which i and j are neighbors and zeroes everywhere else.

However i must also check whether it is the "highest" neighbor of some other node j<i. This can be done by checking the received connectivity rows from the lower numbered neighbors. If node i is the highest in some row j<i, node i must become a cluster head for at least node j. Thus node i is able to broadcast in the ith slot of the 2nd frame his status. This information is needed for the linking of the clusters as will be seen later.

At the end of the second frame each node knows all head nodes that are one hop away and some, but perhaps not all, heads that are two hops away. Thus, by the end of the second frame, clusters have been formed and the data base necessary for the second logical function of the algorithm (the linking of the clusters) is available. The linking is accomplished by the introduction of gateway nodes. Every non-head node is a candidate to become a gateway. There are three cases to be considered. The first case is shown in Figure 4a. Here there is no need for gateways since the heads of the clusters are directly linked. The second case is depicted in Figure 4b. Here exactly one node is needed to link up the two heads. Clearly the candidates are the nodes in the inter-section of the two cluster regions. The third case is pictured in Figure 4c. Here at least two nodes (one from each cluster) are needed. It is, of course, assumed that suitable such nodes exist; otherwise the net cannot be connected. In the sequel we describe the procedures used to achieve the link-up in the last two cases. Every node, which is not a cluster head, is a candidate gateway node. Each pair of heads in a node's list of heads that are one hop away corresponds to a pair of overlapping clusters. To avoid the formation of unneeded gateways, candidate gateway nodes first

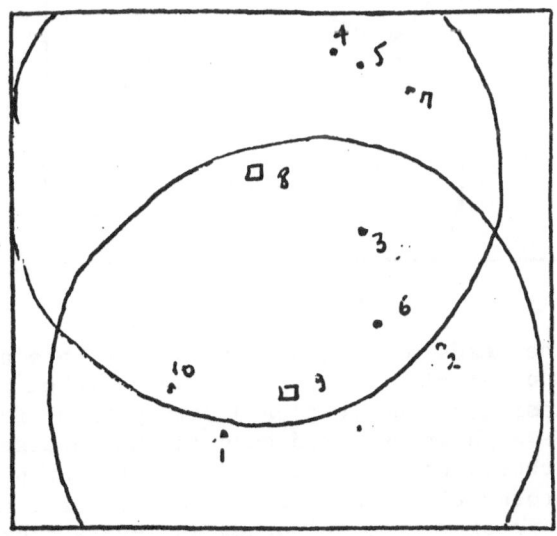

Figure 4a

294

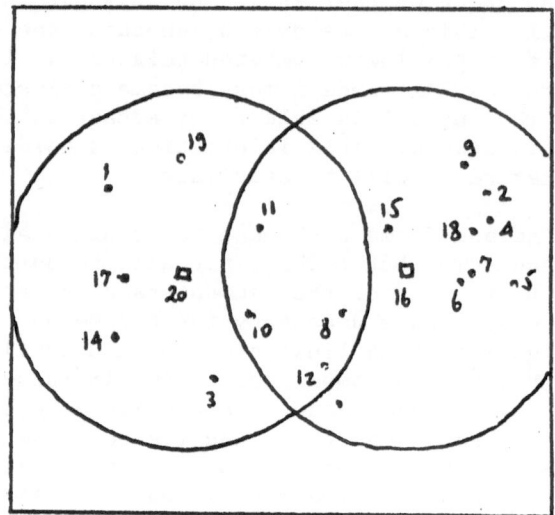

Figure 4b

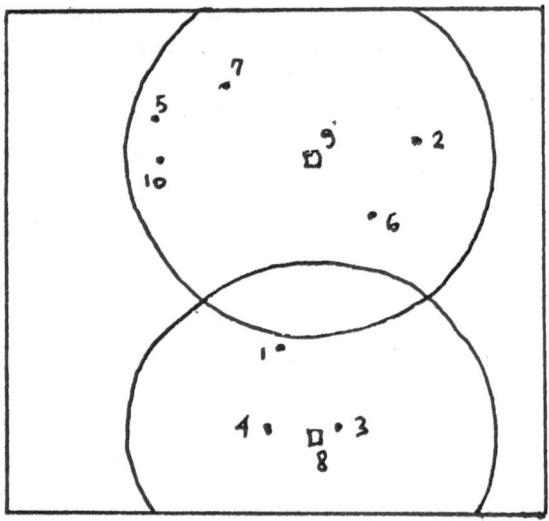

Figure 4c

test to see if the heads of overlapping clusters are already
linked through another cluster head. If an unlinked pair of heads
is found, the examining node is a candidate gateway for linking
these heads. The highest numbered node in the intersection of the
two clusters is chosen to become a gateway for that pair. All
nodes in this intersection are aware of each other since each can
be at most two hops away from the other and every node possessses
the connectivity row information for every one of its neighbors.

Thus there is no ambiguity in the selection of the gateway node.
Again error-free transmissions are assumed.

Case of non-overlapping clusters: Cluster head pairs consisting of
a node's own head plus a head from the node's list of heads that
are two hops away identify non-overlapping clusters. For linking
up two clusters that do not overlap, at least one node from each
cluster must become a gateway. Each node examines every possible
pair of nodes, the first member of which is its own cluster head
and the second member of which is a cluster head in its own list
of heads that are two hops away. To avoid creation of redundant
gateways the node attempts to ascertain the need for the creation
of a gateway by checking, for each such pair, whether a path may
be created through another cluster head. Thus it seeks nodes
among its neighbors that may include in their connectivity rows
the second member of the head pair it is examining, together with a
head from its list of nodes one hop away. If no such circumstance
is established however, the node proceeds to assume that its services are needed for this linkage. There may be several pairs of
potential gateway nodes that can link two clusters. Each node may
be aware of several of those, but perhaps not all of them. The
(arbitrary) deterministic rule chosen for resolving the ambiguity
is to select the pair with the largest sum of identity numbers.
In case of a tie the pair involving the node with the highest
number is chosen. Unlike the previous case here we may end up
with extra gateways with two or perhaps more pairs becoming
gateways. It is worth noting however that such multiple linkage
outcomes are not very likely for most topological configurations.
In some cases only one of the two potential gateway nodes in a
pair may decide to become a gateway while the other may find that
it is not needed if another pair of higher numbered nodes is
available and known to it. The existence of such a pair may not
be known to both partners of the first pair, and thus asymmetric
situations can arise. Such outcomes, however, are rare and only a
harmless nuisance. They need not affect the network's operation
and cannot be avoided without substantially increasing the data
bases available at each node by additional message exhanges.

2nd Method (Distributed Version)

The two methods have nearly identical implementations. Both use
the same data structures, and both follow the same transmission
schedule described earlier. The only difference is that the rules
for forming the clusters and for assigning cluster heads are different. The difference has already been described in the centralized version. In the distributed implementation, the rule is
simply that a node becomes a cluster head unless it has a lower
numbered cluster head for a neighbor. Thus instead of announcing
its node status, each node broadcasts the identity of its own

cluster head during Frame 2 transmissions. This enables nodes to fill in both their lists of heads one and two hops away.

III. THE ACTIVATION OF THE LINKS

Once the clustering has been formed the "potential" links between neighbors must be activated in a coordinated but distributed way.

Essentially, each node attempts to make an arbitrary assignment of slots to its neighbors and thus form a TDMA schedule for its communication with them. Obviously, each neighbor will attempt to do likewise, but as it may have a different number of neighbors than the original node and as these neighbors set up their own schedules also, inconsistencies will in general arise that must be resolved in order to arrive at consistent, conflict-free assignments. A systematic method in which these schedules are set up and the conflicts resolved is needed.

The method consists of two activities, namely, allocating slots and resolving scheduling conflicts. Slot allocating occurs during frame 1 and conflict resolving occurs in frame 2. These frames are the ones corresponding to the previous algorithm. Broadcasts of information relevant to the activation of links take place during the same slots that information relating to the previous algorithm is sent. Alternative methods have also been examined but seem to have serious disadvantages [6]. We briefly describe how the activation proceeds. The first slot of each node's own TDMA schedule is arbitrarily selected as a broadcast slot*. Other time slots are allocated for the activation of specific links. If the link between a pair of nodes is bidirectional, the highest numbered of the two nodes is responsible for allocating the slot. If a link is unidirectional, the receiving node must allocate the slot. When a node allocates a slot, it chooses the earliest slot available for that link. At the end of frame 1 every link has been allocated one slot.

During frame 1 each node broadcasts, in turn, its current link activation schedule, which will contain only those nodes heard from, thus far, during frame 1. As other nodes receive the broadcasts of these schedules, they either allocate a slot, if this is a unidirectional link, or simply update their own schedule, if this link is bidirectional. An update involves a possible change in one's own schedule to make it consistent with information just received.

---

*There are special potential uses of such a slot. A node can monitor its own clusterhead during that slot and/or can access it via contention methods.

At the end of frame 1, each node has its own version of the link activation schedule. During frame 2 the conflicts that may exist can be resolved following a simple procedure described in 4 . The results of the use of this activation procedure for a specific topology are shown in Figures 5 and 6.

Thus we have some means of connecting nodes that were previously disconnected and of providing a backbone structure for their communication. The method is distributed and thus relatively secure. Many issues remain of course to be addressed in order to fully describe the operation of the network. The most important step however has been taken. The backbone structure can be utilized in a variety of ways in order to disseminate control information that will allow the implementation of any set of desired rules of permanent operation.

In a highly volatile communication environment in which established connectivities are changing relatively rapidly the described procedure must be repeated periodically. There is even an asynchronous way of performing such updates [7]. In the sequel we would like to discuss the issues of routing, inherent jamming protection, and analytical tools for performance evaluation.

IV. ROUTING

There are two extreme approaches to routing in a mobile user radio network. One is flooding and the other is some form of path routing for known topology. Flooding in a radio environment involves in general many redundant transmissions which can be costly in terms of bandwidth expenditure and/or delays due to contention. However, if connectivities change too rapidly there may be no other alternative. On the other hand if connectivities are stable over long periods of time it is best to exchange sufficient information to allow each node to learn the full topology of the network and implement some sort of "best path" routing, as is done for non-radio networks. It is of course the intermediate case that presents an interesting challenge, namely when the time scale of topological changes is not fast enough to warrant flooding nor slow enough to permit ordinary path routing.

We have considered two alternatives for the intermediate case. Both are based on the concept of "flooding" short query messages that search for and locate the destination node, followed by transmission of the main message over a "discovered" path. This concept is related to the one used by Merlin and Segall [8] and Segall and Aurebuch [9] for failsafe routing and reliable broadcasting respectively.

The basic idea is the following:

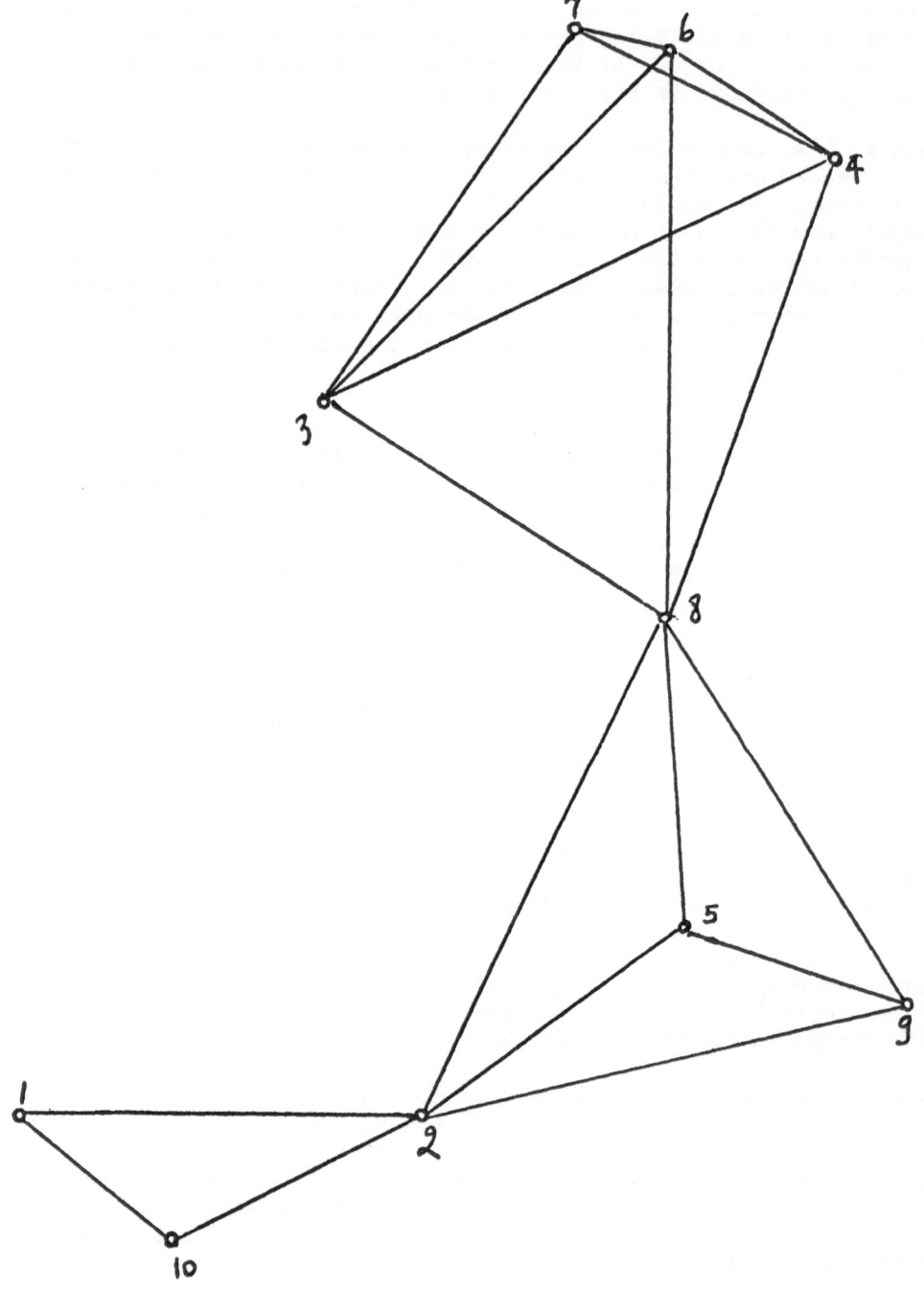

Figure 5

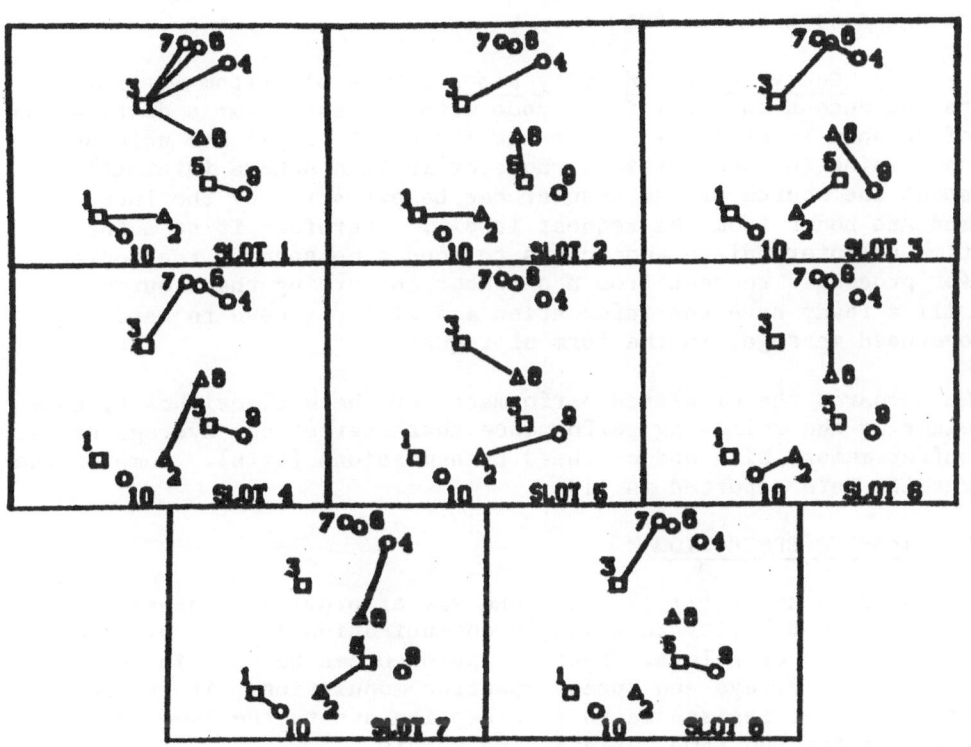

Figure 6

1) Query/source identity unknown - In this algorithm a node
with a message for a destination the location of which is unknown,
will send out a much shorter message to all of his neighbors
requesting routing information about the ultimate destination.
The request will propagate through the network until it reaches a
node that knows a route to the destination. At this point the
request is converted to a reply and is sent to every neighbor who
requested information about the destination. As the reply is
passed through the network each node updates and uses the information about the destination, thus decreasing the number of
requests sent in the future.

2) Query/source identity known - This algorithm is similar
to the second one in which a node with a message for a destination
of unknown location will send out a request to all his neighbors
concerning the destination. However in this scheme information
about the source of the request can be extracted by the intermediate nodes from the request itself. Therefore if at a future
time an intermediate node needs to send a message to the source
(or process a request from a neighbor concerning that source) it
will already have the information and will not have to send more
overhead messages in the form of requests.

We compared the simulated performance of these algorithms by considering the following performance characteristics: Average delay,
buffer memory size and overhead transmissions (bits). Some of the
results were reported in [10].

V. JAMMING PROTECTION

It has been known [11,12] that one way of providing interference
rejection capability to a single communication link is via the
introduction of relays. Such an approach can be used in addition
to adaptive arrays and spread spectrum modulation. It provides
additionally, resistance to interception due to the lower power
required for the same anti-jamming margin.

A network has an inherent capability of providing strong
resistance to threats due to its natural structure that includes
multiple relays. We can demonstrate that capability for the network architecture considered in the earlier sections.

Consider the case of a network as shown in Figure 5. Assume that
all nodes use the algorithm described earlier to organize themselves into overlapping clusters.

In the presence of a jamming source the resulting architecture
"discovers" the right nodes that can serve as relays. The selection or relays is done automatically and without the need for a

central controller. In addition, the network structure is con-
tinually self-adapting in response to a changing jamming environment (moving jammer).

To illustrate this self-organizing capability, we used a digital
computer simulation model. In the simulator, node connectivities
are computed based on some fixed propagation model [5]. The radio
connectivities shown in Figure 5 are for the case when no jammers
are present.

In the absence of jamming, the network self-organizes into the set
of node clusters shown in Figure 7a. Three clusters are formed
with heads at nodes 1, 3, and 5. Nodes 2 and 8 have become
gateways (relays) to join together the cluster heads, forming the
"backbone" network.

The sequence of frames or Figure 7 illustrates how the network re-
structures itself in response to a changing jamming threat.
Figure 7b illustrates how the network re-structures as the jammer
prevents node 7 from hearing node 3 and prevents node 6 from
hearing either 3 or 8. Consequently, nodes 6 and 7 are no longer
bidirectionally connected to cluster head 3 as they were in the
case of Figure 7a. The network responds by forming an additional
cluster head at node 6; also, node 4 becomes a gateway to link
this new head to the rest or the backbone network.

As the jammer gets nearer to nodes 4,6, and 7, the link from 3 to
4 is lost resulting in node 4 becoming a cluster head, (see Figure
7c). This in turn causes 7 to become a head and 6 to become a
gateway.

Approaching node 3, (Figure 7d), the jammer is successful in iso-
lating this node, however, the rest of the network remains
connected.

In Figure 7e, the jammer is shown near the "critical" node 8.
This is a critical node in the sense that its loss will split the
network, as can be seen by examining the connectivities shown in
Figure 7a. That this splitting does in fact occur is shown in
Figure 7e.

When the jammer moves away from nodes 3 and 8 to the position
shown in Figure 7f, the network is able to return to the connected
state. However, as the jammer approaches 2, the ability to use
this node for relaying is lost and the network becomes disconnected again. As illustrated in Figure 7h, the network is able to
recover to the connected state after a moderate separation between
the jammer and node 2 occurs. Finally, with no jammer present,
the network returns to its original configuration.

302

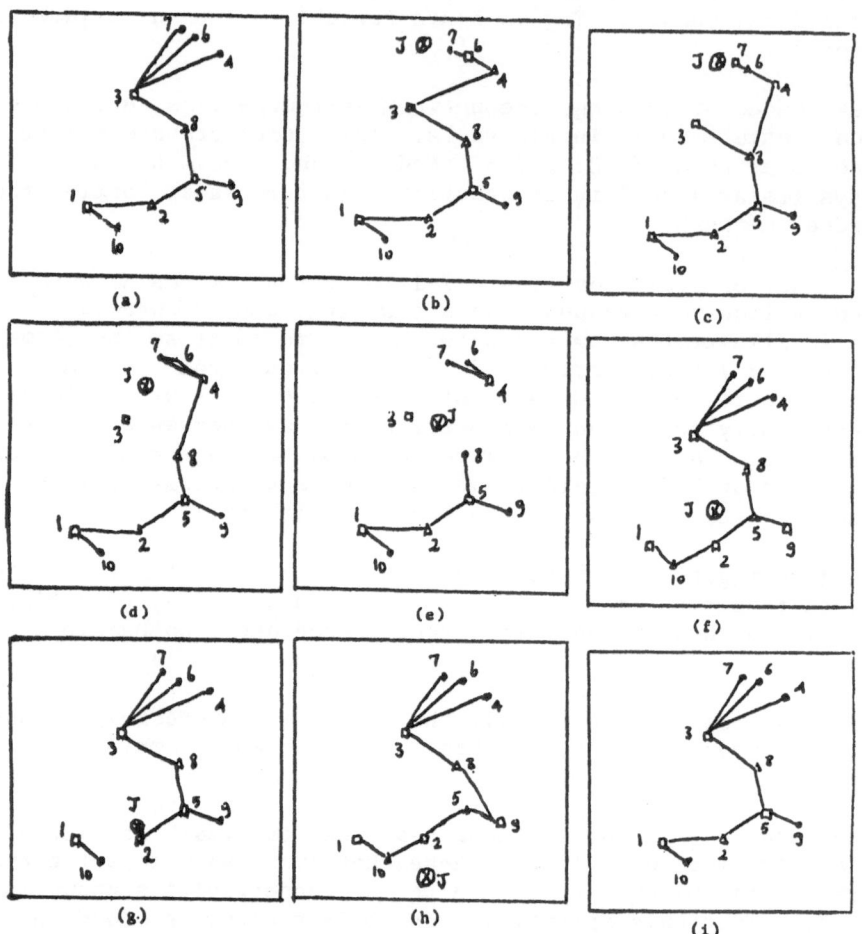

Figure 7

Thus we see that the basic algorithms described earlier allows the set of nodes to form an organized, connected network, without the use of any central controller or coordinator. Furthermore they allow the nodes to activate their "discovered" links in a distributed manner and maintain consistency in this activation. Finally we saw how by using the principle of relaying these same algorithms offer increased protection to interference by allowing the network to reconfigure itself in an adaptive, automatic fashion.

VI. ANALYTICAL TECHNIQUES

The analytical evaluation of performance in a multi-hop radio environment is difficult, if not impossible, even in the presence of many simplifying assumptions. Partially successful results in throughput evaluation have been recently obtained by Boorstyn et al [13] and by Tobagi et al [14]. A phenomenon that cannot be ignored in multi-hop radio is that of capture. Due to a variety of reasons (spread-spectrum signaling, transmitter power variations, differential propagation delays, etc.), it is possible that among several simultaneously transmitted messages one is successfully received. This is known as capture. We have developed an analytical (combinatorial) approach to the study of the capture effect that may prove useful to the overall analysis of multi-hop radio networks. This approach has already been used in some preliminary analysis attempts [15, 16]. We present here briefly the basic idea.

Suppose that in an asynchronous system a fixed length message is successfully transmitted if the duration of the time overlap with any number of other messages is less than some fraction ρ of its own time length. Although this model is not quite "capturing" the capture phenomenon it does represent realistically the case of frequency-hopped M-ary FSK modulation in which single tones are transmitted during each hop. Assuming that m other users transmit at the same time and n of them transmit independently in the same frequency bin one of M tones and assuming a time off-set in synchronization uniformly distributed over one hop duration we find that the conditional probability of success for a given tone is given by

$$P_m = \sum_{n=0}^{m} \binom{m}{n} \left(\frac{2}{q}\right)^n \left(1-\frac{2}{q}\right)^{m-n} Q_n$$

where q = number of frequency bins

Q_n = Probability of a given tone succeeding given that n of the m other users transimit in the same frequency bin = $\Sigma P[n_1,\ldots,n_n] \cdot Pr[success/n_1,\ldots,n_n]$

where

$$P[n_1,\ldots,n_n] = \frac{n!}{\prod_{j=1}^{n} (n_j)!(j!)^{n_j}} \cdot \frac{M!}{(M - \sum_{j=1}^{n} n_j)!(M)^n}$$

and

$$Pr[success/n_1,\ldots,n_n] = \prod_{j=1}^{n} (j+1)(\frac{\rho}{2})^j]^{n_j}$$

where

n_j = number of tones each of which is shared by j users = number of distinct bunches of j users sharing same tone

and

$$\sum_{i=1}^{n} in_i = n$$

Next we can calculate the probability of success of a packet depending on what assumptions we make on the frequency hopping pattern and on the error correcting code used. Details can be found in forthcoming publications [16,17].

A different technique described in [18] can be used to predict throughput performance for the case of no feedback random access transmission. Such situations can arise when ACK messages cannot be transmitted due to receiver radio silence requirements or cannot be received due to jamming conditions at the transmitter, and one-way connectivities.

VII. CONCLUSIONS

We have reviewed some of the problems in designing and evaluating the performance of a mobile radio network. We have shown that the proposed solutions have merit in that they seem to achieve several of the design objectives. Emphasis is placed on distributed techniques that provide survivability features and that take advantage of the speed and efficiency of modern processing techniques.

305

REFERENCES

1. J.E. Wieselthier, D.J. Baker, A. Ephremides, "Survey of Problems in the Design of an HF Intra Task Force Communication Network," NRL Report 8501, October 1981.

2. D.J. Baker, A. Ephremides, "The Architectural Organization of a Mobile Radio Network via a Distributed Algorithm," IEEE Trans. Communications, Vol. 29, pp. 1694-1701, November 1981.

3. D.J. Baker, A. Ephremides, J. Wieselthier, "An Architecture for the HF Intra-Task Force (ITF) Communication Network," NRL Report 8638, December 1982.

4. D.J. Baker, A. Ephremides, "Distributed Algorithm for the Activation of Radio Links in a Mobile User Network," Proc. of ICC, Philadelphia, June 1982.

5. D.J. Baker, J.E. Wieselthier, A. Ephremides, D.N. McGregor, "Distributed Network Reconfiguration in Response to Jamming at HF," MILCOM Proc., Boston, October 1982.

6. D.J. Baker, A. Ephremides, J.E. Wieselthier, "Distribute Assignment of Links in Mobile Radio Networks," 19th Allerton Conference Proc., October 1981.

7. D.J. Baker, "Distributed Control of Broadcast Radio Networks," INFOCOM Proc., San Diego, April 1983.

8. P. Merlin, A. Segall, "A Failsafe Distributed Routing Protocol," IEEE Trans. on Communications, Vol. 27, pp. 1280-1287, September 1979.

9. A. Segall, B. Awerbuch, "A Reliable Broadcast Protocol," INFOCOM Proc., San Diego, April 1983.

10. M.L. Weber, A. Ephremides, "A Simulated Performance Study of Some Distributed Routing Algorithms for Mobile Radio Networks," Proc. of 1983 Conference on Information Sciences and Systems, Johns Hopkins, March 1983.

11. C.E. Cook, "Optimum Deployment of Communications Relays in an Interference Environment," IEEE Trans. on Communications, Vol. 28, pp. 1608-1615, September 1980.

12. C.E. Cook, "Relay-Augmented Data Links in an Interference Environment," Proc. of NTC, November 1980.

13. B. Maglaris, R. Boorstyn, A. Kershenbaum, "Extensions to the

Analysis of Multihop Packet Radio Networks," <u>INFOCOM</u> <u>Proc.</u>, San Diego, April 1983.

14. F.A. Tobagi, J.M. Brazio, "Throughput Analysis of Multihop Packet Radio Networks Under Various Channel Access Schemes," <u>INFOCOM</u> <u>Proc.</u>, San Diego, April 1983.

15. J.E. Wieselthier, A. Ephremides, "A Scheme to Increase Throughput in Frequency Hopping Multiple Access Channels," <u>Proc. of 1983 Conference on Information Sciences and Systems</u>, Johns Hopkins, March 1983.

16. J.E. Wieselthier, A. Ephremides, "A Distributed Reservation Scheme for Spread Spectrum Multiple Access Channels," <u>Proc. of GLOBECOM</u>, November 1983 (submitted).

17. J.E. Wieselthier, A. Ephremides, "A Combinatorial Technique for the Analysis of Framed Contention-Based Multiple Access Protocols," (in preparation).

18. A. Ephremides, "Random Access without Feedback," (in preparation).

SEAL FUNCTION IN TRANSMISSION AND STORAGE

S. HARARI

Unité de Recherche et d'Enseignement de Mathématiques
Université de Picardie
33, rue St Leu
80039 - Amiens Cedex - France -

I. - Introduction.

When important data is written on paper some precautions are taken to ensure authenticity. Special paper is used to detect forgery, and a seal is put at the end.

Visual inspection before using the document gives a proof of its authenticity and the absence of forgery. In case of alteration, the document becomes useless.

A similar procedure would be interesting for the same kind of data, digitized and stored in data banks ; or transmitted on unprotected channels. More specifically one would like to determine a digital quantity, called seal, which would have the following properties.

1) The seal should be very short. Ideally for a given text or a given tape, one would like to have to compute one seal. Corollary to that condition is that the seal computation should be a very fast operation.

2) The seal should be unforgeable. This means that given a set of data, the probability of computing an illegitimate seal should be as low as possible.

3) The authenticity of the seal must be able to be checked by any one.

This condition implies that the seal algorithm should use only simple operations used on general purpose computers, and should not rely on any specific device of any specific computer.

4) The seal should be "transparent" : Data should be able to be processed independantly of seal authentication. One must allow for off line seal checking.

One immediately sees that the usual techniques used in data processing do not satisfy at least one of the conditions. More specifically error detecting or correcting codes are inadequate because they do not satisfy conditions 1) and 2) simultaneously. In any case condition 2) is not met by the general codes used : they are linear, and therefore given a document and its seal, it is quite easy to find another document having the same seal.

II. - Possible solutions.

Another method one may think of is cryptography. If a good cipher is used then the objective of unforgeability is attained, but the tansparency condition is not met, and the ease of implementation and speed condition is also not met.

Cryptography is a solution to the problem when the environment is made of a small set of users, having the same kind of processing equipment. But then these users do not need any seal between themselves.

An intermediate solution between cryptography and error detection, is to use a "signature scheme" like the one proposed by M. Rabbin in [1] . In this case, transparency of the data is ensured, and checking the signature involves only arithmetic operations which are found on all general purpose computers. However, one of the basic assumptions of this signature scheme is that the signa-

ture be as long as the message that is to be signed. This constraint eliminates this method which comes closest to meeting all
the conditions.

III. - Functional problem constraints.

A seal function is a function of all possible messages called text to the set of all possible seals. Let F be a finite
field of order more than 36. A typical length of a text is about
350.000 characters. The seal length is chosen long enough so that
a random choice of a seal for a given message has a low probability
of being correct. If such a probability is taken to 10^{-20} ; then
a seal length of 20 symbols is an adequate length.

A seal function is therefore a function of the type.

$$\phi : F^{350.000} \to F^{20}$$

The set of messages having the same seal is of the order F^{349980}.

A consideration to be made is that an eventual modification of the text will not only disobey a maximum likelihood hypothesis, but will also have little chance of being random : The
objective of an eventual forger will be to transform a meaningful
text into another meaningful text.

Unlike the problem of error detection on a transmission
link, no hypothesis of maximum likelihood, can be made. Two meaningful texts are not close in the Hamming distance sense. Therefore the probability of a text having a different seal from a given
text must be uniform among all the texts.

One last remark to be made is that if the seal is computed
in an algebraic manner, then the chance of the forger finding another meaningful text having the same seal, by dynamic analysis is
almost certain. Therefore the seal must have a random aspect for
an eventual forger. One way to ensure this apparent random character, while having an algebraic fast algorithm to compute it is to

involve a secret quantity in the seal function.

IV. - The seal arichitecture.

The seal function must depend on every digit of the text (One can assume without loss of generality that the text consists only of digits). The fastest way to do this is to use arithmetic operations, i.e. operations for which general purpose computers are best fitted to operate.

Operations using secret quantities being usually slower, one can conceive two distinct parts for the seal function : a "fast" part, consisting of arithmetic operations, involving only public data, and having all digits of the next. Having obtained the result, a "secret" function is applied to it

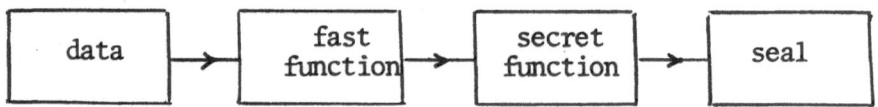

figure 1. Flow chart of a seal function.

An example of such a function could be the modular sum of all the digits for the fast part, and then any cryptographic function for the secret function.

One immediately sees that all the previous arguments developed against a function having no secret part, apply to the fast part.

In this example, permutations of any of the rows intervening in the result would yield the same seal, whatever the chosen secret function is.

The same could be said if instead of taking the modular sum, one takes the modular product of all the numbers in the text. Therefore the "fast function" must already contain secret data, and the preceding dichotomy is not necessary.

Another scheme would be a "stream cipher" function used in a recursive manner

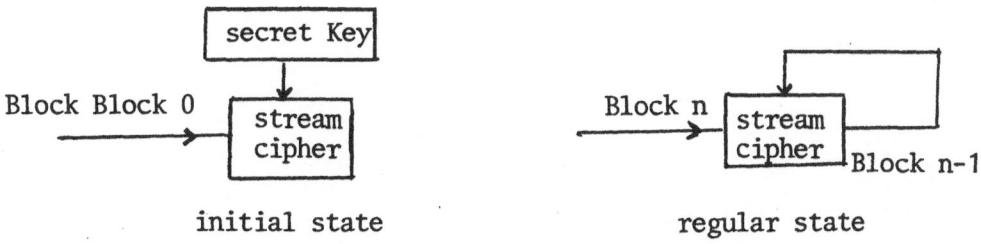

initial state   regular state

Figure 2. Stream cipher used for a seal function.
The resulting cipher of the last block would be the seal.

This function meets all the requirements except one ; speed of operation. Most good stream ciphers like D.E.S., are slow when used in software, computers performing poorly on boolean functions which are inherent part of D.E.S.

One can modify this scheme slightly, so that the slow operations be done beforehand, only fast arithmetic ones being done "on line".

Conclusions. :
The problem of a seal function, arising from the applications, has a class of specific solutions which are functions which are not to be found in related areas. However, ideas developped there allow us to determine satisfying solutions.

Bibliography. :
1. M. Rabbin. - Digital signatures as untractable as factorisation. MIT LCS n° 80, 1979.

PANEL DISCUSSION

ON

LOCAL AREA NETWORKS

AND ON LAYERED ARCHITECTURES AND PROTOCOLS

Chairmen and Organisers:	Professor J.K. Wolf	University of Massachusetts, U.S.A.
	Professor A. Ephremides	University of Maryland, U.S.A.
Panel Members:	Dr. I. Cidon	Technion – ITT, Haifa, Israel
	Mr. M.A. Padlipsky	Mitre Corporation, Bedford, MA., U.S.A.
	Professor J.W.R. Griffiths	University of Loughborough, U.K.
	Dr. A.B. Cooper	U.S. Army, Aberdeen Proving Ground, MD., U.S.A.
	Professor J.L. Massey	ETH-Zentrum, Zürich, Switzerland.
Contributors:	Professor J. Alves	Universidade de Aveiro, Portugal.
	Dr. M. Beale	University of Manchester, U.K.
	Professor T.M. Cover	Standford University, CA. U.S.A.
	Professor P.G. Farrell	University of Manchester, U.K.
	Professor D.G. Messerschmitt	University of California, Berkeley, U.S.A.
	Dr. P.J.W. Rayner	University of Cambridge, U.K.
	Dr. H. Robinson	British Telecom, Martlesham, U.K.

<u>Professor Wolf</u> – This discussion concerns local networks and
layered protocol architecture. Let me remind you that we have here
people who have come to this Institute specially to tell us
something of these subjects. Altogether there are six speakers.
Their contributions will vary in duration from five to fifteen
minutes. We can have general a discussion after their
presentations, but short remarks or comments are welcome during
these. I have a question!

<u>Dr. Rayner</u> – Can you or someone tell us first what are local
networks and what is the layered architecture?

<u>Professor Wolf</u> – I will try to tell you what are local networks,
and perhaps Professor Ephremides will tell you what is layered
architecture. This is my view of the world before local networks
term was introduced (Figure 1). Here in fact are two non-local
networks; they are local only in the sense of being limited in a
geographical area. Some people are talking of satellite networks
as local ones. In Figure 1a we have a set of users with
dedicated lines running from one to another. This is a very
standard way of interconnecting users. Another way is shown in
Figure 1b. I call it a star, but there are various versions of it.
It is a switching circuit where everybody has lines running to a
switch which takes care of interconnecting users when they want to
talk to each other. There are other versions of course.

Now what about examples of local networks. There are many of
these and I will take two to give you some idea of what is going
on. The first is called a ring, or a token ring (Figure 2a).
First of all there are few wires here – the wires are running only
to your next neighbour. As a matter of fact, the transmission goes
only in one direction. The name 'token ring' comes from the idea
that there is 'something' that gets passed around and only the
person that has it can transmit. Say at the moment the 'hatched'
person has it and so he is allowed to transmit. He passes his/her
message to the next user, as well as his token, and then it is
passed on to the address. 'ETHERNET' is a network (Figure 2b)
where you have one channel and users are just connected to that
channel, and have some rules of getting on to it. This is a very
specific type of system that the Xerox Company has put forward.
This involves the concept of 'carrier sensing', where your are at
least polite enough not to talk when somebody else is broadcasting.
This common channel does not have to be a wire, but you can think
of its as a coaxial cable. When somebody is transmitting, you keep
quiet. Then at last when it sounds like it is quiet, you put your
signal on. Unfortunately, because of delays one person may think
that it is quiet, and the next one may think so too and they both
begin to transmit. This leads to collisions. Once you have
collisions, you have to consider collision resolution algorithms.
Some collision resolution algorithms are known to lead to an
eventual deadlock whereupon no messages get transmitted
successfully. Others appear stable but have not be proven so.

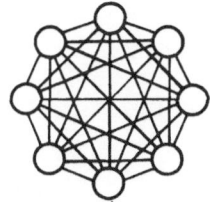

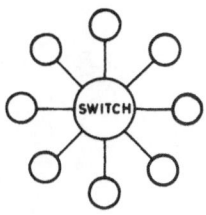

a.   Fully Connected Network   b.   Star Network

Figure 1.   The World Before Local Networks

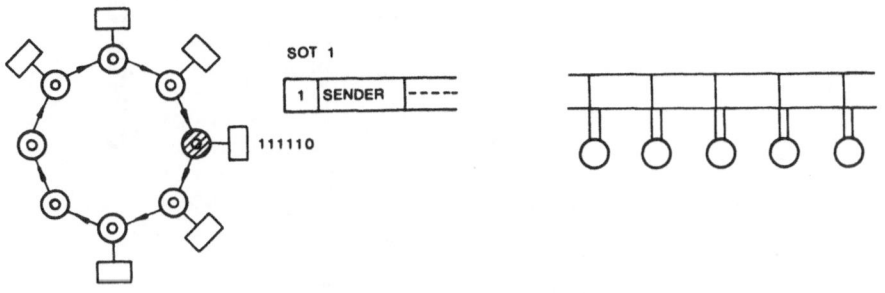

a.   Token Ring (ETHERRING)   b.   ETHERNET

Figure 2.   Local Area Networks

Many unanswered questions exist with respect to these collision resolution algorithms.

I am going to save the rest of my remarks on this problem to my lecture [1] when I deal with group testing. But I would now like to mention other speakers in this panel discussion on the subject of local networks, namely Dr. Cidon, Professor Griffiths and Professor Massey. When I asked Dr. Cidon for the title of his presentation, it was so long that I decided to tell you what he is not going to talk about. He is not going to talk about a single local area network. Professor Griffiths has a very modest title - it is concerned with the universe! This is one example of the Cambridge ring, or rather the Cambridge ring is one example of it. Professor Massey is going to talk about something beyond the

316

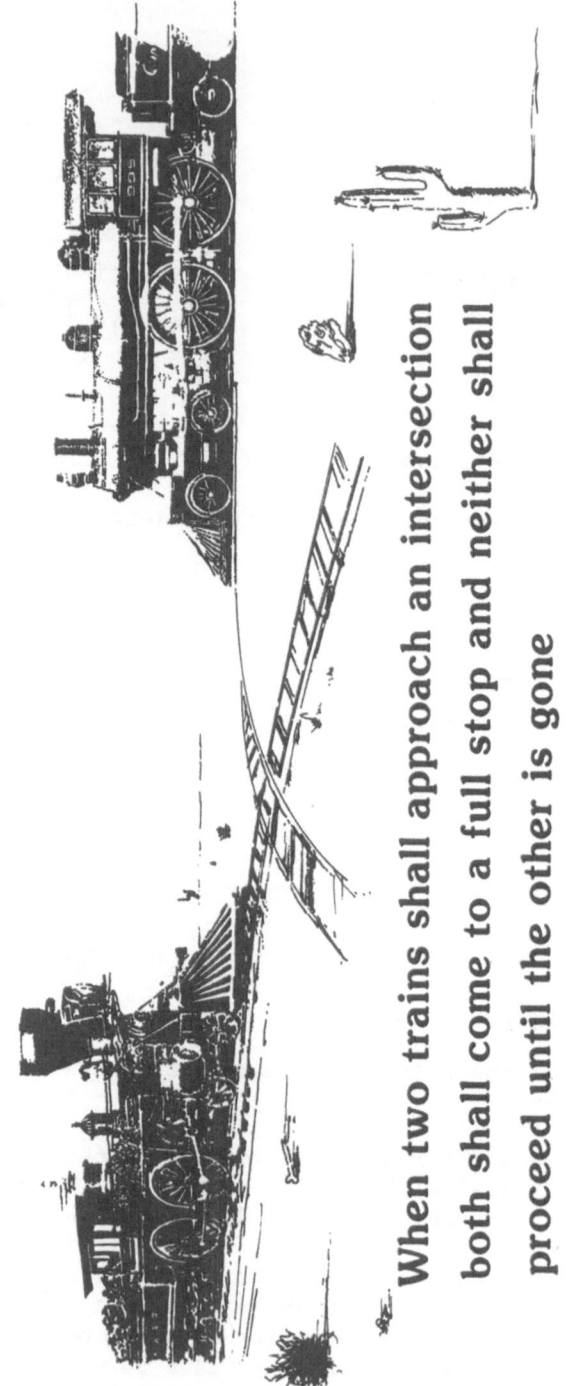

When two trains shall approach an intersection both shall come to a full stop and neither shall proceed until the other is gone

State of Kansas ordinance

Figure 3.   Trains in Kansas

ETHERNET. With that existing promise I will now pass you over to
Professor Ephremides.

Professor Ephremides - My introduction is going to very brief - I
have only one point to make at this late hour! The other three
speakers will address the problem of protocols and they will tell
you what they are. I will just raise the question: 'Should there
be standards? If yes, what are the real technical issues that
cause disagreement?' Are these economical, social or technical
reasons that have caused disagreements? There have been
disagreements ranging from trivial confusion to some real and
difficult technical problems. For instance in the State of Kansas
in the last century, when the rail crossings first appeared (Figure
3) they developed a protocol which is still in the ordinance of the
State of Kansas and which reads: 'When two trains shall approach
at intersection, both shall come to a full stop and neither shall
proceed until the other is gone'. With this example I will invite
our first speaker, Mr. Padlipsky.

Mr. Padlipsky - I am responding to Professor Ephremides' request to
address the technical, economic, and aesthetic bases for the
controversy between the evolving ISO approach to intercomputer
networking and the in-place approach developed on the ARPANET. My
comments should, by the way, be viewed as a necessarily concise
treatment of what are, it should be noted, my personal views on the
topic.

The ISO "Open System Interconnection" Reference Model (ISORM)
is fairly well known. The ARPANET Reference Model (ARM) is less
well known, though it antedates the ISORM by a number of years and
the ISORM is rather clearly indebted to it (see Figures 4 and 5).

In response to Professor Ephremides' question, it is my
contention that the ISORM and the suite of protocols being
developed for it (ISORMS) are not, by and large, a good art,
whereas the ARM Suite (ARMS) is. This is by no means meant to
suggest that the ARMS is held by me to be without its own flaws,
but rather that my little list of ARMS flaws is both shorter and
less profound than my little list of ISORMS flaws. Given the
charter for the talk, I will, of course, be emphasizing ARMS(S)
strengths and ISORM(S) weaknesses, but it does seem important to
mention the fact that in other contexts I am capable of addressing
some ARMS(S) weaknesses and ISORM(S) strengths.

By "good art", I do not limit myself to a technico-aesthetic
judgment only, but also suggest, and will support the allegation,
that the ISORM has built-in inefficiencies which are wasteful of
both cpu cycles and memory - the very resources which a
"resource-sharing network" (the ARM phrasing of what the ISORM
seems to mean by OSI) are meant to make shareable; also as will be
noted, the ISORM appears to lack certain features common to the
present state of the art as instantiated by the ARM and the ARMS.

318

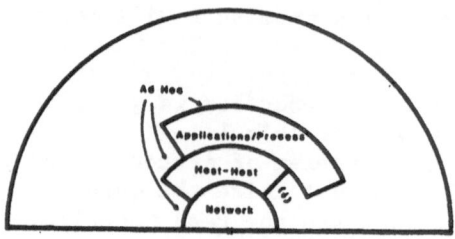

Figure 4.   ARM in the Abstract

Notes:
A.   X marks the egress (and the ingress)
B.   The whole picture either lives in or is flexibly attached
     to a Host
C.   (↓) indicates it is at least imaginable to use the
     network layer without-in-lieu-of a Host-Host protocol in
     some circumstances
D.   (And if comm subnet processors are present, there will
     doubtless be a CSNP-CNSP protocol in play - which is
     rendered uninteresting for present purposes).

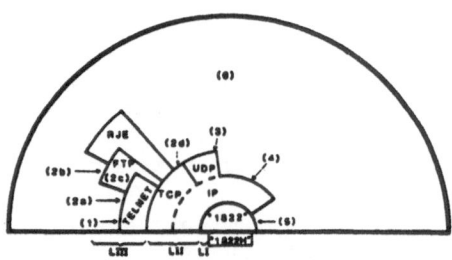

Figure 5.   ARMS Somewhat Particularised

Legend
(1)   Where Terminal-Terminal Protocol would go*
(2a)...(2d) Where Mail Protocol could go*
(3)   Where muxed packetized speech goes
(4)   Where unmuxed packetized speech could go
(5)   Where you could "pump bits at a peripheral" on a LAN
(6)   Still ad hoc land
●   "1822" for concreteness only (need not be taken
      literally)
●   RJE not endorsed, merely suggestive of using two PLP's in
      support-of-doing-virtulization for a third in-support-of/
      doing-virtulization for a third.
*   Found via well-known socket

The "Architectural Highlights" (Table I) rather cryptically
deals with most of the points I want to raise here (the
architecture reference plays on the visual resemblance of the ISORM
to two high-rise apartment houses with parking garages between
them), but before turning to it a few words should be said about

---
ARM ARCHITECTURAL HIGHLIGHTS

- CUSTOMIZED PARKING GARAGES

- PLENTY OF ROADS BETWEEN PARKING GARAGES

- ELEVATORS AS WELL AS STAIRCASES

- MULTIPLE APARTMENTS PER FLOOR

- TOWNHOUSE STYLE APARTMENTS AVAILABLE

- STAIGHT CLOTHES LINES

- FULL GUARD SERVICES

- MANNED CUSTOMS SHEDS

- READY FOR IMMEDIATE OCCUPANCY

Table I

the "Axioms" chart (Table II). Although some observers would hold
that most if not all of the cited Axioms can be applied to both the
ISORM and the ARM, I submit that there are two extremely
significant differences in how they are construed. For one thing,
the ISORM view of layering is extremely prescriptive: "all
n-entities must communicate with other n-entities via n-1 entities"
or words very close to that; the "style" is, to my ear, rather
Confuciansist: extremely concerned with proprietry and order – an
Establishment view, as it were. The ARM view of layering is, in
contrast, merely descriptive (of what we observed "worked" in
practice): layer-transversal is not mandatory; the style here,
again to my ear, is rather Taoist: being at one with Nature,
"going with the flow", dominates abstract strictures. The other
point I'd like to raise in the present context is that the final
two axioms, Efficiency and Equity, seem to me to be absent from the
ISORMS, regardless of whether they are or are not avowed in some
version of the Model I haven't yet seen. That is, because those of
us who developed the ARM and the ARMS were mainly system
programmers and implementation-   orientated computer scientists,
we were particularly concerned with not chewing up the resources of
our home systems (Efficiency) and with not introducing mechanisms
and/or common representations that were extremely awkward to
implement on all "reasonable" operating systems even if they seemed

```
 _____
|                                                |
|                 ARM AXIOMS                     |
|                 _____                      |
|                                                |
| •   DO RESOURCE SHARING                        |
|                                                |
| •   VIA INTERPROCESS COMMUNICATION             |
|                                                |
| •   OF COMMON INTERMEDIATE REPRESENTATIONS|    |
|                                                |
| •   OVER LOGICAL CONNECTIONS                   |
|                                                |
| •   IN A LAYERED FASHION                       |
|                                                |
| •   EFFICIENTLY                                |
|                                                |
| •   AND EQUITABLY                              |
|_____|
```

Table II

"natural" on some operating operating systems (Equity). I find no evidence of either concern in either the ISORM or any of the candidates for inclusion in the ISORMS I've yet reviewed.

Turning to the cryptic headings on the "Highlights" chart, without going on at too great a length ([2] has more to say, and it should be noted that even there the list is by no means exhaustive), I'd offer the following brief glosses (in order of the "bullets"):

1. To those ISORM advocates (some, not all) who hold that X.25 "is" L1-3, I observe that X.25 is not widely held to be a useful interface for Packet Radio or even an efficient one for LAN's, irrespective of its other deficiencies; the ARM accommodates whatever LI is presented (see Table III.)

2. An X.25 "virtual circuit" view at L1-3 is usually construed as mandating fixed routing through the communications subnetwork/internetwork; alternate routing (far preferable for "survivability") comes naturally with the ARM view that the commsubnet processor – comm subnet processor protocol is outside the model.   (Believing in datagrams also helps.)

3. In the ARM, protocol interpreters in a given layer are free to invoke directly other such PI's "on the same floor of the building" without metaphorically (and inefficiently) climbing downstairs and then back up. As noted, the ISORM is rather fierce about n-entities' use of n-1 entities.

4. Again as noted, layer traversal is not mandatory in the ARM;
    until and unless the rumored "null layers" are incorporated
    into the ISORM, however, it does seem to be mandatory there.
    (It may be even afterwards, depending on how "null layers" get
    specified at least to the extent of wasted time on calling
    sequences to entities which will merely pass data-units on
    "down" or "up", and even some wasted space on nominal entity
    headers.)

5. ISORM L3-7, as nearly as I can tell, all have "connections"
    between "peer entities". At the least, that's a lot of table
    entries. In the ARM, logical connections are an L II
    artifact - a "straight clothes line" between the metophorical
    buildings (though it never occurred to me before that if the
    ISORM has high-rises, the ARM has geodesic domes).

6. If desirable ISORM L6-7 protocols eventuate, I think they'd
    "fit on top of" ARM L II. I'm not at all convinced that ARM L
    III protocols fit on ISORM L4 (or 5) (or 6). If I'm right,
    it's another flexibility point for the ARM. (As an aside, the
    ordering of L5/6 militates against offloading/doing outboard
    processing/front-ending ISORMS: you'd like Presentation
    functionality not to burden your Hosts, but Session
    functionality associates logical connections with processes
    and seems to have to be inboard. Again, the ARM is more
    flexible in my view.)

```
            INTERCOMPUTER NETWORKING PROTOCOLS

 • LIII      PROCESS/APPLICATIONS LAYER: TELNET, FTP, T-TP, (RJE)
             - USE COMMON INTERMEDIATE TEP'Ns ("VIRTUALIZATIONS")
             - OPERATE OVER LOGICAL CONNECTIONS

 • LII       HOST-HOST LAYER: TCP, (UDP), IP
             - ALLOW INTERPROCESS COMM. (TCP, UDP)       "HEADERS"
             - ESTAB. RELIABLE FLOW-CONTROLLED LOGICAL
               CONNECS. (TCP)
             - MANAGE INTERNET ADDRESSING, ROUTING (IP)

 • LI        NETWORK INTERFACE LAYER: "1822"
             - PHYSICAL XFER OF BITS TO/FROM HOST.          WIRES
             - PROXIMATE NET ADDRESSING, STATUS          "LEADERS"
             - HOST-IMP FLOW CONTROL

 • "OTHER": CNSP-CNSP: H-FP
```

Table III

7. A fairly subtle point: the ISORM "expedited data" notion was
    not, the last time I looked, defined in such a fashion as to
    make it clearly usable to mechanize the sort of "out-of-band
    signal" a virtual terminal protocol needs. You can get a
    message of your peer rapidly, but how the peer disposes of it
    "upwards" isn't stated. (The chart's bullet refers to taking
    an SST to a foreign country only to find no one on duty at the
    Customs Shed.)

8. A key point: Whatever kind of metaphorical buiding the ARM
    is, in reality the ARMS has been working for years now, while
    the ISORM itself is still being argued over, much less the
    ISORMS.

I've often been told, in response to some specific criticism
or other of mine to the ISORM or a candidate for the ISORMS, "But
even bad standards are better than no standards at all". I can't
agree. Bad standards in intercomputer networking can be extremely
expensive, both to implement and to execute. At best, bad
standards can be gotten around by upgrading the participating Host
systems (cpu power and/or memory complement). At worst, bad
standards can destroy the field by taking "forever" to be arrived
at, for example, or by not delivering the promised functionality.
In response to another question of the chairman's (as to whether we
need networking standards), then, I'd say that above the level of
physical connections/connectors, I don't want bad standards, I want
good protocols.

Finally, returning to the opening question, I hope I've
managed to sketch here a case for the contention that there are
indeed grounds for controversy between the ISORM and the ARM, and
even that although the ISORM is winning (or even has won) the
"P.R." battle, it is by no means a clear advance over the ARM,
which represents, despite its lack of public recognition, the real
proof-of-concept for most of the ISORM's notions.... and the real
state of the art. It's also the case that eventually
implementations of the ARMS, the ISORMS, the Xerox suite, DECNET,
SNA, and whatever other idiosyncratic suites eventuate, will become
cheap enough that they will be employable in parallel, depending on
where a desired application has been implemented, and the
controversy will/should become moot.

Dr. Cidon - I am going to talk about multi-station, single-hop
ratio-networks, a work I have done with my colleague Dr. Moshe Sidi
at the Technion Institute. I believe this work presents new
directions in the research on multi-access schemes and offers some
practical solution to radio networks with large number of
terminals.

A tremendous amount of work has been done on single-station single-hop packet-radio networks. The main property of such networks is a large number of terminals which send packets to a central station. Since naturally a terminal has bursty type traffic demands, dedicating a separate channel for each terminal is not efficient. Typical solution to this problem is to use a common shared channel, and some dynamic access scheme. In this work we will consider a slotted channel and low rate feedback channel used to transmit from the station back to the nodes the results of the last slot.

How should we extend the single-station network when the area and the number of terminals becomes larger? The simplest solution is to use higher rate channel and to increase terminals power while keeping a single station. We will present different and more efficient solution. Let us consider a multi-station packet ratio network (Figure 6). In this network terminals access several stations using a single shared channel. For example terminals, 1, 2, 3, 6, 7, 8 are all at the receiver range of station 1. What can we gain from such network in comparison to the single station one? For example, at our network terminals 1, 5, 10, 16 can transmit at the same time (better utilization of common channel). The second advantage is that we can still use low power terminals and do not have to increase their power. Thirdly we have here an hierarchic structure which is easily expandable. We mean by this that stations themselves can use a common channel to access some super-stations. However, there are new problems with this model. Consider first the types of traffic that we can have in this network:

The first is the terminal to station communication problem. In our case terminals use a slotted shared channel and hence collisions may occur. For this we would use some random access and a low rate feedback channel. This type of communication is the hardest to accomplish due to the new problem of interference between terminals accessing different stations. One should be very careful in implementing known access schemes in the new model since as shown in [3] some protocols may deadlock. We will use the slotted ALOHA scheme and will show that it indeed works with our model.

The second is the inter-station communication system which is simpler. Since only several stations exist and they are concentrators of large numbers of terminals its communication demands are not bursty. In this case a dedicated channel for each station along with a high power transmitter (allowing direct connection between stations) will be a reasonable solution. We propose to use for this task TDMA or FDMA or even wires if stations are not mobile (for example in a cellular data radio network).

324

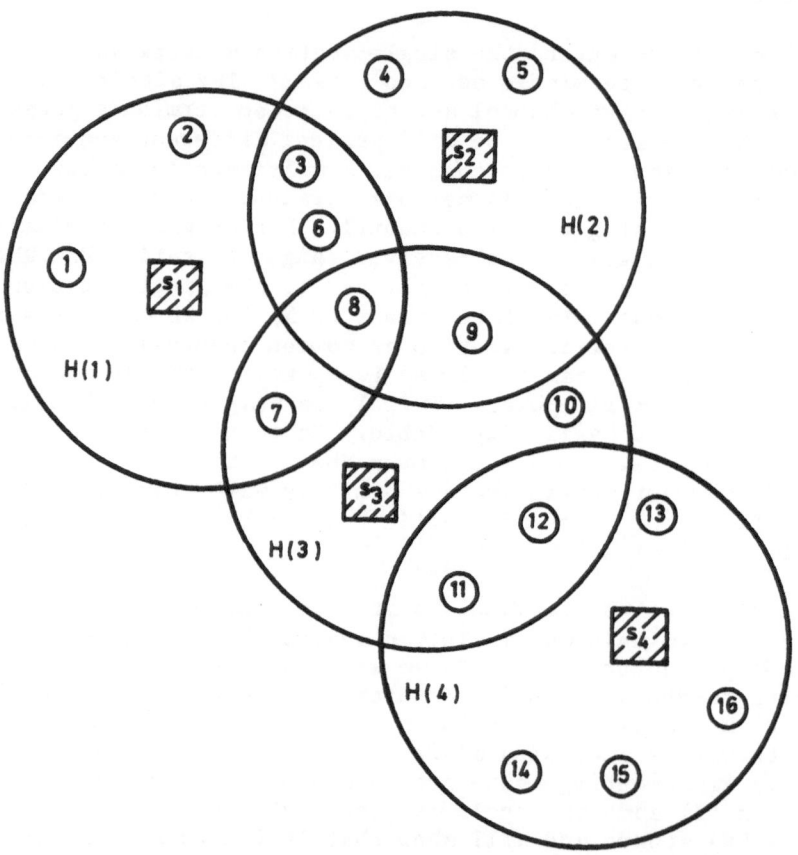

Figure 6.   Multistation Packet Radio Network

The third is the stations to terminals communication which is simple as the previous one and very similar to it. Thus the last two cases can be solved using the same mechanism (by letter nodes to listen to the inter-station communication system).

We will distinguish between two forwarding schemes which can be taken by nodes.

The first is the Fixed Forwarding Scheme (FFS) in which a packet is considered to be successfully received if it has been received correctly at a single predetermined station i.e. a node forwards packets to a fix predetermined station.

The second is the Random Forwarding Scheme (RFS), where a packet is considered to be successfully received if it has been received correctly at any station.

The advantage of RFS is higher throughput; its disadvantage is that we have no fixed routing and duplicates can be generated when a packet is received simultaneously at more than one station. Let us consider the FFS scheme. In Figure 7 we see a typical network. here $H(i)$ denote all nodes heard by station s_i, $B(i)$ denotes the set of all nodes which forward their packets to that station. In our case we have: $B(1)=(1,2,3,8)$, $B(2)=(4,5,6,9)$, $B(3)=(7,10,11)$ and $B(4)=(12,13,14,15,16)$ $I(i)=H(i)-B(i)$ is the interference set of s_i i.e. nodes heard at station s_i which forwards their packets to other stations. And finally $I_j(i)=I(i)\cap B(j)$ as the interference subset of s_j at s_i i.e. the set of nodes that transmit packets to station s_j and are also heard by s_i.

I wish to show some preliminary results on the FFS throughput for slotted ALOHA access scheme using the infinite population assumption and the equilibrium hypothesis of Abramson.

Let TH_i = new packet generation rate for nodes from $B(i)$

λ_i = combined rate of new and retransmitted packets from $B(i)$.

ϕ_{ij} = fraction of λ_i which is heard by s_j, i.e.

$$(0 < \phi_{ij} < 1 \; ; \; \phi_{jj} = 1)$$

G_i = total number of packets heard by s_i per slot.

TH = total network throughput

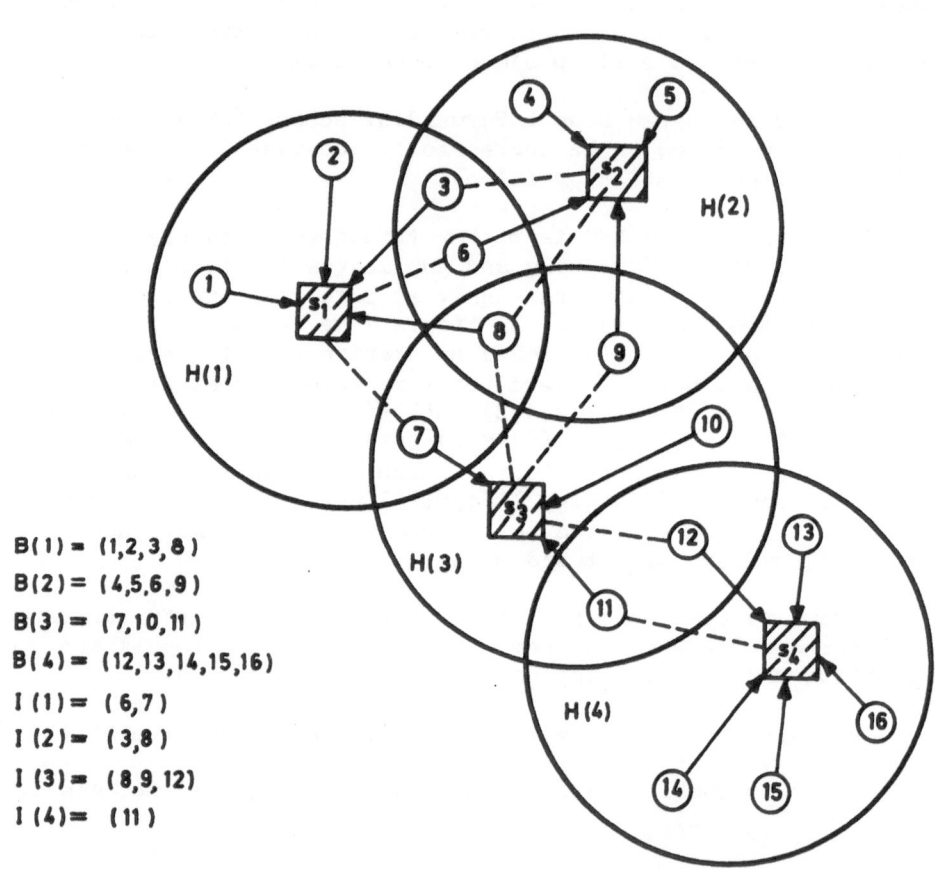

B(1) = (1,2,3,8)
B(2) = (4,5,6,9)
B(3) = (7,10,11)
B(4) = (12,13,14,15,16)
I(1) = (6,7)
I(2) = (3,8)
I(3) = (8,9,12)
I(4) = (11)

Figure 7. A Multi-Station Packet-Radio Network; Fixed Forwarding

Then:

$$G_j = \sum_{i=1}^{m} \lambda_i \, \phi_{ij}$$

$$TH = \sum_{i=1}^{m} TH_i = \sum_{i=1}^{m} \lambda_i \, \exp(-G_i)$$

(making the Poisson assumption).

Let me show you some examples. Consider m stations and a symmetric case.

$$\lambda_i = \lambda \quad \text{for all } i$$

$$\phi_{ij} = \begin{cases} 1 & i = j \\ I & i \neq j \end{cases}$$

Then $TH = m\lambda. \exp\{-[I(m-1) + 1]\lambda\}$

$TH_{max} = m.\exp(-1)/[I(m-1)+1]$

Therefore

$$TH_{max} = \begin{cases} m\cdot\exp(-1) & I = 0 \\ \exp(-1) & I = 1 \end{cases}$$

Figure shows the throughput as a function of I for the case of m = 2, for both RFS and FFS schemes.

Finally, it is well known that the slotted ALOHA protocol policy might cause instabilities, unless a proper algorithm is added in order to control the retransmission of packets. This drawback has led us to consider a Collision Resolution Algorithm (CRA) which has been investigated thoroughly in the literature (3,4,5). The network that we have simulated consists of two stations s_1 and s_2. Two sets of 50 nodes B(i), i = 1, 2 construct their tree according to station s_i respectively. The number k of interfering nodes defined by k = | H(i) − B(i) | i = 1, 2 is a parameter in our simulations. We have simulated both RFS and FFS schemes again here and have obtained the throughput for each scheme as a function of k. The results are shown in Figure 8.

328

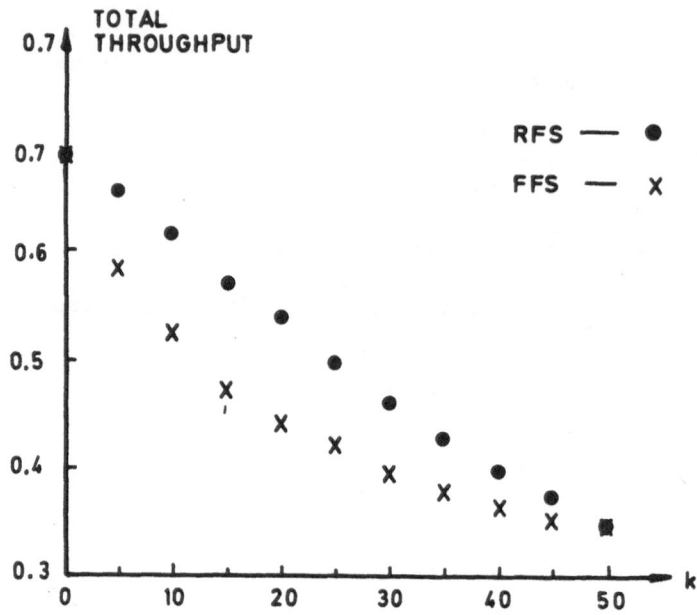

Figure 8.   A Two-Station Symmetric Network – tree CRA.

To summarize this discussion: we presented here some new preliminary results for a new extended model of packet-ratio network. It is very interesting problem to extend other known multiple access scheme to this new model.

Professor Griffiths. – I wish to talk about Project UNIVERSE, which has been mentioned once or twice during our deliberations. The project UNIVERSE (UNIVersities Expanded Ring and Satellite Experiment) is a major project in the U.K., involving universities and industry. It is founded by the SERC (Science and Engineering Research Council), government and industry. The universities involved are Cambridge, Loughborough and UCL (University College of London). Also, referring to the last speaker, we have a connection with ARPANET in the U.S.A. The industry participants are GEC (particularly the Marconi Research Centre), Logica (who manufacture amongst other things the components of rings), and the British Telecom. The SERC laboratory (Rutherford Appleton Labotoratory) is also involved in this network. The total budget is £4 million; and the project started in 1981. The aim is to extend the local area technology into the wide area by the use of satellite communication and other high speed links. The project is mainly based on the Cambridge ring, which is a fairly high speed ring. The basic layout of the system is shown in Figure 9. Actually six sites are communicating directly with the satellite. The other site is communicating via high speed line links. The satellite we are

329

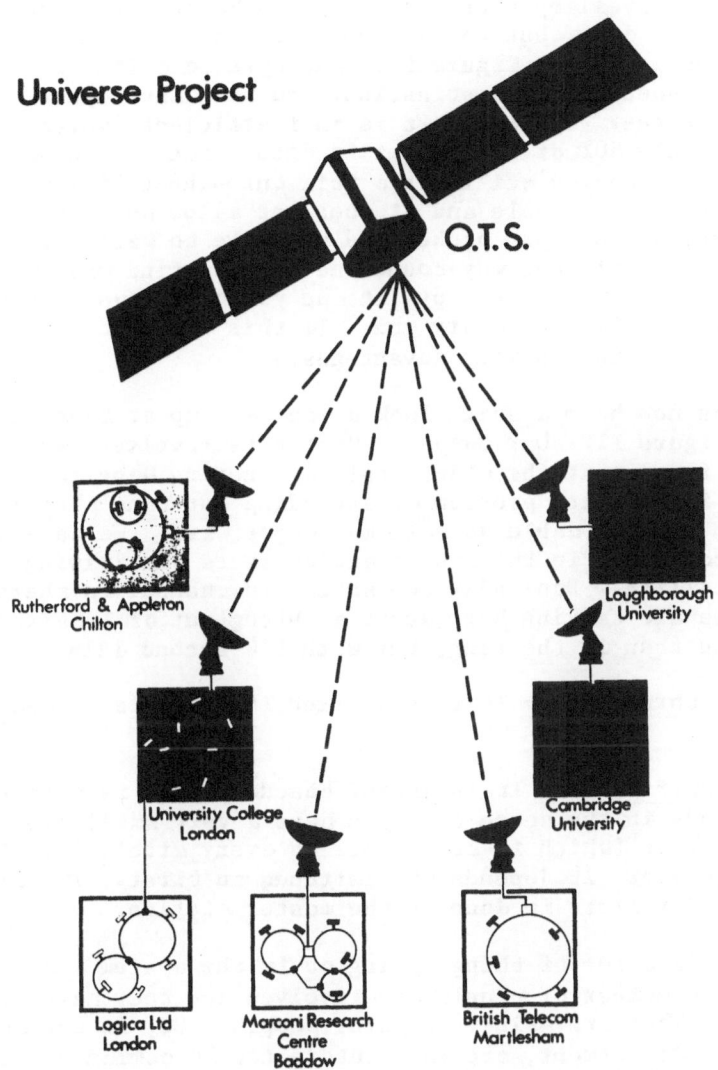

Figure 9.   Project Universe

using is the OTS stationary satellite. It is liable to go out of operation fairly soon - it is now beyond its expected life! Later on we shall use the ECS satellite.

Basically, the Cambridge ring is a mini-packet system. Packets are travelling round the ring. A basic packet could have a fair amount of data, but for various reasons the standard amount is two bytes of data (see Figure 10), one byte to define the source, one byte to specify the destination, and few other bytes for control purposes. Basically it is an inefficient system in the sense that only 50% of the packet is data. But we are travelling at 10 Mbit/sec so you still get a fair throughput of data. The protocol is fairly simple and it does not allow hogging because you can only access an empty packet and you have to wait till that packet has been all the way round the ring to find what has happened to it. Then you empty it and you cannot use another packet till you find an empty one. In this way nobody can hog the system and this is its main advantages.

Let us now have a quick look at our set up at Loughborough, as shown in Figure 11. Our Computer Centre is involved, which is one mile away from us at the Electrical Engineering Department. Thus we had an interesting problem of extending our ring that distance. We used an optical cable as well as copper wire. We have about 10 to 15 computers in the system and there is one driving computer to the satellite. Basically the satellite channel is shared by TDMA and we are talking here about a throughput of 1 Mbit/sec which rather less than in the ring, but with 1/4 second delay.

Professor Ephremides. - Is this a fixed TDMA, or is it demand based?

Professor Griffiths. - It is demand based. There is a little bit of fixed TDMA in the sense that you have a very small slot in the main time frame (which is accessible to every site). Any site can enter the master, it depends who switches on first. The allocation of these major slots is done by the master station.

There is a lot of things going on in the system. A considerable number of people are involved and they have different interests. They are mainly computer people. We, in the Electrical Engineering Department, are more interested in communication aspects, particularly computer systems. The work on infra-structure was no mean task - such as digging trenches across the Campus! One of our main interests is in the moden design and in the satellite bridge design for the next stage of the system. At present we deal with 1 Mbit/sec system over the satellite and we want to go to a higher rate. Professor Farrell is involved with the coding side of the improved version of 8 Mbit/sec. We are interested in speech and image transmission across the network. We have developed a slow scan system for images (Figure 12), which will allow us to capture a frame and then to send this across the ring. At present we are using a normal 64 Kbits PCM signal for the

331

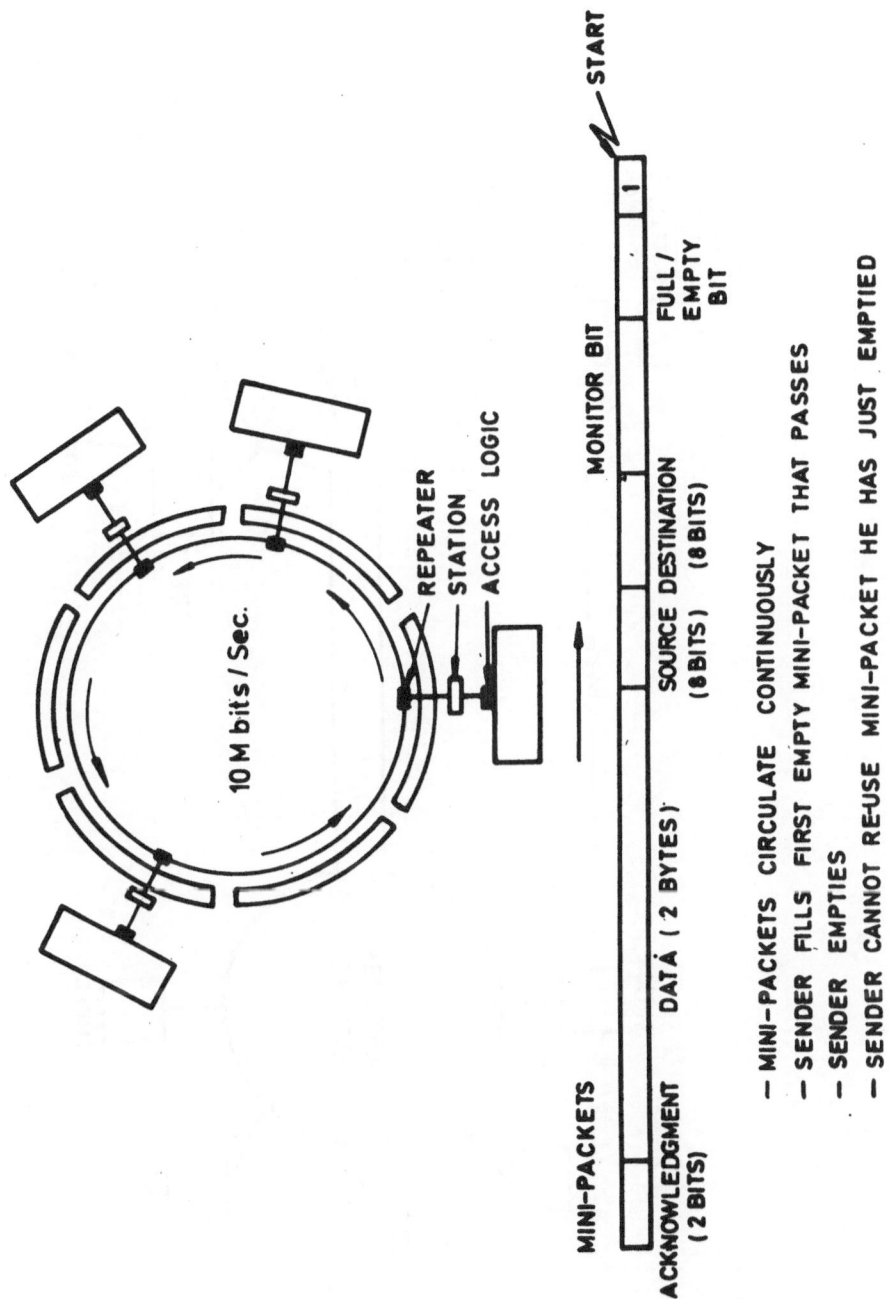

Figure 10.   Cambridge Rings

332

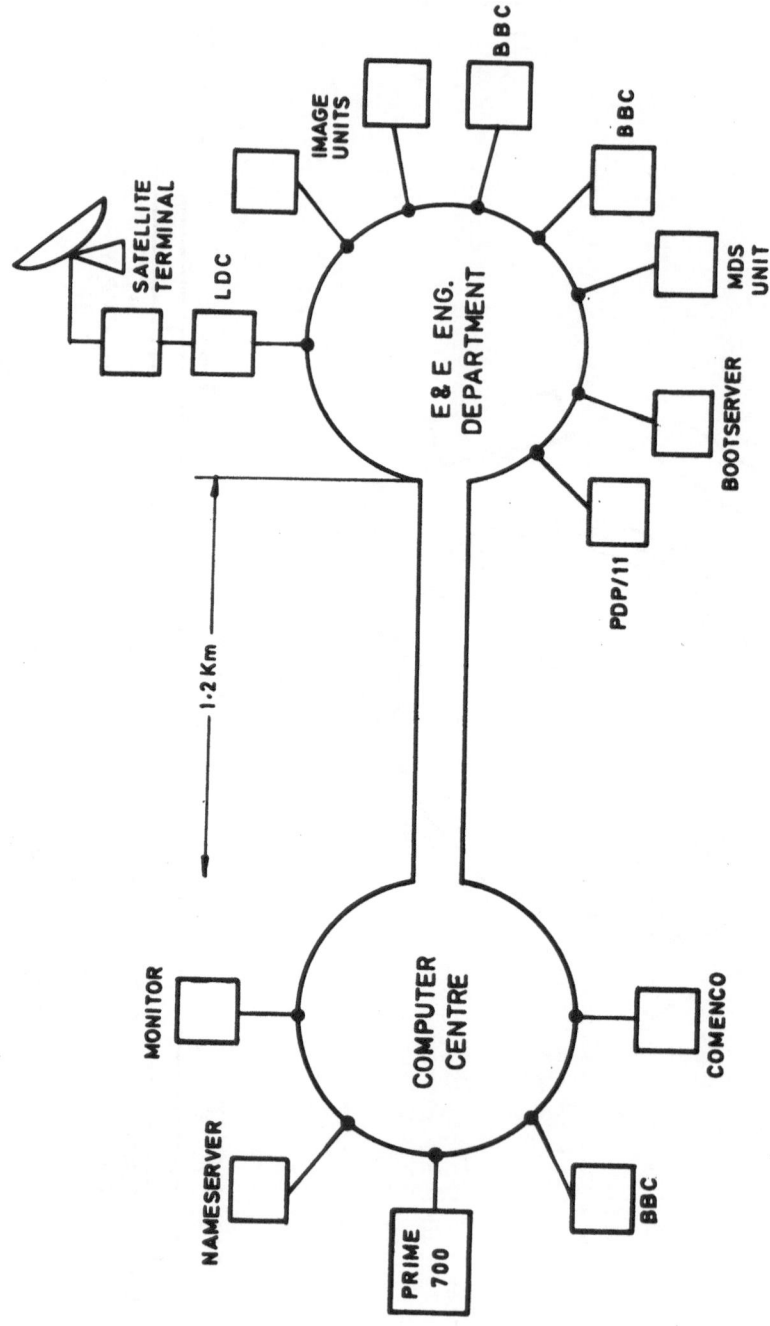

Figure 11.   Loughborough University of Technology Ring Network

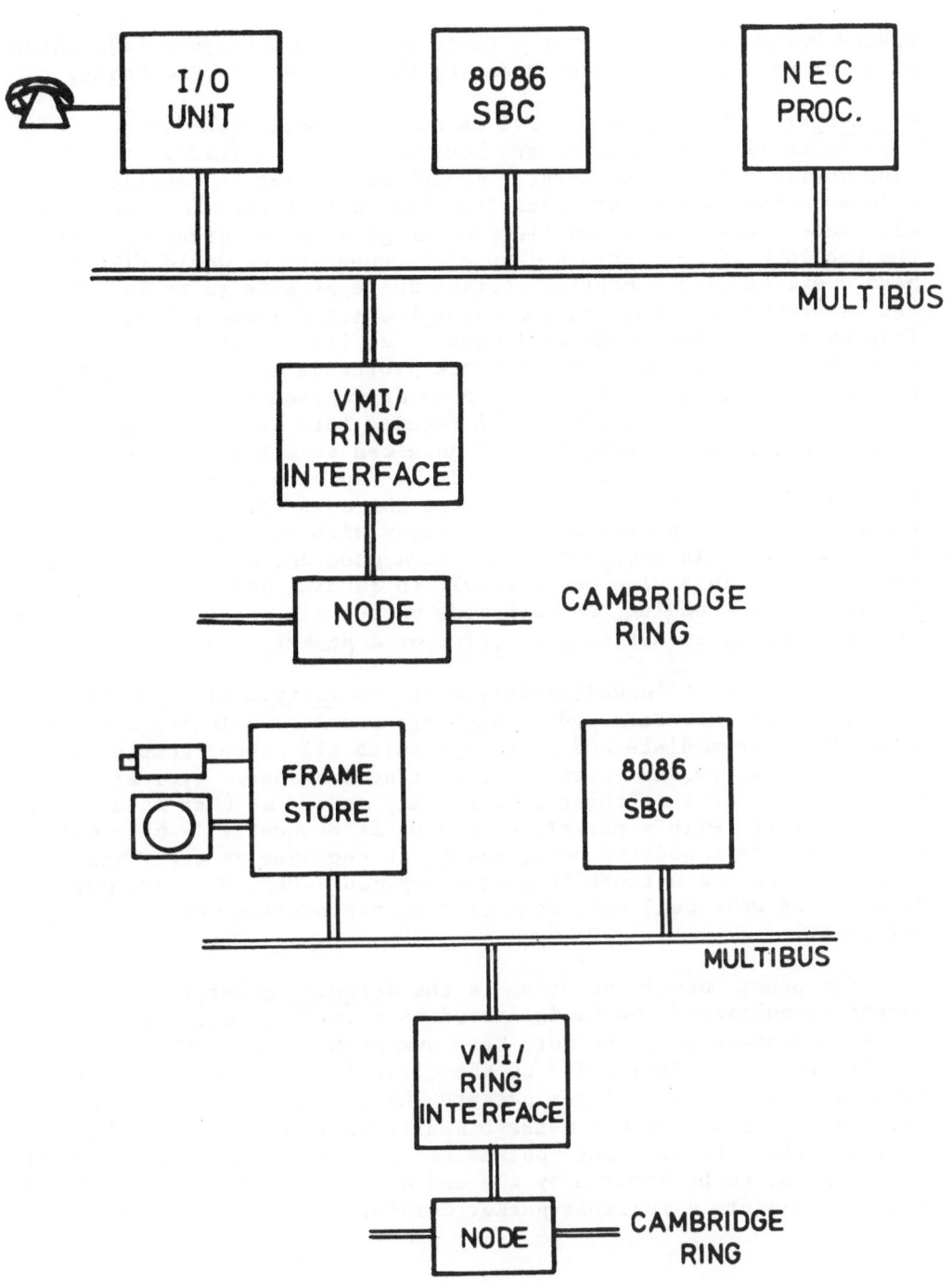

Figures 12 and 13

334

speech but we are developing a voice coding unit (Figure 13), which
will act as a coding server reducing the bit rate to 2.4 Kbits.

Dr. Cooper - The problem I have is that I cannot draw nice
topologies like stars or rings, because I am concerned with
land-mobile computer networks, 'nodes' are riding in vehicles
across country while communicating with each other. Thus we deal
with unreliable nodes and links which go down and disappear, while
the topology of our network changes frequently. We would like to
have a system with a control station where packets go to and fro,
but we have no central station through which all control passes.
This means that each node will have to do its own processing and be
quite 'smart'. In addition, network protocols must be more robust
to remember who is up and who is down at a given instant. The
frequent changes of topology will require lots of processing.
Nodes are not just users, for any node can be either a relay of
packets from one user to another, or it can be a user, or both.
Every packet received at node, except the ones that are intended
for it, has to be passed on to everybody with whom it can
communicate. This does not result in a good use of links; they are
always busy. This also could result in queuing delays over links.
The result is flooding, as each node passes all received packets to
all neighbours, except to the source of a packet.

The effective 'usual' solution is the virtual circuit, where
for every pair of nodes and an exchange packet, we define a set of
effective intermediate nodes through which all these packets pass.
This is not a physical circuit, but it acts as one. Typical
schedules are kept at these intermediate nodes, so that when
someone starts with a packet, he passes it sequentially over the
same route, from node to node, always in the same order. This
works well when a network is stationary and everybody stays put.
It does not work well when some of the intermediate nodes suddenly
disappear.

The other 'usual' solution is the datagram concept, where each
packet is submitted and is delivered as a distinct message, even
though a message could be more than one packet long. This solution
is offered as an alternative to the virtual circuit concept. The
idea is that a user will get packets which come by many routes; the
user will re-assemble the message and it could happen that the
packet 2 will arrive before packet 1. In datagram system that sort
of thing has to be handled by the end a user. Therefore a protocol
must here assure a reliable packet delivery.

At this stage let us compare the processing impact of our case with that of static networks:

1) We have a 'central-less' system, and this means that we have to do much more processing at each node.

2) We need more robust protocols, since in our case each node must remember the acutal state of the network.

3) We have frequent topology changes and so we need frequent updates of the stored network data; also we need additional buffering for in-transient packets.

Let us now consider some new solutions to our problem. Since coming to this Institute I have been made aware of the new by Dr. Cidon, who with Professor A. Segall has submitted a paper of network re-synchronisation protocols [5]. The idea is to define a set of network re-synchronisation protocols which allow each packet to see the network as if its was stationary and unchanging. The basic idea is that when a node fails, all remaining nodes are re-synchronised. This is called 'Communiations-Type Re-Synchronisation Protocol' (CRP). I will try now to describe this. Each node in the network stores a count R_i which is the number of times that node i has been re-synchronised since the time begun. Suppose that the state of node i is now m_i.

The state varible m_i tells you whether a node is in a normal mode, sending or receiving packets, or whether it is in a re-synch mode, at which time no packets are coming in and no packets are going out.

Suppose now that the re-synchronisation count at the node i is $(R_i - 1)$. Let us now assume that the node i detects a failure of its neighbour, namely the node j; for instance, something was sent to it and no acknowledgement was received. Then we proceed as follows:

$$\text{resynch count} \leftarrow R_i$$
$$m_i \leftarrow \text{RESYNCH}$$

$$\text{For all adjacent nodes } k\colon R_k \leftarrow R_k+1$$
$$m_k \leftarrow \text{RESYNCH}$$

(If this is not possible for some node k, then this node is disconnected from the node i).

Thus summarising, when the RESYNCH count is updated by one, a node goes into the RESYNCH mode and all adjacent nodes do the same thing: they update the count and go to the RESYNCH mode.

In other words, what happens at node i is the following:

If R_i = R and m_i = NORMAL, then the packet sent from node i to node j is:

$$\text{ACCEPTED} \quad \text{if } R_j = R \text{ and if } m_j = \text{NORMAL,}$$
$$\text{REJECTED} \quad \text{if } m_j = \text{RESYNCH}$$

I am assuming here a reliable data link protocol, also error detection and correction and re-transmission of non-acknowledged packets, so that you know with some confidence whether the nodes adjacent to you are up or down.

Dr. Cidon's solution goes back to that due to Dr. Finn who recently published a paper of 'fail-safe' algorithm [5]. Here a node i sends a packet n to node j <u>only after</u> receiving the acknowledgement ACK(n-1) from the node j. The node j sends ACK(n) to the node i <u>only when</u> m_j = NORMAL and the packet n was last correctly received before going into the RESYNCH state.

Dr. Cidon has shown that, with changing topology, this algorithm is loss-free and is also duplication-free. It never looses a packet in transit, no matter what that packet is. When a node fails, it never generates the same packet and sends it out twice. I would like to point out that although the count R_i appears to grow without bounds, there are conditions under which you can count R_i mod(n) (where n is some large number), and this can be bounded stochastically by introducing additional counts when nodes go into RESYNCH. The R_i numbers will never be very far from each other throughout the network, so you can easily count back. To conclude, I just want to state that there are several other algorithms in Dr. Cidon's paper.

<u>Professor Massey</u> - Professor Wolf told us something in his lecture about the ETHERNET system. I would like to describe to you something which is called ETHERRING. The idea is due to Dr. M. Schlatter who about two years ago had finished his thesis at our Institute in Zürich [6]. What he did was to take the best features of token rings and of the ETHERNET. He combined these together to devise a ring with a multi-access protocol in which there is a sort of natural resolution of collisions, without any question as to whether a system is stable of not. In Professor Wolf's ring, and in the case of Kansas trains of Professor Ephremides there is no mention in which direction the packets or trains should go. We can easily visualise a case when one station sends a train in one direction and another station will send it in the opposite direction and there will be no collision. Here we assume, that our packets go in the clockwise direction, and this is standard, except possibly in the U.K.! One of the problems in the token ring is that when you have a protocol, which says that when a token comes, you can seize it, and then you can put a packet straight into the

network so that someone else can receive it. This works quite well, except when tokens are lost or duplicated accidentally. Thus token rings can cause crashing of protocols.

Let me now explain the basic ideas of the ETHERNET ring. Suppose that nothing happens in this ring at the time zero. Then '0' signals are sent round the ring. This is the first bit that arrives at any station, and this bit is sent along without any delay. The total delay is then proportional to the total number of stations. Suppose that there are 100 stations in a token ring, and suppose that the 1st station has to send something. Then he has to wait for the token, and in our case of 100 stations he will wait on the average up to 50 bits. This might be slightly inefficient, for why can this station not send his message immediately? He does it by first sending a signal bit '1'. Then the other stations know that the ring is no longer free. Then he sends his own name (of the sender), followed by other stuff, perhaps the address of his message. In this system, unlike in the Cambridge ring, packets are removed from the ring by their senders.

The packets make round trip so if there is automatic feedback in the ring. In our ring, when the sender first receives a message, he checks the sender's address and in this way he knows if it is his own address that he has received a successful package. If he gets someone else's address, he knows that a collision has taken place, or that something else has happened.

Suppose now that we consider a message, say 1000 bits long. At the end of it one sends an 'end-of-transmission' message which could be a flag of let us say 6 'ones'. This would be the end of transmission. In case of a fixed length packets this flag could be allowed to appear in data, but for several reasons it is preferable to use bit-stuffing in this system to avoid the appearance of flags anywhere in data. By increasing the length of the flag economically you reduce the overhead due to bit-stuffing, but you may increase the overhead due to the flag at the end of the message. As I said, the flag is assumed to consist of all 'ones'. If that flag goes by another user and he sees 6 'ones' sitting in front of him when he has a packet to send, he then changes the last 'one' to 'zero'. Then instead of being the end of transmission, this flag becomes the end of message which is an automatic start of another message. This user would then chain this message immediately, just like in a token ring. So, once traffic gets going in this network, packets are chained each behind another, until finally the last packet has come and no one has any messages left. Therefore, after a maximum length of your message you must allow someone else to send a packet, and each station in turn has a chance to do it.

338

The last question is what are you going to do in case of
collisions? Say you send a packet at one node of a ring and then
suppose what comes back to you is someone else's address and some
'ones', then you go into the collision option. This is a rule that
if anyone sees a '1', he realises that there is a message in the
network. In the algorithm that Dr. Schlatter has worked out this
is actually the same as a conventional ALOHA type protocol. All
the users that are involved in the collision retransmit with small
probability in each subsequent bit time. If two or more transmit,
this causes a second collision. The size of the collision set will
then be reduced and this is the essential feature of this protocol.
So collisions tend to thin out quickly. By doing a fairly simple
queuing analysis, Dr. Schlatter has shown that for this network
under a heavy traffic, its performance is much the same as that of
a token ring; it is virtually indistinguishable from the token
ring. Under light traffic conditions the performance of our ring
is similar to that of ETHERNET, although of course we have a delay
in our network which is not present in ETHERNET.

So mine was one more contribution to the area of local
networks, the last one that you have to listen to. By the way,
this network has never been built, as neither have most of the
networks that people are talking about here. It took me a long
time to realise that these things do not exist - they are abstract
since most of them will never be actually built; I have never seen
a packet in my life!

Mr. Padlipsky - There are plenty of LAN's in my neighbourhood but
even so I can see no packets.

Professor Massey - Neither do I! What I do know is that there
seems to be many more names than actual networks. This is all I
want to say.

Professor Ephremides - Now time has come to have some questions
from you all!

Dr. Robinson - Usually people define networks in terms of stars, or
rings and busses; and usually, as a matter of course, the star
network is not being regarded as a promising one. Professor Wolf
just referred to them at the beginning of this discussion. Yet a
lot of work needs to be done on this - there is the question of
crashes, of better protocols etc.

It seems to me that there is an enormous disadvantage of
rings, as opposed to stars. There is their tremendous inefficiency
in the use of tranmisssion links. All the nodes in a ring share
each part of the link, and provided bandwidth is free and
unrestricted, this is fine. But otherwise it seems to be me that
we are going overboard with rings. Perhaps we should start
redesigning star networks? I would be interested in your
comments.

Professor Ephremides - Exactly and precisely what your have stated to be the disadvantage of ring topology, seems to me to be its main advantage. You have a flexibility in adding new users. It is a shared medium and a high speed medium. It is its flexibility that is its main attraction.

Mr. Padlipsky - It is a kind of received wisdom among local area network people that I deal with, particularly at the MIT, that access to rings is almost free, while stars are more expensive to subscribers.

Dr. Robinson - I agree that this is true traditionally, but is it likely to be true for ever?

Mr. Padlipsky - Can you show me that stars are cheaper now?

Dr. Robinson - No, but possibly you could put more effort to re-design stars. We are told that rings provide free bandwidth, but we are already running out of bandwidth, for we now have broad-band version of these rings, and where should we go next?

Dr. Cooper - I have considered this question, but still am not positively either on one side or the other. I still assume that users in a ring can occupy only a small part of available bandwidth. It was stated here by the bandwith available. There is a fixed number of packets running around and they at most use about 80% of the total bandwidth.

Professor Mersserschmitt - The bandwidth is a relatively free commodity, in the sense that we can anyhow get a lot of bits running round a ring. But the problem with high bandwidth rings are busses is that interfaces are expensive, i.e. the things that load bits and take them off. It is not clear to me at all whether in ETHERNET the ring is a cheaper medium. I think tht the main limitation of a star, which is synonymous to the PBX, is that traditionally we send speech at a fixed bit rate, and fixed bit rate has not been adequate for file transfer facilities. This however is fairly easily adjusted by redesigning the PBX to be a widened PBX. We shall probably do this anyway for digital services. Then we shall have available both wideband and narrow band services.

Professor Wolf - I think that an answer to this question, at least from the military standpoint, is that few people like the notion of one very critical node in a network, through which everything goes in and out.

Professor Farrell - I just want to support Dr. Robinson, because, as a communicator, I find rings and ETHERNETS abominations. They waste bandwidth and they create problems. For instance, with simple rings, it is very difficult to see how to do voice transmission between users. The whole concept of providing some form of control over that is very difficult indeed. Engineering means for providing such a control just do not exist.

So I think the future lies with some sort of more hierarchical networks, not necessarily starred, but something in-between. In fact the Computer Science Department of the University of Manchester (U.K.) they are working on something like this, called CENTRENET, which is based on that sort of concept.

Professor Massey - I do not necessarily disagree with the conclusions expressed, but perhaps with the premises. I think a ring network is not a complicated thing to study, but it is the only simple topological structure. As soon as the topology becomes more complex I do not know how to deal with standard problems, such as reliable protocols. Thus, we should think of structures not in terms of whether they are simple or complex, but rather whether they are simple to analyse.

Dr. Gopinath - I think that the difference between star and ring networks boils down to the size of data bus that is required. The data bus is not that big in the ETHERNET, but in a star network it would have to be very large indeed.

Mr. Padlipsky - Another fine point of publicity for MITRENET, or for that matter for any other broadband network, is I guess that you can use their broad bandwidths. If you have a broadband you can use more than one channel over your link. It still boils down to the economy of using a large star which tends to be a 'single-point-of-failure' example.

Dr. Beale - I just want to clear up what you mean by broadband?

Dr. Padlipsky - I do not claim any expertise in hardware, but I understand that in broadband transmission, as opposed to baseband one, you can employ the whole 400 or so MHz that can be put on a CATV cable, discriminating between different portions of the frequencies spectrum. In other words, you can have many channels on some broadband cable. You can use more than one of them for data, and you can still get more bandwidth just by changing your modems. There are even people who are able to use "frequency agile modems", as they say.

Professor Alves - This controversy between rings and stars misses one point, in that as you look at a bus or ring, and its size shrinks, the subscribers are not lost, for their connections are then extended. Then geographically you see a star. The argument is about catastrophic failures of a star, but these do not exist in such growing stars because the stability of the original ring is due to the distributed nature of processing and this is maintained when a ring grows into a star. The distribution of switching equipment is still maintained. So there are no catastrophic failures.

Professor Ephremides - You can ask the same question about any
other type of network. The networking problem can be settled by
providing two-way dedicated links between any possible pair of
users, but this is uneconomical. Thus it is more efficient to
provide a shared bandwidth to users who are bursty.

   Now we must conclude our discussion. Thank you very much
everybody.

REFERENCES

[1] J.K. Wolf, "Principles of Group Testing and an Application to the Design and Analysis of Multi-Access Protocols", (in this volume).

[2] R.A. Padlipsky, "A Perspective on the ARPANET Reference Model", Proc. 1983 IEEE INFOCOM.

[3] J.L. Massey, "Collision Resolution Algorithms and Random Access Communications", University of California, Los Angeles, Tech. Re., UCLA-ENG-8016, April 1980; also in Multi-User Communications (CISM Courses and Lectures Series), G. Longo, Ed., New York: Springer-Verlag, 1981.

[4] C.C. Siegel, "Multiple-Coupled Random Access Techniques for Packet-Radio Networks", MIT Electronic Systems Laboratory, Report ESL-TH-824, June 1978.

J.I. Capetanakis, "The Multiple Access Broadcast Channel: Protocol and Capacity Considerations", MIT Electronic Systems Laboratory, Report ESL-R-806, March 1978.

B.S. Tsybakov and V.A. Mikhailov, "Slotted Multiaccess Packet-Broadcasting Feedback Channel", Problemy Peredachi Informatsii, Vol. 14, No. 4, Oct-Dec. 1978.

J.L. Massey, "Collision Resolution Algorithms and Random Access Communications", University of California, Los Angeles, Tech. Re, UCLA-ENG-8016, April, 1980; also in Multi-User Communications (CISM Courses and Lectures Series), G. Longo. Ed., New York: Springer-Verlag, 1981.

[5] I. Cidon and A. Segall, "Network Resynch Protols", Submitted to IEEE Trans. on Comm.

S.G. Finn, "Resynch Procedures and a Fail-Safe Network Protocol", IEEE Trans. on Comm., Vol. COM-27, No. 6, June, 1979. pp.840-845.

[6] Dr. M. Schlatter, "Ring Systems for Digital Communications – New Concepts and Results", (A dissertation submitted to the Swiss Federal Institute of Technology Zurich, ETH No. 6949), 1981.

Part 4.

APPLICATIONS

Organised by:

J.K. Skwirzynski

G.E.C. – Marconi Electronics, Ltd.,
Chelmsford, Essex, U.K.

THE IMPACT OF PROCESSING TECHNIQUES ON SPREAD SPECTRUM COMMUNICATIONS

George R. Cooper

School of Electrical Engineering
Purdue University
W. Lafayette, Indiana, USA

1. INTRODUCTION

The theoretical aspects of using spread spectrum techniques to improve the performance of communication systems operating in a strong interference environment have been known for over thirty years (1). It is only in recent years, however, that it has become feasible to implement such systems on a practical, operational basis. The ability to accomplish such implementation is almost totally the result of the development of advanced processing techniques. As the capability to implement even more complex processing techniques is developed, the usefulness of spread spectrum techniques in everyday communications will be enhanced. Thus, it is anticipated that spread spectrum communications will play an increasingly important role in the overall communications environment.

This paper outlines some of the fundamental concepts of spread spectrum communication systems and reviews the role that processing techniques play in the generation and detection of spread spectrum signals. The circumstances under which the performance of both military and civilian communication systems can be improved by using spread spectrum techniques is described and the degree to which this improvement is dependent upon complex processing techniques is outlined. Because of space limitations and security requirements the discussion is necessarily general and non-specific.

The impact of processing techniques on spread spectrum communications is not limited to the communicator. Those who would intercept or interfere with spread spectrum communications must also rely upon sophisticated processing techniques. Indeed, the techniques employed by such non-users are often more complex than those employed by the spread spectrum communicators themselves. Such techniques are also discussed here.

2. BACKGROUND

A brief survey of the major types of spread spectrum systems is desirable in order to make clear the role that processing techniques play in the performance of such systems. This survey emphasizes those aspects of spread spectrum that highlight the processing techniques required and is not intended to be complete. More complete discussions of spread spectrum are available in the literature (2),(3).

2.1. Types of Spread Spectrum Systems

A spread spectrum system is by definition one in which the bandwidth of the transmitted signal is much greater than the bandwidth of the message. Furthermore, the bandwidth of the transmitted signal is determined by some modulating function that is independent of the message, but is known to both the transmitter and the receiver. The most common spectrum spreading techniques are direct sequence (usually denoted as PN, for pseudonoise), frequency hopping (FH), and time hopping (TH). Modern spread spectrum systems usually employ some hybrid combination of these techniques.

In a direct sequence system the spreading is accomplished by biphase modulating a carrier with a high rate pseudorandom binary sequence. Each element of this sequence is a "chip" and the rate at which they are produced is the "chip rate". The bandwidth of such a system is frequently taken to be the separation between the spectral nulls closest to the carrier and is, thus, twice the chip rate. Binary message modulation is accomplished by multiplying the pseudorandom sequence by the message, the message bit rate being a sub-multiple of the chip rate.

In the receiver it is necessary to despread the incoming signal by multiplying it by a replica of the original spreading sequence. In order to do this successfully the locally generated replica must be in time synchronizm with the received sequence. Problems associated with the acquisition and synchronization of direct sequence signals form an important area in which complex processing techniques are required.

The simplest form of frequency hopping results from frequency- shift keying (FSK) a carrier with the binary message and then hopping this carrier over a wide range of frequencies in accordance with some pseudorandom frequency pattern. This same pseudorandom hopping pattern must be generated at the receiver in synchronism with the received signal in order to despread it to a normal binary FSK signal.

Time hopping is accomplished by storing a set of messages bits and then transmitting them as a burst in a much shorter time interval. The location of this burst within a time frame that spans the message interval is determined by pseudorandomly selecting one time slot within the frame. The received signal is despread by gating the receiver on during the proper time slot in each frame.

Many different hybrid combinations of these three basic spreading techniques are possible. By appropriately combining techniques it is possible to minimize the disadvantages of each while utilizing the advantages.

For example, a PN/FH hybrid system can achieve a very large bandwidth with a modest number of frequencies and modest PN chip rates.

2.2. Areas of Concern

There are numerous aspects of spread spectrum operation for which the use of complex processing techniques either have improved system performance or show considerable promise of doing so. Of particular importance in this respect is possibility of reducing acquisition time. As noted above, in any type of spread spectrum system it is necessary that the receiver generate a reference signal that is a replica of the spreading portion of the received signal and is in time synchronism with it. In some types of spread spectrum systems the process of acquiring the necessary degree of synchronization may require many seconds. Any processing technique that can substantially reduce this time will be a great advantage.

Another area of concern for the spread spectrum communicator is that of improving the ability of the system to resist the effects of intentional jamming. Again, there are processing techniques available that will aid in this regard. Similarly, the spread spectrum communicator is often concerned with the need to reduce the probability that an unauthorized receiver can detect the existence of a transmission. Beyond this need, however, there is often the additional requirement that the message cannot be recovered by an unauthorized receiver even when it is aware that a transmission is taking place. These concerns are discussed in the sequel under the titles of low probability of intercept (LPI) and security.

The potential interceptor has a converse set of problems. He must be able to detect the presence of transmissions under conditions in which he knows very little about the nature of the signal. Naturally, he would also like to be able to decode those transmissions that he intercepts.

The concerns described above pertain primarily to military communication systems. If spread spectrum is to ever to find widespread use in civilian applications, there are an additional set of concerns that must be addressed. Of primary importance in this regard is the possibility of operating spread spectrum systems and conventional narrowband systems that are geographically and spectrally collocated without significant mutual interference. This mode of operation is currently referred to as an "overlay" and holds great promise for improving the spectral efficiency of civilian communications. Whether or not overlay systems can ever be feasible is in a large measure dependent upon the processing that can be implemented.

In addition to the potential for overlay systems, there are several civilian applications in which the use of spread spectrum is inherently more efficient. An outstanding example of this is mobile communications, in which it has been estimated that the use of spread spectrum can improve the usage of the mobile communication channels by more than an order of magnitude (4). Another civilian application that appears very promising is the cordless telephone, the demand for which is rapidly increasing. Depending upon regulatory considerations, this application may be accommodated either as an overlay or in an exclusively assigned portion of the spectrum.

For either military or civilian spread spectrum systems, an entirely different set of concerns arise when the system involves multiple access; i.e., more than one spread spectrum user attempts to access the same receiving point at the same time. In addition to the problems described above, there is the added problem of separating a set of similar spread spectrum signals into its individual components. This task is frequently complicated by the fact that some users of the system may have signals that are significantly stronger or weaker than others. This is usually referred to as the "near-far" problem and can be handled either by power control or by signal design. Either approach requires a significant processing capability.

3. ACQUISITION OF SPREAD SPECTRUM SIGNALS

The pseudorandom sequence forms a fundamental part of any type of spread spectrum system. All practical pseudorandom sequences are periodic in nature in the sense that the same sequence of values repeat with some fundamental period. When low probability of intercept or secrecy are important issues, it is necessary to make this period very long, frequently hours or even days. Although the intended receiver knows the sequence exactly, and can generate a local replica of it, there is always some uncertainty as to the precise time of arrival. Thus, the receiver must always search in time in order to acquire the incoming spread spectrum signal before any sort of synchronization between the local reference and the received signal can be established.

In some cases there may also be an uncertainty as to the frequency of the received signal. This uncertainty may be due to doppler shift resulting from relative motion between the transmitter and the receiver, or it may be due to frequency drift at either the transmitter or the receiver. If this frequency uncertainty is larger than the coherent processing bandwidth of the receiver (which is on the order of the message bandwidth), then it is also necessary for the receiver to search in frequency in order to acquire the spread spectrum signal.

Thus, in any spread spectrum system there is an uncertainty region in the time-frequency plane that extends from $-\Delta T$ to ΔT along the time axis and from $-\Delta F$ to ΔF along the frequency axis. The magnitudes of ΔT and ΔF can usually be determined from a knowledge of the system parameters. The acquisition phase consists of searching this uncertainty region until the receiver locates the particular time and frequency that correspond to the desired spread spectrum signal. The rate at which this search can be carried out, and the time required to complete it, depend upon the type of spread spectrum system, the search strategy employed, and the amount of parallel processing that is possible.

3.1. Direct Sequence Systems

The long-code direct sequence system poses the most severe problems with regard to acquisition time because of the extremely large number of pseudorandom chips that must be searched over. For example, if the maximum time uncertainty is only one millisecond and the code chip rate is 40 million chips per seconds, the position uncertainty in the sequence may be as great as 40,000 chips. In the presence of large amounts of noise or interference, it may be possible to search only at a rate of a few hundred chips per second. Thus, it could take many seconds to search over the total time uncertainty at just one frequency. Since the search may have to be repeated at more than one frequency, the total acquisition time can become very large indeed.

An obvious approach to reducing the acquisition time in direct sequence systems is to use parallel processing. The total time uncertainty can be divided into N_t time intervals and the total frequency uncertainty divided into N_f frequency intervals. All of the resulting smaller uncertainty regions are then searched at the same time by an appropriate set of parallel correlation operations. This mechanization of the acquisition phase will reduce the maximum acquisition time by a factor on the order of $N_t N_f$. It is clear from this that the reduction in acquisition time is limited only by the amount of parallel processing that it is feasible to implement. With advances in large scale integration, acceptable amounts of parallel processing are steadily increasing to the point where it is reasonable to consider hundreds of parallel correlators and a corrsponding reduction in acquisition times to a few seconds. However, a simultaneous trend to higher chip rates tends to negate the gains made with parallel processing because of the increased number of chips contained in the uncertainty region. Thus, the use of hybrid techniques to gain bandwidth rather than using higher chip rates represents a more favorable trade-off.

Another obvious approach to reducing acquisition time is to use a shorter code, since in this case the acquisition time never needs to exceed the period. While the use of short codes may be acceptable for most civilian applications, it is usually not acceptable for military applications because of LPI and security requirements.

3.2. Frequency Hopping Systems

Although parallel processing techniques can greater reduce the acquisition time in direct sequence systems, it is unlikely that "push-to-talk" can be achieved. For systems with this requirement, frequency hopping is a more reasonable choice because it involves fewer symbols per second. However, even in this case it is usually necessary to utilize some special processing techniques in order to accomplish push-to-talk operation, for which the acquisition time must be less than 0.2 seconds.

In most frequency hop systems the signal energy per hop is sufficiently large that single hop detection is possible. Under these circumstances an approach to reducing acquisition time involves storing short segments of the long frequency hopping sequence and updating these segments in accordance with the receivers estimate of time of day (TOD).

The receiver listens on a frequency that is a function of the TOD and upon hearing that frequency compares subsequent received frequncies with the stored segments. Although this approach is simple in concept, it does involve a substantial amount of data storage and processing. Furthermore, it has been demonstrated to provide acquisition sufficient for push-to-talk operation when the receiver's TOD is known only with wristwatch accuracy.

4. RESISTANCE TO JAMMING

One of the important reasons for the application of spread spectrum techniques to military communication systems is its ability to resist the effects of intentional jamming. How this resistance is achieved depends upon the type of spread spectrum system under consideration. Direct sequence systems are in effect "averaging" systems in that narrowband interference is spread over the entire receiver bandwidth by the despreading operation and then averaged out by the message recovery operation. Frequency hopping systems, on the other hand, are "avoidance" systems in that narrowband interference is simply avoided most of the time by the hopping pattern. A hybrid PN/FH system employs both of these mechanisms to reduce the effects of jamming.

4.1. Direct Sequence Systems

Under ideal conditions a strong narrowband jamming signal in a direct sequence system is suppressed by a factor that is called the "processing gain". This processing gain is on the order of magnitude of the ratio of the spreading sequence chip rate to the message bit rate. Thus, if the chip rate were 40 million chips per second and the message were 2400 bits per second, jamming should ideally be suppressed by 42 dB. In an actual situation, however, the suppression may be considerably less than this because of residual carrier in the locally generated reference signal and because of code unbalance over that portion of the code corresponding to one message bit.

In order to restore the jamming resistance to near its ideal value, it is necessary to filter both the local reference signal and the received signal with narrowband notch filters to remove the unwanted frequency components. Although the reference signal filter can be fixed, the received signal filter must be adaptive in order that its notch frequency can be placed at the jamming signal frequency regardless of where it is in the receiver bandwidth. A technique for accomplishing this employs the fast Fourier transform to estimate the jamming signal frequency, from which the filter notch is adjusted. The time permitted to accomplish both of these tasks may be on the order of one millisecond. Thus, a significant amount of processing is required just to make the system perform more nearly like an ideal system.

The situation becomes even more complex if partial band jamming is involved. In this case the jamming signal consists of a number of narrowband signals that cover only a part of the signal bandwidth. Fortunately, only those frequencies close to the carrier frequency cause the greatest concern. Even so the amount of processing required to restore the ideal jam resistance my be quite substantial.

4.2. Frequency Hopping Systems

When narrowband jamming signals exists in one or more of the frequencies used by the transmitted signal, all message data carried by those particular hops will in all likelihood be destroyed. The type of processing needed to restore this data depends upon how the message data rate compares with the hop rate. If there is exactly one message bit per hop, a common situation, then error correction coding may be employed at the transmitter. The receiver is then able to recover the bits that are destroyed by the jamming signal. In a fast hop system, for which there are several hops during each message bit, there may be sufficient information from the hops that are not interfered with to recover the message without any forward error correction. Alternatively, in a slow hop system, in which there are several message bits per hop, it may be possible to include burst error correction in the transmitted signal and, thus, recover the lost information.

When partial band jamming is present, the situation becomes much more difficult. If a substantial fraction of the hop frequncies are jammed, say on the order of one-half, forward error correction may be inadequate and it may be necessary to resort to some form of retransmission, such as ARQ. The processing that is required in this case calls for the receiver to detect the existence of errors and inform the transmitter. Ideally, the transmitter should utilize some knowledge concerning the jamming frequencies to select a retransmission time that minimizes the number of jammed hops in that particular message segment. Note that the transmitter cannot modify the hopping sequence, because the receiver would be unable to follow an unknown sequence, but it is possible to compute a transmission that is optimum, provided that random delays are tolerable.

5. LOW PROBABILITY OF INTERCEPT

When the objective is to reduce the probability that a spread spectrum transmission can be detected by an unauthorized receiver it is desirable to select a transmission method that makes the instantaneous bandwidth of the transmitted signal as large as possible. This reduces the signal spectral density everywhere and, in many cases, makes it possible for the system to operate below the receiver noise level. Thus, direct sequence systems are much more favorable than frequency hopping systems because their instantaneous bandwidth is the same as their signal bandwidth. Under these circumstances the interceptor, who does not know the spreading sequence, is compelled to rely upon detecting small changes in energy in his receiver.

Under the assumption that the interceptor is using an energy detector, the obvious strategy for the transmitter is to reduce its signal energy to a value that is just sufficient for the intended receiver to recover the message. This calls for some sort of estimation of path loss and power control, both functions requiring a measure of processing capability. Over and above this, however, the transmitter can employ error correction coding to further reduce his transmitted energy per bit. It is not true, as is frequently believed, that the increased transmitted bit rate required by error correction for a given message bit rate, and the corresponding drop in effective processing gain, cancel out the improvement due to coding (5). In fact, transmitted signal energy can be reduced by factors on the order of 5 to 10 dB by proper use of error correction (6).

6. SECURE COMMUNICATIONS

A secure communication system is one in which it is impossible for an unauthorized receiver to recover the message even when it knows that a transmission is taking place but does not know the code. Several techniques, all requiring some degree of processing capabililty, can enhance the security of any spread spectrum system. However, they are most effective in a direct sequence system because of the increased difficulty in observing the code on a chip by chip basis unless the signal is well above the noise level. Almost any frequency hop system can be observed on a hop by hop basis with a suitable spectrum analyzer.

In the first place, the system should employ a code sequence that is much longer than any possible message sequence. Sequences that are several days long are frequently employed for this purpose since the amount of computation required to break the code increases with the length of the code. Secondly, it is desirable to change the code at intervals that are significantly shorter than the period of the code. The hope here is that by the time a given code is cracked, it will no longer be in use.

The pseudorandom sequences most often discussed in connection with spread spectrum systems are maximal length linear shift register sequences because of their desirable correlation properties and their ease of generation. However, if security is a major issue then nonlinear sequences should be employed. The time required to break appropriate nonlinear sequences may be orders of magnitude greater than for a linear sequences of comparable length.

Other processing techniques can also be employed to enhance the security of a spread spectrum communication system. For example, message bits can be interleaved in accordance with some pseudorandom pattern before transmission. Thus, the interceptor must not only break the spreading sequence code, but must also break the interleaving code.

7. INTERCEPTION OF SPREAD SPECTRUM SIGNALS

The unauthorized receiver that wishes to intercept a spread spectrum signal must also engage in a great deal of signal processing if it is to be successful. Since it does not know the spreading code, it must detect the presence of a signal on the basis of an energy measurement only. The common ways of accomplishing this rely upon a spectrum analyzer, an instanteneous frequency measurement, a compressive receiver (which is a very fast scanning frequency analyzer), or Bragg cell electro-optical Fourier analysis.

Having detected the presence of a spread spectrum signal, the next step is to attempt to break the spreading code. This task is extremely processing intensive and requires techniques that are beyond the scope of this discussion. As noted previously, the processing involved in this task is probably orders of magnitude greater than that required by any spread spectrum communicator. Nevertheless, with the development of more efficient algorithms and increased computing capacity, truly secure communications will become more difficult to achieve.

8. SPREAD SPECTRUM OVERLAY SYSTEMS

There are many unresolved problems in connection with operating a spread spectrum system and a narrowband system in an overlay mode, and very little practical experience to rely upon. It is necessary, of course, that the two systems not interfere with one another in any significant way. There are numerous analyses that spell out the interference caused by narrowband signals to spread spectrum systems, since this is strongly related to the jamming problem. These analyses suggest that frequency hopping systems are more likely to be successful in an overlay operation because they are avoidance systems and less susceptible to the near-far problem.

A particular class of frequency hopping systems seems most favorable in this regard. This system employs short sequences that are either phase or frequency modulated by the message. For example, a 32 hop sequence would be modulated with 5 bits of message information. Reception is accomplished with a matched filter that includes separate limiters for each frequency component of the sequence. Very tight bounds have been obtained on the performance of this system when every hop frequency has one or more narrowband interfering signals associated with it (7). These bounds suggest that satisfactory performance of the spread spectrum system can be achieved even when each narrow band signal is significantly stronger than the spread spectrum signal.

Some analytical results are also available on the interference caused to narrowband systems by spread spectrum signals (8). These results suggest that direct sequence systems affect narrowband systems in about the same way as wideband noise would, and that frequency hop systems appear like periodic pulse interference.

The basic question that remains unanswered at this point pertains to designing an overlay system such that the overall performance is optimized. Some tentative steps in this direction have been taken and suggest that there are indeed optimum signal forms for both the spread spectrum system and the narrowband system that will minimize the mutual interference. In order to achieve these signal forms in a real system, it is necessary to make estimates of the spectral density of all channels in the narrowband system and then synthesize a spread spectrum signal that is optimum with respect to this.

Since most communication channels are dynamic in nature, it is necessary that the spectral measurements and the optimization be carried out on a realtime basis with frequent updates. This calls for a great deal of processing capability, but can result in a significant increase in the efficiency with which the spectrum is used.

9. CIVILIAN APPLICATIONS OF SPREAD SPECTRUM

There is recent interest in applying spread spectrum techniques to civilian communication systems. In order for this to be feasible, spread spectrum must offer some desirable characteristics that narrowband systems do not possess. Two such characteristics that are of particular interest in the civilian sector are increased spectral efficiency and privacy. Although spread spectrum is not generally considered to be more spectrally efficient than narrowband communications, there are some situations in which it does appear to be so. An outstanding example of this is the potential application to cellular mobile communications.

Several other applications of spread spectrum in the civilian sector have also been identified. These include the cordless telephone, data communication within buildings, remote control of multiple processes, and medical monitoring systems. Two of the most promising applications are discussed below from the standpoint of the processing required.

9.1. Cellular Mobile Communications

In a cellular mobile communication system, the geographic area to be served is divided up into small regions ("cells"), each having its own base station, and the base stations are interconnected by wire lines or point-to-point microwave links. In a narrowband FM system the total number of channels is divided among the cells so that adjacent cells never have any frequencies in common. Cells that are sufficiently far apart may use the same frequency and it is this frequency re-use that accounts for the increased spectral efficiency of the cellular system as compared to a single-cell system.

The mobile channel is a very difficult one, however, and encounters a great deal of multipath fading when narrowband FM is employed. This fading requires the transmitted signal powers be 18 to 20 dB greater than they would have to be in a non-fading environment, and greatly increases the minimum frequency re-use distance. The net result is that the narrowband FM cellular mobile communication system is not as spectrally efficient as it might at first seem to be. Furthermore, it does require a great deal of processing to cause signals to be "handed off" from one base station to another as a mobile unit moves from cell to cell. The narrowband FM system does not provide for any privacy unless scrambling circuits are used, in which case both ends of the channel must have this equipment. Finally, there is no overload capability since no additional users can be accommodated when all channels are in use.

The situation is quite different when spread spectrum is employed. If a fast frequency hopped system is used, there is a significant amount of frequency diversity that effectively combats multipath fading (9). Thus, the transmitted powers can be substantially less for a given cell size, and the interference powers are reduced by a corresponding amount. Since all users in all cells utilize the entire allocated bandwidth, it is not necessary to switch codes (which is analogous to switching channels in the narrowband system) as the mobile unit moves from one cell to another - although control must still be handed off from one cell to the next. The spread spectrum system inherently provides privacy because a unique code, in the form of a unique frequency hopping sequence, is assigned to each user. Finally, there is limited overload capability because an increased number of users simply degrades the signal-to-noise ratio for all users.

Valid comparisons of the number of users that can be accommodated by narrowband FM systems and spread spectrum systems are difficult to make because of the large number of parameters involved. However, on the basis of some reasonable and conservative assumptions, it has been claimed that the use of spread spectrum in mobile communications would increase the number of potential users by factors of thirty to fifty (4). However, in order to achieve improvements of this order of magnitude a great deal of processing is required. Since the assigned signal codes are unique for each user, and do not change when the user moves from one cell to another, every base station must be capable of receiving and transmitting every code in the entire system in order to accommodate movement between cells. Although this is not as formidable a task as it might seem, it does require a fair amount of computer capability at each cell base station site in addition to the receiver hardware requirements. Processing is also required to accomplish the hand-off from one base station to another as a mobile unit moves from one cell to another, but this task is no more difficult in the spread spectrum case than it is for narrowband FM.

9.2. Cordless Telephone

The application of spread spectrum tecniques to the cordless telephone does not improve the spectral efficiency, but it does provide privacy - an important consideration for many users that is not accommodated by current narrowband systems. Furthermore, it does provide the possibility of accommodating many more simultaneous conversations in a given small geographic area, although current systems are limited by regulatory considerations rather than technological ones.

Two different approaches to the spread spectrum cordless telephone have been identified. One of these is an overlay system in which the cordless telephone utilizes a frequency hopping technique among the guard bands in the UHF television channels. Such a system is expected to accommodate about forty simultaneous users as compared to present systems that can only accommodate two simultaneous users. Furthermore, it would provide complete privacy against the casual eavesdropper.

The second approach proposes to use a 9 MHz band of frequencies that is dedicated to this type of service. Although spread spectrum does not provide greater spectral efficiency in this type of system, it does provide privacy. Frequency hopping is suggested for this application in order to minimize the near-far problem.

Although the processing involved in the application of spread spectrum techniques to the cordless telephone is not as complex as it is for mobile communications, it is still significant. Processing tasks that are required are the generation of the appropriate frequency hopping sequences for dialing, ringing and conversation.

10. SPREAD SPECTRUM MULTIPLE ACCESS

Although the use of spread spectrum techniques in a multiple access system may not be as effective as some of the more conventional approaches, there may be other objectives that dictate its use. These objectives include the need for random access on a non-scheduled basis, the need for privacy, or the need for large bandwidths to achieve more precise time of arrival measurements. Such needs might arise in a military satellite communication system or in a satellite navigation system for either civilian or military applications.

In a multiple access environment a major source of interference for any one user are the signals intended for other users. Most analyses of such systems tend to treat the other user signals as either noise or as essentially unknown interference. This is an unduly pessimistic approach, however, because these other signals can, in fact, be known to each user. With sufficient processing capability the interfering signals at each receiving point can be detected and cancelled out. This would be an extremely difficult task if the total number of interfering signals is large, but could quite conceivably be carried out when that number is small. This approach also solves the near-far problem, quite possibly without the need for power control.

The possibility exists for designing signal sets that exhibit crosscorrelation properties that are more favorable for multiple access than signals selected on a more ad hoc basis. Methods are known for creating such signal sets that contain very nearly the theoretical maximum number of signals (10), (11). Such signals create mutual interference that is much less than the equivalent noise power and, hence, make it possible for the multiple access system to accommodate an increased number of users. They do not solve the near-far problem, however, and power control may be needed to cope with this.

11. CONCLUSIONS

Although the discussion contained in the previous sections has been very cursory, it does emphasize the point that almost all applications of spread spectrum in modern communication systems are critically dependent upon the processing requirements. Furthermore, as the use of spread spectrum becomes more prevalent, the complexity of these processing requirements is bound to increase. It is difficult to predict the extent to which spread spectrum communications will supplant the more conventional narrowband techniques, but it is safe to say that more sophisticated processing techniques hold the key to this development.

1. Scholtz, R. A. The Origins of Spread Spectrum Communications. IEEE Transactions on Communications, vol. COM-30, no. 5, May 1982, pp. 822-854.

2. Dixon, R. C. Spread Spectrum Systems (New York, Wiley,1976)

3. Holmes, J. K. Coherent Spread Spectrum Systems (New York, Wiley,1982)

4. Cooper, G. R. and R. W. Nettleton. Cellular Mobile Communications. IEEE Spectrum, June 1983.

5. Viterbi, A. J. Spread Spectrum Communications-Myths and Realities. IEEE Communications Magazine, vol. 17, no. 3, May 1979, pp. 11-18.

6. Chandler, E. W. and G. R. Cooper. Error Correction Coding Performance Bounds for Spread Spectrum Systems. Proc. 1982 IEEE Military Communications Conference, pp. 8.4-1 - 8.4-5.

7. Chapman, W. W. and G. R. Cooper. A Preliminary Error Analysis of a FH Spread Spectrum Overlay With Uniform Narrow Band Interference. 1983 Conference on Information Sciences and Systems.

8. Farber, L. Spread Spectrum Interference to Voice Communications Systems. Electromagnetic Compatibility Symposium Record, 1978, pp. 282-287.

9. Cooper, G. R. and R. W. Nettleton. A Spread Spectrum Technique for High Capacity Mobile Communications. IEEE Transactions on Vehicular Technology, vol. VT-27, no. 4, Nov. 1978, pp. 264-275.

10. Cooper, G. R. and E. H. Cooper. Design of Spread Spectrum Signal Sets with Small Mean-Square Crosscorrelation. Proceedings of 20th Allerton Conference on Communication, Control and Computing, October 1982, pp. 955-956.

11. Cai, K. V. and G. R. Cooper. Relationships Between the Number of Signals and Their Correlation Properties. IEEE International Symposium on Information Theory, June 1982.

DATA COMMUNICATION APPLICATIONS OF INCREMENTAL SIGNAL PROCESSING*

L. E. Franks and F. S. Hill, Jr.

Department of Electrical and Computer Engineering
University of Massachusetts
Amherst, Massachusetts 01003

1   INTRODUCTION AND BACKGROUND

From the standpoint of VLSI implementation, the ideal architecture for digital signal processing devices is one which exhibits
a high degree of modularity, along with a low degree of interconnectivity between modules. Such a structure not only permits concurrent localized processing, but also avoids the 'bottleneck' problem of time-sharing and communicating data to and from a single,
high-speed arithmetic processor. Of course, this "parallel processing" approach can be applied to conventional digital signal
processing, where typically 8- to 12-bit words are used to represent signal and filtering parameters, but the circuit complexity of
multipliers is a deterrent to implementing a large-order system in
this manner. Another major problem is simply the amount of chip
area that must be devoted to the 8- to 12-bit data buses. These
problems are alleviated in direct proportion to the amount of reduction in word length that can be accomplished by alternative
methods of encoding the signal and system parameters.

Word length reduction by differential encoding (e.g., DPCM)
has been extensively studied from the standpoint of reducing
channel capacity requirements in digital data transmission systems.
It is well known that word-length reductions can also be accomplished when sampling rates are increased. However, specific
system implementations which exploit this property of time-quanti-
zation vs. amplitude-quantization tradeoff seem not to have been
very thoroughly investigated. One interesting proposal and some

*This work has been supported by the National Science Foundation
under Grant No. ECS-8005688.

specific results are reported by a group at Philips Research Laboratories, where the performance of the A/D conversion operation is improved with an elevated sampling rate (1).

As we pass to the limit where sampling rate is sufficiently high that "one-bit" words accurately characterize incremental signal quantities, we find several forms of signal processing structures which appear very attractive for VLSI implementation. One version of the signal format is linear delta modulation, familiar to communications engineers and famous for its extremely simple implementation. In another context, the concept is familiar to computer engineers in the form of the digital differential analyzer (DDA). This machine can be regarded historically as bridging the gap between analog computers and general-purpose digital computers (2-5). Since that time, the DDA has received little attention until it was recognized that it could be an ideal structure for the implementation of special-purpose signal processing functions on an IC chip (6-7).

The principle of the DDA is that of interconnecting a large number of basic functional blocks (as in the analog computer) which operate on coarsely quantized (e.g., binary-or ternary-valued incremental signal values (another name for the DDA was 'digital incremental computer' (DIC)) at a rate sufficiently large to achieve the desired accuracy. A final integration operation is used to convert the results into absolute, rather than incremental values. DDA processing techniques also permit signal multiplication and non-linear operations, suggesting communications applications in the design of modulators, data detectors, and timing recovery circuits. We shall limit our attention here, however, to linear, time-invariant filtering or equalization operations. The overall structure, as shown in Figure 1, involves a delta modulator for generating the incremental signal quantities as input to the processing device; the output of which is demodulated with a discrete-time digital integrator. The delta modulator is characterized by the equations:

$$\delta x(n) = Q(x(n) - \overline{x}(n))$$

$$\overline{x}(n) = \overline{x}(n-1) + s \cdot \delta x(n-1)$$

(1)

where s is the step size and the quantizer, Q, may be binary, $Q(\cdot) = \text{sgn}(\cdot)$, or Q may produce a ternary incremental signal, $\delta x(n) \in \{-1, 0, +1\}$. The use of ternary signals is not entirely inconsistent with conventional logic design because, for example, the digital integrator is basically just an up/down counter and the input signal can be used to indicate whether register contents should be incremented, decremented or left unchanged.

The first proposal for using delta modulation for implementation of filters was made by Lockhart (8) about ten years ago. His

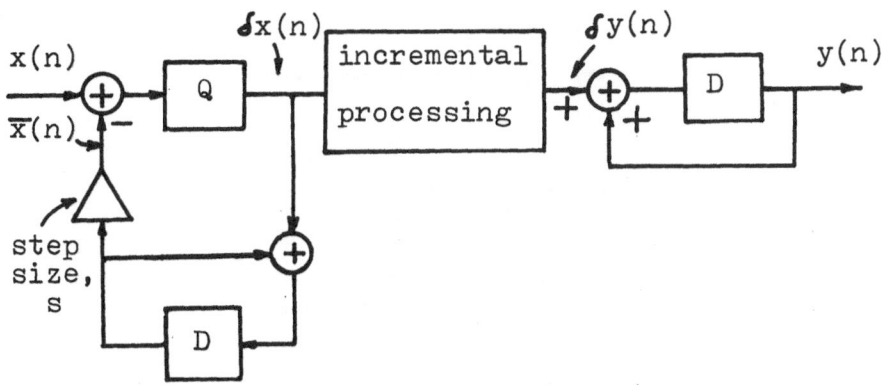

Figure 1. Digital Filtering Based on Delta Modulation

structure was that of a transversal filter, forming the incremental output signal,

$$\delta y(n) = \sum_{k=0}^{N} h(k \cdot K) \, \delta x(n - k \cdot K) \qquad (2)$$

Since the $\delta x(\cdot)$ is a binary sequence, the delay lines in the earlier analog transversal filters could be replaced by a shift register. Since the sampling rate, $1/\tau$, for delta modulation is usually substantially higher than the Nyquist rate, $1/T$, for the analog signal source, and since transversal filter taps usually use a T-second delay interval, only every Kth term in the discrete convolution of Equation 2 needs to be used; i.e., the shift register is tapped every Kth stage. The parameter $K = T/\tau$ is called the "oversampling parameter" for delta modulation. In Lockhart's structure, the scaling (coefficient multiplication) and summing in (2) were performed by analog circuits. The same structure, called a binary transversal filter, was also studied as a generator of bandlimited pulse waveforms by Voelcker (9), Lockhart (10), and Hill and Lee (11, 12). Franks (13, 14) presented schemes for digital scaling by modulating the widths of the delta modulation pulses or by time-multiplexing pulses of the same binary weights from different taps into a single sampling interval. Summation was, in effect, provided by analog means.

In considering all-digital implementations, the requirements of Equation 2 are sometimes regarded as "multiplier-free" because of the one-bit nature of the $\delta x(n)$ word as a factor (15). At each tap in the structure, it is only necessary to add (or subtract) the coefficient word from the accumulated sum up to that tap. Viewing the computation in this manner leads to a different viewpoint of the nature of the transversal structure. The structure can be re-garded as a simple one-dimensional array of two-input, two-output modules of the form shown in Figure 2. Using D-transforms for the $u_i(\cdot)$ and $v_i(\cdot)$ sequences, where $D = \exp[-j2\pi\tau f]$, we characterize

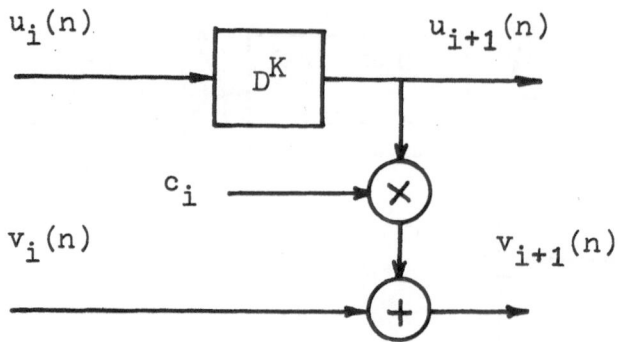

Figure 2. Transversal Filter Module

the module by the matrix equation,

$$
\begin{bmatrix} U_{i+1}(D) \\ V_{i+1}(D) \end{bmatrix} = \begin{bmatrix} D^K & 0 \\ c_i D^K & 1 \end{bmatrix} \begin{bmatrix} U_i(D) \\ V_i(D) \end{bmatrix}
\tag{3}
$$

where c_i is the ith tap factor. If we make the connection $V_0(D) = c_0 U_0(D)$ at the input module of the array and take the output of the array as $V_N(D)$, we get the familiar form of the non-recursive filter transfer function,

$$
[V_N(D)/U_0(D)] = H(D) = \sum_{i=0}^{N} c_i D^{K \cdot i}
\tag{4}
$$

It is interesting to note that, from this viewpoint, the transversal filter is a special case of the lattice filter.

We shall show later that the module can be greatly simplified by restricting the coefficients to ternary values; $c_i \in \{-1, 0, +1\}$. Hence, in addition to the shift register delay device, we only require a "one-bit" summer to form the $v_{i+1}(\cdot)$ output sequence.

A device for performing the logical summation of two delta modulation sequences, and its application to transversal filtering for very specialized sets of tap coefficients, was given by Engel and Steenaart (16). Later, Kouvaras (17) gave a different implementation of this summer, requiring only a binary full adder device and a flip-flop. This type of summer provides a delta-modulation sequence corresponding to one-half the sum of the original analog sequences. Unfortunately, the 'one-half' factor makes the device unsuitable for the modular form of the transversal filter. Kouvaras also showed how b of the 'half-sum' adders should be interconnected to implement an arbitrary b-bit scaling operation. A novel realization of the delta modulation transversal filter is due to Peled and

Liu (18). Their scheme uses the binary tap signals to address registers of a ROM, each register containing the required value of the
sum-of-products for that particular output sample, thus completely
avoiding multiplication and summing circuitry.

From this discussion, we appreciate that there are a great
number of structural alternatives for an all-digital realization of
the delta-modulation based transversal filter. We shall examine
some of the more promising of these and also consider some recursive
filter structures in the following sections.

2 THE DDA AS A CIRCUIT ELEMENT

The DDA can be regarded as an interconnection of basic blocks
of the form shown in Figure 3, each block consisting primarily of
two registers. Depending on external connections, this basic block
can function as an integrator, a scaler, or a summer (2-5). In
Figure 3, R and Y are b-bit registers. Y contains the accumulated
sum of the ternary increments δy. Hence its contents are a signed
integer, usually in two's-complement form. The R register holds the
remainder, or quantization error, $r(n) = \Delta z(n) - \delta z(n)$. As the remainder is saved for the next computation cycle, errors due to the
ternary quantization Q have only a limited effect on the integrated
output signal. The contents of R are interpreted as an unsigned
fractional quantity; $0 <= r(n) < 1$. The device marked TFR (transfer register) simply adds or subtracts the contents of the Y register to the R register on each machine cycle (each τ seconds) according to whether $\delta x(n) = +1$ or -1. If $\delta x(n)$ is a ternary sequence,
then $\delta x(n) = 0$ causes no transfer to be made. The basic DDA block
is thus characterized by

$$y(n) = y(n-1) + \delta y(n)$$
$$\Delta z(n) = 2^{-b} y(n)\ \delta x(n) + r(n-1) = Q(\Delta z(n)) + r(n) \tag{5}$$
$$\delta z(n) = Q(\Delta z(n)).$$

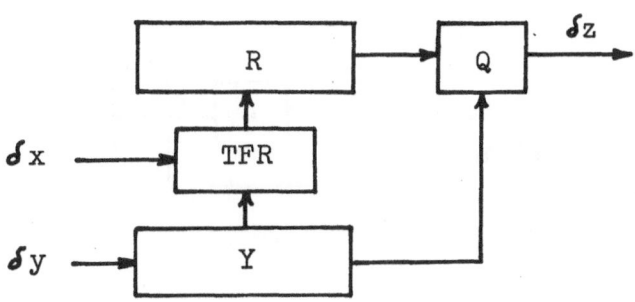

Figure 3. Basic DDA Block

The ternary output sequence $\delta z(n)$ is formed as follows: $\delta z(n) = +1$ if the result of combining the Y and R registers exceeds 1. $\delta z(n) = 0$ if the result is between 0 and 1, and $\delta z(n) = -1$ if the result is negative. Thus the quantizer can be implemented by simple logic operations on the carry (overflow) bit from the R register and the sign bit from the Y register. In the binary in- cremental logic case, the structure is quite similar if a different form of numerical representation is used for the R and Y registers. One form that works well is to use two's-complement with the sign bit only complemented in both registers.

To use the device of Figure 3 as an integrator, $\delta y(n)$ is the input signal and $\delta x(n) = 1$ for all n. For a scaler, however, $\delta y(n) = 0$ and Y contains the scaling constant C. The input is $\delta x(n)$ and $\delta z(n)$ is the ternary delta modulation sequence for the analog signal $C \cdot x(n)$. A rearrangement of the equations (5) leads to an equivalent block diagram for the scaler, Figure 4, which is recognized as the structure for a delta-sigma modulator. An obvious application for the DDA scaler is implementation of the tap factors on a transversal filter. We have tested this idea by simulation of a 32-term nonrecursive approximation to the response of a fifth- order analog Chebyshev lowpass filter with a passband ripple of 0.1 db. (19). A cuttoff frequency of $(14T)^{-1}$ and an oversampling of K = 200 was selected. Tap coefficients were rounded to 7-bits, as shown in Table I, so the size of the R and Y registers was cor- respondingly limited. Neglecting the 0 and 1 coefficients, only 28 DDA scalers needed to be simulated.

A DDA summer is implemented by "multiplexing" a number of $\delta x(n)$ inputs in the basic scaling configuration. In the present example, 28 register transfers would take place during each τ-sec- ond machine cycle, according to the increment signals at the var- ious shift-register taps. The summer was implemented with a scaling constant of C = 1/32. The gain frequency response of the actual filter and the target response are shown in Figure 5.

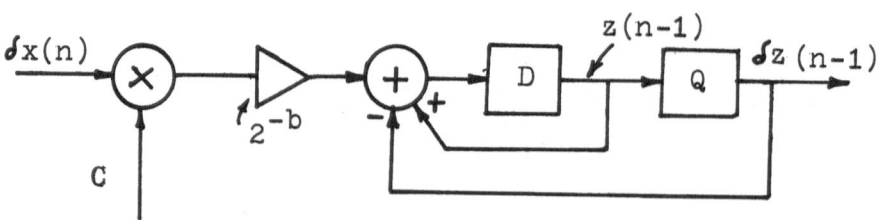

Figure 4. Delta-Sigma Version of the DDA Scaler

$c_0 = 0.0001011,$ $c_1 = 0.01,$ $c_2 = 0.0111110,$
$c_3 = 0.1011110,$ $c_4 = 0.1110111,$ $c_5 = 1.0,$
$c_6 = 0.1110110,$ $c_7 = 0.1011100,$ $c_8 = 0.011100,$
$c_9 = 0.0010010,$ $c_{10} = -0.0001101,$ $c_{11} = -0.01,$
$c_{12} = -.0100110,$ $c_{13} = -0.01,$ $c_{14} = -0.0010011,$
$c_{15} = -0.0000111,$ $c_{16} = 0.0001001,$ $c_{17} = 0.001,$
$c_{18} = 0.001001,$ $c_{19} = 0.0001110,$ $c_{20} = 0.0000111,$
$c_{21} = 0.0,$ $c_{22} = -.0000110,$ $c_{23} = -0.0001001,$
$c_{24} = -0.0001,$ $c_{25} = -0.0000110,$ $c_{26} = 0.0,$
$c_{27} = 0.0,$ $c_{28} = 0.0000100$ $c_{29} = 0.0000101,$
$c_{30} = 0.0000100,$ $c_{31} = 0.0000011.$

Table I.   List of Tap Factors For Transversal Filter Design Example
of Figure 5

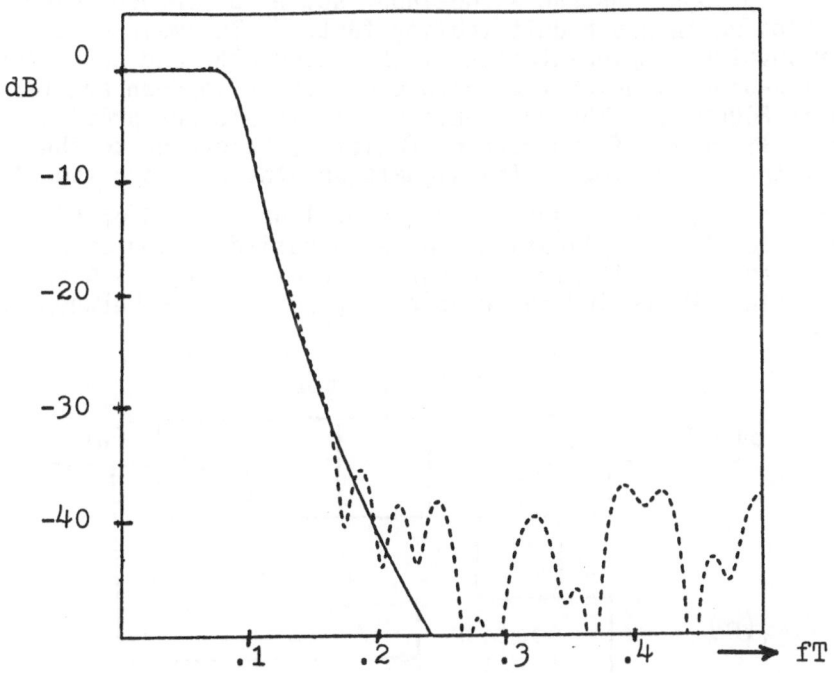

Figure 5.   Gain vs. Frequency Response: Target Function and 32-term
Nonrecursive Approximation

3 THE TERNARY-COEFFICIENT TRANSVERSAL FILTER

If we incorporate two modifications to the foregoing transversal filter design approach, we can gain a remarkable simplification in filter implementation. First, we employ a much denser
tapping arrangement by replacing D^k by D in the basic module shown
in Figure 2. This is not meant to imply a reduced oversampling
rate, which might typically be a factor of 100, but rather that
taps are made at every stage of the shift register, not every Kth
stage. Second, a low-order recursive filter (such as an integrator)
is placed in cascade with the transversal structure. This allows
us to regard the set of coefficients as incremental values and hence
they may be very coarsely quantized. We examine in particular the
ternary case where $c_i \; \varepsilon \; \{-1,0,+1\}$. In this case, the structure is

not only "multiplier-free" but we require only the addition of two
"one-bit" words in each module of the array.

As mentioned previously, earlier implementations of a two-
input delta-modulation summer (16-17) involved a "one-half" scaling
factor. While this factor can be compensated for in some structures,
it cannot be used in the module of Figure 2, because it introduces
a D-plane scaling which repositions the roots of the transfer function. We propose instead a two-input summer based upon binary DDA
principles which has a unit scaling factor. The main part of the
summer involves implementation of the integrator and quantizer in
the delta-sigma modulator of Figure 4 with an up/down counter as
shown in Figure 6. The combinational logic element provides increments of ±1 or ±3 of the counter register, according to the polari-
ties of the $\delta x_1(n)$ and $\delta x_2(n)$ signals and the feedback signal from

the MSB stage of the counter. Register lengths of 4 or 5 bits seems
adequate for this application. An alternative to the combinational
logic is to time-multiplex the three increment signals within one
clock cycle. We remind the reader that connections between trans-

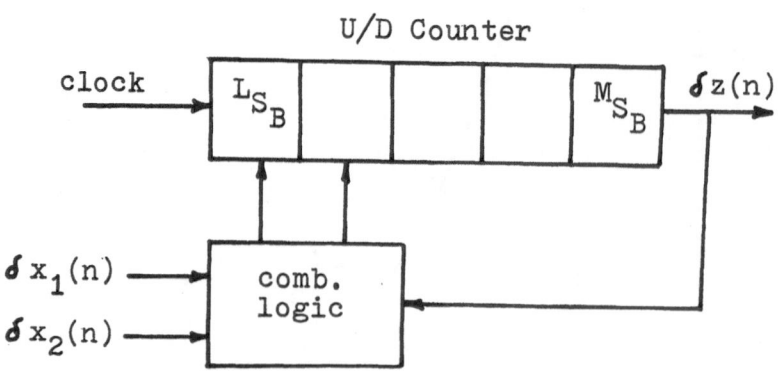

Figure 6. Two-Input Delta-Modulation Summer

versal filter modules are via "one-bit-wide" data buses; a great advantage in chip layout.

Addressing now the question of how to select the set of ternary c_i coefficients, we first consider a very simple method which also gives some insight as to why the approach is effective. Regarding the $h(\cdot)$ sequence in Eqn. 2 as representing samples of the desired analog impulse response, we can "delta-modulate" this sequence to get a ternary c_i sequence from

$$\overline{h}(i) = \overline{h}(i-1) + \sigma \cdot c_i \qquad (6)$$

where σ is the step size in this construction. The running sum of the c_i, which forms the $\overline{h}(i)$ approximations to the desired $h(i)$, is formed by a first-order recursive section (integrator) with function, $(1-D)^{-1}$. On the premise that the bandwidth of the filter is on the same order as the bandwidth of the signal, an oversampling factor which is adequate for one should be approximately right for the other.

This procedure, while very simple, may not yield an optimum set of ternary coefficients. It is possible to employ an iterative procedure to successively modify the coefficients to minimize the mean-squared-error in the overall response function (20). We have developed a dynamic programming algorithm which greatly speeds up the search procedure (21-22). This has permitted us to study filter designs, not only for a broad range of parameter values, but also for different types of recursive reconstruction sections to be incorporated in cascade with the transversal structure. For example, using a double integrator, $(1-D)^{-2}$; i.e., letting the c_i represent second-differences of the $h(\cdot)$ sequence we always find substantial improvement in mean-squared-error for a given oversampling factor. Some slightly more complex "averagers" give even better results, although the "best" seems to depend upon the nature of the target response. As an example, we consider the design of a Nyquist rolloff filter for data communication. The target function, illustrated in Figure 7, is of the "double-jump" variety with 40% excess bandwidth. This rolloff shape is selected for its optimal immunity to intersymbol interference resulting from small symbol timing errors (23). The actual transfer function realized, using a double-integration averaging, is also shown in Figure 7. Figure 8 shows the mean-squared-error obtainable with other recursive averaging sections. These results are presented in another fashion in Table II, showing the possible reduction in array length and sampling rate, for a given mean-squared-error.

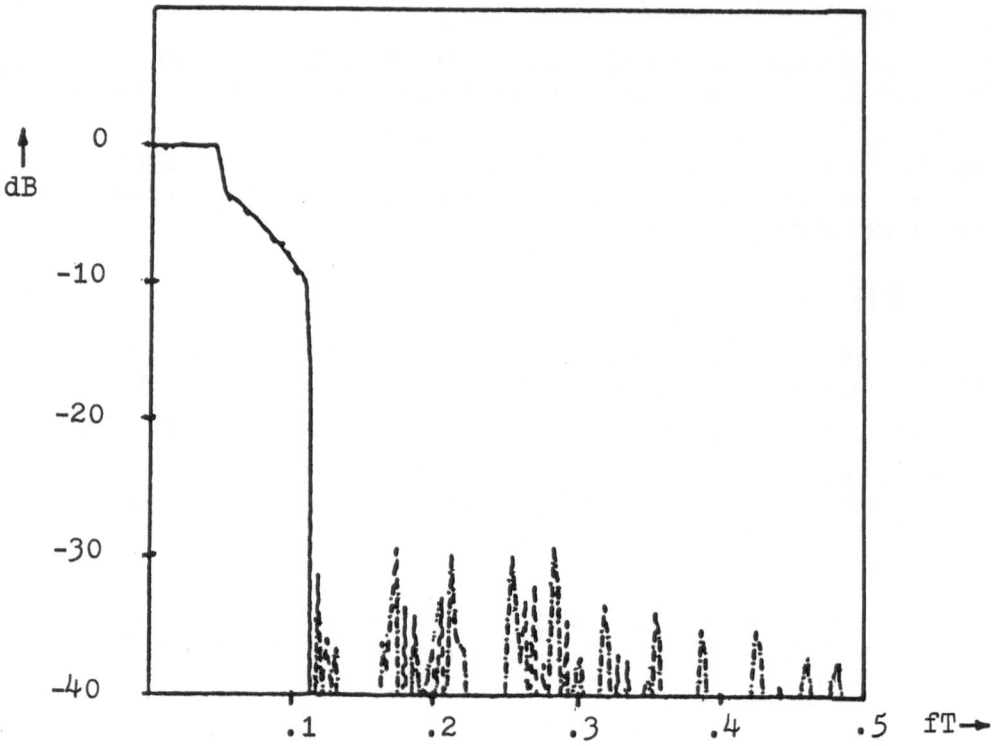

Figure 7. Nyquist-Filter Frequency Response: Target Function and
Ternary-Coefficient Transversal Approximation Using
Double Integration.

F(D)	OVER-SAMPLING FACTOR	# OF TAPS
$1/(1-D)$	26	1317
$1/(1-D)^2$	7	356
$1/(1-D)^3$	5	256
$1/(1+D)(1-D)^3$	5	257
$(1-D^K)/(1-D)^2$	6	299

Table II. Required Oversampling and Filter Length for MSE = 2×10^{-3}
for Nyquist Filter. Various Recursive Averaging Sections.

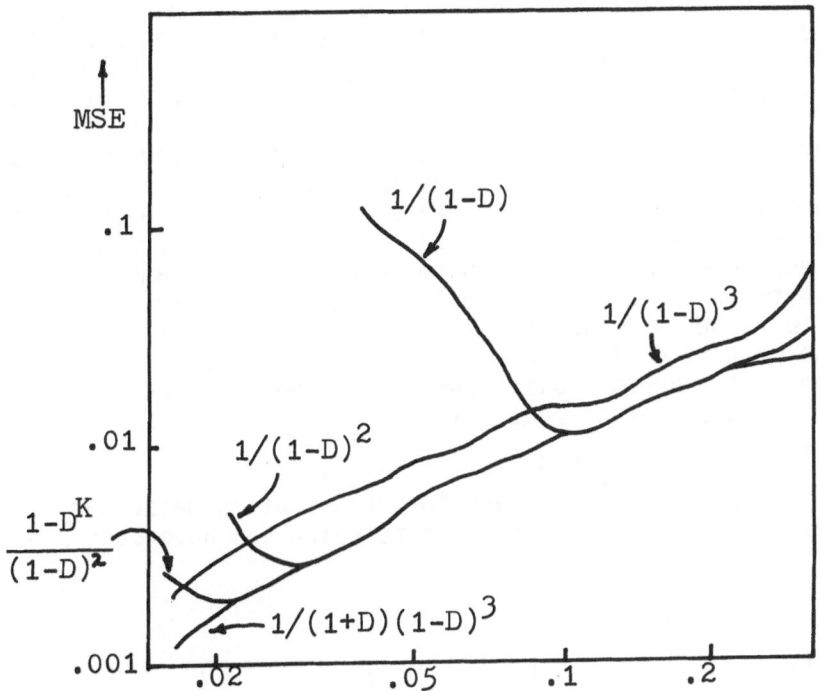

Figure 8. Mean-Squared-Error vs. Step Size for Transversal Realization of Nyquist Filter in Fig. 7. Various Recursive
Averaging Sections. Oversampling Factor = 6.

4 RECURSIVE INCREMENTAL FILTERING

Because the basic DDA block of Figure 3 can be made to function as an integrator, a scaler, or a summer, it is clear that the
appropriate inter-connection of such blocks could be made to realize
an arbitrary transfer function rational in D. The design procedure
would, in fact, resemble some of the procedures employed in active
RC filter design. This approach has been proposed by Abu-El-Haija
et al (6). Because of large oversampling factors, pole locations
tend to cluster about the D = 1+j0 point. This is known to cause
problems with sensitivity and roundoff noise with conventional
digital filters, but the problem is alleviated by using the digital
integrator as a basic element rather than delay as a basic element
(24). As expected, the DDA recursive filter ralizations tend to
perform better as the sampling rate is increased; i.e., as poles
move in closer to the D = 1 + j0 point. It seems natural to inves-

tigate the "biquad" approach to filter design, using a cascade of second-order recursive filters to realize the overall transfer function. The straightforward approach to a second-order lowpass filter would call for the interconnection of five of the basic DDA blocks; two integrators, two scalers, and a summer. This structure would therefore contain five ternary quantizers, each contributing in some degree to the total quantization noise of the second-order section. The quantization noise performance can be dramatically improved by using extended-precision registers in a modified configuration shown in Figure 9 (25-26). This structure will also reduce or eliminate the occurrence of limit cycles (27). The improved performance comes about because higher resolution (more bits) data transfers are permitted between registers within the basic second-order section. We regard the Y and R registers as the concatenation of two registers, with the most significant bits in the left hand part. (This is why the signal flow appears different from normal in Figure 9.) It is sometimes convenient to regard Y_{11} and Y_{21} as comprising the integer parts of the Y_1 and Y_2 registers while Y_{12} and Y_{22} contain the fractional parts. Some expressions for quantization noise and some preliminary guidelines for register size selection have been presented (25).

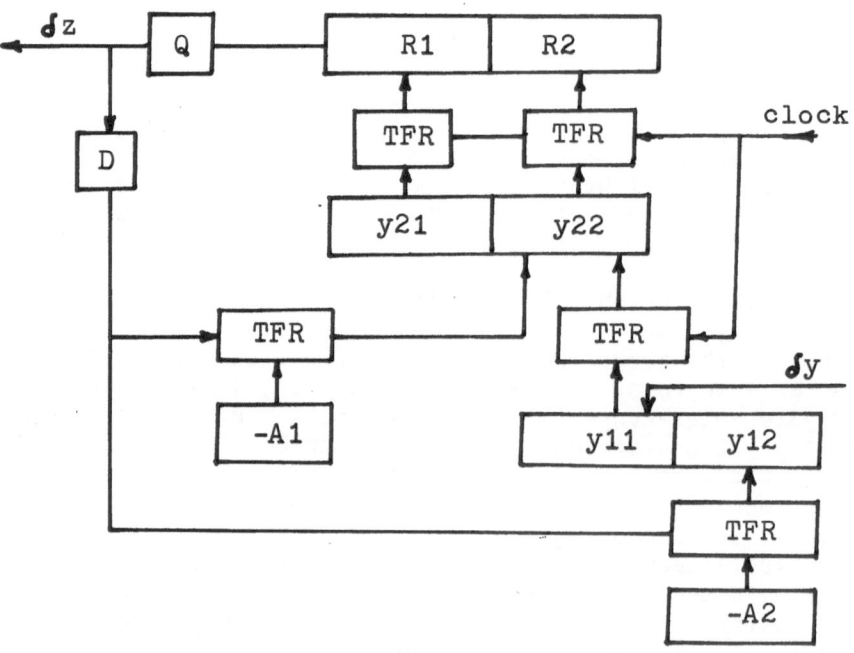

Figure 9. Second-Order Recursive DDA Filter. Parameters Stored in Registers A1 and A2 Determine the Pole Positions.

5 CONCLUSIONS

Some of the background on the evolution of structures for transversal filters based on delta-modulation encoding of the analog signals has been presented. New structures, of a type more suitable for VLSI implementation have been introduced along with some discussion of the design procedures that may be used. Currently under investigation are alternative modular structures (such as the lattice, or ladder structures based on continued-fraction expansions) with the objective of maximizing hardware efficiency under the constraint of ternary parameter values. Also under investigation is the suitability of these structures as adaptive equalizers for data communication channels.

Our approach to the design of recursive filters has taken a rather different path. Simulation studies of a particular second-order configuration indicate the possibility of high-performance digital "biquads" with quite modest hardware requirements. However, questions of optimizing the configuration and allocation of register resources deserve further investigation.

6 REFERENCES

1. Claasen, T.A.C.M., et al. Signal Processing Method for Improving the Dynamic Range of A/D and D/A converters, IEEE Trans., ASSP-28, pp. 529-538, October 1980.

2. Shileiko, A. V. Digital Differential Analyzers, Pergamon Press, 1964, (D. P. Barrett, Trans.).

3. Sizer, T.R.N. The Digital Differential Analyzer, Chapman & Hall, Ltd., 1968.

4. Mayorev, F. V. Electronic Digital Integrating Computers-Digital Differential Analyzers, American Elsevier, 1964.

5. Bartee, T. C., I. L. Lebow and I. S. Reed. Theory and Design of Digital Machines, McGraw Hill Book Company, Inc., 1962.

6. Abu-El-Haija, A. I., K. Shenoi, A. M. Peterson. Digital Filter Structures Having Low Error and Simple Hardware Implementation, IEEE Trans., CAS-25, pp. 593-599, August 1978.

7. Franks, L. E. Signal Processing Structures Based on Digital Incremental Computing Techniques, Int. Conf. on Digital Signal Processing, Florence, pp. 419-431, September 1981.

8. Lockhart, G. B. Digital Encoding and Filtering Using Delta Modulation, The Radio and Electronic Engineer, Vol. 42, pp. 547-551, December 1972.

9. Voelcker, H. B. Generation of Digital Signaling Waveforms, IEEE Trans., COM-16, pp. 81-93, February 1968.

10. Lockhart, G. B. Binary Transversal Filters with Quantized Co-efficients, Elec. Lett., Vol. 7, pp. 305-307, June 1971.

11. Hill, F. S., Jr. and W. U. Lee. PAM Pulse Transmission Using Binary Transversal Filters, IEEE Trans., COM-22, pp. 904-913, July 1974.

12. Hill, F. S. Jr., and W. U. Lee. Implementation of PAM Pulse Generators with Correlative Coding, IEEE Trans., COM-23, pp. 483-485, April 1975.

13. Franks, L. E. Hybrid Implementation of Sampled-Data Filters, IEEE Int'l. Symp. on Circuits and Systems, San Francisco, April 1974.

14. Franks, L. E. Hybrid Implementation of Digital Filters, European Conference on Circuit Theory and Design, London, July 1974.

15. Lee, J. C. and C. K. Un. On FIR Delta Modulation Digital Filters, IEEE Trans., ASSP-29, pp. 1194-1201, December 1981.

16. Engel, L. J. and W. Steenaart. Digital Summation of Delta Modulation Signals, Canadian Communication and Power Conference, Montreal, 1976.

17. Kouvaras, N. Operations on Delta-Modulated Signals and Their Application in the Realization of Digital Filters, The Radio and Electronic Engineer, Vol. 48, pp. 431-438, September 1978.

18. Peled, A. and B. Liu. A New Approach to the Realization of Non-recursive Digital Filters, IEEE Trans., AU-21, pp. 477-484, December 1973.

19. Xiang, J., T. Lu, F. S. Hill, Jr., and L. E. Franks. A New Digital Transversal Filter Implementation Based on Delta Mod-ulation Encoding, Canadian Communications and Energy Conference, Montreal, Canada, October 1982.

20. Bateman, M. R. and B. Liu. An Approach to Programmable CTD Filters Using Coefficients 0, +1, and -1, IEEE Trans., CAS-27, pp. 451-456, July 1980.

21. Benvenuto, N., L. E. Franks and F. S. Hill, Jr. Realization of Finite Impulse Response Filters Using Coefficients +1, 0 and -1, submitted to IEEE Trans. on Communications.

22. Benvenuto, N., F. S. Hill, Jr. and L. E. Franks. Dynamic Pro-gramming Methods for Designing FIR Filters Using Coefficients -1, 0 and 1, submitted to IEEE Trans. on Acoustics, Speech and Signal Processing.

23. Franks, L. E. Further Results on Nyquist's Problem in Pulse Transmission, IEEE Trans., COM-16, pp. 337-340, April 1968.

24. Agarwal, R. C. and C. S. Burrus. New Recursive Filter Structures Having Very Low Sensitivity and Roundoff Noise, IEEE Trans., CAS-22, pp. 921-927, December 1975.

25. Xiang, J., L. E. Franks and F. S. Hill, Jr. New Structures for Delta Modulation Recursive Digital Filters, 25th Midwest Symposium on Circuits and Systems, Houghton, Michigan, August 1982.

26. McGhee, R. B. and R. N. Nilsen. The Extended Resolution Digital Differential Analyzer: A New Computing Structure for Solving Differential Equations, IEEE Trans. Comp., Vol. C-19, 1-9, January 1970.

27. Witten, I. H. and P. G. McCrea. Suppressing Limit Cycle in Digital Incremental Computers, IEEE Trans. Circuits Syst., Vol. CAS-28, No. 7, July 1981.

A WIDEBAND ADAPTIVE FILTER USING "SAW" DEVICES

Avni MORGÜL   and   Peter M. GRANT †

Dept. of Electrical Eng., Boğaziçi University, Istanbul, TURKEY
† *Dept. of Electrical Eng., University of Edinburgh, Edinburgh,*
Scotland, U.K.

INTRODUCTION

This paper will cover the basic principles of a frequency domain
adaptive filter and implementation of surface acoustic wave   (SAW)
analogue Fourier transform processors in order to perform the forward and inverse Fourier transformations. A practical realisation
of this kind of filters which employs a cascade of forward and inverse SAW Fourier transform processors interconnected via a digital
interface will also be presented, Figure 1.

The frequency domain adaptive filtering was suggested by Dantino
et al. in 1978 (3) and there are a number of published reports since
then on this subject (3,4,5). These filters have potential applications in the cancellation of uncorrelated noise and interference
and in channel equalisation in communication systems.   They offer
faster more controllable convergence then conventional time domain
adaptive transversal filters (5) by reducing the eigenvalue spread
of the data autocorrelation matrix and they considerably reduce
the computational complexity. This is obtained at the restriction
of a filter which implements circular rather then linear convolution.

Although it is possible to use frequency domain adaptive filters in
conjunction with any kind of Fourier transformer _such as digital
fast Fourier transformers-, analogue surface acoustic wave (SAW)
chirp Fourier transform techniques permit these filters to be realised with 1-60 MHz bandwidth.

In our realisation the input samples are transformed to yield their
Fourier coefficients and subsequently multiplied by a set of stored
weights before inverse Fourier transformation to yield the normal

time domain output. The filter adapts by comparing the multiplied output, sample by sample, with the Fourier transform of a separately supplied desired response to yield a set of error signals which each update the appropriate stored weight. This minimises the error or discrepancy between the filtered and desired signals.

Computer simulations with fixed convergence coefficient (μ) for each of the multiplier stages and also variable μ, where the precise μ value for each weight is selected dependent on the signal level at the transformed outputs are presented. It is shown that the variable μ gives considerable enhancement in convergence rate over the fixed μ filter for signals where there is a wide spread in the eigenvalues of the input auto-correlation matrix. Practical results are presented along with the simulations for a frequency domain adaptive filter based on 100 point SAW chirp transform processors which give a 4 MHz real time bandwidth capability.

THEORY

In a conventional time domain transversal filter the input signal samples x(n) are multiplied by the appropriate stored filter weights w(n) and summed over the filter points to yield the output y(n):

$$y(n) = \underset{\sim}{w}^T(n).x(n) \tag{1}$$

where $\underset{\sim}{x}(n) = \left(x(n) \ x(n-1) \ \dots \ x(n-M+1) \right)^T$ is the vector of M stored samples and $\underset{\sim}{w}(n)$ is the vector of filter weights. The processor is made adaptive by differencing the desired signal d(n) with the filtered output y(n) to derive an error e(n):

$$e(n) = d(n) - y(n) \tag{2}$$

which is used to update the weights and minimise the error. In the least mean square (LMS) adaption algorithm (1) the updated weight vector for the n + 1'th filter iteration is given by

$$\underset{\sim}{w}(n+1) = \underset{\sim}{w}(n) + 2\mu e(n) \ \underset{\sim}{x}(n) \tag{3}$$

where μ is the step size or convergence coefficient which controls the adaption rate.

In the frequency domain realisation data is collected into M sample blocks and processed in M point discrete Fourier transform (DFT) processor (2) into M contiguous complex outputs. Time domain convolution can now be implemented by multiplication with a set of stored frequency domain weights H(n). Thus the output coefficient corresponding to the frequency k for the n'th iteration of the filter is given by :

$$Y_k(n) = H_k(n) \cdot X_k(n) \tag{4}$$

The error is again calculated by differencing

$$E_k(n) = D_k(n) - Y_k(n) \tag{5}$$

Note that this provides M orthogonal error outputs rather than the single global error of time domain approach. $E_k(n)$ is minimised by using the LMS algorithm (1)

$$H_k(n+1) = H_k(n) + 2\mu E_k(n) \, X_k^*(n) \tag{6}$$

$X_k^*(n)$ is the complex conjugate of $X_k(n)$.

If processor complexity is to be minimised then the complex form of the clipped LMS algorithm (8) can be used and $X_k^*(n)$ replaced by its sign (-1 or +1) information only:

$$H_k(n+1) = H_k(n) + 2\mu E_k(n) \left(\text{sign } X_k^*(n)\right) \tag{7}$$

The frequency domain implementation is initially more complicated because the transforms of real signals yield complex Fourier coefficients and hence all calculations including the LMS algorithm of equation 6 must be performed with complex arithmetic (9). However for real input signals the DFT is conjugate symmetric and hence only M/2 rather then M complex single tap filters are required (4).

As the filter processes the data in M point blocks each weight is updated only once each data block, reducing the number of processor iterations by M. This can be compensated for by increasing μ by M (3) to give equivalent convergence rate performance to the time domain filter (4). However the overall computational complexity is considerably reduced in the frequency domain implementation for long filters (3, 13, 14).

A further difference between the realisations is that a separate error signal is derived for each frequency component in the frequency domain approach, equation 5, which is a significant advantage over the single global error of the transversal filter. When combined with the separation of the signal into orthogonal DFT samples, this ultimately permits the μ_k value for each cell to be adjusted relative to the power estimates of the input and desired signals $p_k \simeq (X_k)^2$ and $q_k \simeq (D_k)^2$ respectively (5), in a manner similar to that which is adapted in gradient lattice filters (6). Thus, if μ_0 is the maximum possible step size to ensure filter stability with input and desired signal power level of unity, then we can select individual μ_k values for each cell as:

$$\mu_k = \frac{2\mu_o}{p_k + q_k} \qquad (p_k + q_k > 0)$$

$$\hspace{2.5cm} = \mu_o \qquad\qquad (p_k + q_k = 0) \hspace{2cm} (8)$$

This gives the possibility of designing these filters to give equal rates of convergence for all components of an input signal even when there is large spread in the eigenvalues of the auto-correlation matrix of the input signal. This variable μ development is functionally equivalent to normalising the DFT output powers before adaption, or prewhitening the input signal.

SIMULATIONS

The frequency domain adaptive filter based on 100 point DFT processor has been simulated. Some of the simulation results have previously been reported (10, 11).

Figure 2 shows the filtering simulations when the adaptive filter is input with sinusoids at 0.23 MHz and 0.93 MHz relative to a 100 point DFT operating at 4 MHz sample rate. This is filtered with a d(t) comprising only the lower frequency sinusoid. Figure 2.a shows the input spectrum of these non-integer input sinusoids and 2.b the converged filter weights. If this is processed with real transforms and only a single interconnecting multiplier, figure 1, then the filtered output is windowed (modulated) due to the loss of phase information. If however fully complex transforms and multiplication is used, Figure 5, then the output is shown to comprise only the lower frequency component without the amplitude modulation across the window, Figure 2.c. Processing with modulus information only does yield the correct results in this example if the signals have integer numbers of cycles in the sample window, i.e. their spectral lines coincide with DFT output samples. However such a penalty is unlikely to be acceptable in practical applications.

Figure 3 shows simulation results when the filter is used as a channel equaliser or echo suppressor. The input, x(t) (Figure 3.a), comprises one sample of a pulse of amplitude unity followed by one sample of simulated echo of relative amplitude 0.5. The spectrum of this signal X(n) is shown in Figure 3.b. The processor was then fed with a desired signal, d(t), comprising only one sample of a pulse of unity amplitude. After convergence the filter weights H(n) Figure 3.c, are shown to be the inverse of the transformed signal x(n) such that the output spectrum is flat, corresponding to the unit impulse y(t)=d(t).

These simulations are further developed in Figure 4 to show the effect of altering the μ_k value in the interface processor relative to the 100 individual adaptive loops which update each of the $H_k(n)$ weights. Figure 4.a shows the convergence of the processor weight values with identical μ values throughout the filter when converging into a signal with a wide band spectrum. In comparison figure 5.b shows the same result with independent μ values selected as defined earlier in equation 8. Note how this provides equal convergence rate for all signals independent of their levels.

PRACTICAL FILTER REALISATION

In our demonstration processor we have used SAW Discrete Fourier Transform processors. These are analog processors which use a development of chirp z-transform algorithm (12) to calculate the DFT coefficients. This can be visualised by expanding the Fourier transform relationship (7)

$$F(\omega) = \int_{-\infty}^{\infty} f(t)\, e^{-j\omega t}\, dt \tag{9}$$

with a substitution $-2\omega t = (t-\omega)^2 - t^2 - \omega^2$ to yield

$$F(\omega)=F(\alpha t)=e^{-j\frac{1}{2}\alpha t^2} \int_{-\infty}^{\infty} f(\tau)e^{-j\frac{1}{2}\alpha\tau^2}\, e^{j\frac{1}{2}\alpha(t-\tau)^2}\, d\tau \tag{10}$$

This implies a premultiplication of the input signal x(n) by a chirp waveform, followed by convolution in a chirp filter and subsequent postmultiplication with another chirp. This multiply-convolve-multiply (M-C-M) operation (7) can be simplified to a M-C operation when only modulus (spectral) information is required by replacing the postmultiplying chirp by an envelope detector.

Our SAW processors each use two linear frequency modulated down chirp devices with centre frequencies of 32 and 22 MHz, bandwidths of 8 MHz and 4 MHz respectively and dispersive delays of 50 μs and 25 μs, arranged in M-C configuration (7). These chirp filters implemented a 100 point transform with ±2 MHz real time bandwith.

The output from the forward Fourier transform processor is fed to an 8 bit A/D converter, operating at a 4 MHz sample rate and the transformed signal is input to the digital interface circuit, Figure 5.

This comprises an 8x8 bit digital multiplier which selects the 8 most significant bits from the 16 bit precision weights stored in the 100 location RAM. The multiplied output is subtracted from the desired signal which is either stored in another RAM or ROM, or is separately supplied, to calculate the error, equation 5. A shifter

scales the error to implement the multiplication by μ which takes
values in the range 2^{-i} where i is an integer in the range $0<i<15$.
The stored weights are updated in sequence by adding the shifted
error to the value stored in the appropriate RAM location and load-
ing the sum back into the same memory location. This operation is
repeated every 250 ns to update each coefficient once each 25 μs
processor cycle and the operation repeats until convergence is
achieved. Adaption then can be frozen to fix the filter characteristic or it can be left to continuously adjust to permit it to track
a time varying signal.

The digital output is converted in an 8 bit D/A into the frequency
domain analog signal, $Y_k(n)$, which is fed to the inverse Fourier
transform processor to obtain the time domain output, y(t).

Figure 5 represents a fully complex processor with 4 real multipliers. We have only constructed a real processor with dual chirp filter Fourier transformers, envelope demodulation and a single real
multiplier in the digital interface unit. In addition spectral information always has a positive polarity so a clipped LMS algorithm
has been accomplished by replacing sign $X_k^*(n)$ in equation 7 by
unity i.e. omitting this multiplication.

Figure 6 shows the adaptive filter operation when input with two
sinusoids at 200 and 930 kHz (upper trace). The lower trace shows
the processor output when the higher frequency component has been
suppressed in the adaptive filter as shown in Figure 2 of simulations. Figure 7 shows the experimental result when the filter is
used as a channel equaliser.

Figure 8 shows results for a single frequency cell in a fixed μ
processor where the convergence coefficient has been changed to
alter the rate of convergence of the filter. The upper trace shows
how the faster convergence with $\mu = 2^{-2}$ compares with $\mu = 2^{-5}$ in the
lower trace. Notice how increased μ increases the loop gain and
also the converged filter errors.

CONCLUSION

We have outlined the principles behind the design of frequency domain adaptive filter and described its implementation with SAW and
digital components. This has shown that our realisation of adaptive
filter provides not only wideband operation, but offers the ultimate
potential of superior convergence rate when compared to an equivalent adaptive transversal filter. Further work can now be performed
to extend our approach to other more sophisticated frequency domain
adaptive filter designs (13, 14, 15) which have the capability of
performing either circular or linear convolution. It can be shown
that fastest convergence requires the calculation of sliding Fourier

381

transforms where all the transformed output samples are updated at
the input sample rate. In that case it is attractive to consider
replacing the SAW chirp transform processors by SAW contiguous filter banks (16). This increases the digital processor complexity but
avoids the problems associated with DFT block processing.

REFERENCES

1. Widrow et al, "Adaptive Noise Cancelling: Principles and Applications" Proc. IEEE 63 No. 12 pp 1692-1716 Dec 1975

2. Bracewell R. "The Fourier Transform and its Applications" McGraw
   Hill New York 1965

3. Dentino M. McCool J. and Widrow B., "Adaptive Filtering in the
   Frequency Domain" Proc. IEEE 66 No 12 pp. 1658-9 Dec. 1978.

4. Reed and Feintuch, "A Comparison of LMS Adaptive Cancellers
   Implemented in the Frequency Domain and Time Domain" IEEE Trans
   ASSP 29 No 3 pp.770-5 June 1981.

5. Shankar and Peterson, "Frequency Domain Least Mean Square Algorithm" Proc IEEE 69 No 1 pp 124-6 Jan 1981

6. Friedlander, "Lattice Filters for Adaptive Processing" Proc IEEE
   Vol 70 No 8 pp 829-876 Aug. 1982

7. Jack, Grant and Collins, "Theory Design and Application of Surface Acoustic Wave Fourier Transform Processors" Proc IEEE 68
   No 4 pp 450-68 April 1980.

8. Moshner, "Adaptive Filter whith Clipped Input Data" Stanford
   Univ. Inf. Sys. Lab Report No 6796-1 1970.

9. Widrow et al, "The Complex LMS Algorithm" Proc IEEE 63 pp 719-
   20 April 1975.

10. Morgül, Grant and Cowan, "Wideband Frequency Domain Adaptive
    Filter Module" Paper No 413 IEEE ICASSP Conference, April 1983

11. Morgül, Grant and Cowan, "Wideband Hybrid Analog/Digital Frequ-
    ency Domain Adaptive Filter" Manuscript submitted to IEEE
    transaction ASSP.

12. Rabiner at al, "The Chirp-z-Transform Algorithm" IEEE Trans AU
    17 pp 86-92 June 1969.

13. Ferrara, "Fast Implementation of LMS Adaptive Filters" IEEE
    Trans ASSP-28 pp 474-5 Aug. 1980.

14. Copeland, "Transmultiplexers used as Adaptive Frequency Sampling
    Filters" Proc. IEEE ASSP Conf. pp 319-22 1982.

15. Clark, Parker and Mitra, "Efficient Realisation of Adaptive Di-
    gital Filters in the Time and Frequency Domains" IEEE ICASSP
    Conf. Proc. pp 1345-8 1982.

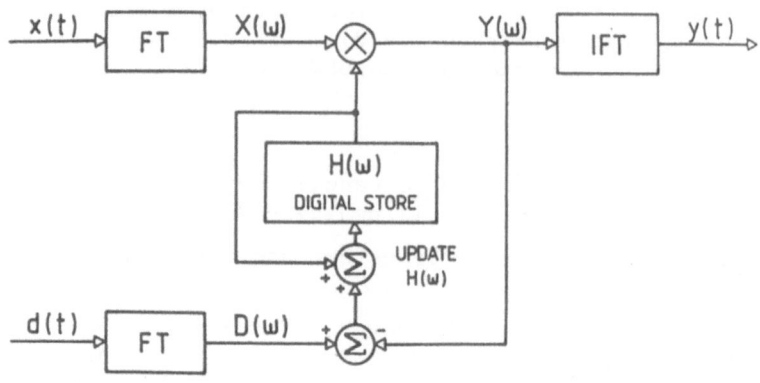

Figure 1. Frequency Domain Adaptive Filter

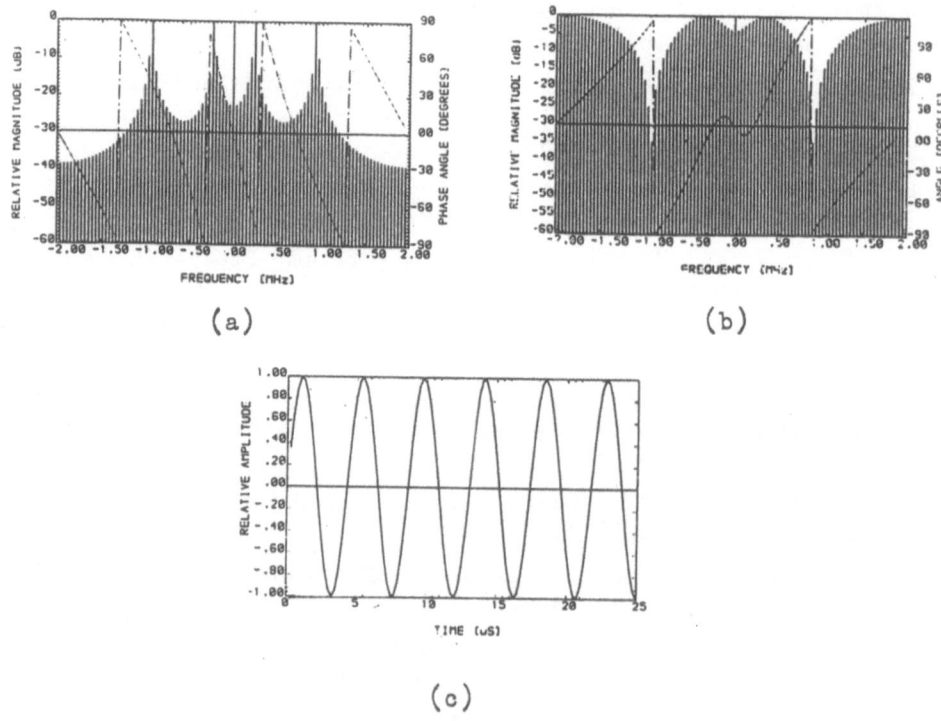

(a)   (b)

(c)

Figure 2.

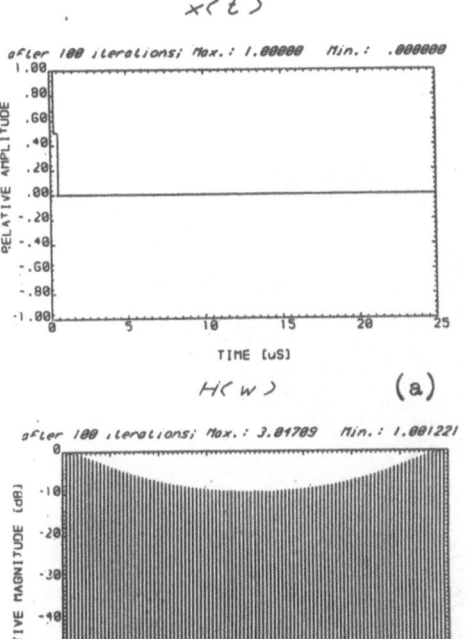

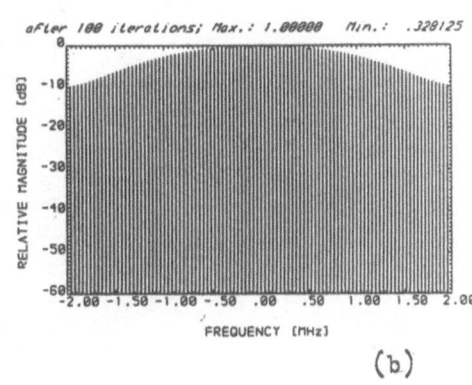

Figure 3.

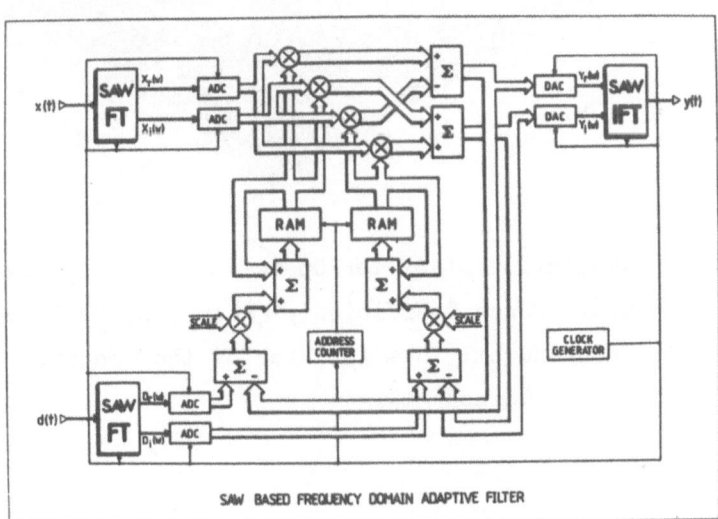

Figure 5. Adaptive Filter Detail.

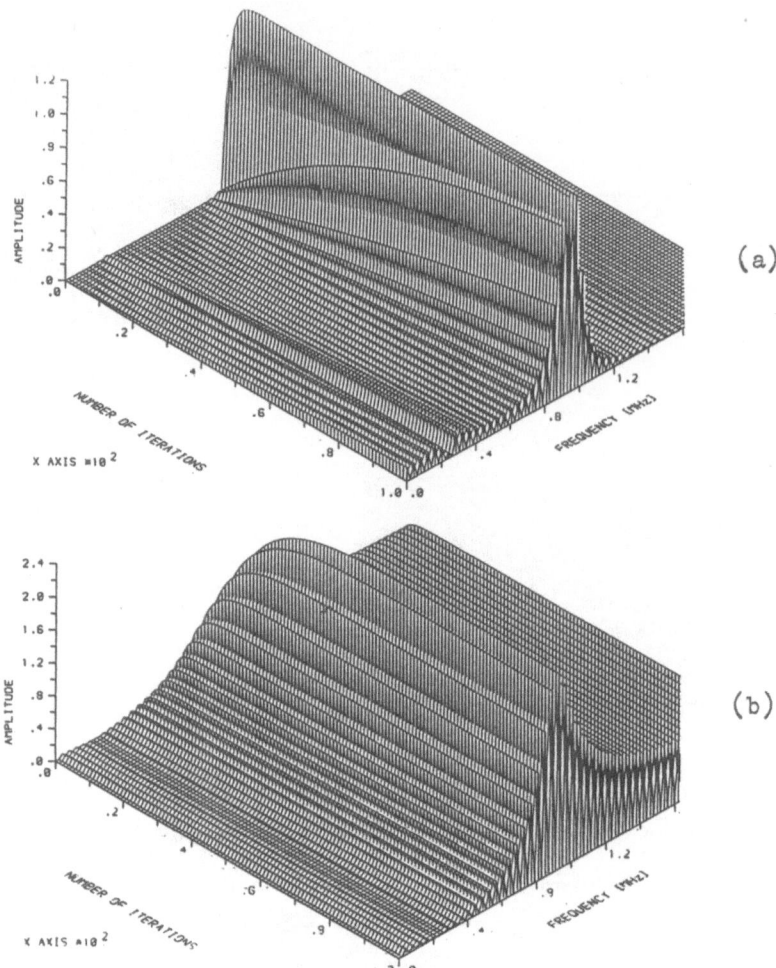

Figure 4.   Simulation of Filter Operation with (a) Fixed
Convergence Coefficient, μ and (b) μ Values
Independently Set for Each of the Transformed
Signal Components.

385

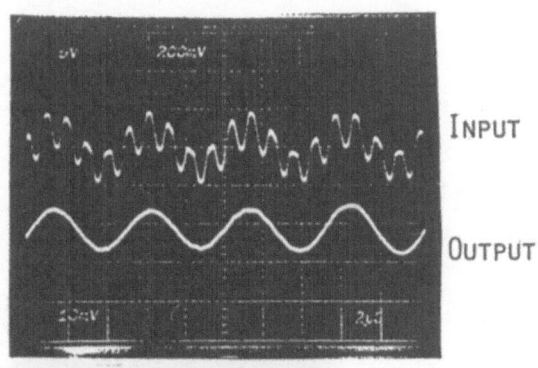
INPUT

Figure 6. Operation of
Adaptive Filter Module
as a Notch Filter Sup-
pression Unwanted CW
Interference.

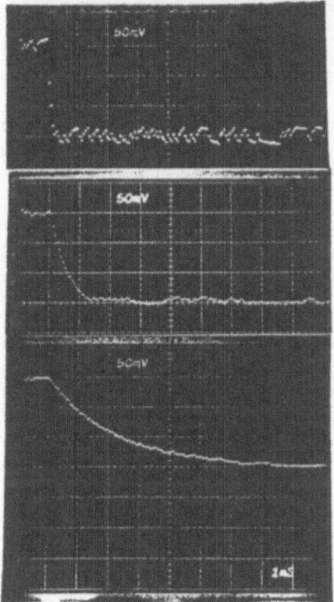

Figure 7.

$\mu = 2^{-1}$

$\mu = 2^{-3}$

$\mu = 2^{-5}$

Figure 8. Practical Results
for Adaptive Filter Operating as an Equaliser which
Compensates for Post Echo.

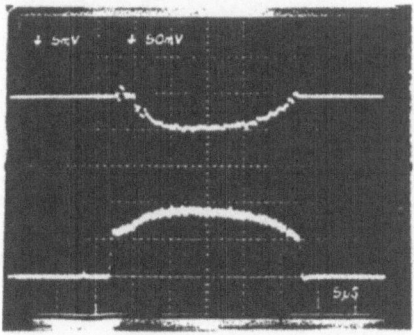
ADAPTED FILTER
WEIGHTS H(w)

TRANSFORMED INPUT
SIGNAL X(w)

PANEL DISCUSSION

ON

FAULT-TOLERANT COMPUTERS

Tuesday 12th July 1983 at 20.30 hours

Chairman and Professor R. McEliece California Institute of
Organiser: Technology, Pasedena,
                                     CA., U.S.A.

Panel Members: Professor D.G. Messeschmitt University of
                                            California, Berekeley,
                                            CA., U.S.A.

Mr. E.A. Palo   MITRE Corporation,
                       Bedford, MA., U.S.A.

Professor K. Steiglitz Princeton University,
                        N.J., U.S.A.

Dr. C. Heegard   Cornell University,
                       N.Y., U.S.A.

Dr. R.M.F. Goodman   University of Hull,
                       U.K.

Contributors: Professor K.W. Cattermole University of Essex,
                                         Colchester, U.K.

Mr. F.P. Coakley   University of Essex,
                       Colchester, U.K.

Dr. A.B. Cooper   U.S. Army Ballistic
                       Research Laboratory,
                       Aberdeen, MA., U.S.A.

Professor S.W. Golomb University of
                       California, Los
                       Angeles, CA., U.S.A.

Mr. B.L. Johnson   MITRE Corporation,
                       Bedford, MA., U.S.A.

Dr. J.A.E. Simons   GCHG, Cheltenham,
                       U.K.

Professor N. Tepedelenlioglu Middle East Technical
                              University, Ankara,
                              Turkey.

Professor McEliece - Our subject is fault-tolerant computers. I have been told by our director that this panel discussion will be recorded and reproduced in the proceedings of this Institute. Therefore each speaker has to give us copies of his viewgraphs, or their copies. Also, if even someone asks a question from the back of the hall, he will have first to give his or her name. We have to-day five speakers in the panel. The first is Professor Messerschmitt.

Professor Messerschmitt - It is my fortune to have been employed by Bell Telephone Laboratories for a number of years. The telephone system is their business, beside satellite systems etc. They have been most concerned with fault-tolerance. So what I intend to do is to describe, from a fault-tolerance point of view, a large switching machine. This might provide some insight of how these things have been arranged. The machine I wish to describe is the No. 4 ESS, which is a large toll switching machine. It was designed in the 1970-ies and first went into service with the Bell North American Telephone in 1976. Now it essentially replaces the whole of the toll network switches in the U.S.A. It deals with about 10^5 trunk switch lines and with about 10^5 telephone calls at a given time. Potentially it is the largest configuration, and of course it is all digital, meaning that the internal switching is done digitally in the machine. There is also a lot of analogue circuitry, in the interfaces, for example to analogue trunks. It is 'stored program controlled', as is everything these days.

I will mention here the overall maintenance objectives which govern the design of a circuit like this, i.e. a development cost of several hundred million dollars, over a period of five years or so. It is quite a large digital machine. Its bulk configuration would fill this château. Its maintenance objective (and this is very standard for telephones) is to have at most two hours of total downtime (when the whole machine is down and inoperational) in forty years. This implies that to meet this short of objective, you have to use lots of redundancy, as we shall see in a moment, and lots of protective switching types of mechanisms, to be used when switching hardware fails. Fault isolation is not actually a part of this. It is more associated with the maintenance of the machine, so you can get it fixed when it fails. Of course, the faster you fix it, the better you are in terms of down-time. Probably the most important aspect of the design of a machine like this is the design of software. This software naturally forms a large program and there is a lot of testing to be done. The major source of down-time, as in any large computer system, is the software. Therefore this software is designed very defensively. In particular, there are various stages of panic when the software goes through one of its difficulties. For instance, if one procedure when you write something on a disk, you do not assume automatically, when you read something from a disc, that it is exactly in the state and in the form that you have put in there.

because there might have been hundreds of other procedures execute
in the meantime that might have bugs and destroyed some of your
data. So you put check samples on the disc and this is a part of
recovery mechanism. Such recovery mechanisms are very elaborate in
data bases, where they try to isolate such problems, so they do not
grow. We would be fairly happy to drop a few of the calls, even
tens or hundreds, as long as it does not crash the entire machine.
In terms of maintenance there is fault isolation, and the objective
is to isolate faults to the card level. Thus, very little training
is required from the maintenance personnel. Then what the system
would do is to print out three or four cards, to attempt to replace
the faulty one. The personnel would replace these cards in the
right order, in the hope that the first card replaced was the one
that was in fault. The average number of cards that is replaced is
two.

I would like now to make a point that this kind of maintenance
is very expensive in terms of overheads associated with it. About
30% of hardware cost that is associated with fault tolerance and
isolation. Probably about 80 to 85% of software is devoted to
maintenance. This is also associated with fault tolerance and with
defensive software strategy, as well as software costs associated
with re-configuration due to hardware faults.

Professor Tepedelenlioglu - Excuse me! That hardware figure of 30%
that you have quoted, does it include the cost of spares?

1A Processor

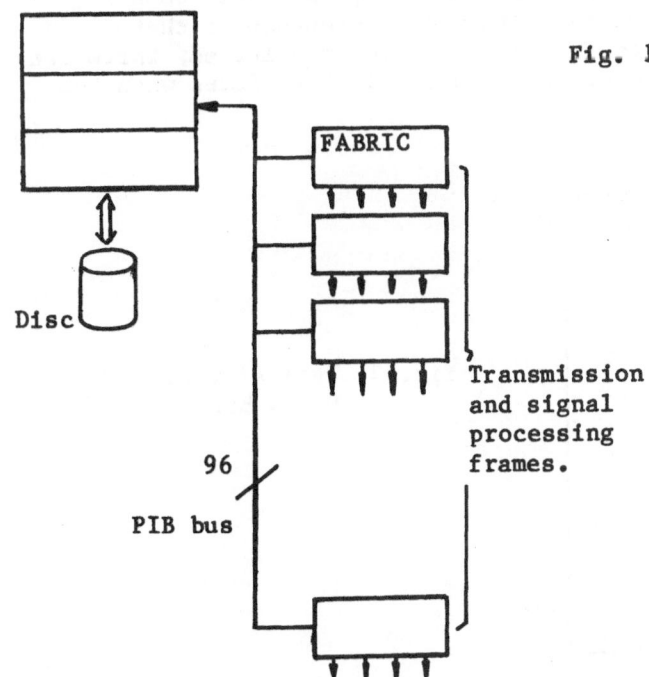

Disc

FABRIC

96

PIB bus

Fig. 1. Maintenance
architecture for
the 1A processor and
transmission/signal
processing
frames.

Transmission
and signal
processing
frames.

Professor Messerschmitt - Yes. Then about 70% of the cost is for the overhead of fault isolation, recovery and tolerance, in the design and effort.

Just to show you very briefly the design architecture of this particular machine, from the point of view of fault tolerance, let us consider the maintenance of the central computer, i.e. the so-called 1A processor (Fig. 1). It has a large disc file. This runs the majority of software, including the software associated with fault recovery. That processor is triplicated; there are three processors which are continuously matched against each other, so that if one processor goes bad, you can tell which one it is and switch it off. From the point of view of the maintenance, there is a large bus that goes throughout the office. It is called the PIB bus. It is 96 bits wide, and is completely redundant, for there are only 48 bits information. That bus is to communicate between the frames. Associated with the actual switching functions are the frames - they form a small portion of the total hardware. One of them would easily fill half of this lecture theatre, whereas the entire switching machine would fill the château.

What is left are the entire transmission and signal processing frames. All these frames are interconnected. Even the frames that are associated with the transmission function are connected to the main processor, and they follow the same maintenance pilosophy.

Let us now look what a typical frame looks like. A typical frame in this switching machine would be about 9 feet high and 3 feet wide. At the top of the frame (see Fig. 2) would be a controller which is associated with the maintenance of that particular frame. That controller is again triplicated; there are three copies and they are cross matched. It interfaces with the

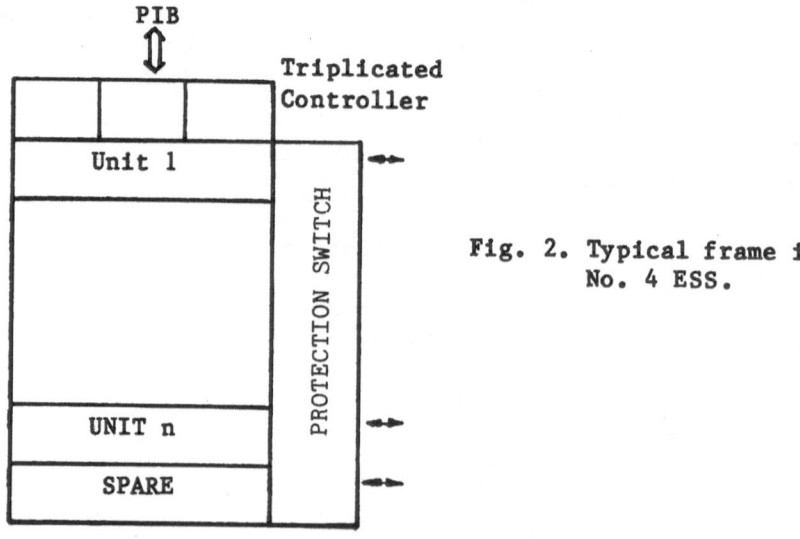

Fig. 2. Typical frame in No. 4 ESS.

PIB and it talks to the units in the frame. We have here a set of
units which actually implement the functionality of that frame.
For example, they can interface with the digital transmission.
There are say n units like that, and there is a spare unit. The
spare unit replaces one of the 'in-service' units if it develops a
hardware failure, on a 1/n basis where n is typically 10. This is
the overhead associated with the spare, relative to Professor
Tepedelenlioglu's question, it is now 10%. Then there is a
protection switch which sits between the actual transmission
facilities. On the other side, between the unit and the actual
hardware of the switch itself, there is regulation by the
controller which selects one of the spares in place of the
non-working unit. The protection switch is implemented with relays
in these days of VLSI, for it still turns out to be appropriate,
since relays have the nice property that if there is a fault in the
power supply, they default to a resonable condition, unlike logic
gates.

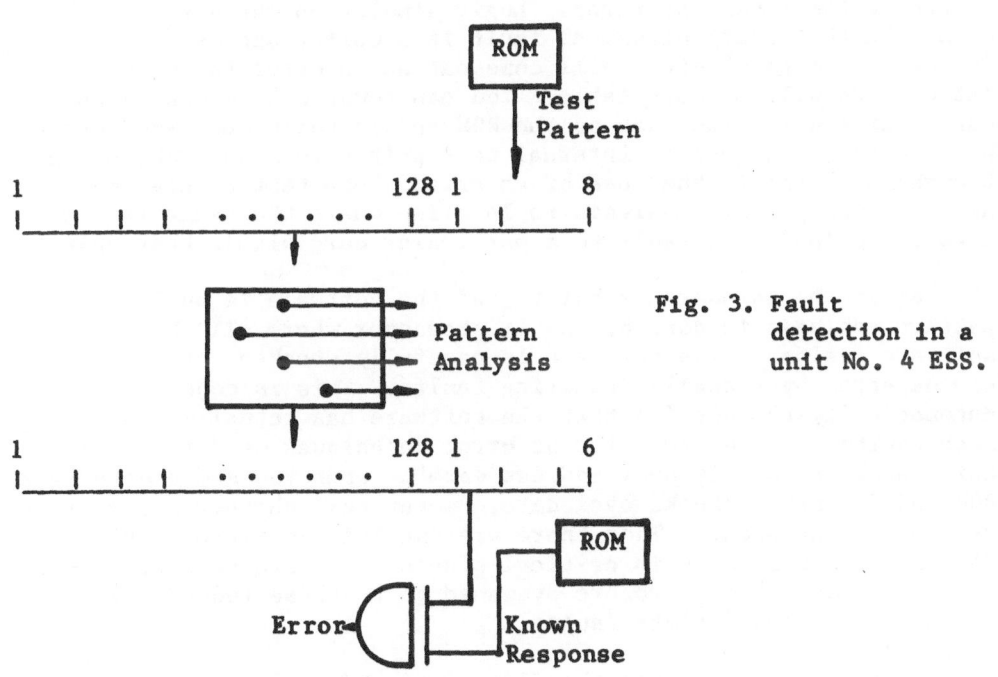

Fig. 3. Fault detection in a unit No. 4 ESS.

Now we shall look at how the fault detection is done in a typical signal processing unit. How do we actually detect that one of these units has failed? A typical unit here (Fig. 3) is actually a set of logic represented by the box in the middle. This logic is time-shared between many channels, and in the case of No. 4 ESS, they are running throughout the machine with 128 time slots. Each time slot corresponds to one voice channel; it is 64 kb/s and the overall bit rate is 8 M bits, except that some units are duplicated, so it turns out to be 16 M bits. Of these 128 time slots, 120 or more are actually used for calls. The other 8 time slots are used strictly for maintenance functions. So we have a set of hardware here which is performing identical functions for all the 128 time slots. The philosophy is that if a piece of hardware is working correctly in one of these 128 time slots, then presumably it is working correctly in all the time slots. There are measures taken in the hardware to ensure that this assumption is valid. Typically we take a test pattern which exercises the functionality of that unit and we put that pattern into one of the 8 maintenance time slots. When something goes wrong, we read what is coming from a unit and compare that against the known response to that stimulus. That known response is stored in a ROM. The actual stimulus is carefully chosen and a lot of effort was devoted to choose these test patterns. Logic simulation was used to establish that every classical fault in a unit (such as 'stuck-at-one gate' etc.) will come out as an error in the test pattern and will be detectable. You can compare or cross-match the output of a unit with that of the ROM and it tells what went wrong. We also have a number of internal test points in a unit which are brought out, and in the case of an error these test points can be used via the pattern analysis to localise where the fault is. We attempt to isolate a fault to a particular card within that unit.

After the hardware is built, and the software is built, the pattern analysis is done by the 1A processor where all the software resides. The software is carefully checked - the errors are detected by actually inserting faults. This is done automatically by checking that the software has actually isolated such faults. There are a lot of error techniques used internal to this, such as 'parity over address/data'. When you store data in a ROM and do parity checks over data, we can tell whether any address mechanism went wrong. Then there are 'activity detectors' to determine that in certain critical places there are correct signals. The 30% of hardware overhead go to these redundancy checks to try and isolate faults.

Finally, let us taste the flavour of this switching unit, by considering the implications of the VLSI revolution, if there are any. I think that when you design a machine like this, with full maintenance and fault detection service, you have a different problem to that of a VLSI chip design. There are some things that jump out at you. One thing that I was puzzled about is the issue of 'per channel' versus 'multi-channel'. Particularly in the context of signal processing functions, before the advent of VLSI, most things were done on a multi-channel basis as I have shown here: One takes a LSI chip off the shelf, one implements the functionality of some signal processing and then you simply time-share that hardware over many time slots. On the other hand, when you get to the area of the VLSI chips, what people tend to do is to build a chip which implements such a functionality for a single time slot or a single channel. So this is quite different philosophy and one of the things that jump out a you is the 'per channel' approach. It is much more expensive to determine fault detection if you use the 'cross-matching' schemes; it is more expensive in terms of software and of hardware overhead to build two VLSI chips, or to put two identical units on a single chip and then to cross-match them. Whereas in the multi-channel type of context this is relatively inexpensive because you simply devote to it an extra time slot for maintenance purposes and such overhead is very small. This might suggest that, from the point of view of fault tolerance and fault detection, the way to build VLSI chips is to utilise extra time slots. You can use one of them for the actual data and the second one for the purposes of maintenance. That is a lot cheaper than duplicating the functionality. There is a lot of hardware considerations that come into that choice, and it has struck me that perhaps it is not the best thing to do things on the 'per channel' basis.

Another question that comes out with VLSI is whether we can put into the system all this fault detection to solve testing problems. Can we put it on a VLSI chip for the purposes of on-going maintenance when a chip is actually deployed in the field? It strikes me that this sort of policy might be quite helpful in the terms of the testing problems.

The other question is, can we also utilise the same circuitry to help us with initial testing during the manufacture? It is obvious of course that fault detection at a chip level does not solve all fault detection problems. The major source of problems, at least in my experience, are the connections and the bounding of wires on chips, then also power supplies etc. These are problems which will not be aided by a fault detection circuitry on a chip.

The other question I have, and I do not know the answer, is with the problem of transient faults. I have been told that they are important with VLSI chips. There we have to deal not only with the static faults, but also with faults that occur randomly. This is a much more difficult problem.

Then the final question is what happens when there is a fault in the fault detection circuitry? There have been for example circuits designed which, with a fault present, they can still be expected to work with the help of cross-matching facilities. There is a lot of work to be done there.

That is all I have to say. Are there any questions?

Mr. Coakley - I was under the impression that in your system you are only duplicating the redundancy and are not triplicating it. So, how do you vote? Also, it has been observed that in practice in SPC system some 25% of down-time faults is caused by technicians making wrong decisions during fault finding, e.g. taking working processor or a component out of service. Also, the majority of the remaining down time is caused by faults within the fault recovery software!

Professor Messerschmitt - I have put the case as I knew it in 1976. They might have improved it now. Considering your other point re technicians, this has also been my experience. Once we had a look at re-frames in the network and we found that almost all the re-frames happened between 8 a.m. and 5 p.m., without exception.

Mr. Simons - Your figure of 2 hours in forty years for system life seems to be very stringent, particularly when you consider power supplies.

Professor Messerschmitt - Yes. You understand however that this figure refers to the whole machine. What actually happens is that one unit may fail and the rest will work. Then you see it is not so stringent. I do not know much about power supplies, but I can say that the system has a battery back-up to the commercial power supply in case of long power failures. The overall system power supply is duplicated, and incidentally, we do not need much power, say a 5 in. by 3 in. box with 48 V and that gives you a lot of current. Each frame unit has an independent power supply, so if power fails in one unit, there is a spare.

Professor Cattermole - At various times in the last 30 years I have been associated with trials of new systems, and my experience bears out what you are saying. First, if you consider faults in the less sophisticated parts, such as power supplies, contacts and things of that nature, in every carefully documented trial they form the majority of failures, as compared to more subtle items of electronics. Then again, faults can be provoked by power surges or by maintenance. Also, test or fault-finding equipment can be a great pitfall, for if it goes wrong it can introduce errors. I have met several systems in which the least reliable part was the test equipment. I recall one equipment in which, because of faults in a tester, there was a tester to test the tester, and that also went wrong. Also, with duplicated or triplicated equipment, there is always some little thing that is critical. I recall an early trial of the IOC switching system in which one of the first things to go wrong in the field was the bus linking the duplicated processors. This sort of problem can be difficult to design out. I am very interested in your comments on fault-tolerant comparison equipment: that might be the key.

<u>Professor Messerschmitt</u> - I might add that in the case of software
it is much more difficult to test fault recovery software, than it
is to test functional software, because it is difficult to create a
lot of realistic fault conditions.

<u>Professor Cattermole</u> - One more comment. The fact is that various
industries have standards of reliability differing by many orders
of magnitude. Some computer installations are thought to be doing
well if they have only a few percent of down time. All telephone
switching systems meet much higher standards, with MTBF of some
decades.

<u>Professor McEliece</u> - Our next speaker is Dr. Palo.

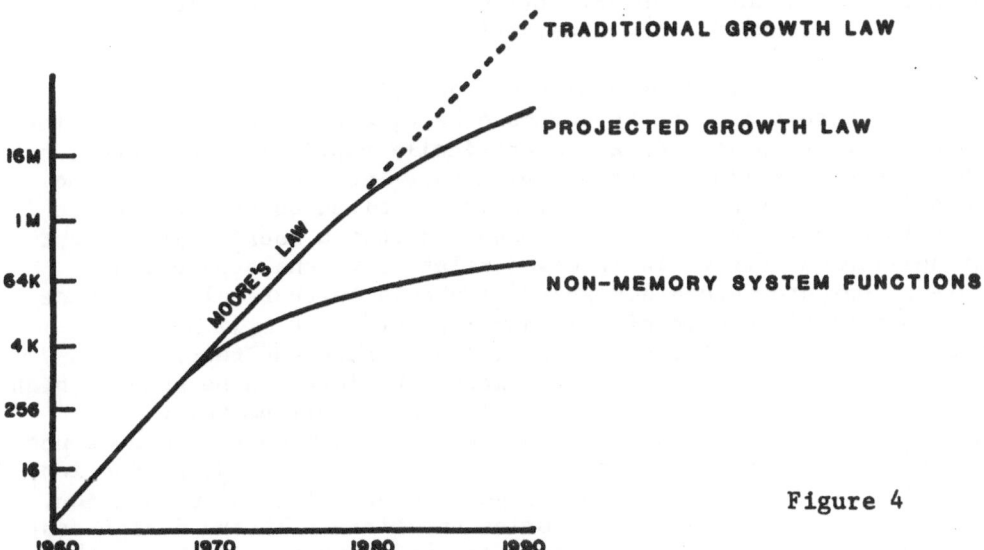

Figure 4

<u>Dr. Palo</u> - The particular theme for this discussion session is
error- or fault-tolerant computing. I will try to cover this in an
organised fashion. First I would like to start discussing the
complexity problem in the VLSI area. Here is a graph proposed by
Gordon Moore ([1], See Fig. 4). It shows the complexity of IC
chips, i.e. the number of components per chip, which had a record
of doubling every year since the 1960's. Then in the late 1970's
Dr. Morse and others observed that may be we are at the end of that
traditional rate of growth, and that for a number of reasons we are
to bend downwards somewhat, as is suggested by the projected growth
law curve. There is some question as to which curve we are still
on. I think that it has been noted that the IC technology
achieving this density has come from a number of factors. One of
these is the method of semiconductor processing: we have been
seeing smaller feature sizes on larger chips. But that has not
been the only factor. It has been in some cases at least the
dominant one. Another reason contributing to the traditional
growth is that we have been getting clever design and clever
architectures and this enables us to realise more components on a

chip. It is the combination of these two factors that realises this exponential growth (please note that the coordinate on graph in Fig. 4 is exponential).

As we move into the era of VLSI, there comes the question of what sort of problems will arise. There are still a number of problems that have not been solved yet. One is how to use this complexity? What function should be put on a chip? Clearly memories have played a dominant role here. This has been because of the regularity of their structure and their broad application. Thus a considerable design effort has gone into them to provide generically useful chips. We are already witnessing a high density and complexity following Morse's Law there. It has been stated that when the era of VLSI has passed and whatever else comes, although it is not yet clear what will happen, we may look back and say only: Thanks for the memories!

Then there are some non-memory generic functions - microprocessors are the key example here, and there are data points on the curves in Fig. 4, are specifically applicable to micros, but these tend to follow a rapid growth curve yet not as high as the memories. There have been some only a limited number of generic functions implemented and the reason is that a substantial amount of development effort is involved relative to the production volume. Computer-aided design (CAD) should further help to reduce the cost in all phases of chip development. CAD will help to identify and to partition the function, design the function and test it, or get it to work. Now more functions can be sold in high volume! The definition of high volume is I believe this: the design cost is small compared to the production cost. It does not matter if the development cost is large as long as the production cost is very large, for then you get a high volume of production. Such has been the case for memories and micros. If the design cost is modest and the production cost is small, there will be in this sense a low volume production. High volume provides the drive in the area of integrated circuits.

The next problem we are running into is testing and reliability. This was just raised by the previous speaker here. First we have to worry about hard errors. I will not deal with this extensively. Some people may say that they are found by design verification, but here we shall assume that we start with a working device. We have a device that has been tested and is assumed to be designed properly and to be implemented properly. Yet it still might develop some hard errors in it; typically they are called 'stuck-at' errors, when logic sticks at '0' or '1'. In VLSI technology there is the question of fault modelling - how do we model the failures? This is not settled, but a lot of work is ongoing in this area, at least in the U.S.A. The reliability models are not yet well defined, and we have to worry both about hard errors and the soft errors, because feature sizes are getting so small. Even ambient radiation is causing soft errors that are there at one clock cycle and not at the next. Then there are data-dependent errors that can appear as soft errors. These can be very hard problems to deal with, and I would like to consider them a bit.

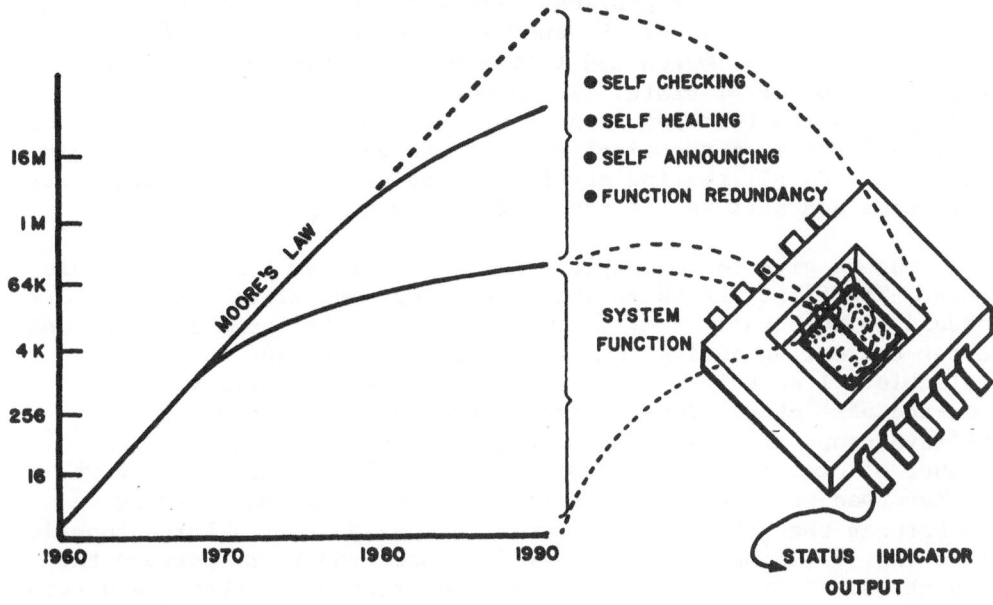

Figure 5

What can we do with this available complexity to get better system functions, when it is difficult to define bigger system functions that will be produced in high volume? We have to think again about our approach to chip design. Then we can hope to march in this direction to follow the exponential scale in Fig. 4. One of the things we can try for is to use some of this curve for fault-tolerance and testing and thus essentially introduce some redundancy on a VLSI chip. Let us again examine the plots of chip complexity (see Fig. 5). I can use the portion of the available complexity, which is large, to implement a function. I can put that function on a chip, and then I can use the remaining available chip complexity to add some fault-tolerance and testing capability. Remember that the ordinate is an exponential scale, then the function part of a chip would use a small amount of the available complexity. What I would like to do is to promote the use of this additional available complexity for fault tolerance and for concurrent testing.

What do I mean by concurrent testing?

I mean that we test the device while we are using it.
Everytime we clock the device, we not only get the answer but we
also do get some form of test which says whether that answer was
good or not. That is the beginning of our attack on the soft error
problem, the transient problem. The chip itself would announce at
every clock time whether the answer was good. It does not do any
good to test by stimulus and response, if we are concerned about
the transient error state. It is possible to do the test and
decide that the function is good at one time and then come back to
operate later in a transient error state where the function would
be wrong. So all testing should be done simultaneously with every
operation and that is the theme of my remaining talk.

Let me give you a couple of examples of how I might use this
idea. The first one is in the area of communication. People here
should be reasonably familiar with the fact that communication and
error-correcting coding tend to go together to make things
reliable. I am going to address the problem of getting
'on-and-off' chip. This is one of the areas that tends to be
failure-prone because of the input/output (I/O) problem. The
number of pins for I/O in VLSI can obviously be unusually large.
We can then put some error-correcting, encoding and decoding
in-between the VLSI chips. Assume that as part of a large function
(see Fig. 6), I have to transfer data back and forth between the
two chips. I want to add some error-correction as shown, and this
is going to do two things for me. One is the error-correcting part
of this code which provides robustness against soft errors and
corrects these errors as they appear. Also, that basically provides
a concurrent testing capability, for error-correcting coding is
particularly suitable for concurrent testing. If an error is found
at some time the decoder will not only correct that error, but it
also gives an indication if something was wrong, and I can use that
answer later on. In particular, I can use the original correction
to correct for hard errors; i.e. errors due to a failure in the
communication medium, that is the data bus. Let me describe this
further.

I am taking a set of data, say k words or symbols of data,
each m bits long. This might be 8 or 16 bit words and I to have k
of them. I encode them by an error-correcting code of n words
long. Thus I have expanded the block of words from the size k to
n. Now, the important feature is that I am going to do a sort of
corner turning (see lower part of Fig. 6) or transfer, as shown.
Thus, I take a symbol, turn it on its side and transmit it bit
serial on a line of my n-bit data bus. I will do this in parallel
with every one of the n words. Thus, the number of words that I
transmit is equal to the number of lines in the data bus. If on
any one of these data bus lines there occurs a fault, say due to a
soft error, the the error-correcting code will correct for this, or
rather its decoder will do it. At the receiving end one uses a
control monitoring of the error correction process. If over a
period of time one notes one line having continuously an error,
then the monitor will say: 'Gee! One of these lines has been in

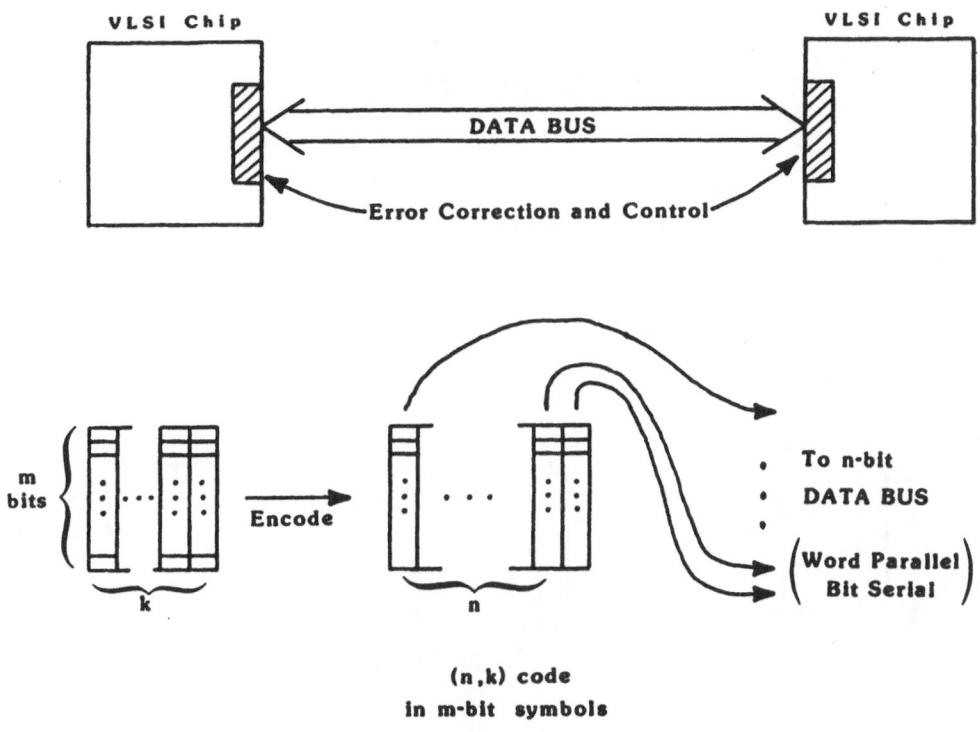

(n,k) code
in m-bit symbols

Figure 6

constant error; that means it must have failed'. One then declares
that line is in the erasure state. After noting failures in a bus
line due to a hard error, the monitor declares that line as an
erased line and stores a signal in the controller, so that later on
one can read this out to decide which chip to change. Thus, we are
invoking the error-correcting coding and decoding on both sides of
the data bus, and the monitoring control is also on both sides.
This transmission method can be called 'word-parallel' and
'bit-serial', since each word is transmitted on a different line in
a bit-serial manner.

Now to a specific example. We have heard of some new
technology on Reed-Solomon coding [2,3,4], so I will advertise it
further! Here is a 16-bit data bus example (Fig. 7) using a
Reed-Solomon (R-S) transform code. We shall deal here with a short
code; data consists of eight 8-bit symbols. The bits are coming
from some memory storage and we have 8 x 8 = 64 data bits. I am
going to encode this eventually into a (16,8) Reed-Solomon code,
i.e. R-S(16,8), but because I am in the transform domain, I am
initially encoding it into a R-S(17,8) code, over the field $GF(2^8)$
and this means that I can define the 17-point transform in that
field. Each word is 8-bits long. Before I transmit it I am going
to subtract one word: R-S(17,8) → R-S(16,8), and then I transmit.
The 16 words are again transmitted (as in Fig. 6) on the 16-line
data bus, with one word per data line. Now I have R-S(16,8) code

16-BIT DATA BUS EXAMPLE

R-S Transform Code

Data 8 8-bit symbols

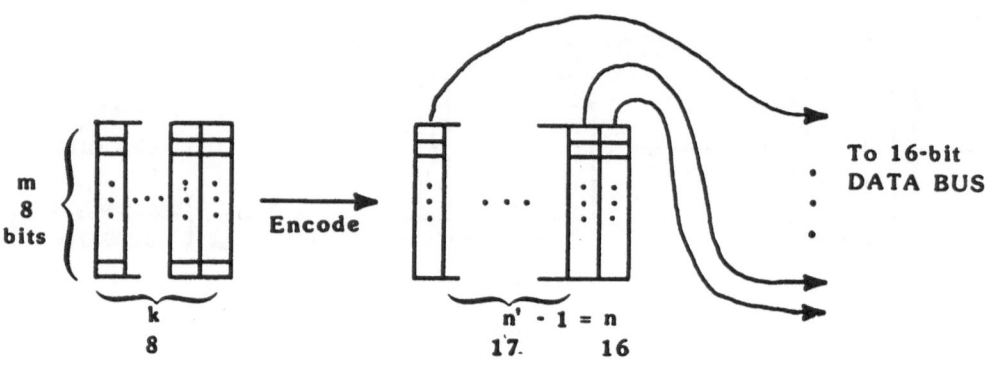

R-S (17,8)   R-S (16,8)

$$d = n - k + 1 = 9 > 2t + e$$

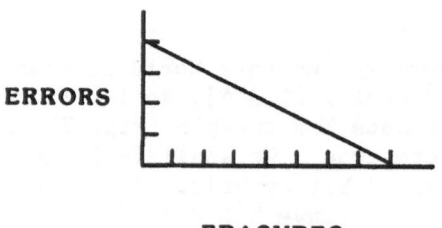

Figure 7

and I have two to one redundancy factor. There is no reason that I
am restricted to that amount of redundancy, but it matches nicely
for a typical bit, byte or word length. The power of this
particular code is in the relative Hamming distance:
d = n-k+1 = 9, which means that I can correct any combination of t
errors, and e erasures such that 9 < 2t+e, as illustrated in the
graph in Figure 7. This means that you can accomodate any 4 data
bus lines in error; they can be totally shorted or totally open,
whatever you wish; they can be either in a transient error or in a
hard error. If the monitor notes consistent hard errors, I can
declare up to 8 of these lines as erasures, and I can still have
reliable data transmission between my two VLSI chips.

So this is a little example of how I might include
error-correction in VLSI chips to increase the reliability of these
chips in particular in the areas where input/output stages are
prone to errors, and thus also to the transient errors.
Reed-Solomon codes are potentially very powerful in these areas.

Let us now take one other issue here, namely the use of
redundancy in computation. I have a function which has an input
and an output (Fig. 8a), a processing function. Assume that this
function can be easily implemented with the present VLSI
technology. I will now propose to use the ever increasing
capabilities of this technology to achieve a more reliable
function. This I present to you in sort of a generic diagram
(Fig. 8b). I will divide this function into several parts, so as
to introduce redundancy and there is more here than just to
subdivide my original function. The input is distributed to
several parallel circuits, each of which eventually contributes to
the output. The most trivial example is shown in Fig. 9. Here I
have a triple modular redundancy, since I have repeated my function
three times. All three outputs from these identical functions go

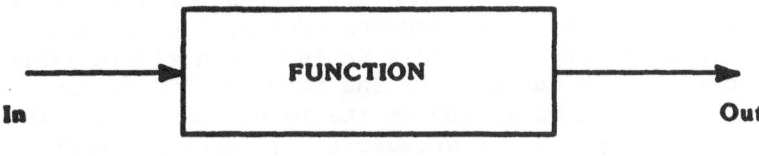

FUNCTION

In
Out

Figure 8a

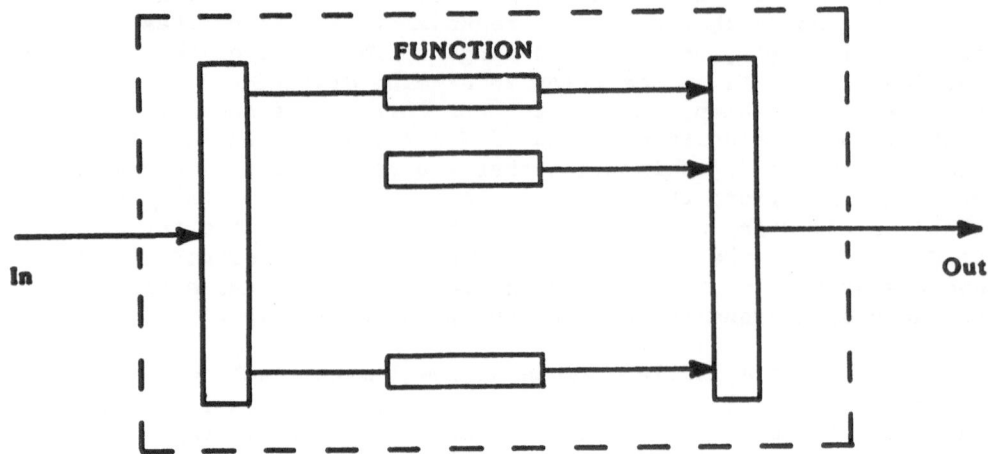

Figure 8b

to a common block, a voter. Then, depending on the voter's
decision, they come out. In this case I assume a reliable voter to
choose the proper function out of the three, particularly if they
differ in their output. The triple complexity may not be so
expensive now with the exponential growth in VLSI capability. This
is one way to achieve reliable functions and concurrent testing.
Another way is shown in Fig. 10. I now subdivide the function f
into several pieces. This can be done quite easily in several
signal processing applications. In this case f is decomposed into
f_1, f_2 and f_3 where f_1, f_2 and f_3 compute different parts of f.
Each part is of lesser complexity than f. There are several
potentialities in this area, particularly in finite field
processing and residue number processing. The parts of the function
are computed in parallel and then are recombined at the output
block in several different and essentially redundant ways. This
gives me monitoring facilities to see that all three modules are
working. I would get a different result if any one of these is not
working properly, and I can accommodate such situations in the
output block. The third case involves redundant computation in
time and is shown in Fig. 11. Here we include a simple switch and
an extra operation. We can compute the function f directly, or
otherwise we can do an operation on its input and output. This
consists of a coder (g) and a decoder (g^{-1}), with f in-between. We
feed both these functions into the output and compare them. They
are the same function there and they should give the same result.
The important thing is that this allows me to go through f
sequentially in two different fashions with the same input, and in
each case I will get a different error coverage. A small amount of
extra circuitry gives me an effective double redundancy.

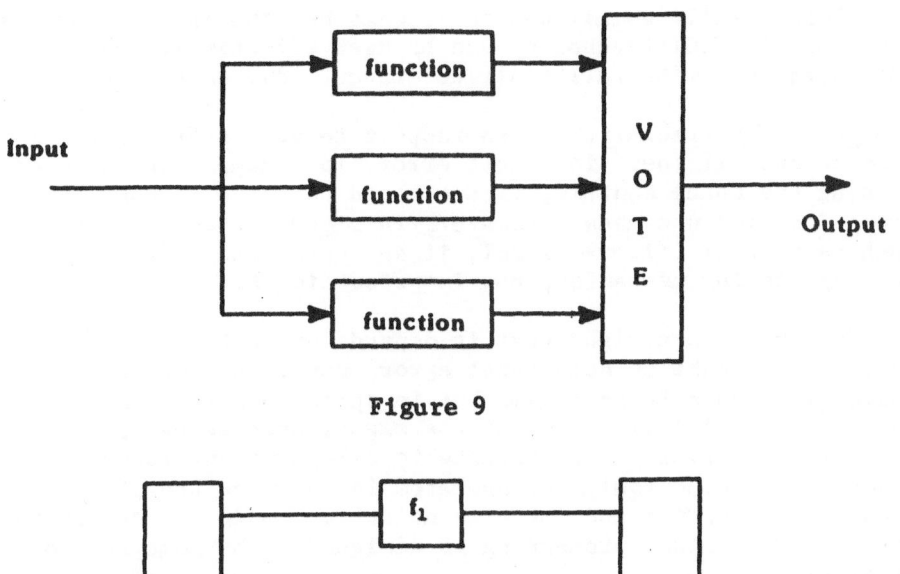

function
function
function
Input
V
O
T
E
Output

Figure 9

Input
f_1
f_2
f_3
Output
subdivide
recombine

Figure 10

Input
g
function
compare
g^{-1}
Output
$f = g * f * g^{-1}$

Figure 11

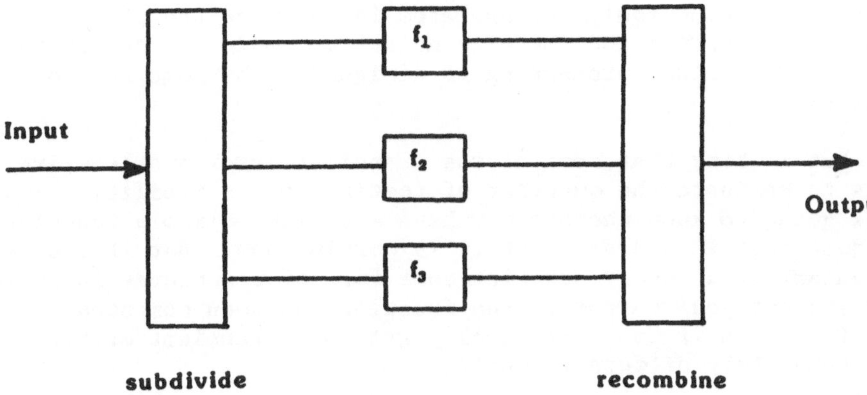

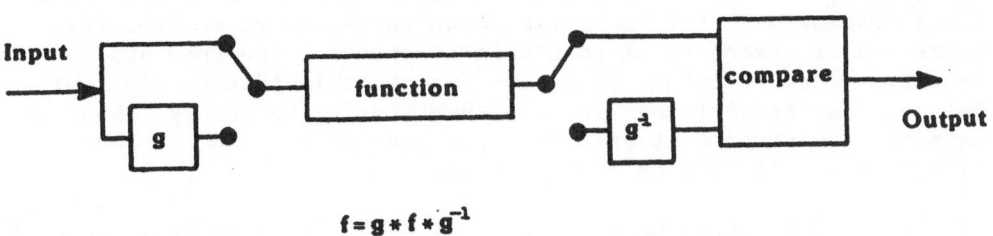

This briefly summarises some ideas how the growing complexity available in VLSI technology can be used effectively in communication, with full fault-tolerance. Any questions?

Professor Messerschmitt - With respect to your soft errors, I would like to know if there is a soft error, for example in the case of landing the space shuttle, if one would not like to know that a wrong decision was made. However, in signal processing examples, such as digital filters or FFT, if an occasional soft error occurs in a processing mechanism, how important is it?

Dr. Palo - You are right that it depends on what you are using it for. If you make an occasional error, say in a radar processor, it could give you a false alarm, but in space work false computation may be very critical. Some of the signal processing applications are error-tolerant. For instance in telephone environment a lot of signal processing equipment operates in the presence of transmission errors and there is no attempt made to correct these errors. The actual processing is designed to be somewhat tolerant of errors.

But I think that some of the transient error problems are going to dominate the question of testing for reliability. You are never going to know whether you have a really reliable function if you just test it and say that it is working now. But if you test it concurrent at every use with some form of redundancy in there, then you get your output if the function has been computed properly, even if you occasionally get some transient error. This is a completely different position.

Professor Messerschmitt - I think this is very important, because for instance I raised this question once with some circuit design people and their reaction was: 'Why worry about soft errors? We have to design and test that the thing is working properly'. That is sometimes the reaction you get.

Mr. Coakley - Is it not possible to actually adopt here some of the ideas developed by the Newcastle group for software reliability, where you implement check points in computation and then use the recovery block technique [5]. You can use this idea in VLSI chips. One part is used for checking and then storing in memory. When you compute, you compare it with what you have stored, and if these differ, you fall back on another block.

Dr. Palo - This pertains more to error detection question, and what I was talking here refers to the actual error correction.

Mr. Coakley - This technique can also be used for error correction, depending how far you want to implement this. In your example you compare the function with something on a different part of a chip.

405

Dr. Palo - If you take the case of my triple modular redundancy (Fig. 8), you can compute there and then compare. If there is an error the voter decides.

Professor McEliece - To keep the timing of the session going at a right pace we should now invite our next speaker, Professor Steiglitz.

Professor Steiglitz - What I am going to tell you, to put it simply in the context of this discussion, has to do more perhaps with what is meant by production testing, rather than finding whether a chip is still functional at a certain time of its operation. The main problem in production of chips is that many of them do not work. Thus I would like to bring out the theme of structure, which in design and in layout has many advantages. One of these is the ability to test, and hence, I suspect, the ability to make that chip fault-tolerant. The work I wish to present was done by a student of mine, Mr. A. Vergis [6] on bilateral arrays which is a sort of generalisation of systolic arrays. The idea is roughly as follows: We have a very large VLSI circuit which looks like this (Fig. 12), and which is made of little cells like the one shown. We organise the circuit in a way that allows us to identify such a little cell among many. What we plan to do is to test that little cell exhaustively by applying all possible inputs and observing all possible outputs. If you like to see something concrete, that little cell might be a part of an array, like the one referred to above. With say three inputs lines we need to apply $2^3 = 8$ inputs and we observe four outputs on two lines. The problem is to test exhaustively such a cell and the initial assumption we make is that we have a fault only in one cell. I will return to this point later on. By the way, the cell might be a single wire, so that covers the question of testing the connections between cells, as well as the cells themselves. Now, when does this paradigm work? It works when we can control the inputs to the cell by manipulating the input bits. Another condition for this to work is that a cell to be exhaustively tested must not be too big. The cell must be controllable in the sense that we need to be able to set every possible input pattern by setting various input vectors. It also needs to be observable in the sense that every output pattern that we need to observe must be somehow obtainable at the chip port. You may say: 'When is this possible?'

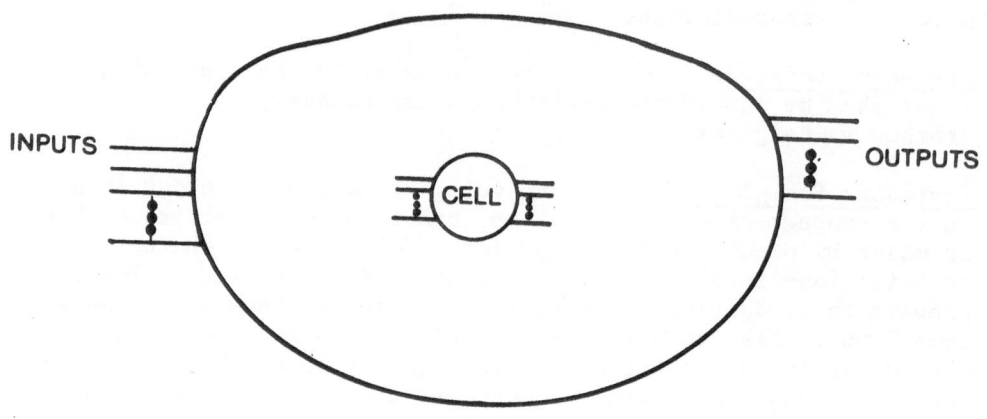

Figure 12

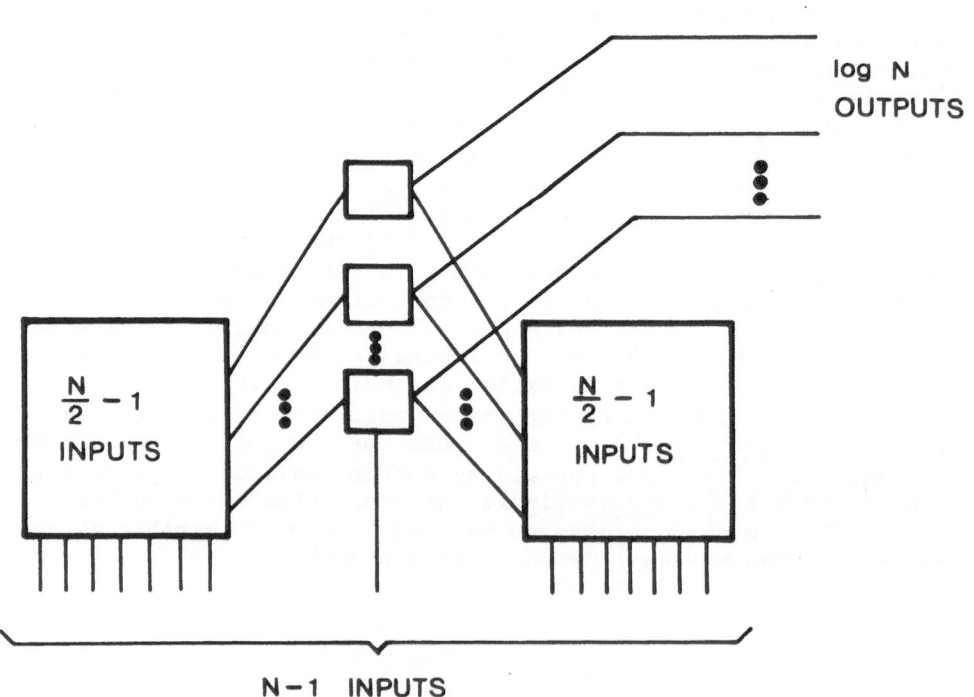

Figure 13

It seems like a pretty stiff job. Let me show you how this is done on a particular example which I have used, namely a tally circuit (Fig. 13) which is mentioned and discussed in my lecture [7]. This was done by my former student, Dr. P.R. Cappello [8]. Let me remind you what this is. It is a recursive layout for a structure that simply tallies the number of 'ones' at the input lines. There are N-1 input lines, and log N output lines. The way it works is that we build a circuit which handles N/2-1 inputs and another circuit which also handles N/2-1 inputs. We add their outputs with a simple array of one-bit field adders. This is built the same way - recursively, and so it is a recursively specified structure. It has a regularity and a structure. It is not repetitive in the sense of being rows and columns. It is repetitive in the sense of having a succinct description. Let us now think about how we would test it. Let us go down each possible input until we exhaust them. Then if we have 63 input lines, we obtain:

$$2^{63} \approx 10^{19}$$

which is clearly too large a number of tests, which would be inpractical. So what we can do is to use simply a recursive procedure; that is, we find out what the test vector must be for each full adder cell recursively. The procedure of the test of this circuit involves first the test of one circuit of half-size, say the left one, setting the inputs to zero at the right side. That establishes a certain output at the centre which will be added to that zero. Thus the policy is:

Test (N/2) on left, 0's on right.

Test (N/2) on right, 0's on left.

Then we also have to test the full adders at the join, where these two inputs come together. This gives: log N cell tests at joining cells. We can do this by choosing the inputs so that the outputs of these half-size circuits establish the right inputs at one of the full adders, with 'zero' elsewhere. The total number of full tests is therefore:

$T(N) = 2T(N/2) + C \log N$, where we need C tests per adder. That is, twice the number of half-size tests plus the number of output adders. I have discussed this expression with Professor Cattermole this morning and he has shown me that this number $T(N)$ is asymptotically linear with N. We have:

$$T(N) = (\text{Number of cells}) \cdot C$$

$$= (N - \log N - 2) \cdot C$$

$$\approx 56 \text{ cell tests (if } N = 64).$$

Thus our program would in this case generate 56 sets of test vectors, each being designed to test the full adder. The point that I am trying to make is that if one has a very clear idea and a succinct description of the circuit, then it becomes much easier to deal with the testing.

Now I should go back to Fig. 12, because I want to consider the question of the single-fault hypothesis. When I said that I was going to test this chip exhaustively by setting its inputs and by observing and measuring its outputs, I have assumed that there are no faults elsewhere. With this assumption, if you are testing for single faults, and in fact you have two faulty cells, or seven faulty cells, then in the course of testing each of the seven cells you might detect in fact that something is not right. So the general issue here, on which I would like you to comment, is as follows: When is this single-fault modular testing exhaustive and when is it not?

Mr. Coakley - The smallest cells that you test are usually not small enough, because you might have two faults in a cell.

Professor Steiglitz - Two faults inside a cell presumably count as a single-fault as far as the input/output relation is concerned, and that would be discovered by an exhaustive test of a cell. If I have nominal input/output relations in every cell then the circuit is functioning correctly.

Professor Messerschmitt - If you use your idea for testing software, then it certainly is not right, because even if you test every single output in your program, it is still not enough.

Professor Steiglitz - There is a difference here, for I assume that the design is correct, so I am not really looking for errors in connections. I am simply worrying whether each element that I have placed in the chip is functioning as it should. I had to extend my test to wires in this case, but, as I said, I considered them as cells.

Mr. Johnson - I think the potential problem is, when I look for the identity of errors, whether a cell that is working affects the state of the whole machine. For every given state of the machine a cell may operate differently. This is a typical problem in testing memory devices.

Professor Steiglitz - That certainly might be the case - that the single fault model is not valid in many practical situations.

Professor Messerschmitt - There is a whole new dimension involved, particularly when you consider timing errors.

Professor Steiglitz - My technique is so to speak embedded in static situations. I could consider a sequence of waveforms, but I would not be testing for all the transitions. This is a very limited technique, I would be first to admit. It could be used for static production testing.

Professor Messerschmitt - The problem with static testing is that you have a data sequence going through a VLSI chip. Then the typical static 0-stuck or 1-stuck faults do not necessarily behave combinatorially. Basically we have here a capacitive circuit and what you are actually getting are high impedance faults. In that case I can have a single-bit finite state machine which would depend not only on the actual set of data, but also on the preceeding data. So this is not a simple case. But I do admit that in any case you would benefit from your structural approach, or a hierarchical approach.

Professor Cattermole - If in fact you are considering data-dependent test results on a particular subset of cells in a chip, then I imagine your procedure can also apply to a set of overlapping cells, rather than only to completely disjoint cells. So you should take into account not only a basic cell, but also its neighbours, and then you move on.

Professor Steiglitz - Of course, one can apply this method to all pairs of cells if one has enough time. This would not be unreasonable for the case shown in Fig. 13. Then I would have a chance to test the interaction of every cell with every other cell.

Dr. Cooper - I am very interested to know what is the nature of the proof that the equation you gave for the number of tests is linear asymptotically with N. Can you say how it is done?

Professor Steiglitz - You have Professor Cattermole to thank for this. First, if you can guess the answer, you can verify it easily by induction, going from $N/2$ to N. To derive the asymptotic linearity from scratch, let $N=2^k$ and $U(k) = T(n)$. Then the recursive relation becomes:

$$U(k) = Ck + 2U(k-1)$$

$$= Ck + 2[C(k-1) + 2U(k-2)]$$

continuing in this way, we get

$$U(k) \ < \ C \ \sum_{i=0}^{k} (k-i)2^i$$

$$< \ C \ \sum_{j=0}^{k} j \ 2^{k-j}$$

$$< \ C \ 2^k \sum_{j=0}^{\infty} j \ 2^{-j}$$

The infinite series converges to a constant, so $U(k)=O(2^k)$ and therefore $T(N)=O(N)$. Our chairman is signalling that we should now stop.

Professor McEliece - Indeed this is the case, and now we have something quite different which is presented by Dr. Heegard.

Dr. Heegard - I would like to discuss the use of algebraic codes for optical data storage. I started on this when I read an article by Rivest and Shamir [9]. This inspired two papers one by myself [10] and the other by Wolf, Wyner, Ziv and Körner [11]. They derive the potential capacity of an optical disc. Now that we know the capacity, the problem is how can gains be achieved in practice. I will start with a simple model. An optical disc is like a key-punch card. It has the property that after you write a '0' on a card, it is easy at a later stage to punch a hole or '1'. But, once you punch a hole or '1', there is no way to return to a '0'. Actually you might try and glue something on a card, but you cannot do that on an optical disc. There are two possible channel states S, a '0' channel state and a '1' channel state. If X and Y are input and output respectively, what you get in each state is shown in Fig. 14.

```
MODEL

0 → 1

1 ↛ 0
```

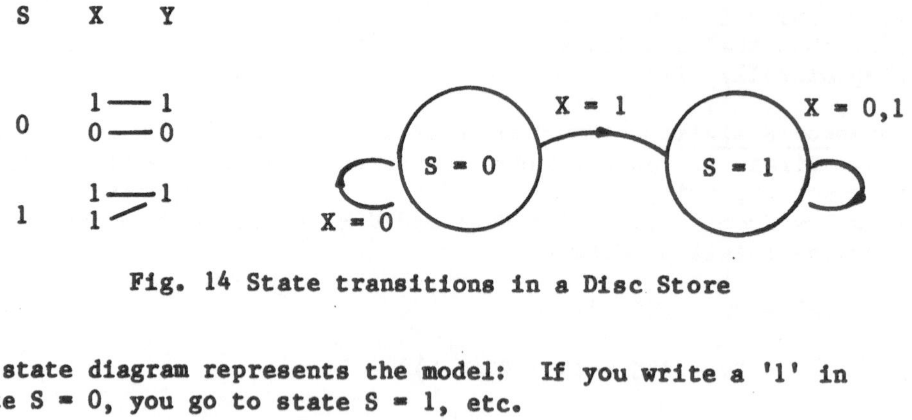

Fig. 14 State transitions in a Disc Store

The state diagram represents the model: If you write a '1' in state S = 0, you go to state S = 1, etc.

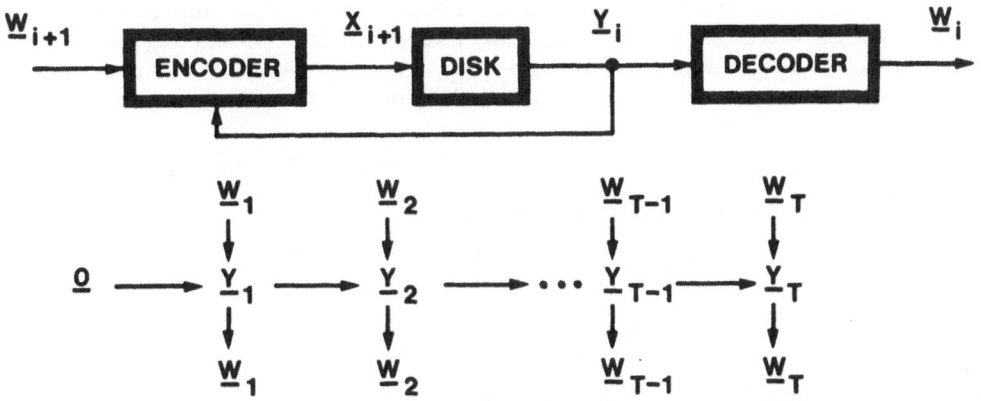

Figure 15

The problem we are considering is the following. Suppose there is a sequence of messages represented by the $\underline{W}_i$ sequence in Fig. 15, that you wish to store in the disc, where at time i the message is $\underline{W}_i$. At some time in the future I shall want to update the disc, so I look what is on the disc ($\underline{Y}_i$), take the current new message at time (i+1) and I generate a new word $\underline{Y}_{i+1}$ to put it in the disc. I want to do it in such a way that I get a sequence $\underline{Y}_i$ of vectors that are comparable with one another in the sense that they satisfy the write once constraint.

I am going to give you three examples. The parameters of the first code are: (n=3; k=2; T=2), where: n = number of cells,
        k = number of bits of data stored,

    T = number of times the storage is done.

Now we define the rate: R = k/n = 2/3

and the efficiency: RT = kT/n = 4/3 > 1 bits/cell.

Let us now consider what the decoder does. The decoder is simply a matrix multiplication. It has a nice structure and is easy to implement:

$$\underline{W}_n = \underline{Y}_n \ H^t \ \text{where} \ H = \begin{bmatrix} 1 & 0 & 1 \\ 0 & 1 & 1 \end{bmatrix}$$

It is a low complexity decoder; it is simply a linear transformation. In this case H is the identity matrix with two 1's added. The encoder's job is as follows: given the new message $\underline{W}_{n+1}$, given what is written in the three cells, namely $\underline{Y}_n$, we try to minimise the distance measure $d(\underline{Y}_n, \underline{Y}_{n+1})$, subject to the sequence that you wish to be decoded properly:

$$\underline{W}_{n+1} = \underline{Y}_{n+1}\ H^t.$$

the distance matrix is not symmetrical:

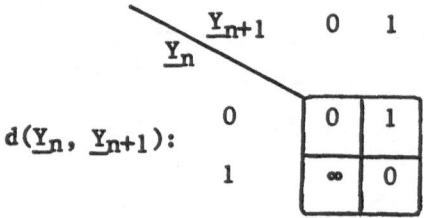

$$
d(\underline{Y}_n, \underline{Y}_{n+1}):
\begin{array}{c|c|c|}
\multicolumn{1}{c}{} & \multicolumn{1}{c}{0} & \multicolumn{1}{c}{1} \\
\cline{2-3}
0 & 0 & 1 \\
\cline{2-3}
1 & \infty & 0 \\
\cline{2-3}
\end{array}
$$

You will note here that if '1' is written in a cell, you cannot next write '0' there. Thus I have put ∞ there, since this is not allowed.

We try to minimize the distance measure d. We try to find a sequence that is associated with the current message and minimizes the creation of a new one. To see how this code works, we write all the possible two-bit messages $\underline{W}$ and all the sequences associated with each message:

$\underline{W}$		Table 1	Table 2	
0	0	0 0 0	1 1 1	e.g. $\underline{W}_1 = 1\ 0,\ \underline{Y}_1 = 1\ 0\ 0$
1	0	1 0 0	0 1 1	$\underline{W}_2 = 0\ 1,\ \underline{Y}_2 = 1\ 0\ 1$
0	1	0 1 0	1 0 1	
1	1	0 0 1	1 1 0	

the idea is that for the first time you go around you use essentially Table 1 and here you find at most single 'ones'. That leaves two 'zeros' that you can play with. The matrix H has a nice property that every 2 x 2 sub-matrix is invertible. This property assures that T = 2.

Now to the second example. This algebraic code by itself is not very good, but it will become useful in the next example. The parameters are:

(n = 5; k = 2; T = 3)

and it is over the field: $F_4 = \{0, 1, \alpha, \alpha^2 = \alpha + 1\}$. Over the binary field it takes 10 cells, writes 4 bits of data and does it three times. Its rate is R = 2/5 and its efficiency is RT = 6/5 > 1, bits per cell. Here again we have the decoder matrix:

$$H = \begin{bmatrix} 1 & 0 & 1 & 1 & 1 \\ 0 & 1 & 1 & \alpha & \alpha^2 \end{bmatrix}$$

Perhaps you can recognise this as the parity check matrix of a Hamming code. We can again make up the distance measure

$d(\underline{Y}_n, \underline{Y}_{n+1})$:

$\underline{Y}_n$ \ $\underline{Y}_{n+1}$	0	1	α	α^2
0	0	1	1	1
1	∞	0	∞	∞
α	∞	∞	0	∞
α^2	∞	∞	∞	0

Note that this is not a binary distortion measure. We can write the sequences three times and the reason for this is that the matrix H has two properties:

1) Given any $\underline{W}$, assume we start with all zeros on the disc. I can find a disc vector that has only one 'non-zero' component. Thus, if $\underline{W}_1 = [1 \ 0 \ 1 \ 1] = [\alpha \ \alpha^2]$, I can obtain the vector $\underline{Y}$ which has a 'non-zero' element in the fourth position, namely:

$$\underline{Y}_1 = [0 \ 0 \ 0 \ \alpha \ 0] \ (=[0 \ 0 \ 0 \ 0 \ 0 \ 0 \ 1 \ 0 \ 0 \ 0]).$$

Similarly: $\underline{W}_2 = [0 \ 0]$, $\underline{Y}_2 = [\alpha \ \alpha^2 \ 0 \ \alpha \ 0]$

and: $\underline{W}_3 = [\alpha \ \alpha^2]$, $\underline{Y}_3 = [\alpha \ \alpha^2 \ 1 \ \alpha \ \alpha^2]$.

Now we come to the third and last example. Here we concatenate the two codes used already in examples 1 and 2. Such a code stores four bits in 15 cells. The code parameters are:

(n = 15; k = 4; T = 6).

The rate is R = 4/15, and the efficiency RT = 8/5 = 1.6. Again it is easier to explain by seeing what the decoder does. It simply takes the 15 bits and breaks them into five three-bit long sequences:

$\underline{Y}$: |⎵⎵⎵| |⎵⎵⎵| |⎵⎵⎵| |⎵⎵⎵| |⎵⎵⎵|
  $\underline{Y}_1$ $\underline{Y}_2$ $\underline{Y}_3$ $\underline{Y}_4$ $\underline{Y}_5$

$$\underline{Z}_1 = \underline{Y}_1 H_1^t \text{ where}$$

$\underline{Z}$: |⎵⎵| |⎵⎵| |⎵⎵| |⎵⎵| |⎵⎵| $H_1 = \begin{bmatrix} 1 & 0 & 1 \\ 0 & 1 & 1 \end{bmatrix}$
  $\underline{Y}_1$ $\underline{Y}_2$ $\underline{Y}_3$ $\underline{Y}_4$ $\underline{Y}_5$

$$\underline{W} = \underline{Z} H_2^t \text{ where}$$

$\underline{W}$: |⎵⎵⎵⎵| $H_2 = \begin{bmatrix} 1 & 0 & 1 & 1 & 1 \\ 0 & 1 & 1 & \alpha & \alpha^2 \end{bmatrix}$

First we multiply by H_1^t and that gives us 5 blocks each 2-bit
long. By multiplying it by H_2^t we get a 4-bit long $\underline{W}$ sequence. An
example is given in the table below. Note that in any efficient
coding scheme when you have done you should have mostly 'ones' on
the disc.

Time	W	Z	Y
1	1 0 0 1	0 0 0 0 α	0 0 0 0 0 0 0 0 0 0 0 0 1 0 0
2	0 0 0 0	α 1 0 0 α	1 0 0 0 1 0 0 0 0 0 0 0 1 0 0
3	0 0 0 1	α 1 α α α	1 0 0 0 1 0 1 0 0 1 0 0 1 0 0
4	1 0 0 0	α 1 α α 0	1 0 0 0 1 0 1 0 0 1 0 0 1 1 1
5	1 1 1 1	α² α α α 0	1 1 0 0 1 1 1 0 0 1 0 0 1 1 1
6	1 0 0 0	α² α 0 1 0	1 1 0 0 1 1 1 1 1 1 0 1 1 1 1

<u>Professor McEliece</u> - Thank you Dr. Heegard. Any questions?

<u>Professor Golomb</u> - What is the purpose of using these long
code-words?

<u>Dr. Heegard</u> - Basically we know exactly how much gain coding can
provide. It turns out that if you fix T (the number of writes)
then for a rate $[\log (1 + T)]/T$ you get an efficiency as high as
$\log(1 + T)$. So if you encode 1023 times, potentially you get 10
bits per cell. That is why the code-word is that long.

<u>Professor McEliece</u> - We now have the last speaker here,
Dr. Goodman.

<u>Dr. Goodman</u> - I am going to talk about memory coding for VLSI
RAM's. At the last ASI organised by our director in Norwich in
1980, while I was sitting next to Professor McEliece, instead of
listening to a lecture I was baffling on some horrible
integral - it was just too horrible for me. So I asked him: 'Can
you solve that?'. He could not do it instantly, so I was a little

bit pleased. That integral was not that trivial; it was to do with memory coding. Since then we generated more integrals which have surpassed my wildest dreams of horrible things, for I am just an engineer myself. Professor McEliece has taken control of this business. He has four students working on this problem and this problem is: In a typical computer memory using VLSI D-RAM's, these days 64 K-RAM's, we arrange them in computer memory like shown in Fig. 16. We have a row of chips, the width is for your data words and if you want to correct any errors, we have some parity checks at the end. Depending how big you want your memory to be, you just keep replicating the rows of chips and increase your M. The data width is k and we have (n − k) parity checks. If you want it cheap, you do not have any of these and you just stick up with your data width. That is not a very good thing to do as I am going to show you. We know that chips fail, and they may fail catastrophically; they blow up or they go short or their input/output lines fail. Let us get a bit mathematical. If they fail at certain rate then the reliability chaps tell us that they fail at a constant failure rate, say $\lambda = 10^{-6}$ failures/hour. Then the reliability is the probability of a chip operating correctly at time t and we write $R = \exp(-\lambda t)$. The integral of that is called the Mean-Time-Between-Failures (MTBF) and this is:

$$(MTBF)_{chip} = \int_{o}^{\infty} \exp(-\lambda t)\ dt = 1/\lambda.$$

This, by the way is not that horrible integral I was telling you about. This means that for a single chip the MTBF is 10^6 hours, or 100 years. Not bad, but then in our case you are replicating these things. For instance, with an uncoded memory of 32 bits times 1 M words, the total number of devices is 32 x 1024 k/64 k = 512. Then MTBF = $1/\lambda$ Mk = (100/152) years = 2.5 months. That is not good at all. Furthermore, the 1 month reliability R_{month} = 0.7 and that means that only 0.3 of the system will last one month. That means that 30% of your system, if it is mass-produced, will fail in one month. The first time I said this in front of some VLSI people, their faces went white! This is because they were designing a multi-M-byte memory. So what should we do? Can you stop the processor? That is the question number one. In all these things about error-correcting codes on the system level people want you to hold the processor time. If you say to them that you can correct all these errors and that you will put there some fault-tolerant stuff, then they will say: 'How many nanoseconds do you want?' You say to them 10 and they say no, you cannot have more than 5 nanoseconds at most. So this is the problem. If you can stop the processor, then fine, we can do a lot of wonderful things, but if you cannot stop the processor then it has to be done in time. If you take a standard memory system, then what you do is to use something like a Hamming code, single error correction and double error detection (SEC-DED), because this is about as complicated as you can get, and do it very quickly and very cheaply; which is all they will allow you to do.

416

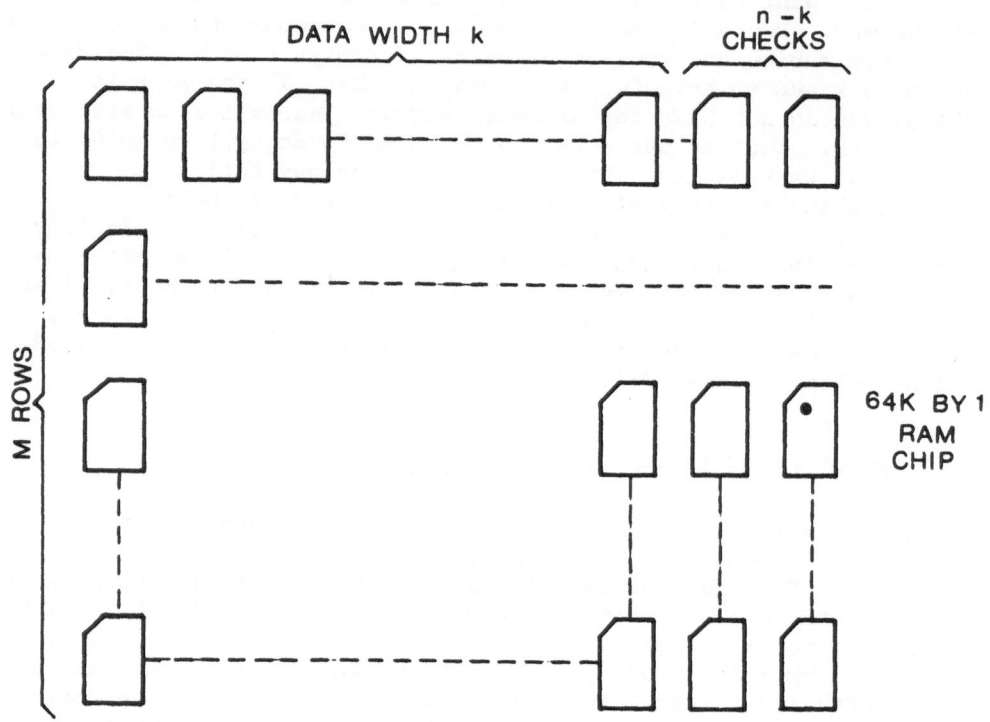

Figure 16

417

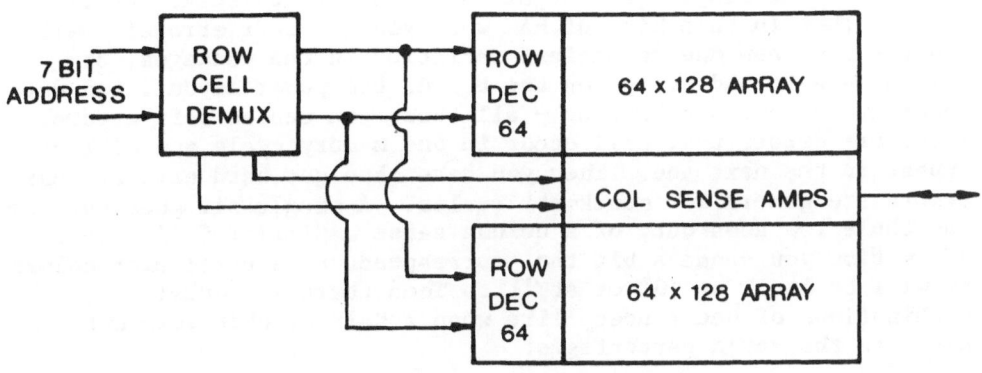

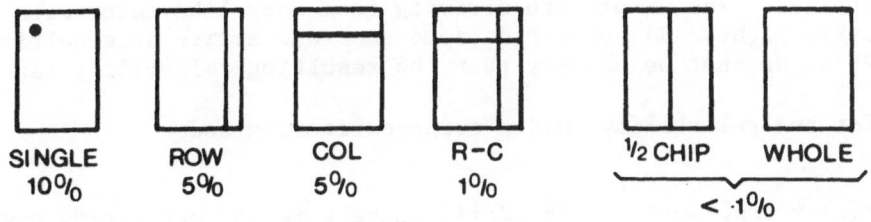

| SINGLE 10% | ROW 5% | COL 5% | R-C 1% | ½ CHIP | WHOLE |
| < 1% |

Figure 17

Things like erasure decoding and multiple error correction are acting differently here than in communication systems. When you get an error in a memory, it is stuck there. If it is not a soft error, but a hard one, then it stays where it is. When it stops and we can remember where it is in the memory, we can rub it out, and maybe when another error comes along, we can decode it later using erasure decoding techniques. Here is a typical old-fashioned D-RAM 16K (Fig. 17). I have assumed 16 K here because all others have more memories. It has basically two symmetric memory arrays, with the thing that took five years to design, namely the column sense amplifier in-between. Row decoders are fed from a multiple address bus which puts a row address on, while the column address is shared in the middle, and at the intersect out pops an output bit out of the single bit output that you want to read. What happens then in this kind of RAM when you get soft errors? Well, you can get them due to ambient radiation in the packages, or perhaps due to bad design on the board, bad power layout, bad data bussing, bad address bussing - all these can causes soft errors. These are errors that will occur in one memory cycle and will not appear in the next one. Then you have also got hard errors. Hard errors are generally 'stuck-at' faults. A single bit goes out, or the whole row goes out, of a column sense amplifier fails. So, every time you sense a bit that corresponds to a particular column it will be stuck at '0' or at '1'. Then there are other combinations of occurences, like when a half of chip goes out. Here are the rough percentages:

80% due to soft errors,

20% due to hard errors.

Then in Fig. 17 you see rough percentages of failures for different structures in RAM memories. I have modelled these.

You can model this situation by using the Poisson process technique. The errors are arriving in memory like motor cars at a traffic light. So you can do some sums and arrive at resulting MTBF's, so that we can say that the resulting reliability is:

- for row reliability, using error-correcting code:

$$R(t) = \exp(-\lambda nt) \sum_{j=0}^{r} (\lambda nt)^j / j!, \text{ where r is the correcting power}$$

of the code.

- for M rows in the memory, the MTBF becomes:

$$MTBF = \int_{0}^{\infty} \exp(-\lambda nMt) \left[\sum_{j=0}^{r} (\lambda nt)^j / j! \right]^M dt$$

Then for M large and r = 1: MTBF $= \sqrt{\pi/2M} \cdot (\lambda n)^{-1}$

We can now define the 'coding gain', which is the ratio of MTBF with coded memory to MTBF without coded memory: 'Coding Gain'

$= \sqrt{\pi M/2} \cdot (k/n)$, where k/n is the coding rate function. In the example which we are considering, i.e. for M = 16 and (n,k) = (38,32) we have: 'Coding Gain' = 4.2. This ratio refers only to catastrophic errors. In fact, if you have a row of memory with a Hamming code, what kills you when two errors occur is the Hamming code error along a certain row. We shall forget here about double error detection, but even so the Hamming code will not be able to deal with it. Furthermore, inside the chip, if a column goes out (on one particular chip in a row) and then a row goes out, then the intersect is in error with probability of one. So we get two errors down a particular bit line and that particular row is going to fail. There there are more complicated things, like when you are getting a row and a column failing and then a soft error occurs. Professor McEliece's and my students are working on it and are modelling this situation. Other things that we are looking at is error correction inside the chip at refresh time. Basically what you are doing here is to read out what is on the whole row into the column sense amplifiers, feed them up with some charge and then stick them up into the row. That happens with the refresh time; this is a cyclic round. We can do the error corrections, all in one refresh time. We can read it in, do the free charging and do the error correction, since the refresh time still goes on. So we can still decide what is wrong. Then we can read it back up to that time; and then, if needed we can switch in the spares.

One final thing is when two errors occur in a particular row. Professor McEliece works on it now, and it has an exact parallel with the number of people in this room which have the same birthday on the same day. Thus you have M rows of memory, and the probability of memory failing is given by M days, or rather 365 days, and given that we have here 60 people, the probability is up to 80%, that two people here have the same birthday. This is all I wish to say.

Dr. Johnson - What is the VLSI technology that you are using here?

Dr. Goodman - One thing about this game is to get the reliability figures out of semiconductor manufacturers. It is impossible. The only exception are the INTEL people; these in my experience are the only people who publish their reliability figures. So I have taken their figures. Other are up and down by 15% from what we hear inside and outside; they are even worse sometimes. This is because of the standard trick: You buy a 16K D-RAM but it is really a 64K D-RAM and only a quarter of this works!

420

Professor McEliece - I should mention that the newly advertised
89.000 microprocessor has single error correction and double error
detection with automatic spares, which refresh the memory
automatically. This is technically a very difficult problem, even
given a simple Poisson mode, to calculate what the expected
life-time is, given a set of spares etc. Anyhow, thank you
everbody, this concludes the panel discussion.

References

[1] G. Moore, "VLSI" Some Fundamental Challenges", IEEE Spectrum
     pp 30-37, (April 1979).

     See also:
     G. Moore, "Progress in Digital Electronics", 1975 IEEE
     International Electron Devices Meeting, Washington, D.C.,
     December 1975.

[2] R.J. McEliece, "Some Applications on VLSI RAM's to Coded
     Communications Systems" (in this volume).

[3] E.T. Cohen, "Special Purpose Digital Hardware", (in this
     volume).

[4] B.L. Johnson, "Design and Hardware Implementation of a
     Versatile Transform Decoder for Reed-Solomon Codes (in this
     volume).

[5] T. Anderson and P.A. Lee, "Fault Tolerance: Principles and
     Practice", Prentice Hall (ISONO-13-308254-7).

[6] A. Vergis and K. Steiglitz, "Testability Conditions for
     Bilateral Arrays of Combinational Cells", 1983, IEEE
     International Conference on Computer Design: VLSI in
     Computers, New York, Oct. 31-November 3rd, 1983.

[7] K. Steiglitz, "Hierarchical Parallel and Systolic Array
     Processors", (in this volume).

[8] P.R. Cappello and K. Steiglitz, "AVLSI Layout for a Pipeline
     Dadda Multiplier", ACM Transactions on Computer Systems,
     Vol. 1, No. 2, May 1983, pp 157-174.

     See also:
     E.E. Swartzlander, Jr., "Parallel Counters", IEEE Transactions
     on Computers, Vol. C-22, No. 11, pp.1021-1024, November 1973.

[9] R. Rivest and A. Shamir, "How to Reuse a Write-Once Memory,"
     to appear in "Information and Control", in 1983.

[10] C. Heegard, "On the Capacity of Permanent Memory", to appear
     in IEEE Transactions on Information Theory, 1984.

[11] J.K. Wolf, A. D. Wyner, J. Fio, J. Körner, "Coding for a
     Write-Once Memory", private information.

SPECIAL PURPOSE DIGITAL HARDWARE

Earl T. Cohen

Cyclotomics, Inc., Berkeley, California, U.S.A.

ABSTRACT

This paper explores the implementation of special purpose digital
hardware devices, using Reed-Solomon decoders as its primary example. A design methodology is presented for implementing "special
purpose microprocessors" -- microprogrammed devices designed to
carry out specific tasks. Several Reed-Solomon decoder designs with
differing throughputs and costs are discussed in light of this
methodology.

Parts of this paper are based on the Ph.D. dissertation "On the
Implementation of Reed-Solomon Decoders", U.C. Berkeley, 1983.

1. INTRODUCTION

There is an increasing need in the data processing and communications industries for specialized hardware devices designed to carry out specific tasks. Some examples of applications for such devices are disk controllers, network interfaces, and error-correcting decoders.

The conventional approach for designing such devices employs a single-chip microprocessor (SCMP) to obtain the needed computational power, and surrounds that chip with enough additional hardware to obtain the desired performance. This paper proposes an alternative design approach which does not use SCMPs and obtains much higher throughput as a result.

The application used to demonstrate the effectiveness of this design methodology is Reed-Solomon (RS) decoding, a task that requires specialized computations such as Galois Field multiplication.

Three actual implementations of RS decoders are presented. These implementations range from a conventional SCMP based design to a very special purpose Galois Field Processor -- the GF1^{TM} Decoder*. As the complexity of these designs increases, so does their performance.

The hardware design methodology expressed in this paper differs in some respects from that commercially practiced. By describing the hardware designs and tradeoffs for the particular problem of decoding Reed-Solomon codes, this paper will shed additional light on the digital hardware design process in general.

2. HARDWARE DESIGN PHILOSOPHY

2.1. Motivation

Many projects call for the design of Special Purpose Microprocessors (SPMPs) of one sort or another. We distinguish such devices from "straight" hardware by the inclusion of PROMs that are defined by some higher level representation of their contents (such as microcode or assembly language).

SPMPs are distinguished from conventional microprocessors by their dedicated nature -- they are designed to carry out a single task as rapidly as possible using the minimum amount of hardware.

Some common examples of SPMPs are:

(1) disk controllers, which require a large amount of intelligent processing.

---

* GF1^{TM} is a trademark of Cyclotomics, Inc.

(2) dedicated graphics processors, as are found in video games

(3) terminal controllers, such as those in intelligent, bit-mapped terminals.

Devices such as those above are frequently implemented according to a conventional design methodology. A single-chip microprocessor (SCMP) is chosen as the computational unit, and sufficient interfacing is added to allow this processor to communicate with and control other devices.

For applications that do not require high speeds, a SCMP is probably the right choice. If the application requires speeds beyond the capability of a SCMP, the conventional solution is to equip the microprocessor with enough outboard, special purpose hardware that it can keep up.

The view expressed in this paper, on the other hand, is that a design which chooses a SCMP as the heart of the system concedes a large factor in speed and performance. For applications where high throughput is important, a SPMP implemented in discrete hardware may run significantly faster than a SCMP based system.

2.2. Conventional Microprocessor Design

Given some task requiring computational power, such as implementing a disk controller, the conventional approach is to use a SCMP and appropriate interfacing. The microprocessor is programmed in assembly language. The code it runs is most likely developed on either a development system for that microprocessor, or a larger computer with cross assemblers and simulation facilities. If the project is small enough, and the amount of software to be written is not very large, it is possible to develop the software without either of these two facilities; but this is not usually practical.

From the early days of the 6502 and the 8080 to the advent of the Motorola 68000 in the last few years, the power of SCMPs has grown enormously. The difference in computational power between these early processors and more modern ones is at least an order of magnitude. Furthermore, the number of features in these chips is also increasing. The 68000 is a full 32-bit processor, with 16-bit input/output data paths, and 16 registers on chip [10]. Compared to this, the 8-bit 8080 seems to be a child's toy.

But one pays a penalty for all these advanced features. Interfacing to the 68000 is more complicated than interfacing to the 8080 -- the 68000 supports many features (such as memory mapping, provisions for caches, etc.) that are needed only in the design of full-fledged computers.

A conventional processor built from one of these off-the-shelf microprocessors generally has the following components:

(1) a memory, containing a PROM store for the code the processor runs and RAM for the use of the processor (and possibly the

external world),

(2) a means of communicating with the external world, possibly via DMAs,

(3) a SCMP, which plays the role of Arithmetic Logic Unit (ALU), program counter and branching logic.

If there are any needed features that cannot be implemented easily or efficiently by the microprocessor (such as communicating with the external world, special computational features, etc.), they can always be added by "memory-mapping" them in, so that references to specific memory addresses activate the special features.   Almost all interesting applications require some kind of memory-mapped extra feature.

Even conventional design with bit-slice based microprocessors follows similar lines. The only difference is that the bit-slice(s) are used as just an ALU; an external program counter and memory con- trols are added to make the system equivalent to a microprocessor. Often, there are families of chips designed to support these bit- slice ALUs [1]. Such a system offers slightly more flexibility than a SCMP based system, but conventional design methodologies try to force bit-slice based processors into a very standardized framework. The advantages of a conventionally designed bit-slice based system over a SCMP system are a (slight) speed improvement and greater sim- plicity in interfacing to external I/O.

2.3. Special Purpose Microprocessor Design

Given a problem such as the rapid decoding of Reed-Solomon codewords (a task that uses Galois Field arithmetic, not binary arithmetic), the choice of a SCMP as the central processing unit of the system greatly limits its overall throughput since the micropro- cessor will spend too much of its time simulating operations that can be done more efficiently with a small amount of hardware. (This claim is substantiated in succeeding sections of this paper).   An 8086 implementation of a Reed-Solomon decoder has shown the throughput obtainable by a SCMP to be more than two orders of magni- tude less than that obtainable with a special purpose microproces- sor. Even building outboard hardware that the SCMP can access does not improve matters much. The time to access such hardware is still non-trivial, and the additional hardware defeats one of the prime motivations for choosing a SCMP in the first place: keeping the parts count low.

Each SPMP must be created for the task at hand, but there are many features all SPMPs share. Such designs generally include:

(1) A PROM store for the microcode,

(2) A Program Counter (PC) used to address the PROM store, and branching logic (based on results obtained from other parts of the machine) to determine whether the PC is incremented or is

loaded with a new value,

(3) One or more ALUs (made either out of discrete components bit-slices) together with registers,

(4) A memory (possibly with some kind of Memory Address and Memory Buffer registers),

(5) A means of communicating with the external world, possibly via DMAs, possibly by dedicated data paths.

The creation of such a specially designed microprocessor also requires the implementation of support software (assemblers, simulators, etc.) in order to develop and test the microcode the SPMP runs. Using conventional techniques, the cost of creating this software is quite large, but there are techniques that minimize this cost.

The cost of developing a simulator is minimized by making the microcode appear to be a higher level language so that it is directly executable in that language. Special purpose assemblers for such a restricted language can be implemented with editor scripts (and other system-level programs), avoiding the high cost of writing a full-fledged assembler in a conventional programming language.

Using this high-level language approach, the firmware for a SPMP can be developed entirely on a large mainframe computer in a simulation environment that parallels the hardware to such an extent that when PROMs are finally burned, the code is practically bug-free.

2.4. Comparison

2.4.1. Hardware

It must be conceded that designing a system around a SCMP (very likely) requires less hardware than a SPMP built out of discrete components. The added cost of building a SPMP is non-trivial; hardware must be designed to do things that the SCMP does automatically, such as branch control, instruction fetching, etc.

But there is a large performance gain as a result, partially from increased parallelism, partially from the higher speed of the components, and partially from having an instruction set that is tailored to the problem at hand. For example, any SCMP has to share its I/O pins for both instruction and data fetching, but a SPMP might have separate memories (which function in parallel) for these tasks. Although the speed of the SCMP system is generally limited by the speed of the microprocessor, a SPMP is quite often limited only by memory speed.

2.4.2. Software

In addition to the microcode/assembly language, there is a need for support software, such as assemblers, linkers, and loaders. The advantage of the SCMP approach is that large amounts of this software already exist. The disadvantage is that all this software was developed elsewhere, often does not do quite what is needed for the current project, and cannot be easily fixed if bugs are found.

With the SPMP approach, the micro-assembler must be developed, and it is unlikely that it will be as "friendly" as a commercially available assembler written for a SCMP. But in this case, all of the software development is done in a higher level language with the ability to use all the tools and facilities of a large mainframe computer to aid in the development and debugging. In fact, the development of the software may go significantly faster because of this.

2.4.3. Integration

In either system the hardware and software must be integrated and tested. In the SCMP system, the software is usually not fully tested until it is burned into PROMs; in this mode, both hardware and software are debugged simultaneously. This procedure generally results in longer debugging times, since when problems are found, it is not clear whether they are caused by hardware or software.

With the SPMP approach, the software is fully simulated before PROMs are ever burned. This simulation can be as detailed as is desired (including simulating random DMAs, for example). The advantage of this is that when PROMs are burned, the designers can be fairly sure that all bugs are hardware-related.

Although it is possible to simulate assembly language for the SCMP systems in equal detail, it is not as convenient. The directly executable high-level language approach used with the SPMP greatly enhances debugging and simulation capabilities.

2.5. Summary

When high performance is a major system goal, high throughput can be obtained by designing a special purpose microprocessor for the task at hand. The added software problem of designing a simulator and micro-assembler can be minimized by using microcode that is "directly executable" in a high level language. A processor designed this way may have throughput an order of magnitude or more better than a system designed around a SCMP, and does not have significantly more hardware or software development costs. In some cases, the development costs may be significantly less.

3. RS DECODING

3.1. Implementation Issues

The three RS decoder designs presented in this paper are designed to be efficient at decoding high rate RS codes with algebraic decoding techniques. Another method of decoding RS codes called transform decoding [5] has certain asymptotic advantages for decoding low rate RS codes, but the areas of applicability of low rate block codes are limited. The decoding of high rate RS codes with algebraic decoding techniques is the appropriate choice in a wide variety of circumstances.

All the given designs are restricted to fields of characteristic two, due to its practicality of implementation with digital hardware.

3.2. Algebraic Decoding Procedure

The worst case decoding of RS codes involves several steps:

(1) computing the power sum symmetric functions for the received word, which may involve computing the remainder of the received word divided by the generator polynomial of the code

(2) solving the Key Equation to find the error locator polynomial

(3) finding the inverses of the roots of the error locator polynomial

(4) computing the error evaluator polynomial from the error locator polynomial and the power sums

(5) finding the error values (The error evaluator polynomial may be computed at the same time as the error locator polynomial via an algorithm such as the Key Equation Solver [4], but the method described above which computes the error evaluator polynomial via a separate step is slightly more efficient).

3.3. Algebraic Decoding Complexity

For an RS code of length n and redundancy r, the decoding process requires roughly

$$\frac{3nr}{2} + n + \frac{15r^2}{8}$$

memory references and

$$\frac{3nr}{2} + r^2$$

Galois Field multiplications in the worst case [6].

If a Reed-Solomon Encoder (RSE) is used to generate the remainder directly from the received word, the number of memory references and multiplications are

$$\frac{nr}{2} + n + \frac{23r^2}{8}$$

$$\frac{nr}{2} + 2r^2$$

respectively. Bit-serial implementation techniques for RS encoders make the cost of their inclusion in an RS decoder very small [3]. There is no additional data delay caused by the RSE.

The most frequent operation in RS decoding is evaluating a polynomial at a point. The inner loop of this operation involves one Galois Field multiplication (multiplying the previous result by the evaluation point), one memory reference (fetching the next coefficient of the polynomial from memory), and one Galois Field addition (adding the next coefficient to the newly computed product). We call this operation, involving one memory reference and one Galois Field multiplication, a prototypical inner loop, since it occurs so frequently in the RS decoding procedure. The time required for the entire decoding process with a given decoder design is very closely related to the time required to execute this prototypical inner loop [6].

4. A SINGLE-CHIP MICROPROCESSOR BASED RS DECODER

4.1. Implementing Galois Field Operations

Reed-Solomon decoders implemented on binary computers require a method for doing Galois Field operations. There are two common table lookup methods for implementing addition and multiplication in small Galois Fields (of characteristic two). The first method maintains all Galois Field values in an m-bit binary form and uses log and antilog tables to do multiplication; addition is done via exclusive-OR. The second method stores all Galois Field values by their logarithm, performs multiplication by modular addition, and performs addition via Zech's logarithms (first described by Conway [7]). If throughput is the primary consideration, there are some implementation advantages to the log/antilog table method [6].

4.2. Basic Hardware

An Intel 8086 running at a clock rate of 6 Mhz was used to implement a RS decoder for the (255, 243) RS code over $GF(2^8)$. This decoder ran at data rates up to 18.5 kilobits per second (kbps). The code run on the 8086 was written in carefully optimized assembly language; little speed improvement could be obtained by improving the code. A block diagram of this system is shown in Figure 4-1.

Hardware based around an 8086 to decode the (255, 243) RS code over $GF(2^8)$ requires ROM (approximately 12K bytes for both code and fixed data such as the log and anitlog tables), RAM (512 bytes, assuming only single codewords are to be decoded at a time), and a small number of additional chips for controlling memory access and

Microprocessor Based RS Decoder

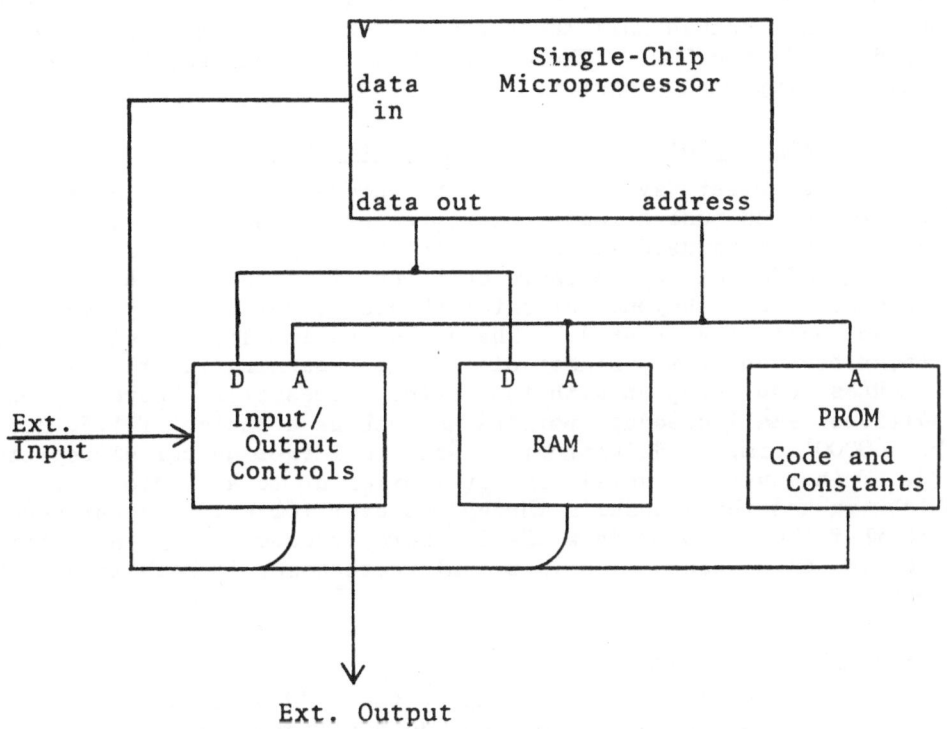

Figure 4-1

handling external Input/Output via DMAs or memory mapping. This entire decoder might have as few as a dozen ICs. Over larger fields, the amount of RAM and of data ROM will increase, but the amount of code ROM will stay the same.

4.3. Improvements?

The question that remains is: can the decoding time be significantly improved with the addition of a small amount of hardware? (The addition of more than a small amount of hardware is pointless, since the bit-slice microprocessor based design described in the next section requires less than twice the hardware of this

430

implementation and runs much faster).

The problem with a SCMP implementation is a lack of flexibility -- the microprocessor becomes the center of the design. It takes on the roles of program counter, memory controller, and arithmetic unit, and insists on handling all these functions by itself and in its own way. There is no obvious way to pipeline or improve the design because all the parts suitable to such improvement are buried inside the SCMP. There are several approaches, however, that might speed up such a system.

4.3.1. Using a Faster Single-chip Microprocessor

The simplest way to improve the throughput of a SCMP based decoder is to use a faster microprocessor. Assume a 24 Mhz 68000 microprocessor is used in place of the 6 Mhz 8086. (The fastest existing 68000 microprocessors only run at 16 Mhz, so this alternative candidate is beyond current technology, but probably achievable in the next few years). There is an immediate factor of four speedup because of the higher clock rate (assuming that the PROMs and RAMs can keep up with the faster processor). There may be an additional small speedup, perhaps as much as a factor of 1.5, due to the 68000's improved instruction set and larger number of registers [9]. This gives an overall improvement of up to a factor of six, which pushes the decoder's throughput over 100 kilobits per second. Note that the 68000 being a 32-bit microprocessor makes no difference in this case -- a 16-bit microprocessor is more than sufficient.

4.3.2. Additional Hardware

Another approach to speeding up the SCMP based decoder is to take some frequent operation and add additional hardware to implement it directly. Unfortunately, one quickly comes to the realization that the easiest things to implement in hardware can be done almost as efficiently in software.

Implementing Galois Field addition in hardware clearly does not save any time -- microprocessors already have exclusive-OR instructions and can implement Galois Field addition directly. Consider implementing Galois Field multiplication in hardware. In software with table lookups, this operation requires four additions and three memory references. (Three of these additions add the base address of an array and an index; the remaining addition is the sum of the two looked-up logarithms). This can be done in only three instructions, since (as is the case in a carefully coded inner loop) the base addresses of the log and antilog tables can be stored in registers, allowing the addition of the index into these tables to be done as an addressing mode instead of as a separate instruction. The 68000 code for Galois Field multiplication over $GF(2^8)$ is shown in Figure 4-2. (Note that instructions dealing with logarithms must be word instructions, since the log table contains 9-bit values).

68000 Code for Galois Field Multiplication

```
                              ; in-line code to GF multiply
                              ; registers R1 and R2
     MOV.W   R1@log, R0       ; log[R1] in R0
     ADD.W   R2@log, R0       ; log[R1] + log[R2] in R0
     MOV.B   R0@alog, R0      ; antilog of sum = result in R0
```

Figure 4-2

Implementing Galois Field multiplication in hardware could be done by memory mapping in the new function. One way to do this might be to have two special memory addresses X and Y. Writing a 16-bit value into address X causes a hardware Galois Field multiplier to multiply the high 8 bits of the 16-bit quantity written by the low 8 bits. A subsequent read from address Y returns the result of the previous "write" at address X. The improvement over the software method is not as great as might be hoped. First, both multiplicands must be packed together in a single register, then a memory write and a memory read are required -- three instructions are still needed. These instructions are shorter (and faster) than the original three, but the improvement is not as great as might be hoped because of the high overhead of instruction execution.

As another hardware speedup alternative, consider implementing a more complex operation, such as the evaluation of a polynomial at a point. The implemented 8086 code spends about 30% of its time evaluating polynomials. Implementing this kind of operation in hardware would save at most 30% of the total running time, and would require a (comparatively) massive amount of additional hardware -- in addition to memory mapping hardware to control the polynomial's base address, length, and evaluation point, alternate paths to allow the additional hardware to access memory (to fetch the coefficients of the polynomial being evaluated) must be provided. External Galois Field addition and multiplication hardware must also be added. Once this has been done, the external hardware is almost capable of replacing the SCMP altogether! This interesting alternative will be further investigated in the next section.

5. A BIT-SLICE BASED RS DECODER

5.1. Design

Unlike the microprocessor based system in the previous section, a bit-slice based system uses VLSI components (the bit-slices) only as an ALU. Separate addressing and branching controls are added which give the system some parallelism. The addition of registers on critical data paths makes the system pipelined and decreases the clock cycle time. An entire decoder for the (255, 239) RS code over $GF(2^8)$, shown in block diagram form in Figure 5-1, requires less than 30 ICs.

5.2. Clock Rate

Note that although the computational power of SCMP based systems can, to a large extent, be compared just by comparing the clock rates of the processors, a comparison of the clock rate of this bit-slice based system with a SCMP based system might lead one to think the microprocessor system was more powerful. The error in this argument arises from the fact that the bit-slice based system executes one instruction on each clock cycle; the microprocessor based system requires many clock cycles for each instruction.

A typical microprocessor requires four clock cycles for a simple instruction, such as the addition of two 16-bit registers [11]. Thus, a bit-slice based decoder with a 6 Mhz clock might very well be competitive with the 24 Mhz 68000 system discussed in the previous section. In fact, as will be seen, the bit-slice based system of Figure 5-1 achieves more than a factor of two higher throughput than the microprocessor based system, even though its actual clock rate is only 4 Mhz. The higher throughput is due both to more parallelism and pipelining in the architecture, and an instruction set better suited to the problem of RS decoding. One example of this is the pipelined branch instructions of the bit-slice based decoder. Due to a stage of pipelining flip-flops on the output of the microcode PROM, the effect of a branch instruction is delayed by one cycle. This helps keep the clock cycle time short at the cost of making the microcode more difficult to write.

5.3. Implementing Galois Field Operations

The key to implementing a fast decoder around a bit-slice ALU is the ability to do Galois Field operations quickly and easily. This is accomplished by the addition of a log/antilog table to the input data path of the bit-slices.

All bit-slices provide an external data input path and an output that is the result of the bit-slice's ALU operation. By putting a PROM and a flip-flop between the ALU output and the data input, a fast, efficient means of taking logs and antilogs can be implemented in a small number of integrated circuits. (Without at least one flip-flop in this data path, there is nothing to latch the ALU

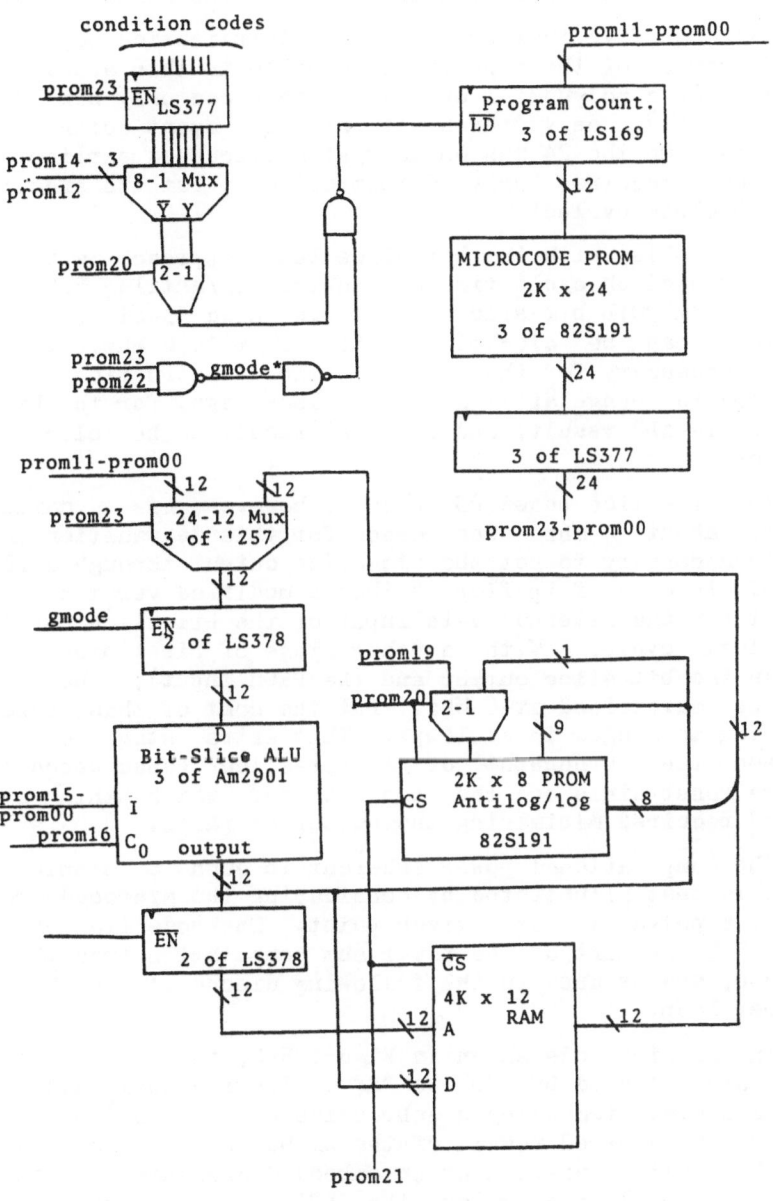

Figure 5-1

output of one instruction and thus allow the data input of the next instruction to be a function of the previous result). With the binary ALU of the bit-slice and the log/antilog table method of doing Galois Field multiplication, two numbers may be multiplied in only five instructions (including fetching the numbers from memory).

Although five instructions per multiplication may sound excessive, because of the pipelining possible in this architecture, the inner loop of a polynomial evaluation requires only six instructions (clock cycles), only three of which are spent doing Galois Field operations. On the 24 Mhz 68000 system discussed earlier, this same inner loop requires only 5 instructions, but its execution takes almost 60 clock cycles!

One problem with the bit-slice based approach is the degradation in the clock cycle time due to the log/antilog PROM. A system with three Am2901B bit-slices that is used just for arithmetic operations can be clocked at rates of up to 6 Mhz. This reflects the time necessary for the current micro-instruction (held in a flip-flop) to change after a rising clock edge, for the bit-slice to generate its ALU result, and for that result to be clocked into a flip-flop.

The bit-slice based RS decoder, however, has a maximum clock rate of about 4 Mhz. The reason for this degradation in speed is the time necessary to get the bit-slice output through a PROM and a MUX and into a flip-flop so that a modified version of it can be re-input (on the external data input of the bit-slice ALU) on the next clock cycle. With another stage of pipelining on this path (between the bit-slice output and the PROM input), the clock rate could be maintained at 6 Mhz. But the cost of this extra stage of pipelining is longer inner loops. This extra stage of pipelining increases the throughput of a bit-slice based decoder, but the hardware constraints on the project for which this system was designed required minimizing the number of parts.

The computational power inherent in such a simple processor design is best illustrated by considering the microcode required to evaluate a polynomial at a given point. The code fragment shown in Figure 5-2 is part of the power sum computation from the decoder's microcode, and is used in the following discussion as the prototypical inner loop.

In the microcode shown in Figure 5-2, the 16 registers of the Am2901 are denoted by "RO" to "RF". Instructions that control the Am2901 are simulated using a subroutine call -- AS() -- whose argument is the desired result of the Am2901 ALU for that instruction. This ALU result is written as an embedded assignment statement with a function call to carry out the ALU's opcode on the desired arguments. These ALU controls can be followed by additional subroutine calls that affect the disposition of the Am2901 output (and hence the value input through the Am2901's D input on future clock cycles). Examples of this are the MAR() subroutine call, which

Power Sum Evaluation Microcode

```
SYNLOOP:
    AS( R3 = OR(0, R9) );      RAM();
    AS( OR(D, 0) );            LOG();

syninner:
    AS( R2 = ADD(D, R4) );     ALOG();
    AS( R2 = OR(D, 0) );
    AS( R1 = ADD(R5, R1) );    MAR();
    AS( R3 = BUS1(0, R3) );    RAM();
    BR( !ZC );
    AS( R2 = XOR(D, R2) );     LOG();   if(h)goto  syninner;

    AS( R6 = ADD1(0, R6) );    MAR();
    AS( OR(0, R2) );           WE();
    AS( R4 = ADD(RA, R4) );
    AS( R8 = BUS1(0, R8) );
    BR( !ZC );
    AS( R1 = OR(0, R7) );      MAR();   if(h)goto  SYNLOOP;
```

Figure 5-2

loads the Memory Address Register with the output of the bit-slice's
ALU, and the LOG() subroutine call, which loads the flip-flop at the
bit-slice's D input with the logarithm of the current bit-slice ALU
output. Another instruction format -- BR() -- makes branch decisions. The CQ() instruction format (not shown in the code fragment
in Figure 5-2) allows constants in the PROM to enter the bit-slice's
external data input. As can be seen, the inner loop of a polynomial
evaluation requires only six instructions.

Each of the instruction lines shown in Figure 5-2 is executed
in a single clock cycle by the bit-slice based decoder. The width
of the microcode PROM for the implemented Am2901 based decoder was
only 24 bits.

5.4. Improvements

The simplest improvement to increase the performance of the
above architecture is to increase the width of the PROM in order to
provide more facilities. One step in this direction might be to
allow each instruction to have its own constant field, branch op-

436

code, and branch destination (as opposed to the implemented design which has separate instruction formats for branching and for con stants). This does not have a large effect, however, since almos all inner loops on the implemented decoder are five or six instruc tions in length, and constants are not used in inner loops. separate branch field/address for each instruction saves only on instruction per inner loop (on the average) and greatly increase th width of the microcode.

For a slightly higher cost in hardware, more bit-slices can b added. By keeping the memory addressing bit-slices separate from the Galois Field ALU bit-slices, memory addressing can be done i parallel with Galois Field operations. (Note that this als requires a wider PROM in order to control both kinds of bit-slice in parallel). These additional bit-slices save about two instruc tions per six cycle inner loop, since both the loop test and th memory reference can be done by one set of bit-slices in paralle with the needed Galois Field arithmetic operations.

With a microprocessor based decoder, the addition of Galoi Field multiplication hardware does not offer any significan improvement because the time needed to access such hardware is no significantly shorter than the time needed to do the multiplicatio directly by table look-up in software. In a bit-slice base decoder, however, the addition of Galois Field multiplicatio hardware saves time -- one instruction per Galois Field multiplica tion (as explained below). With this hardware improvement and the additional addressing arithmetic bit-slices described above, the inner loop of a polynomial evaluation takes only three instructions.

Hardware that can do Galois Field multiplication in times com parable to that of the bit-slice based decoder's clock cycle is shown in Figure 5-3. This hardware, which requires less than 2 ICs, has been designed for a minimal parts count, not for speed.

The Galois Field multiplication hardware is used by a bit slice based decoder in the following manner: on the first clock cycle, the bit-slice ALU outputs the first multiplicand, which gets latched in a flip-flop. On the second clock cycle, the bit-slice ALU outputs the second multiplicand, and on the third clock cycle the desired product may be input through the bit-slice's external data input. When doing a polynomial evaluation, however, the first multiplicand may be set once for the entire inner loop and the mul tiplication itself requires only two clock cycles.

5.5. Summary

The initial bit-slice based design could decode the (255, 239) RS code at speeds up to 300 kilobits per second. By adding more bit-slices to do the addressing arithmetic and Galois Field opera tions in parallel, and by adding a Galois Field multiplier, the throughput can be increased to 600 kilobits per second. This improved design has less than 60 ICs, and outperforms the 24 Mhz

Simple Galois Field Multiplication Hardware

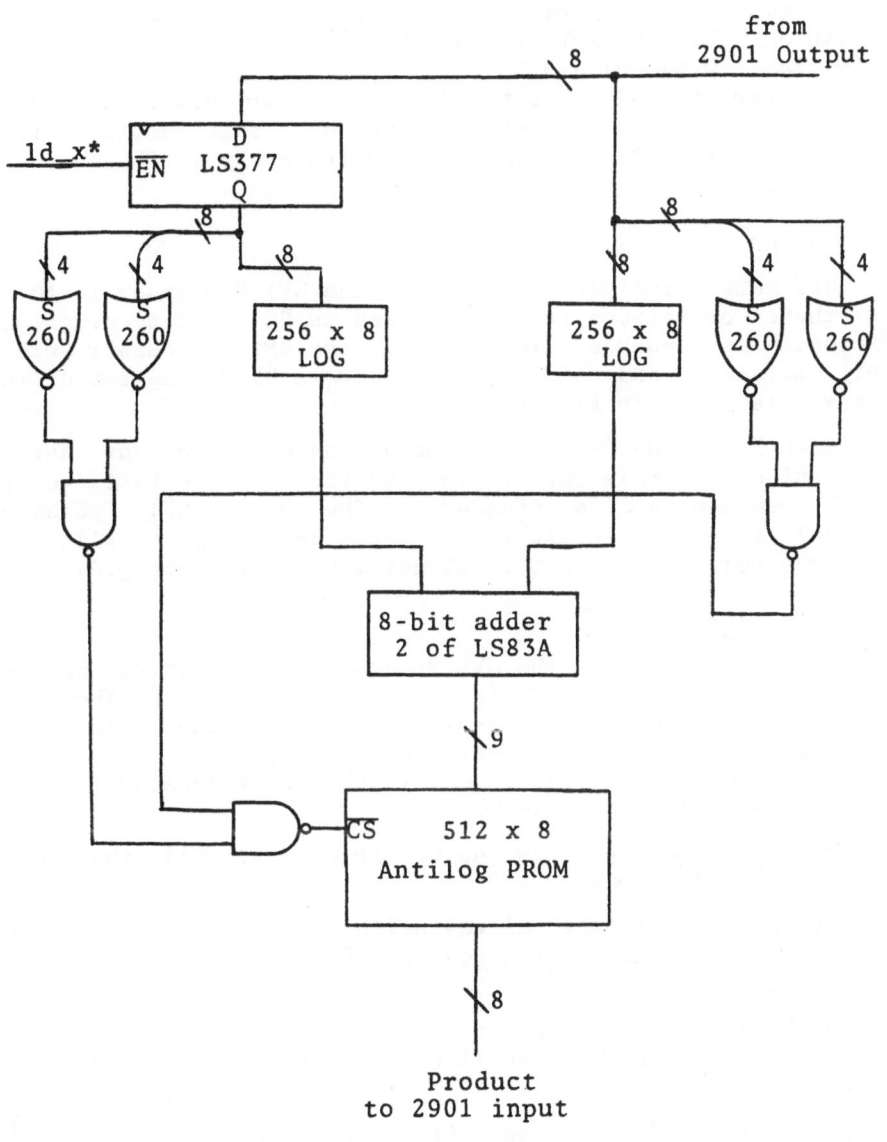

Figure 5-3

68000 system by a factor of four.

6. A HIGH SPEED RS DECODER

The GF1[TM] Processor is a special-purpose, pipelined architecture designed by E. R. Berlekamp especially for decoding Reed-Solomon codes [2]. A 5-bit version of this architecture (which operates on 5-bit characters over $GF(2^5)$) requires less than 70 ICs, all (vintage 1975) SSI and MSI, and operates at speeds up to 200 kilobits per second at rate 1/2. At higher code rates, its worst case speed approaches 1 megabit per second. This section discusses the design of the GF1 Processor and analyzes why it is so successful at RS decoding.

6.1. Design

The main principle of design of the GF1 Processor is to implement those operations frequently used in RS decoding as simply and efficiently as possible. To this end, the GF1 Processor was made highly parallel and highly pipelined. A block diagram of the GF1 Processor is shown in Figure 6-1.

Both 5-bit and 8-bit versions of the GF1 Processor have been built out of discrete parts. This section uses an 8-bit version of the GF1 Processor as its standard. The 8-bit GF1 Processor is identical in architecture to the 5-bit one; only the widths of the data paths between the functional units have been changed.

6.1.1. Parallelism

The parallelism in the GF1 Processor allows it to do in one clock cycle what other designs do in several. Like the bit-slice based design in the previous section, separate addressing and arithmetic units are employed to carry out memory referencing and Galois Field operations in parallel. But in the GF1 Processor this parallelism is carried even further.

On each clock cycle in the GF1 Processor, all the following operations (may) happen:

(1) The contents of the micro-instruction PROM at the address selected by the Program Counter (PC) are clocked into a set of flip-flops,

(2) The PC either increments or loads to a new address (under control of a branch condition selected on the previous clock cycle),

(3) Any of the ALU's registers (x, y and z) may be loaded with one of several different possible values (depending on the register),

(4) The ALU carries out a Galois Field operation, either a multiplication (x*y) or a multiplication and an addition (x*y + z),

GF1 Processor Block Diagram

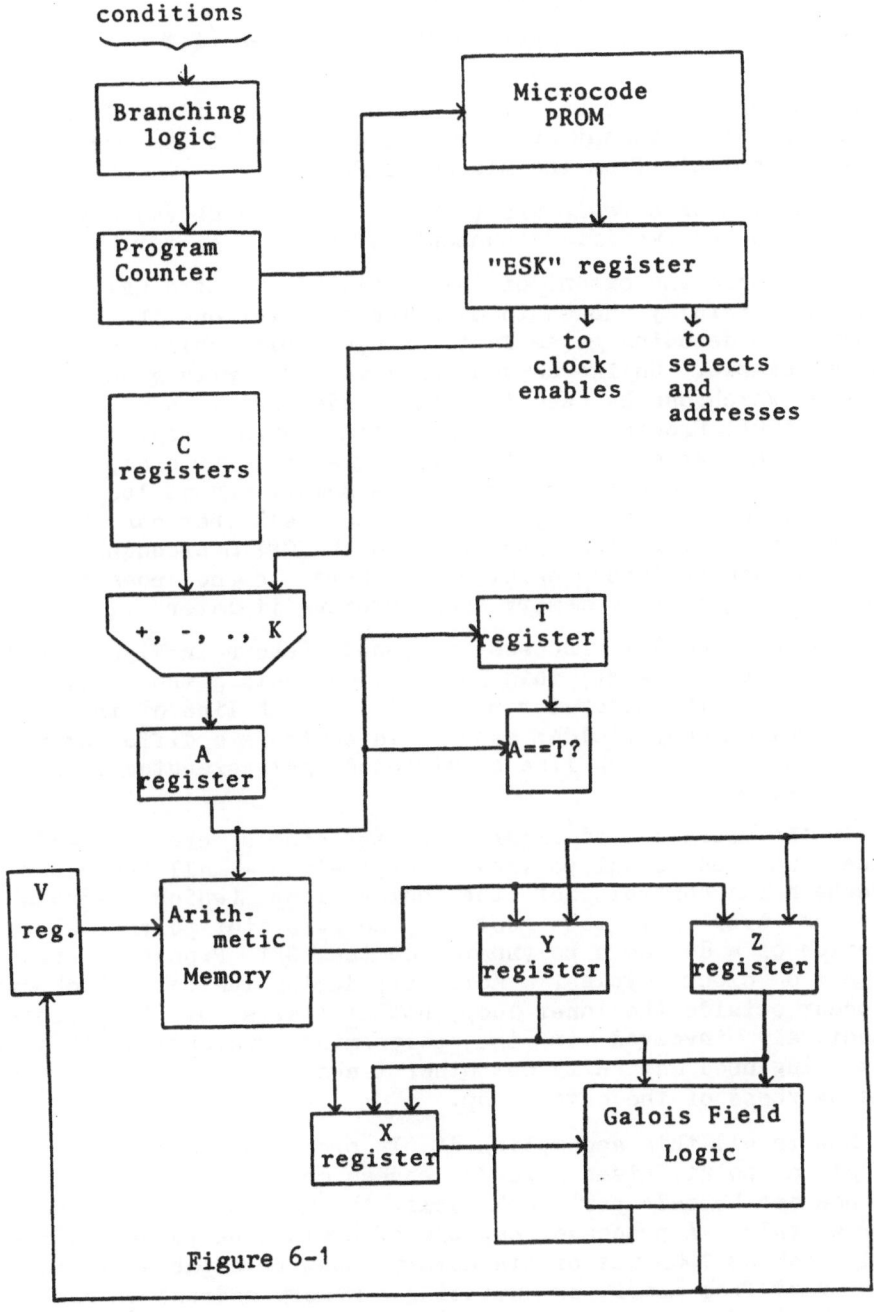

Figure 6-1

(5) The v register may be loaded with the result of the ALU, or the value in the v register may be written into memory at the address controlled by the a register,

(6) The t register may be set to the a register (for a future a==t branch test),

(7) The a register may be set to any of the 32 c (index) registers, or to the increment or decrement of any of these registers, or to a constant (from the microcode PROM),

(8) One of the 32 c registers (the one selected above) may be written with the (modified) contents of the a register.

The nature and extent of the parallelism in the GF1 Processor has been carefully chosen to make those operations that occur frequently in RS decoding go as fast as possible while keeping the hardware simple. Unlike conventional machines with general purpose registers (which can be used for both addressing and for arithmetic), the GF1 Processor has only special purpose registers. The a (address) register and the c (index) registers, for example, are all 12 bits wide and are used to address memory and as loop counters. The memory, however, is only 8 bits wide -- all that can be stored there is data for the ALU. Woe to the GF1 microcode author who wants more than 32 index registers -- there is no possibility of "dumping" a register to memory and restoring it later.

Consider the GF1 Processor microcode (shown in Figure 6-2) for calculating power sums; this involves evaluating the received word at the roots of the generator polynomial. Each line of this microcode contains many fields; each field controls a different part of the GF1 Processor. Each line of microcode gets executed in a single clock cycle.

In the microcode of Figure 6-2, the single cycle "sel" loop carries out the actual polynomial evaluation -- all the other code is overhead and control for the outer loop (which repeats the evaluation over all the roots of the generator polynomial). The evaluation of a degree n polynomial on the GF1 Processor requires just $n + 6$ clock cycles. Moreover, during the "+6" clock cycles that occur outside the inner loop, not all parts of the processor are actively involved in the polynomial evaluation; those that aren't being used may carry out other tasks, such as those relating to the overhead of the outer loop.

How is all this accomplished? In order to evaluate a polynomial at a point, given a random access memory from which only one reference can be made per clock cycle, the most efficient means is Horner's rule. A processor capable of evaluating polynomials efficiently (making 100% use of its memory) must be capable of retrieving a word from memory and updating a memory address on each clock cycle. Furthermore, the processor's ALU must be able to multiply the previous result by the evaluation point and add in the new coefficient just read from memory on each clock cycle. Lastly, there

Evaluating a Polynomial at a Point

```
sps:
uf();g=0        ;x=y;y=m[a];              a=c0=a   ;if(h)goto hl;
uf();g=1        ;                         a= --c0  ;if(h)goto hl;
uf();g=a!=t     ;         z=m[a];         a= --c0  ;if(h)goto hl;

sel:
ff();g=a!=t&&g;y=r     ;z=m[a];           a= --c0  ;if(h)goto sel;

ff();g=0        ;y=r   ;z=m[a];           a=kpsf15;if(h)goto hl;
ff();g=0        ;              v=r  ;t=a;a= ++c2  ;if(h)goto hl;
uf();g=a!=t     ;              m[a]=v;    a=kRTINC;if(h)goto hl;
uf();g=0        ;y=m[a];                  a=c1     ;if(h)goto hl;
lf();g=0        ;y=r   ;                  t=a;a=cusd-1;if(h)goto sps;
```

Figure 6-2

must be a test and a (potential) branch on each clock cycle. The GF1 Processor does all these operations on every clock cycle.

The GF1 Processor is also very efficient at other polynomial operations that are frequent in RS decoding. These operations include:

polynomial addition

polynomial multiplication

polynomial convolution

All of these operations run as fast as the data can be retrieved from and the results stored back into memory.

6.1.2. Pipelining

Being able to evaluate a degree n polynomial in n + 6 clock cycles is no major accomplishment if the clock cycle time is a snail's pace. The clock cycle time on an ECL 8-bit GF1 Processor, however, is about 60 nanoseconds (16 Mhz). The worst case decoding time of a (255, 239) RS received word on this processor is about 10000 clock cycles; the worst case throughput of such a processor is about 3.4 Mhz.

442

The pipelining of various data paths in the GF1 Processor keeps the clock cycle time to a minimum. One example of this is the pipelining of branches. Each micro-instruction has a branch code that controls (through a multiplexor) the setting of the "goto"-bit at the end of the corresponding clock cycle. If the goto-bit was set by the previous instruction, then the constant field of the current instruction is loaded into the program counter at the end of this second clock cycle. But, because of a stage of pipelining on the outputs of the micro-instruction PROM, the effect of this change in the PC is not seen for an additional clock cycle. Thus, when a branch test is successful, the two instructions following the successful test are executed before the destination of the branch is reached.

6.1.3. Timing Bottlenecks

The timing bottleneck on the GF1 Processor is the read-modify-write access on the c (index) register array. On any clock cycle, an element of the c register array may be accessed, fed through an exclusive-OR gate, a four-to-one multiplexor, and a latch, and may then be written back at the same address. Note that this is one of the most efficient ways to use a RAM -- the address access time is equal (and large) whether one is reading or writing; paying this penalty only once (and reading and writing at the same address on each clock cycle) makes efficient use of the memory. The auto-increment and -decrement feature this modification scheme implements for the index registers is highly effective for stepping through consecutive locations in memory (as when accessing the coefficients of a polynomial).

Because this path is known to be the bottleneck, the increment/decrement of the c registers is done by a shift register sequence instead of by binary addition, saving about 20 nanoseconds on each 60 nanosecond clock cycle. That is, instead of having a 12-bit binary increment (with resulting ripple-carry delays), the GF1 Processor uses a 12-bit shift register to increment and decrement. When the register is shifted one way, it "increments"; when shifted the other way, it "decrements".

There is no physical shift register needed to accomplish this shifting -- the four-to-one multiplexor mentioned above is wired with normal, shifted right, and shifted left versions of its input (as well as with a constant from the PROM). The shifting is done at no cost. A schematic of the c register modification scheme is shown in Figure 6-3.

6.1.4. Microcode

Although the hardware of the GF1 Processor is extremely important, great care must be devoted to the microcode as well. It is easy to give up a factor of two or more in running time through inefficient microcode.

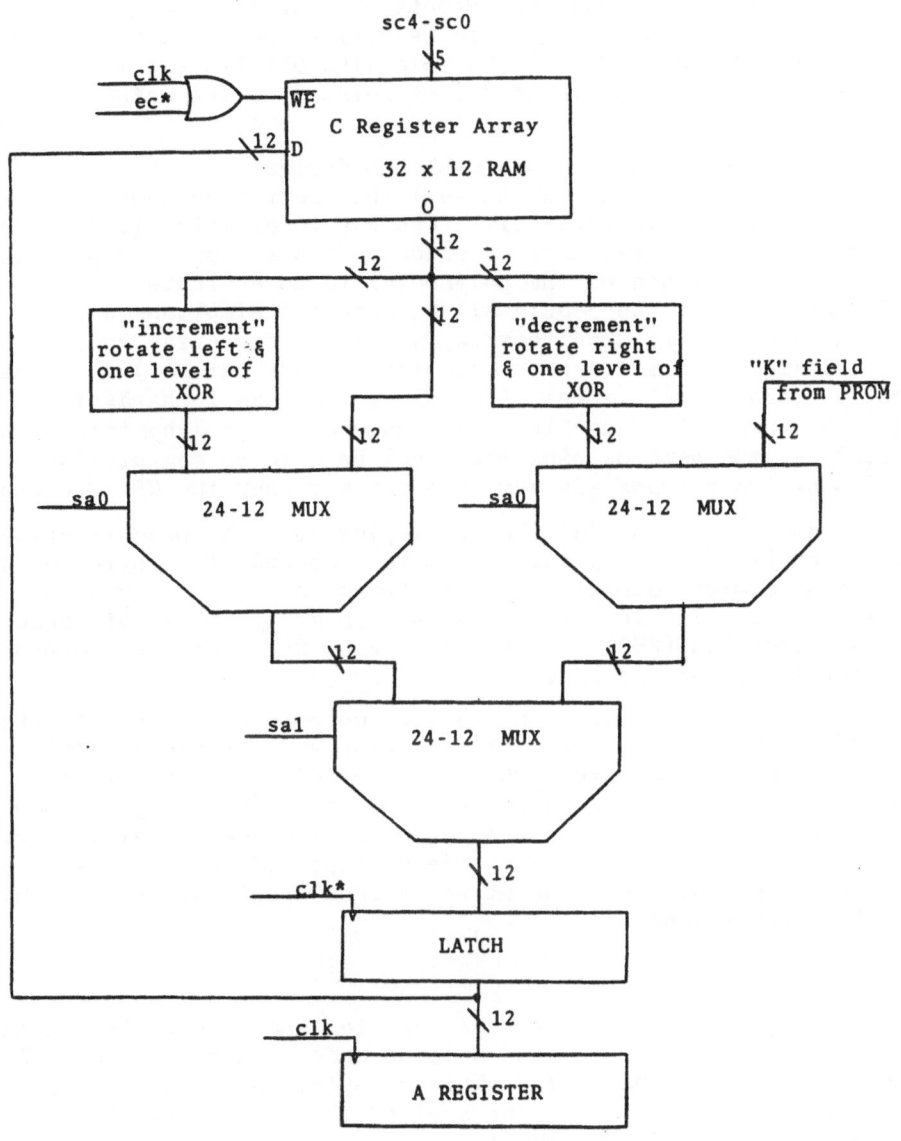

C Register Modification Scheme

Figure 6-3

6.2. Alternative Designs

Given a random access memory from which only one word can be read per clock cycle, the evaluation of a polynomial of degree n must require at least n + 1 clock cycles -- one clock cycle to read each coefficient. The GF1 Processor's ability to do a polynomial evaluation in n + 6 clock cycles is very close to this bound. In fact, the extra five clock cycles the GF1 Processor requires are mostly for outer loop overhead (not related to the actual polynomial evaluation).

A processor could conceivably be designed with a "multiple-port" memory, one in which all the coefficients of a polynomial could be read out in parallel. (One way to do this is to segment the memory into banks, each of which can be accessed in parallel; if all the coefficients of the polynomial to be evaluated are stored in different banks, they could all be accessed simultaneously). Given this kind of memory architecture, a large logic array could evaluate a polynomial of fixed degree (limited to some maximum) at a point in just one clock cycle. But, the cost to do this in hardware is huge and the result is a machine that can evaluate polynomials at points quickly -- the same machine might not be able to compute the convolution of two polynomials any more quickly than the GF1 Processor.

Some recent work at CMU has implemented a programmable systolic chip. Each systolic chip is composed of 64 bytes of RAM, a 60-bit wide microcode store, some branching logic, and an 8-bit binary ALU. It is estimated that an array of 112 of these chips could decode the (255, 223) RS code over $GF(2^8)$ at rates approaching 8 megabits per second [8].

The complexity of each systolic chip is comparable to that of the GF1 Processor. If the GF1 Processor was implemented on a single chip, it could decode the (255, 223) RS code at rates of more than 500 kilobits per second in the worst case (assuming a 25000 clock cycle decoding time with a 3 Mhz processor clock). But an array of 112 GF1 Processors that was time-multiplexed to handle a high data rate could do this decoding at worst case speeds of greater than 50 megabits per second.

7. CONCLUSIONS

Our goal in this paper has been to show that the algebraic decoding of high rate Reed-Solomon codes can be implemented efficiently in digital hardware. The presentation of various implementations has shown that the cost of the decoder can be traded off against the desired throughput.

By examining possible hardware solutions to the problem of Reed-Solomon decoder design, light has been shed on the hardware design process in general. In systems where high performance is necessary, a special purpose microprocessor design may represent a significant cost savings over more conventional approaches involving single-chip microprocessors surrounded by massive amounts of additional hardware.

8. REFERENCES

[1] "Bipolar Microprocessor Logic and Interface Data Book", Advanced Micro Devices, 1982.

[2] Berlekamp, E. R., "Galois Field Computer", U.S. Patent No. 4,162,480, July 24, 1979.

[3] Berlekamp, E. R., "Bit-Serial Reed-Solomon Encoders", IEEE Transactions on Information Theory, November, 1982. Related to U.S. patent application No. 2,15-361 (approved).

[4] Berlekamp, E. R., "Algebraic Coding Theory", Agaean Park Press, 1983. (Also, McGraw-Hill, 1968).

[5] Blahut, R. E., "Transform Techniques for Error Control Codes", IBM J. of Res. Dev., Vol. 23, May, 1979.

[6] Cohen, Earl. T., "On the Implementation of Reed-Solomon Decoders", Ph.D. dissertation, University of California, Berkeley, 1983.

[7] Conway, J. H., "A Tabulation of Some Information Concerning Finite Fields", in Computers in Mathematical Research, R. F. Churchhouse and J.-C. Herz, eds., North-Holland, Amsterdam, 1968.

[8] Fisher, Allan L., Kung, H. T., Monier, Louis M., and Dohi, Y., "Architecture of the PSC: A Programmable Systolic Chip", Proceedings of the 10th Annual Symposium on Computer Architecture, June, 1983.

[9] Kane, G., "68000 Microprocessor Handbook", Osborne/McGraw Hill, 1981.

[10] "16-Bit Microprocessor User's Manual", third edition, Motorola, 1982.

[11] Osborne, A., and Kane, G., "Osborne 16-bit Microprocessor Handbook", Osborne/McGraw Hill, 1981.

DESIGN AND HARDWARE IMPLEMENTATION OF A VERSATILE TRANSFORM DECODER FOR REED-SOLOMON CODES

Bruce L. Johnson

The MITRE Corporation*

Bedford, MA USA

ABSTRACT

The design and hardware implementation of a versatile Reed-Solomon encoder and decoder based on a transform decoding algorithm is presented. The decoder is electronically reconfigurable to accommodate a large number of different code parameters; the symbol fields range from 4 to 8 bits, the code lengths range from 3 to 255 symbols, and the rates are programmable downward to a minimum of one-half. The discrete transform used for encoding and syndrome computation is implemented with an algorithm that minimizes the number of extension-field products. The error locator uses a modified version of the Berlekamp-Massey feedback shift register synthesis algorithm to correct both errors and erasures. In the decoder's implementation, extension-field operations are separated to a maximum extent from the normal binary operations to promote reconfiguration. The decoding algorithm results in a highly repetitive architecture intended for VLSI implementation. This VLSI architecture is described in detail. The design and measured performance of a TTL breadboard and key-function VLSI macrocells that have been implemented as custom designed LSI circuits are described.

*The work reported was supported by the Rome Air Development Center, Electronic Systems Division, AFSC, under Contract F19628-82-C-0001 and by the MITRE Corporation's Independent Research and Development Program.

I. INTRODUCTION

The hardware implementation of Reed-Solomon [1] codes has lagged far behind their mathematical development for two fundamental reasons – insufficient circuit complexity and high cost.   At this time, even the increased circuit complexity afforded by very large scale integration (VLSI) has not been sufficient to implement existing algorithms [2]. New algorithms must be developed that take maximal advantage of the regularity and parallelism afforded by VLSI technology. Also, the cost of designing a VLSI circuit dedicated to a particular forward-error-correction code is prohibitive. A cost-effective error correction processor must be capable of accommodating a wide range of codes with little throughput penalty incurred for versatility. The cost of design and fabrication then could be amortized over a wide applications base.

Discrete transform techniques can be applied to decoding Reed-Solomon codes [3,4,5], affecting a substantial computational savings over conventional BCH decoding techniques [6,7,8,9]. The computational savings can be directly transferred to circuitry, allowing decoders to be built using existing integrated circuit technology.

The particular adaptation of transform decoding techniques described here results in a design for a versatile encoder and decoder with high circuit regularity. Key features include:

- a computationally efficient finite-field transform algorithm developed to reduce the number of finite-field multiplications [10],

- a modified version of the Berlekamp-Massey(B/M) minimal-length linear feedback shift register (LFSR) synthesis algorithm, for correcting erasures as well as errors [4,5,10], and

- a repetitive decoder architecture designed for VLSI implementation to be electronically reconfigurable to accommodate a wide range of forward-error-correction codes [10,11,12,13].

The architectural design of the transform-based encoder and decoder consists of two major sections, the $GF(2^m)$ transformer and the errata locator. The transformer is electronically reconfigurable over $GF(2^m)$ with m = 4,5,6,7, and 8. Within the errata locator's architecture, the decoding of erasures and errors requires no more hardware than is required for decoding errors only. The encoder/decoder concept has been proven by means of a breadboard implemented with off-the-shelf TTL integrated circuits.

II TRANSFORM METHODS FOR REED-SOLOMON CODES

To define an (n,k) Reed-Solomon code over $GF(2^m)$, where n divides 2^m-1, let $\bar{m} = (m_0, m_1, \ldots, m_{k-1})$ represent the information symbols. Represent message sequence $\bar{a} = (a_1, a_2, \ldots, a_{n-1})$ as

$$a(x) = \sum_{i=0}^{n-1} a_i x^i = a_0 + a_1 x + \ldots + a_{n-1} x^{n-1} \qquad (1)$$

where $a_i = 0$ for $0 \le i \le n-k-1$ and $a_i = m_{i-n+k}$ for $n-k \le i \le n-1$. The Reed-Solomon codeword corresponding to $\bar{m}$ is the n-tuple

$$\bar{A} = (A_0, A_1, \ldots, A_{n-1}) = (a(1), a(\alpha), a(\alpha^2), \ldots, a(\alpha^{2^m-2})) \qquad (2)$$

where α is a primitive n-th root of unity in $GF(2^m)$. Note that $a(\alpha^i) = a_0 + a_1 \alpha^i + a_2 \alpha^{2i} + \ldots + a_{n-1} \alpha^{i(n-1)}$. The encoding operation can be viewed as matrix multiplication

$$\bar{A} = \bar{a}H \qquad (3)$$

where $H_{ij} = \alpha^{ij}$.

The first of two steps in the decoding algorithm is to compute the inverse transform of the received sequence. Assume that $\bar{A}$ (equation 3) is the transmitted codeword and $\bar{E} = (E_0, E_1, \ldots, E_{n-1})$ is the channel error sequence. $\bar{R} = \bar{A} + \bar{E}$ is therefore the received sequence. The inverse transform of R is $\bar{r} = \bar{a} + \bar{e}$ where $\bar{e}$ is the transform of the channel error sequence and $\bar{a}$ is the original message sequence. The matrix H (equation 3) is invertible, with the (i,j)th entry in H^{-1} equal to α^{-ij} which is equal to $\alpha^{(n-i)j}$, making H^{-1} essentially the same matrix as H with its rows permuted. The first step in decoding, calculating $\bar{r}$ from $\bar{R}$, is computed in the same manner as the encoding procedure. The same transform hardware can be used for encoding and decoding.

$$\bar{r} = \bar{a} + \bar{e} = \bar{R}H^{-1} = \bar{A}H^{-1} + \bar{E}H^{-1} \qquad (4)$$

When $\bar{E}$ is zero, $\bar{R} = \bar{A}$ and $\bar{r} = \bar{a}$ -- the original information sequence has been recovered. When $\bar{E}$ is not equal to zero, then $\bar{r} = \bar{a} + \bar{e}$; $\bar{e}$ must be determined in order to extract $\bar{a}$ from $\bar{r}$.

The error syndrome, $\bar{s} = (s_0, s_1, \ldots, s_{n-k})$, is defined as

$$s_i = \sum_{j=0}^{n-1} R_j \alpha^{-ji} \qquad\qquad 0 < i \leq n-k-1 \qquad\qquad (5)$$

Since $a_i = 0$ for $0 \leq i \leq n-k-1$ and

$$a_i + r_i = \sum_{j=0}^{n-1} R_j \alpha^{-ji} \qquad\qquad 0 \leq i \leq n-1 \qquad\qquad (6)$$

$s_i = r_i$, and the syndrome values are the first n-k points in the n-length inverse transform. The error syndrome is used to determine the location of the errors in $\bar{E}$. Several procedures are available for determining the error locations, including a method of continued fractions [14], Euclid's algorithm [15] and an algorithm attributed to Berlekamp and Massey [6,9]. Common to all of these methods is the determination of a polynomial whose distinct roots designate the error locations or their multipicative inverses.

We choose the Berlekamp-Massey algorithm, because it is available, computationally efficient, and potentially attractive for VLSI-based circuit implementation. This algorithm has been discussed thoroughly by its authors [6,9], and its application to transform decoding of Reed-Solomon codes has been examined [5,7]. For our purposes, it may be regarded as an iterative algorithm that uses as inputs the error syndrome values (equation 5) and synthesizes the shortest LFSR whose feedback - connection polynomial, $\sigma(z)$, is the desired error-locator polynomial. If the bound of the code has not been exceeded, a unique relationship can be established between the error-syndrome values and the coefficients of the error-locator polynomial,

$$s_i + s_{i-1}\sigma_1 + s_{i-2}\sigma_2 + \ldots + s_{i-\nu}\sigma_\nu = 0 \qquad\qquad (7)$$

This relationship will be valid for all $\nu < i < n-k-1$. There exists a polynomial $e(x)$ (with coefficients e_i, $i = 0,1,\ldots,n-1$) of degree less than n that satisfies the linear recursion with $\sigma(z)$ for all $1 < i < n$, and $e(x)$ is uniquely specified by ν consecutive values of its transform values $E(\alpha^{-i})$ (where for example $i = 0,1,\ldots,\nu-1$) and $E(x)$ has no more than ν non-zero coefficients [3]. In this case, $E(x)$ is the channel error pattern and $E(\alpha^{-i})$ form its transform values. The error-locator polynomial can then be used to extrapolate $\bar{e}$, which is subtracted from $\bar{r}$ and decoding is complete.

The decoding algorithm may be modified to accommodate erasures [4,5,12]. We initialize the B/M algorithm with a feedback-connection polynomial computed from the known erasure location and then continue the algorithm normally, synthesizing an errata-locator polynomial that is the product of the error-locator polynomial and the initial erasure-locator polynomial. When the errata-locator polynomial has been synthesized, no further distinction need be made between errors and erasures, and the inverse transform of the channel errata pattern can be extrapolated by free-running the synthesized LFSR as before. A block diagram of the encoder and decoder is shown in Figure 1.

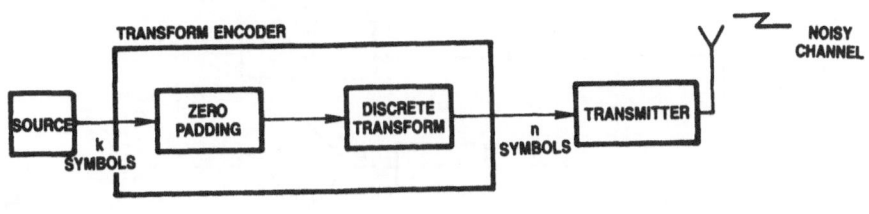

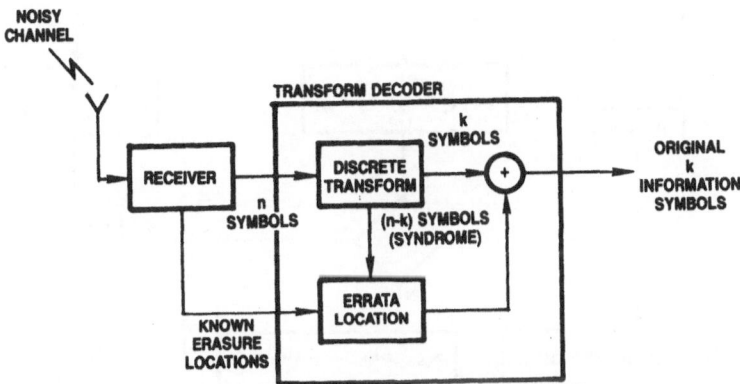

Figure 1. Transform Encoder and Decoder

The key features of the decoder's algorithm are:

- Errata-location and symbol correction require serial input data; each point in the transform may be calculated sequentially, thereby reducing the hardware complexity of the transform section.

- The circuitry required to iteratively generate the erasure-locator polynomial is contained within the hardware required to implement the B/M algorithm.

452

- The decoding of errata requires the same amount of hardware and time as the errors only case.

The complete transform-based decoding algorithm for both errors and erasures is shown in Figure 2.

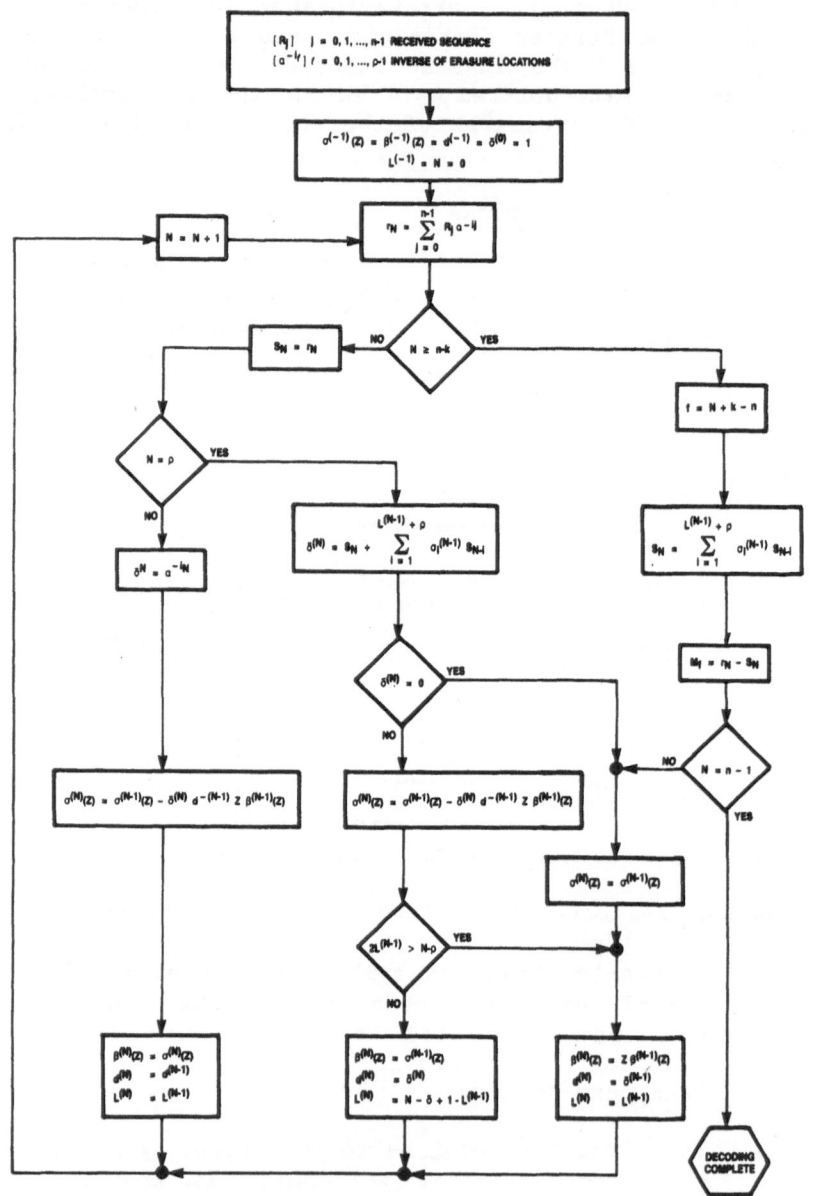

Figure 2. Transform-Based Decoding Algorithm with Modified B/M Algorithm for Correcting Errors and Erasures

III TRANSFORM ENCODER AND DECODER ARCHITECTURE

The natural partitioning of the transform decoding algorithm separates the decoder´s architecture into a transform section, an errata-location section and a control section. The encoding algorithm partitions the encoder into a transform section and a control section. The transform section was designed to calculate both a forward and an inverse discrete transform over the fields of interest; the hardware required for encoding is contained within the decoder. The control sections provide data management to the transform and errata-location sections. Control can be implemented by using standard logic or a software programmable microprocessor.

The transform section calculates both forward (equation 3) and inverse (equation 4) discrete transforms defined on $GF(2^m)$ with m = 4,5,6,7, and 8. To implement either vector-matrix multiplication directly would normally require n^2 multiplications in $GF(2^m)$. However, efficient algorithms that mimic FFT algorithms can be developed to reduce the number of required multiplications and result in hardware-efficient architectures. Note that $A_j = a(\alpha^j) \equiv a(x)$ mod $x-\alpha^j$. The values for A_j could be determined by dividing $a(x)$ by the first-degree polynomials $x-\alpha^j$, $0 \le j \le n-1$, but this technique would require n^2 $GF(2^m)$ multiplications. Instead, divide $a(x)$ by a smaller set of polynomials of higher degree, which contain distinct factors of the form $x-\alpha^j$. If the set of divisor polynomials is the set of minimal polynomials of the nonzero field elements, then their coefficients are binary and only additions are required for division. The equivalence of the two methods is shown by factoring x^n-1 over $GF(2^m)$:

$$x^n-1 = \prod_{j=0}^{n-1} (x-\alpha^j) = \prod_{i=1}^{M} m_i(x) \qquad (8)$$

where M is the number of irreducible factors of x^n-1 and each $m_i(x)$ is a binary irreducible polynomial. To complete the transform, the residue polynomials are evaluated at the conjugate roots of the associated minimal-polynomial divisor. This step requires multiplications in $GF(2^m)$, but the number is substantially reduced and approaches $O(n\log_2 n)$.

The operation of the transform section can be partitioned into two functions. The transformer simultaneously divides the (n-1)th degree input polynomial, $a(x)$, by all minimal polynomials of the non-zero field elements of $GF(2^m)$. Each point in the transform is sequentially calculated by evaluating the appropriate residue polynomial at the corresponding element in the field. The order of evaluation determines whether the transform is forward or inverse. The transform section (Figure 3) consists of a polynomial residue calculator, a multiplexer, a polynomial residue evaluator, and an arithmetic controller.

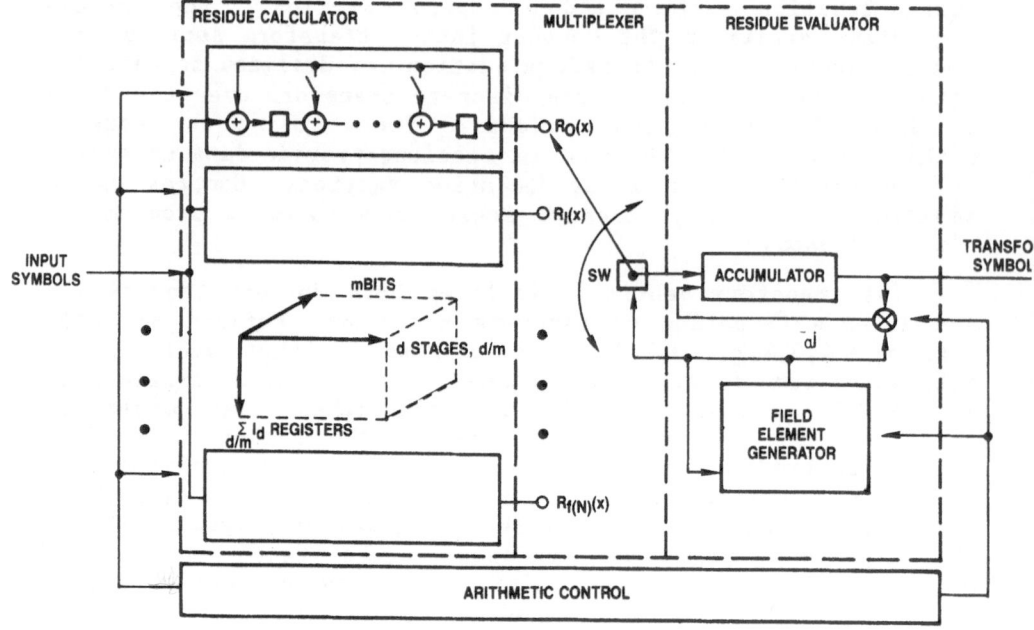

Figure 3. Transform Section

The polynomial residue calculator simultaneously divides the polynomial representing the data to be transformed by all minimal polynomials of the nonzero elements of $GF(2^m)$ where m = 4,5,6,7 and 8. There are 66 minimal polynomials for the five different extension fields. However, only simultaneous division by the polynomials from the field of operation is required for calculating a particular transform. The Galois field $GF(2^8)$ has the most minimal polynomials -- 35 divider circuits to be implemented for transformation. Polynomial division is implemented with LFSRs whose feedback-connection polynomials are defined to be the divisor polynomials. Since the divisor polynomials are irreducible over $GF(2)$, each divider circuit consists of eight identical <u>binary</u> feedback shift registers (BFSRs). The maximum length of any BFSR is eight stages. The 35 divider circuits are electronically reconfigurable to provide for division by the remaining 31 minimal polynomials required for division in the other four finite fields.

The requirement to operate with different symbol sizes recurs throughout the design of the transform and errata-location sections. Our approach is to define a standard symbol size of eight bits, then design all hardware to accommodate this symbol size and to be programmable for smaller fields. Any symbol from $GF(2^m)$ where $m \leq 8$ can be represented as an eight-bit symbol whose$(8-m)$ most significant bits have been set to zero. When the polynomial residue calculator is operating with symbols from $GF(2^m)$ where $m < 8$, zeros are fed into $8-m$ slices corresponding to the bit-positions greater than $m-1$. The output of the corresponding BFSRs is zero.

The polynomial to be divided is fed sequentially into all 35 divider circuits. Division is completed after the last coefficient of the data polynomial has been entered. At that time, the calculated residue polynomials are stored within the delay stages of the divider circuits. These residues are transferred into a temporary holding memory, and the divider circuits become available for processing the next codeword. In this manner, the polynomial residue calculator is pipelined, simultaneously operating on two contiguous blocks of n symbols, thus accepting a continuous-input data stream.

The polynomial residue evaluator implements the second portion of the division-evaluation transform algorithm. The residue evaluator sequentially calculates each point in the transform by selecting (via the multiplexer) a predetermined residue polynomial from the residue calculator and evaluating that residue at an appropriate root of its corresponding divisor polynomial. It implements polynomial evaluation using a continued-product expansion algorithm in a single multiplier and accumulator architecture. This structure is multiplexed to sequentially calculate each point in the transform. The residue evaluator must be capable of operation in the five binary-extension fields of interest. The accumulator is easily implemented using eight two-input Exclusive-Or gates. (The standard symbol produces correct zeros for fields with $m < 8$.) The multiplier structure is an asynchronous-array programmable $GF(2^m)$ multiplier. This structure was selected because the multiplier is the processing bottleneck within the transform section, and the array-type multiplier offers the fastest multiplication rates. However, a penalty is paid for this speed because the hardware implementation of the array-type structure requires the maximum number of gates of all $GF(2^m)$ multiplier structures in- vestigated. However, the $GF(2^m)$ multiplier is the only extension-field multiplier in the entire transform structure, and it will be shown that exactly the same $GF(2^m)$ multiplier structure is used to implement a critical portion of the errata-location section. A block diagram of the programmable $GF(2^m)$ multiplier is shown in Figure 4.

The arithmetic controller provides all timing and control signals required to operate the transform section. The

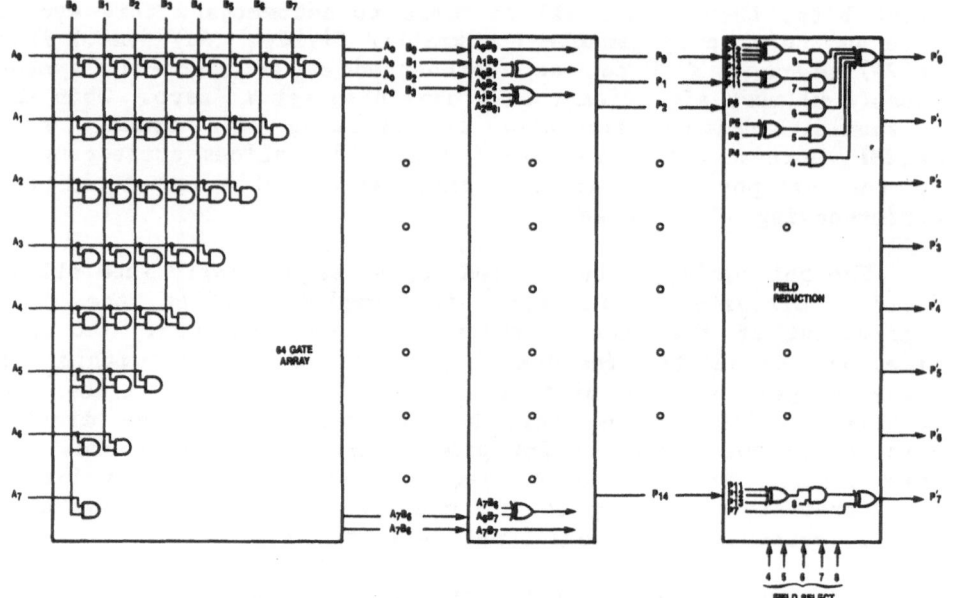

Figure 4. Programmable Extension-Field Multiplier

calculations implemented by both the polynomial calculator and evaluator are independent of either a forward or an inverse transform. The arithmetic controller determines the order of evaluation and therefore dictates the type of transform to be computed. Many practical implementations of the controller are feasible. However, high throughput warrants a special purpose processor. A sequence generator (counter) with local memory or programmable logic arrays implement a suitable high-performance controller.

The errata-location section implements a modified version of the B/M minimal-length LFSR synthesis algorithm as shown in Figure 1. First, this section uses the known erasure locations to calculate the erasure-locator polynomial. The same hardware then uses the error-syndrome values to iteratively calculate the errata-locator polynomial. Finally, the errata-locator polynomial is used to generate the transform of the channel errata pattern which is subtracted from the transform of the received codeword in order to recover the original message.

A block diagram of the errata-location section is shown in Figure 5. The decision and control circuitry implied by the decoding algorithm are not shown but are implicit in the circuit operation. Three major computational steps (see Figure 1) implement the decoding algorithm:

- calculation of the present discrepancy $\delta^{(N)}$

$$\delta^{(N)} = s_i + \sum_{i=1}^{L^{N+}+\nu} \sigma_i^{(N-1)} \, s_{N-i} \qquad (9)$$

- calculation of the present feedback connection polynomial $\sigma^{(N)}(z)$

$$\sigma^{(N)}(z) = \sigma^{(N-1)}(z)$$
$$- \delta^{(N)} \, d^{-(N-1)} z \beta^{(N-1)}(z) \qquad (10)$$

- update of the previous feedback connection polynomial $\beta^{(N)}(z)$ to one of the three values $\beta z^{(N-1)}(z)$, $\sigma^{(N-1)}(z)$, or $\sigma^{(N)}(z)$.

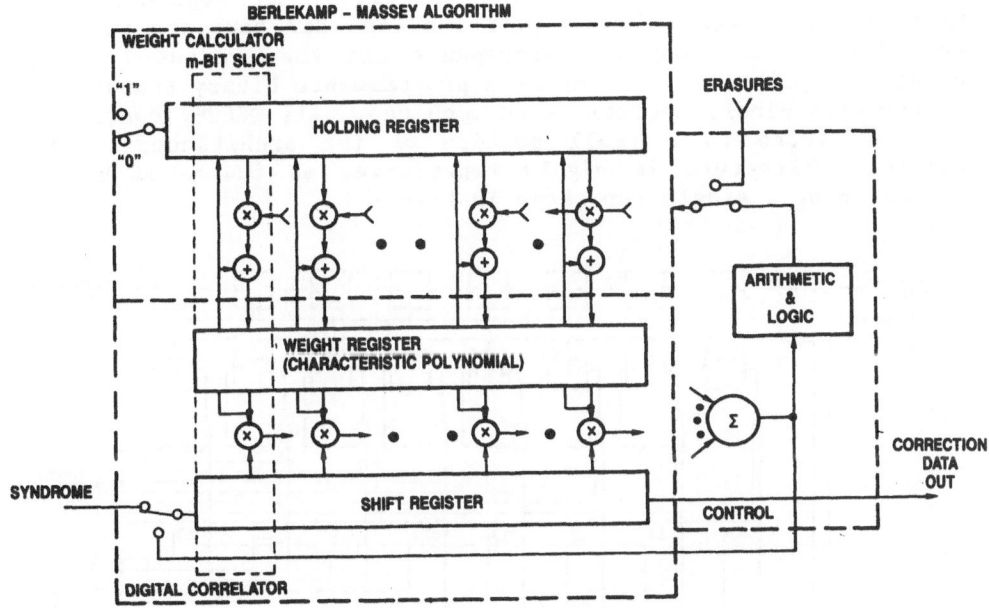

Figure 5. Errata-Location Section

The calculation of $\sigma^{(N)}$ is equivalent to the convolution of the first N error syndrome values with the coefficients of the (N-1)th feedback connection polynomial. The circuit for computing this resembles a convolver whose operations are defined over $GF(2^m)$. This operation can be decomposed so that only binary operations are implemented at each convolver tap, partial sums are accumulated, and serial $GF(2^m)$ multiplication is

458

implemented in the arithmetic and logic section (see Figure 5). Therefore, the circuitry required to implement the calculation of $\sigma^{(N)}$ resembles a binary convolver with all $GF(2^m)$ operations restricted to a small portion of the overall circuit.

The calculation of the present feedback connection polynomial can be thought of as the summation of two polynomials. One polynomial is fixed; the other is multiplied by a constant. The constant $\delta^{(N)}d^{-(N-1)}$ is formed using an asynchronous array $GF(2^m)$ multiplier that is identical to the one used in the transform section. The product $\delta^{(N)}d^{-(N-1)}z_{\beta}^{(N-1)}(z)$ is formed using bit serial $GF(2^m)$ multiplication. The summation of the two polynomials (equation 10) is formed using partial summation so that only binary operations are required at each tap to calculate the present feedback connection polynomial.

The circuitry required to update the previous feedback connection polynomial is a temporary register and some control signals. In this manner, $GF(2^m)$ operations are separated to a maximum extent from binary operation, and the hardware required to calculate the present discrepancy and the present feedback connection polynomial resembles a programmable binary transversal filter with binary operations at each tap. All $GF(2^m)$ operations are restricted to a small portion of the architecture. The resulting structure is highly repetitive, as indicated by the schematic of a single tap shown in Figure 6.

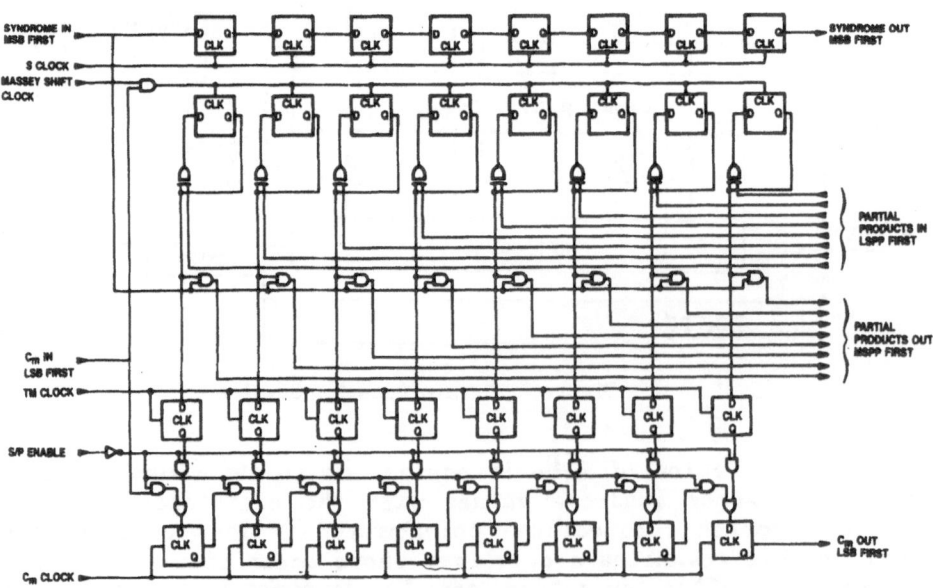

Figure 6. Errata Locator: 8-Bit Slice

IV CODING CAPABIITIES

The errata-location section has a bit-slice architecture that is expandable to accommodate any code rate; each symbol used for redundancy requires a corresponding hardware slice (Figure 6) within the decoder. The maximum number of syndrome symbols that can be accommodated depends on the number of bit-slices integrated onto one chip the number of chips used in a system. If the errata-location section contains 128 identical slices, the decoder can correct all combinations of errors, t, and erasure, r, provided the inequality

$$2t + s \leq n-k \leq 128$$

is satisfied.

Table I shows the range of Reed-Solomon codes that can be processed by the transform encoder/decoder. These codes represent a large number of both maximum ($n = 2^m-1$) and submaximum ($n|2^n-1$) length codes over $GF(2^m)$ where the symbol representation, m, ranges from four to eight bits.

Table I. Reed-Solomon Codes

PARENT CODE		PARAMETERS		CODES
$GF(2^8)$	$(255,k)$	$127 \leq n' \leq 255,$	$127 \leq k \leq n'$	8,256
	$(85,k)$	$51 < n' \leq 85,$	$1 \leq k \leq n'$	2,329
	$(51,k)$	$17 < n' \leq 51,$	$1 \leq k \leq n'$	1,173
	$(17,k)$	$15 < n' \leq 17,$	$1 \leq k \leq n'$	33
	$(15,k)$	$5 < n' \leq 15,$	$1 \leq k \leq n'$	105
	$(5,k)$	$3 < n' \leq 5,$	$1 \leq k \leq n'$	9
	$(3,k)$	$1 \leq n' \leq 3,$	$1 \leq k \leq n'$	6
$GF(2^7)$	$(127,k)$	$1 \leq n' \leq 127,$	$1 \leq k \leq n'$	8,128
$GF(2^6)$	$(63,k)$	$21 < n' \leq 63,$	$1 \leq k \leq n'$	1,785
	$(21,k)$	$9 < n' \leq 21,$	$1 \leq k \leq n'$	186
	$(9,k)$	$7 < n' \leq 9,$	$1 \leq k \leq n'$	8
	$(7,k)$	$3 < n' \leq 7,$	$1 \leq k \leq n'$	22
	$(3,k)$	$1 \leq n' \leq 3,$	$1 \leq k \leq n'$	6
$GF(2^5)$	$(31,k)$	$1 \leq n' \leq 31,$	$1 \leq k \leq n'$	496
$GF(2^4)$	$(15,k)$	$5 < n' \leq 15,$	$1 \leq k \leq n'$	105
	$(5,k)$	$3 < n' \leq 5,$	$1 \leq k \leq n'$	9
	$(3,k)$	$1 \leq n' \leq 3,$	$1 \leq k \leq n'$	6
				22,260

A class of shortened Reed-Solomon codes is also included in Table I. Encoding is accomplished in a manner identical to normal transform encoding techniques: a k-length message sequence is padded with zeros to produce an n-length sequence where n divides 2^m-1. An n-point transform is computed and an (n,k) Reed-Solomon codeword is produced. A shortened codeword is defined by removing j symbols where $j \leq n-k$. The new codeword has length $n-j = n'$ and dimension k. It can correct t errors and s erasures provided

$$2t + s \leq n'-k \leq 128$$

These shortened codewords are decoded using prior knowledge of the position of the symbols removed during encoding. The tranform decoder identifies the missing symbols as predetermined erasures; j zero value symbols are added to the received sequence, and the decoding procedure continues as for a full-length codeword.

The transform decoding algorithm need not be restricted to nonsystematic Reed-Solomon codes. In some cases a systematic encoder is desirable because the transmitted codeword actually contains the unaltered information symbols. In this case, the discrete transform is not part of the encoder. To decode, simply repeat the entire procedure followed by an inverse transformation to recover the information.

With only minor modifications to the decoder architecture, the transform decoder can be used to decode the more general class of Alternant codes [16]. These codes may well be the largest and most general class of codes available with an easily implementable decoding algorithm. Subclasses of Alternant codes include Goppa, BCH, Generalized BCH, Hamming, and Reed-Solomon codes [17]. Using some of these codes, the system designer has a much greater choice in matching modulator with coding schemes.

V. BREADBOARD AND VLSI MACROCELLS

A TTL breadboard that encodes and decodes a large number of Reed-Solomon symbol-correction codes was designed and fabricated. The breadboard implements the transform encoding and decoding algorithm described in section 2, and is a scaled-down version of the architecture described in section 3. The breadboard can process codes with up to 51 symbols, with each symbol represented by eight bits. The breadboard operates over $GF(2^m)$ m = 4,5,6 and 8, and serves as a proof-of-concept verification of the transform-based decoder architecture. It comprises approximately 600 Schottky SSI logic devices and operates with an internal clock rate of 10 MHz. Based on breadboard performance we project the VLSI implementation (operating at a conservative 10 MHz) of the decoder should be able to decode an (n,k) Reed-Solomon code in 4n microseconds. With the pipeline in the transform section, the decoder should produce a block of decoded data every 2n microseconds.

Using the same breadboard as a basis for complexity, we estimate that the transform section would require 4.5k bits of shift registers and 13k logic gates. The errata-location section should also require approximately 4.5 k bits of shift registers and 15 k logic gates. Both the transform and the errata locator have hardware complexities that suggest fabrication as single VLSI devices.

Direct extrapolation from TTL to VLSI technology is inaccurate. Therefore, two VLSI macrocells have been designed and fabricated in NMOS technology to support performance and complexity estimates. The two macrocells, the polynomial residue calculator and the programmable $GF(2^m)$ array multiplier, were chosen because they are both generically useful and significant to the operation of the encoder and decoder.

The polynomial residue calculator represents approximately 80% of the hardware within the transform section and therefore provides an accurate measure for complexity estimates. A section consisting of eight programmable BFSRs and output multiplexer was designed in 5μm NMOS using conservative design rules [18]. The macrocell implements one of eight identical slices used in the TTL breadboard. The design was completed at MITRE and fabricated through a commercial silicon foundry. The resulting circuit used dynamic shift registers and required approximately 1,100 NMOS transistors. The 5.2 mm^2 chip dissipates 100 milliwatts, while operating at a maximum clock rate of 12 MHz. If the minimum feature size is assumed to be 1.25μm, we estimate that the entire transform section (consisting of 35 divider circuits and the residue calculator) could be implemented in a chip consisting of approximately 41,000 transistors, with an area of 0.12 cm^2 and dissipating approximately 0.5 watts. A minimum feature size of 1.25μm would permit implementing the full transform section on a single NMOS chip. The design experience gained from the NMOS design of the polynomial residue calculator indicates that the errata locator, which is also shift-register extensive, can be implemented as a single VLSI device.

The processing bottleneck within both the transform section and the errata-loction section is the programmable $GF(2^m)$ array multiplier. A VLSI implementation of this macrocell would give an accurate indication of the speed at which a VLSI encoder and decoder could be expected to operate. A complete programmable $GF(2^m)$ asynchronous array multiplier was designed and fabricated in 5 μm NMOS technology. The speed at which the multiplier operates is directly related to the speed at which the encoder or decoder function . The 5 μm NMOS multiplier chip has an area of approximately 8.89 mm^2, contains about 1800 transistors, and dissipates about 120 milliwatts while multiplying at a maximum frequency of 9.6 MHz. It is not unreasonable to assume that smaller feature size will result in a faster multiplication times. If 1.25μm technology can produce a multiplier that will operate at 20 MHz, then an n-length codeword would be decoded in 2n microseconds and a decoded block of data would be available every n microsecond. Block decoding times and throughput data rates (assuming 20 MHz clock) are shown in Table II.

Table II. Block Decoding Time
and Data Throughput

CODE	BLOCK DECODING TIME *		DATA THROUGHPUT *	
(255,127)	510	μ sec	3.98	Mb/sec
(127,63)	254	μ sec	3.47	Mb/sec
(63,31)	126	μ sec	2.95	Mb/sec
(31,15)	62	μ sec	2.42	Mb/sec
(15,7)	30	μ sec	1.87	Mb/sec

* ASSUMING 20 MHz INTERNAL CLOCK

VI SUMMARY

Algorithms and architectures can be developed concurrently to provide a versatile error-correction capability suitable for VLSI implementation. Key features of the transform-based encoder and decoder include:

1) highly repetitive circuitry that reduces costly
   design time,

2) electronically reprogrammable circuitry to
   accommodate a wide range of coding parameters,

3) an efficient discrete transform algorithm that reduces the number of GF(2^m) muliplications, and

4) the ability to accommodate erasures and errors within the same hardware.

   The resulting decoder is an excellent example of how VLSI technology can impact communication systems. The two-chip codec is programmable over a large set of code lengths, code rates and clock rates. It would furnish communication systems engineers with a new design tool -- a powerful codec with sufficient flexibility to be applicable to a wide variety of system problems.

REFERENCES

[1] Reed, I. S., and Solomon, G., "Polynomial Codes over Certain Finite Fields," J. Soc. Ind. Applied. Math., Volume 8, pp. 300-304, June 1960.

[2] Berlekamp, E. R., "The Technology of Error-Correcting Codes," Proceedings of IEEE, Vol. 68, No. 5, May 1980.

[3] Murakami, H., Reed, I. S., and Welch, L. R., "A Transform Decoder for Reed-Solomon Codes in Multiple-User Communications System," IEEE Transaction on Information Theory, Volume 23, pp. 675-682, November 1977.

[4] Blahut, R. E., "Transform Techniques for Error Control Codes," IBM J. Res. Development, Volume 23, No. 3, May 1979.

[5] Haggarty, R. D., Palo, E. A., Carhoun, D. O., and Meehan, S. J., "High Speed Signal Processing for Error Correction Coding," presented at 1979 IEEE International Symposium on Circuits and Sytems, Tokyo, Japan: 1979.

[6] Berlekamp, E. R., Algebraic Coding Theory, McGraw Hill, New York: 1968.

[7] Peterson, W. and Welson, N., Error Correcting Codes, 2nd Edition, The MIT Press, Cambridge, MA: 1972.

[8] Forney, G. D., "On Decoding BCH Codes," IEEE Transaction on Information Theory, IT-11, pp. 549-557, October 1968.

[9] Massey, J. L., "Shift-Register Synthesis and BCH Decoding," IEEE Transaction on Information Theory, IT-15, No. 1, pp. 122-127, January 1969.

464

[10] Carhoun, D. O., Johnson, B. L., Meehan, S. J., "Design and Hardware Implementation of a Versatile Transform Decoder for Reed-Solomon Codes," presented at the 1981 IEEE International Symposium on Information Theory, Santa Monica, CA, 1981.

[11] Carhoun, D. O., Johnson, B. L., Meehan, S. J., "VLSI Architectural Design for a Reed-Solomon Transform Decoder," presented at the 1981 IEEE International Symposium on Circuits and Systems, Chicago, IL: 1981.

[12] Carhoun, D. O., Johnson, B. L., Meehan, S. J., "Transform Decoding of Reed-Solomon Codes Volume I: Algorithm and Signal Processing Structure," ESD-TR-82-403, Vol. I, November 1982, ADA123953.

[13] Johnson, B. L., Bequillard, A. L., Meehan, S. J., "Transform Decoding of Reed-Solomon Codes Volume II: Logical Design and Implementation," ESD-TR-82-403, Vol. II, November 1982, ADA123977.

[14] Reed, I. S., Scholtz, R. A., Truong, T. K., and Welch, L. R., "The Fast Decoding of Reed-Solomon Codes Using Fermat Theoretic Transforms and Continued Fractions," IEEE Transactions on Information Theory, IT-24, pp. 100-106, 1978.

[15] Mandelbaum, D., "Construction of Error Correcting Codes by Interpolation," IEEE Transaction on Information Theory, Vol. I, pp. 27-35, January 1979.

[16] MacWilliams, F. S., Sloane, N. J. A., The Theory of Error Correcting Codes, North Holland, New York, 1977.

[17] Ferguson, T. J., Johnson, B. L., Carhoun, D. O., "Implementation of a Transform Decoder for Reed-Solomon and Alternant Codes" presented at the 1983 IEEE International Conference on Communications, Boston, MA, June 1982.

[18] Paczan, M. W., and Johnson, B. L., "LSI Design of a Programmable Polynomial Residue Calculator for a Reed-Solomon Transform Decoder," presented at the 1982 IEEE Custom Integrated Circuits Conference, Rochester, NY: 1982.

PROCESSING TECHNIQUES IN PUBLIC KEY CRYPTOSYSTEMS

Rod Goodman

Department of Electronic Engineering
University of Hull, Hull HU6 7RX , U.K.

Abstract

The increasing use of cryptographic techniques in business and
commercial data communications systems will only come about if
cheap and fast hardware LSI devices can be designed to implement
the algorithms. This has already happened with the DES but in the
case of public key cryptosystems the process is only at the
development stage. This is due to this nature of the algorithms
and to the fact that the algorithms are themselves under
suspicion. The paper examines public key cryptosystems and their
modifications from an implementation point of view.

1. Introduction

In modern cryptography, the security of a transmission rests in
the secrecy of a "key" rather than in keeping the algorithm used
by the cipher machine a secret. In a conventional (or symmetric)
cryptosystem (CC) such as the DES (ref.1), the algorithm is in
fact an international standard. In such a system the cipher
machine is "primed" with a secret key which tells the machine
which transformation to apply to the plaintext, out of the many
possible transformations within the algorithm, to turn it into the
ciphertext. The receiving cipher machine uses the same (or a
directly related) key to effect the inverse transformation from
ciphertext to plaintext. An enemy with an identical machine can
only try to find or steal the key, given that the cryptosystem has
been well designed in that it is computationally and statistically
infeasible to deduce the key even with known plaintext ciphertext
pairs. A fundamental limitation of this system is therefore that
the users must have previously securely set up a common key.

This "key distribution problem" is severe in a networking or
electronic mail environment. If a packet or local area network
has n users, any pair of whom may wish to communicate, the number
of potential keys rises as n squared. There is a further
limitation of the CC when authentication is considered. An
enciphered order may contain a weak authenticator such as an order
number and date, but because the receiver has the ability to
create a ciphertext from any messagetext desired, disputes between
the two parties cannot be resolved by a judge.

In 1976 Diffie and Hellman (ref.2) and independently Merkle
(ref.3) proposed a new cryptographic scheme called a
Public-Key-Cryptosystem (PKC) that is essentially asymmetric
(ref.4). The elegance of the method stems from the use of
different keys for encryption and decryption (EK and DK), and that
it is infeasible to derive one from the other. Thus if network
users generate key pairs and make their encryption keys public in
a secure (say printed) directory the need to distribute keys does
not arise. In order to send a message M to a user, we generate
the ciphertext $C = E(M)$ by encryption with his (public) encryption
key. The user keeps his decryption key secret so that only he can
invert the procedure to give $M = D(C) = D(E(M))$. An enemy is
faced with deriving DK from EK, which we have said is "hard".
Note that in this scenario prior authentication is assumed, i.e.
the very fact that our encryption key is public enables any user
to send us messages. We must be sure that the person sending the
message is who she says she is.

PKC schemes also permit us to devise authentication procedures
such as signing a contract which we cannot later deny, showing
that a message has not been tampered with, and establishing
identity beyond doubt. In order to achieve this we require the
additional property that for all (or nearly all) cryptograms
$E(D(C)) = C$. That is, "decryption" of a message makes sense
because most messages are also cryptograms. We can sign a
document as follows. User A sends an message M to user B,
encrypted with B's public key for security. A then forms a
pre-signature which is some function of the plaintext M and is
also a valid cryptogram in A's encryption algorithm (ref.5). A
decrypts this using his secret key to form the signature $S = D(C)$
which only he can do. This is then sent to B (via encryption for
secrecy if neccessary). B uses A's public key to form
$C = E(D(C))$. B operates on the already received message M to
form the pre-signature which she then compares to C. If they
match then B is sure that M came from A. Furthermore, as no one
other than A (including B) could have produced S, A cannot later
deny that he signed M. Also B cannot alter M or S without
destroying the correspondence $C = E(S)$. There are situations in
which keeping the message secret is not desirable in the
authentication process. Consider that 'WE' have an ambassador in

an alien country who has to use 'THEIR' PTT to send 'US' messages.
'THEY' will not allow him to send coded messages which 'THEY'
cannot read for fear of espionage. 'WE' must be sure that 'THEY'
are not tampering with the ambassador's messages. With asymmetric
encryption both parties can be satisfied. 'WE' give 'THEM' the
decryption key but keep the encryption key secret. The ambassador
gives 'THEM' a message M and the resultant cryptogram C. 'THEY'
decrypt C and check that $D(C) = M$. 'WE' receive M and C and check
that the message is authentic if $E(M) = C$.

We can concieve of a PKC system working in conjunction with a CC,
particularly in an electronic mail environment. The PKC is used
to securely distribute session keys and to authenticate users. A
disadvantage of the PKC is that it is essentially one-to-one.
That is, we set up a two user secure channel. If we wish to send
a broadcast to several users we need to encrypt the same message
several times. This redundancy is particularly severe in a packet
switched network where group addressing is usually built in.
Several authors have attacked this problem (ref.36,40,41,42).

The secrecy of the PKC resides in the one-wayness of the
operations involved. The investigation of suitable one way
functions has been the subject of intense research since Diffie
and Hellman's paper.

2. Trapdoor One-Way Functions

A function $y = F(x)$ is said to be one-way if 1) there is a
one-to-one relation between x and y, 2) given x it is 'easy' to
compute y, and 3) given y it is 'hard' to compute x. Furthermore
in a trapdoor one-way-function it is easy to compute x from y
given some secret side information. Diffie and Hellman (ref.2 and
6) describe a key distribution system based on the one-wayness of
the discrete exponential and logarithm functions. If p is a prime
and a is a primitive element, then for x and y in the range 0 to
p-1

$$y = a^x \mod p \qquad \text{and} \qquad x = \log_a y \text{ over } GF(p)$$

It is easy to compute y given x (ref.7) in about $2\log p$ (base 2)
multiplications (ref.8). For example (ref.9) noting that
$35 = 10011$ in binary, we have

$$a^{35} = a^{32} \cdot a^2 \cdot a^1 = ((((a^2)^2)^2)^2) \cdot a^2 \cdot a$$

which requires seven multiplications. Even if the x's are several
hundred bits long it is still easy to evaluate the exponential as

given in (refs.5 and 10). Evaluation of the logarithm is
conjectured to be much more difficult requiring of the order of
root p steps (ref.8). Tighter bounds are known (refs.11,12,13)
but for numbers of the order of 500 bits, it is still
computationally infeasible to find the log.

The distribution method works as follows. Given

$$U = a^u \bmod p, \quad I = a^i \bmod p, \quad K = a^{ui} = a^{iu} \bmod p$$

'You' think up a random number u and tell me U. 'I' think up a
random number i and tell 'you' I. 'You' raise I to the power u,
'I' raise U to the power i and we have both calculated the key K.
'They' only know U and I and our one-way function. To find K
'they' need to find either u or i and 'they' are up against a
one-way function.

The generation of secure PKC depends on the finding suitable
one-way functions with hidden trapdoor information to make the
inversion feasible. Such schemes have been proposed by
Merkle-Hellman (ref.5), Rivest et al. (ref.14), McEliece
(ref.10), Lu-Lee (ref.15), Kravitz and Reed (ref.16),MITRE
(ref.17), Gordon (ref.18), etc. These schemes have been subjected
to intense scrutiny and some have fallen as a result. Indeed it
may prove impossible to devise workable PKC's given the rate at
which new 'holes' in the techniques are found, and the rate at
which new modifications are proposed to overcome some of the
disadvantages. Notwithstanding this, let us now consider the two
most popular systems.

3. The Merkle-Hellman Trapdoor-Knapsack PKC

The knapsack problem is a combinatorial problem in which one is
given a vector a of n integers (the weight of each possible object
in the knapsack) and an integer S which is the sum of a subset of
the a's (the actual weight of the knapsack). The problem is to
solve for the subset, that is the binary vector x corresponding to
S = a * x (ie find which objects are in the knapsack). The
general knapsack problem is one in which the coefficients of x are
integers instead of 0 or 1, and this problem is known to be
NP-complete and therefore 'hard'. However, some knapsacks are
easy to solve. For example if a = (1,2,4,8,16...) ie the powers
of 2, then x is the binary representation of S. Merkle and
Hellman (ref.5) use a knapsack vector a' which is superincreasing.
That is, each integer is strictly greater than the sum of all
previous integers. For example a' = (171,196,457,1191,2410).
Given S' = 3797 we can easily compute S' = 2410+1191+196
ie x = 11010. This 'easy' knapsack is then disguised by k
iterations of modular multiplication to produce a trapdoor

knapsack vector a that is 'hard'.Thus

$$a = (((a'_1 * w_1) \bmod m) * \ldots) * w_k \quad \bmod m$$

where w_j is invertable modulo m, that is $gcd(w_j , m_j) = 1$.

if the vector a is made public then anyone wishing to transmit a message x would calculate the hard knapsack S = x * a , which the recipient would transform into the 'easy' knapsack S' = x * a' using the secret m and w, Thus:

$$S' = (((S * w_k^{-1}) \bmod m) * \ldots) * w_1^{-1} \quad \bmod m$$

Using the example from ref.5, chose m=8443 and w=2550, then w^{-1} =3950 by Euclid's algorithm (ref.8). The published knapsack is now a = (5457, 1663,216,6013,7439). Given S = 1663+6013+7439 =15115, we compute S' = 3950.15115 mod 8443 = 3797 as before. The scheme is attractive because encryption is fast, requiring only addition, and also fast decryption schemes have been proposed (ref.28). The original Merkle Hellman scheme proposed n=100 knapsack vectors and a 202 bit modulus thus making the a 202 bit pseudorandom numbers of length 202 bits. The sum S requires a 209 bit representation giving an intrinsic 2.09 data expansion from x to S with k=1 iterations, and a public key size of 20kbits. Attacks on the system have however forced these parameters to be revised upwards. In particular Shamir (ref.18) shows that two or more iterations are definitely needed, and this causes the data expansion to increase by seven bits at each iteration. At present the whole security of the superincreasing trapdoor is in question (refs.20,21, 22,23,24). In particular Desmet et al. (ref.22) find that iterative transformations do not guarantee higher security, and that infinitely many superincreasing decryption keys exist as soon as one exists. Shamir (ref.21) has discovered a technique which will solve a given knapsack with a probability of success that is directly proportional to the density of the subset sums S, where a dense knapsack is one in which nearly all integers in the interval between 1 and the sum of all the knapsack integers are valid subset sums. This makes dense knapsacks unsafe for cryptographic use. McAuley and Goodman (ref.25) have proposed a new trapdoor in a knapsack PKC that is not based on a superincreasing sequence in order to defeat these attacks. However, new results (ref.26) seem to indicate that most cryptographic knapsacks can be solved in polynomial time, even if they are not based on superincreasing sequences. These results further question whether all useful knapsacks can be cracked, and wheather useful ones can be generated and tested.

The inherent expansion in the trapdoor knapsack means that these systems are not well suited to providing public key authentication, because only a small fraction of all possible message words of a typical length lead to a binary solution of the knapsack. Schobi and Massey (ref.27) have proposed a nonbinary solution that overcomes this.

4. The Rivest-Shamir-Adelman (RSA) scheme

The RSA scheme (ref.14) is based on the fact that it is much easier to generate large primes and multiply them together than it is to factor the result. The key generator chooses two large primes p and q which are a few hundred bits long. if n=pq then Euler's function is (p-1)(q-1) that is, the number of integers between 1 and n which have no common factor with n. We then choose a number e relatively prime to (p-1)(q-1) and use Euclids algorithm to find the 'inverse' d via the expression e.d =1 mod (p-1)(q-1). The public key is (n,e) and our secret trapdoor information is d and the factorisation of n. The message text is represented as an integer from 0 to n-1 and the enciphering and deciphering procedures are the modular exponentiations :

$$C = M^e \bmod n \qquad\qquad M = C^d \bmod n$$

We have seen previously that even if the e and d are large the number of multiplications required in the exponentiation is managable, whilst the enemy has a task as difficult as factoring n. For example (ref.6): choose p = 5 and q = 11. Then n = 55 and (p-1)(q-1) = 40. If e = 7 then d = 23 as 7.23 = 1 mod 40. Choosing a message M = 2 :

$$C = 2^7 \bmod 55 \quad = \quad 2^1.2^2.2^4 \bmod 55 \quad = 18$$

$$M = 18^{23} \bmod 55 = 18^1.18^2.18^4.18^{16} \bmod 55$$

$$= 18.49.36.26 \bmod 55$$

$$= 2$$

The RSA method has also been subjected to attacks but has withstood these much better than than the trapdoor knapsack scheme. (refs 20, 29-34). Furthermore the RSA scheme gives digital signatures directly as there is no expansion of the message text. In addition, an elegant probabalistic test (ref.14) gives us a means of generating large primes efficiently, thus makeing the RSA algorithm self-contained and secure.

5. Implementation

Cryptographic algorithms are ideally suited to VLSI implementation because of their computation-intensive nature. In addition, there have been new developments in tamper-proof chips for software protection (ref.35) and these permit the possibility of secure generation of keys, without user intervention. The main implementation of cryptographic systems so far has been the production of DES chips. These are available from several manufacturers in both single chip and chip-set form. For example the Advanced Micro Devices AMZ8068 which gives throughput rates of over 1 Mbyte per second.

The integration of PKC systems is still at the development stage. There are several reasons for this. Firstly the PKC algorithms require more computation than the DES and thus imply lower data throughput rates, but more importantly he algorithms are still under development and their security is still in question. Given this fact it is not surprising that all implementations have been directed towards the RSA scheme or hybrid DES-RSA schemes where the RSA is used for key distribution and the DES for fast encryption.

Rivest (ref.37) has reported a single-chip implementation of the RSA algorithm. The design is essentially a big-number ALU, that can operate on 512 bit numbers and hence perform all calculations needed by the RSA. The chip implements the operations of addition, subtraction, multiplication, division, remainder, and modular exponentiation. In addition other useful functions such as generation of large primes are performed. The 512 bit ALU is organised in a bit-slice manner with 8 general purpose registers, up-down shifter logic, and multiplier (carry-save) logic. The ALU is only capable of performing the operations A.B+C, shift-left, shift-right, test least significant bit. All the higher functions are implemented by the microprogram stored in the internal PLA. With a feature size of 2 microns the chip is large measuring 5.5 mm by 8 mm, and the 4MHz gives an encryption rate of 1200 bits per second. The complexity of this chip however raises doubts as to the yields obtainable.

The essential operation in the RSA algorithm is that of modular exponentiation. This reduces to modular multiplication at its simplest level. Simmons and Tavares (ref.38) are working on a modular multiplier in NMOS technology using a 6 micron feature size. The design is again bit slice orientated and the two steps of multiplication and then modulo reduction are performed by the same device. That the two operations are essentially the same can be seen as follows. The process of multiplication can be seen as that of conditionally adding together shifted versions of the multiplicand. Thus for each 1 in the multiplier an intermediate product is formed by adding in a version of the multiplicand that has been shifted the same number of places as the 1 in the multiplier. Modulo reduction can be considered as conditional subtraction of the modulus from the product. The modulus is first shifted left until its MSB is aligned with the MSB of the product. If the shifted modulus is less than the present result, we subtract it to form a new result, and then shift the modulus one bit right. If the shifted modulus is greater we do not subtract but just shift. This repeats until the result is less than the shifted modulus. The implicit comparison operation is complex but can fortunately be eliminated as follows. First, a sign bit is required in the result which is initially presumed positive i.e. s=1. Then starting at the MSB end as before the subtractions are replaced by the addition of one of two values: the modulus or its two's complement. Both values are shifted and the sign of the present result determines which value is to be added. If s=1 ie the result is negative we add the modulus, if s=0 add the two's complement. When the LSB's line up one further addition of the modulus is needed if the result is negative, to ensure a positive final result. Thus a single unit consisting of an adder with inputs that may be shifted can perform both multiplication and modulo reduction. This requires that the input to the adder can be selected from one of four values: zero, the multiplier, the modulus, or the two's complement of the modulus. In ref.38 the authors hope that chips will be ready during 1983. They estimate a total multiply modulo time of 250 microseconds for 128 bit inputs. The design appears attractive, particularly with the incorporation of parallel pipelining of the inputs and outputs. That is, as one set of data is being processed by the arithmetic unit, the result of the previous set is being output and the next data set is being input. The major limitation of the device is its 128 bit maximum wordlength, and its slow throughput of about 4Kbits.

McAuley and Parker (ref.39) have been working on an Advanced Cipher Processor (ACP). The device has a mask allocation number MA743 and is being fabricated by the GEC research laboratories at the Hirst Research Centre, Wembley, England. The device essentially performs modular exponentiation of 512 bit numbers, and is to be fabricated in bulk CMOS technology using 2.5 micron feature size. A data throughput rate of 50 Kbits per second is hoped for.

The VLSI architecture group at Hirst have an active systolic array program and the design of the cipher processor reflects this. The systolic 'data pumping' approach is particularly suitable for high performance computing VLSI structures. In general a systolic array is a one or two dimensional array of identical functional modules, typically simple digital circuits, arranged in a regular fashion. Each module is connected only to its nearest neighbours

for₮ the purposes of data transfer. Each module utilises common
control and timing so that all the modules perform the same
function simultaneously but on different data items. The data
streams move at constant velocity over fixed paths and interact
whenever they meet. Multiple use is made of each data item which
results in high computational throughput without the need for high
bandwidth memory links. The precise function of the cell depends
on the problem to be solved.The advantages of the systolic
approach include short interconnects thus giving high speed
transfer with low power drivers and small chip area, easy-to-scale
architecture,small system control overhead, and minimal data
transfers to and from memory (data is input once, used, and
discarded). Note that this also allows parallel pipelining. The
design of the chip builds on previous work at Hirst on an
inner-product step cell. This is a circuit for computing the
function C' = A.B + C and propagating delayed versions of A and Ḃ
to neighbouring cells. This work indicated that although
bit-parallel arithmetic is faster than bit-serial, the latter
approach gives a greater functional throughput per unit chip area.
For these reasons, and because of the high density caused by the
512 bit integers aimed at on this single chip, serial arithmetic
is used.

The ACP communicates with the host microcomputer via an eight bit
bidirectional data bus, and a number of control pins. DMA
transfers are supported. Internally the device consists of the
control unit, the modular exponential unit, four 512 bit registers
which hold the exponent, modulus, inverse and output.
Communication with the host is via a 64 byte I/O stack, a control
register and a status register. The internal and external
operations are essentially asynchronous with communication through
the status register, so that for example input data can be loaded
to the stack whilst the modular exponential unit is operating.
The heart of the unit is the modular exponentiator which consists
of a 512 bit serial parallel multiplier which performs the
exponentiation and a similar 512 bit divider which performs the
modular reduction. The multiplier feeds the divider and
vice-versa so that data is only input once and thus input-output
pipelining is possible. The data flow is thus circular. To
operate the unit the keys are first loaded into the appropriate
registers via the I/O stack, that is the modulus, exponent and
after a precomputation, the inverse register. The input data is
then supplied to the exponentiator which will circulate until the
output is ready and in the output register. During this period
the I/O stack can be loaded with new data if required. If the
result is ready and the stack is still busy, with new data, then
the output waits until the new data is claimed by the
exponentiator before transferring itself to the I/O stack. This
parallel pipelining contributes greatly to the overall speed of
the device.

6. References

1 . "The AmZ8068 Data ciphering Processor", Product Description, Advanced Micro Devices, Sept 1980.
2 . W.Diffie and M.E.Hellman, "New directions in cryptography", IEEE Trans. Inf. Th.,vol IT-22, Nov 1976.
3 . R.C.Merkle, "Secure communication over an insecure channel", Common. Ass. Comput. Mach.,vol 21, Apr 1978.
4 . G.J.Simmons, "Cryptology: The mathematics of secure communication", The Math. Intell., vol 1, no 4, Jan 1979.
5 . R.C.Merkle and M.E.Hellman, "Hiding information and signatures in trapdoor knapsacks", IEEE Trans. Inf. Th., vol IT-24, Sept 1978.
6 . M.E.Hellman, "An overview of public key cryptography", IEEE Comm. Soc. Mag., Nov 1978.
7 . A.G.Konheim, Cryptography: A Primer", John Wiley, 1981.
8 . D.E.Knuth, The Art of Computer Programming, Vol 2, Seminumerical Algorithms, Reading, MA: Addison-Wesley, 1969.
9 . J.A.Gordon, "Recent trends in cryptology", Electronics and Power, vol 26, no 2, Feb 1980.
10. R.J.McEliece, "A public key system based on algebraic coding theory", JPL DSN Progress Rep., 1978.
11. S.C.Pohlig and M.E.Hellman, "An improved algorithm for computing logarithms over GF(p) and its cryptographic significance", IEEE Trans. Inf. Th., vol IT 24, no 1, Jan 1978.
12. L.Adelman, "A subexponential algorithm for the discrete logarithm problem with applications to cryptography, Dept. of Math., MIT.
13. A.Shamir and R.Schroepple, "A TS2 = O2n time/space tradeoff for certain NP-complete problems", SIAM J. Comput, vol 10, 1981.
14. R.Rivest, A.Shamir, L.Adelman, "A method for obtaining digital signatures and public key cryptosystems", Comm. ACM, vol 21, 1978.
15. S.C.Lu and L.N.Lee, "A simple and effective public key cryptosystem", COMSAT Tech. Rev. 9, 1979.
16. D.W.Kravitz and I.S.Reed, "Extension of RSA cryptostructure : a Galois approach", Elect. Lett., 18(6), 1982.
17. B.P.Schanning, "Applying public key distribution to local area networks", Workshop on electronic privacy and authentication, The Hatfield Polytechnic, July 1982.
18. J.A.Gordon, "Public key cryptosystems and related topics", Proc. IEE Conf. on data transmission codes, London, Nov 1980.
19. A.Shamir and R.E.Zipple, "On the security of the Merkle-Hellman cryptographic scheme", IEEE Trans. Inf. Th., vol IT 26, no 3,May 1980.
20. T.Herlestam, "Critical remarks on some public key cryptosystems", BIT, vol 18, 1978.

21. A.Shamir, "Cryptocomplexity of knapsack systems", Symposium on the theory of complexity, Atlanta, Georgia, April 1979.

22. Y.Desmet, J.Vandewalle, R.Govaerts, "A critica analysis of the security of knapsack public key algorithms", IEEE Int. Symp. on Inf. Th., Les Arcs, France, Jun 1982.

23. I.Ingemarsson, "A new algorithm for the solution of the knapsack problem", IEEE Int. Symp. on Inf. Th., Les Arcs, France, Jun 1982.

24. A.Shamir, "New results on public key cryptosystems", to appear.

25. A.J.McAuley and R.M.F.Goodman, "Modifications to the trapdoor- knapsack public key cryptosystem", IEEE Int. Symp. on Inf. Th., St. Jovite, Canada, 1983

26. Lagarias and Odlyzko, "Solving low density subset sum problems", Bell Sys. Tech. J., 1983.

27. P.Schobi and J.L.Massey, " Fast Authentication in a trapdoor-knapsack public key system", IEEE Int. Symp. on Inf. Th., Les Arcs, France, Jun 1982.

28. P.S.Henry, "Fast decryption algorithm for the knapsack cryptographic problem", Bell Sys. Tech. J., vol 60, May–Jun 1981.

29. R.R.Rivest, "Critical remarks on critical remarks on some public key cryptosystems by T.Herlestam", BIT, vol 19,1979.

30. B.Blakley and G.R.Blakley, "Security of number theoretic public key cryptosystems against random attack", Cryptologia, vol 3, nos 1 and 2, 1979.

31. G.Simmons and M.Norris, "Preliminary comments on the MIT public key cryptosystem", Cryptologia, Oct 1977.

32. R.Rivest, "Remarks on a proposed cryptoanalytic attack on the MTT public key cryptosystem", Cryptologia, Jan 1978.

33. H.C.Williams and B.Schmid, "Some remarks concerning the MIT public key cryptosystem", BIT, vol 19, 1979.

34. G.R.Blakley and I.Borosh, "Rivest-Shamir-Adelman public key cryptosystems do not always conceal messages", Comp. and Maths. with Appls., vol 5, 1979.

35. R.G.F. Aitchson, "A cryptographic approach to software", Workshop on electronic privacy and authentication", The Hatfield Polytechnic, July 1982.

36. R.M.F.Goodman and A.J.McAuley, "Broadcast Public Key Cryptosystems", IEEE Int. Symp. on Inf. Th., Les Arcs, France, Jun 1982.

37. R.L.Rivest, "A description of a single-chip implementation of the RSA cipher", Lambda, MIT, 1980.

38. D.Simmons and S.E. Tavares, "An NMOS implementation of a large number multiplier for data encryption systems", IEEE Trans, 1983.

39. A.J.McAuley and N.Parker, "MA743 advanced cipher processor-preliminary data", GEC Research Labs, Hirst Research Centre, 1983.

40. L.N.Lee and S.C.Lu, "A multiple-destination cryptosystem for broadcast networks", COMSAT Tech. Rev., vol 9, no 1, 1979.
41. F.Luccio and S.Mazzone, "A cryptosystem for multiple communication", Inf. Procc. Letts., vol 10, no 4, July 1980.
42. S.T.Kent, "Security requirements and protocols for a broadcast scenario", IEEE Trans. Inf. Th., vol COM 29, no 6, June 1981.

PANEL DISCUSSION

ON

CRYPTOGRAPHY AND SECURITY

Friday 22nd July 1983 at 10.00 hours

Chairman and Organiser:	Professor J.L. Massey	ETH-Zentrum, Zürich, Switzerland.
Panel Members:	Professor I. Ingemarsson	Linköping University, Sweden.
	Dr. S. Harari	Université de Picardie, Amiens, France.
	Dr. R. Johannesson	University of Lund, Sweden.
	Mr. P.G. Wright	Marconi Research Centre, Great Baddow, U.K.
	Dr. R. Blom	Linköping University, Sweden.
	Professor J. Ziv.	Technion – IIT, Haifa, Israel.
	Dr. R.M.F. Goodman	University of Hull, U.K.
Contributors:	Dr. D. Andelman	IAA, Haifa, Israel.
	Dr. B.K. Bhargava	Concordia University, Montreal, Canada.
	Dr. R.E. Blahut	IBM Corporation, Owego, NY, U.S.A.
	Professor K.W. Cattermole	University of Essex, Colchester, U.K.

Professor T.M. Cover	Stanford University, CA, U.S.A.
Professor P.G. Farrell	University of Manchester, U.K.
Professor S.W. Golomb	University of Southern California, Los Angeles, CA, U.S.A.
Dr. B. Gopinath	Bell Laboratories, Murray Hill, NJ, U.S.A.
Dr. C. Heegard	Cornell University, Ithaca, NY, U.S.A.
Professor R. McEliece	California Inst. of Technology, Pasadena, CA, U.S.A.
Professor D.G. Messerschmitt	University of California, Berkeley, CA, U.S.A.
Professor K. Steiglitz	University of Princeton, NJ, U.S.A.
Dr. H. Trzaska	Technical University, Wrocław, Poland.
Professor J.K. Wolf	University of Massachusetts, Amherst, MASS, U.S.A.

<u>Professor Massey</u> - This is the panel discussion on cryptography.
Let me just outline a little bit on the idea that I had while
organising this panel. Most of you are aware that in 1976 Diffie
and Hellman published a famous paper on "New Directions in
Cryptography" [1] in which they proposed for the first time the
possibility of creating a secrecy system, a cryptographic system,
in which there was no need for a secure or a private transmission
of the key to the destination, and this area has become know as
public key cryptography (PKC). Initially that idea caused a
tremendous amount of excitement, stimulating much work both in
cryptography and in related areas of mathematics which were
required for cryptographic problems and there was a tremendous
enthusiasm for this work. However, lately things have been going a
little bit badly for the problem of PKC. The breaking by Shamir
[2] of the Merkle-Hellman uniterated trapdoor knapsack system [3]
shook a lot of people's confidence in the PKC. There were also
suggestions of threats to the Rivest, Shamir and Adleman PKC [4]
and we shall hear a little about those to-day. There seems to be a
feeling by many in our community that perhaps the PKC was a brief
aberration in the long history of cryptography, and that really
there is no way to go further.

Therefore I have asked the panellists to address that question
with their own view-points, either to take the positon 'pro' for
the future of the PKC or to take the position 'contra' for that
future. Of course, I was able to find more 'contras' than 'pros'.
Thus we shall have those who support the PKC outnumbered by
'contras'. Therefore I have tried to interleave the speakers a
little bit in the order of their position on the PKC with a
'contra' starting and alternating in all the odd positions, at
least to some extent.

Let me first introduce Professor Ingemarsson of the Linköping
University in Sweden, who is going to be one of the 'contras', Then
we have Dr. Harari of the Universite de Picardie in France whom you
have already heard speaking on the 'seal functions' [5]. Next is
Dr. Johannesson from the Technical University in Lund, in
Sweden - there is still one more Swede to come! Then we have Mr.
Wright from Marconi Research Centre in England and Dr. Blom, also
from the University of Linköping in Sweden. Professor Ziv is from
the Technion in Israel and Dr. Goodman is from the University of
Hull in England.

As I said already, the first of these, the third and the fifth
speakers will be, as I understand, against the PKC. The second and
the fourth speakers are in favour of it. Whether Professor Ziv is
in favour or against it is little bit doubtful, and Dr. Goodman has
changed his position at least twice since we have assembled here!

The ground rules laid down for this panel discussion are that
each speaker gets five minutes to state his case with a public
statement. After five minutes I will rise from my seat and after

six minutes I will raise my hand and at that point I would like the audience to start clapping. This can be interpreted as applause for the speaker, or as a suggestion for him to end his prepared statement. Let us practice that!

First we hear Professor Ingemarsson and I will time him with my Swiss watch.

Professor Ingemarsson - I will tell you something about the . algorithms as such. As you all know, Diffie and Hellman started this debate in 1976 by their famous paper [1]. One of their ideas that is still with us is the public key distribution (PKD) system, and it may be that this idea is a little bit forgotten. Now everything turns round Rivest, Shamir and Adleman (RSA). But the PKD idea still works very nicely. We have already talked at this Institute how the PKD is done [6], so I will not go into details, but one function that is still useful is:

$$y = \alpha^x \mod p \quad \text{(p is a prime number)} \quad \ldots\ldots (1)$$

Here the relation between x and y is one-to-one. Here x and y are positive integers less than p.

People have also tried to use the Galois field namely $GF(2^r)$ in place of mod p . However this is not so good. It has been shown that in many cases this function can be reversed. So we will ignore it. Maybe Dr. Johannesson will have some comments to make on this later on.

As far as I know, the function in (1) above still stands as secure. We can use it in PKD with a sufficiently large prime number p.

Now let us turn to the problem of public key encryption (PKE). The knapsack problem is very much talked about and we have heard that the knapsack as such is broken. This is not entirely true, but is almost true. One of the last statements will be made by A. M. Odlyzko at the symposium in Montreal in 1983 [10]. It will then be demonstrated that given any sparsely distributed knapsack sum S of the form:

$$S = \sum_{i=1}^{n} x_i \, k_i , \quad\quad\quad \ldots\ldots (2)$$

it can be analysed. That is, if you go through the binary set of x_i and if you have a set of k_i such that the distribution of S is sparse on the real axis, then k_i can be derived. I have a very strong feeling, which is underpinned by my own investigations, that even if we propose a stronger requirement for the knapsack, namely that S (as a function of x_i) is injective, the k_i can still be derived. So unfortunately we have to ignore this case as well.

Let us now turn to the algorithm of McEliece [7]. I do not know whether it is good or bad. It has not been published widely and it suffers from drawbacks such as the demand for extensive computation and for large data blocks involved in it. These are forbiddingly large. So at least from the practical point of view we have to put it into brackets.

Finally we come to the RSA solution [4] which is based on the function:

$$C = M^e \bmod n \qquad \ldots\ldots (3)$$

where n is the product of two primes (n = pq), or it could be something else, but we shall concentrate on the case of n = pq. C and M are positive integers less than n. A function like this relies heavily on the possibility of factoring large numbers, for n should be a large number. The latest result is from Schnorr and and Lenstra. It has not been published yet, but there is an internal report on it [8]. They report factoring times which take the form:

$$\text{const. } \exp \sqrt{\ln(n) \times \ln(\ln(n))} \qquad \ldots\ldots (4)$$

Thus for $n \approx 10^{60} \approx 2^{200}$ it takes 45 minutes

and for $n \approx 10^{154} \approx 2^{512}$ it takes 10^5 years.

Actually it is stated by RSA in the original paper [4] that there is yet another solution to this problem. It was proposed by R. Schroeppel, but it is not correct. This was the result of a misunderstanding between him and R.L. Rivest.

Hence we shall discuss here only the estimated times as proposed by Schnorr and Lenstra. These are the fastest times obtained so far; the method was developed last year. It can factor numbers up to 10^{60} in 45 minutes. Actually they have not as yet programmed for this. The second figure is a bit unfair, and the unfairness comes from the fact that in their algorithm they must do the 'greatest common divisor' operation. It is supposed to be done by hardware which is not yet available. Then in the first case, assuming that you could do it by hardware, you can arrive at a solution in 45 minutes. But what hardware is needed for $n \approx 10^{154}$? Will it ever be available? (Laughter). The extrapolation is that the key size number of 512 bits should be solved in 100,000 years. But that is not the whole truth. The truth is that this algorithm is Monte Carlo – it involves a random walk. So, with the probability of 1/1000 you could do it in only 100 years (Clapping.)

Professor Massey – Speakers get only five minutes for their prepared statement, but the idea is that they will have an opportunity to answer question both from the panellists and from

482

the audience. Who would like to begin with a question to Professor Ingemarsson?

Professor Cover - What are your conclusions?

Professor Ingemarsson - My conclusions are based on this statement:

Requirements in (3) on p and q are:

$\frac{p - 1}{2}$ is prime,

$\frac{p + 1}{6}$ contains high prime factors.

Then the RSA algorithm works well enough, and also the original Diffie - Hellman algorithm [1] works well enough for PKD. However, the suggestion of our chairman that I am a 'contra' has to be modified a bit. I think that these methods have a very limited area of use, and you may see why this is so when you hear Dr. Blom.

Professor Massey - Any other questions?

Professor Ingemarsson - Can I comment further on Professor Cover's question? I might have some doubts just for the reason that new factoring algorithms are popping up and we do not know the limits of their speeds. We know that they are all exponential in time but we do not know their absolute limits.

Dr. Blahut - Concerning the characterisation of your last algorithm, are your limits for the worst cases, or are they for typical average ones?

Professor Ingemarsson - As I said, this is a Monte Carlo case. You could do it much faster if you are lucky. The figures stated are average.

Professor Massey - The chair has just noted that Professor McEliece has joined us now. Before he came to this room, his PKC algorithm was half-eliminated from consideration (Laughter.)

Professor McEliece - What has happened?

Professor Ingemarsson - Would you like to comment on your PKC algorithm? Is it useable?

Professor McEliece - My paper was written several years ago and I have had practically no reactions to it. The few attacks I have heard proposed are based not on finding a new trapdoor function but simply on decoding an almost arbitrary linear code in the presence of some errors. If such an attack can succeed in even 10 percent of cases, then this certainly is not a secure system.

Professor Ingemarsson - Your algorithm needs a lot of data and a lot of computation. Can you comment on that?

Professor McEliece - The public key encoding requires a large fixed matrix, and it also requires matrix multiplications with specific parameters. We are talking here of a 1000 x 1000 binary matrix. I suppose you could put it into a ROM and then do the multiplication, but it is not a simple as one would wish.

Professor Massey - This, to me at least, does not sound much more complicated than RSA. Yours is a binary matrix multiplication so you have no problems with large numbers, etc. Thank you, Professor Ingemarsson, and I trust that later questions will come back to some of the earlier points that you have made. Now Dr. Harari will make his statement.

Dr. Harari - In PKC we have the following order of operations: an encyphering key K_1 and a different decyphering key K_2 (see Figure 1). They are different and by this I mean that you cannot deduce

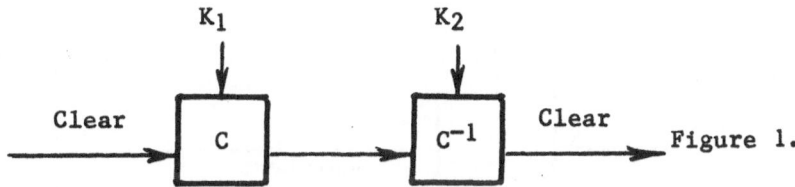

Figure 1.

one from the other. This is a great bonus to the usual cryptographic problems, in that it is easier to implement such a key distribution problem. But then we can still have another situation in which the key K_2 is a secret key and the key K_1 is the public key. You then decipher first the text with the secret key and then you send along a clear text. In other words, you send along M (see Figure 2) and then you decipher part of M with the

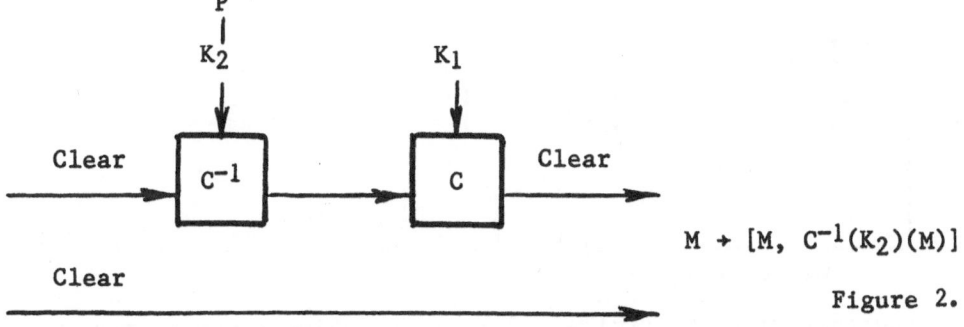

$$M \rightarrow [M, \ C^{-1}(K_2)(M)]$$

Figure 2.

secret key. This gives you a way to check that your choice of the
secret key was a correct one. This is a way to solve this
authentication problem. It allows us to solve it in a very
interesting manner, because it allows any user with a limited
amount of computing power to check that P is the legitimate owner
of K_2, i.e. that P has the right to use such a facility. This
fact has opened a lot of new wide-spread public applications.

One of them is the electronic money experiment using Rabin's
signature scheme [9]. I spoke about this for a few minutes in my
lecture here [5]. The experiment works like this. You go with
your card to your bank and unload your account at the banking
machine. Your card is now loaded with money. Then you go to a
store to buy something. You sign such a procedure and immediately
your card is discharged by the correct amount of money, and at the
same time the store owner has his account correspondingly credited.
This experiment is going on, on a wide scale in France.

The second application I want to talk about is toll
television. This is a more complex situation. TV programs for
which you want a customer to pay are sent encrypted with a key C
(see Figure 3). The key is divided into two parts: Part of data P
is concerned with

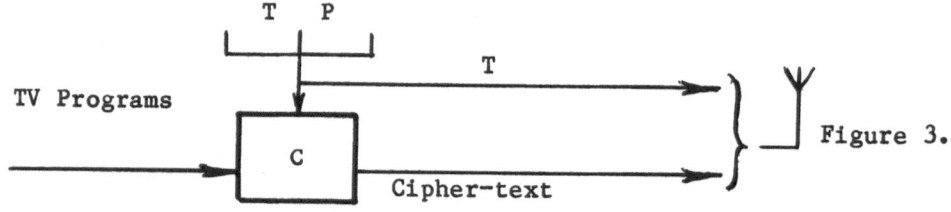

Figure 3.

T = time indication
P = program indication

program indications, while the other part concerns the timimg.
The TV program is sent encyphered. At the receiving end
(Figure 4) if you want to decrypt the program, you can get
everything, but you can only decrypt that part for which you are
paying for. You buy a ticket which contains program information.

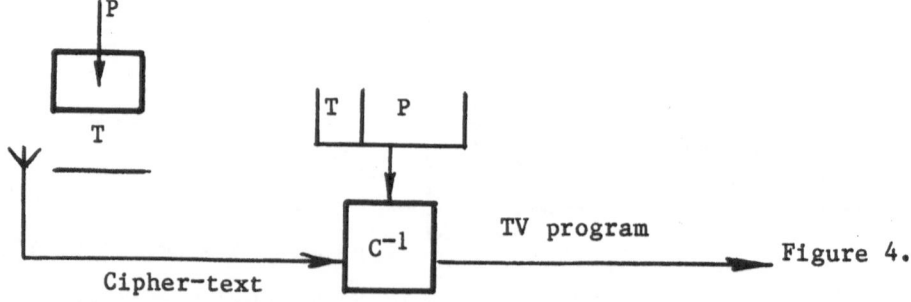

Figure 4.

Then together with the timing information (which comes to you
uncyphered from the channel) you can use your key to decypher the

whole TV signal. This can be done with any kind of cryptographic algorithm.

The publc key idea has thus been used with regular secret key algorithms. This is another application showing that PKC has a very fruitful use. May this influence classical cryptography? That is all I want to say about applications. Other forms of applications are going on, on an experimental scale in France. There is a talk on having all the national network operating in this way.

The last thing I want to argue for PKC concerns the knapsack. You have heard in the preceeding talk that there are indications showing that superincreasing knapsack has been broken by Shamir. This is true, but some work is now going on showing that the superincreasing knapsack is just a total ordering related to some subsets of integers from 1 to N, and there are some stronger order relations on integers which will lead hopefully to harder knapsacks. I do not think that the idea of the knapsack is to be thrown away.

Professor Massey - Thank you Dr. Harari. Any questions?

Professor Wolf - Could you say more about the purpose of the time indicator T in your TV toll example, which is sent to a customer in the clear. Is that used to unload the program so that on the card you bought you are allowed to watch TV for a certain time only?

Dr. Harari - No. There is a time interval in the key data. When you buy a card, there are time data punched on it. Then you are allowed to use your card only for a certain amount of time.

Professor Wolf - Is this an absolute time? Has anyone tried to make a system where you bought a certain number of hours of watching, and as you watch TV you unload your card until you have no more time left?

Dr. Harari - It is theoretically possible to arrange things like that. However, this system is aimed at a very wide public utilisation and you want to keep the cost down for the time being. What you do for customers has to be very cheap.

Dr. Gopinath - I have a question about the knapsack to Professor Ingemarsson. If I understood the theorem he has stated, it looks like that any other totally ordered relation can still be broken.

Professor Ingemarsson - Yes, this is correct. There is an algorithm by Odlyzko [10], (his name is spelled differently in different places) which is applicable to sparse sums, so you do not need the requirement of a superincreasing knapsack.

Professor Massey - If this is true, then this is really a suprising result. Do we know how sparse is 'sparse'? As far as the theory is concerned that the knapsack function is NP-complete, this poses no limitations on whether the knapsack is sparse or dense.

Professor Ingemarsson - Can I comment on this? It is true that the knapsack problem is NP-complete, but it has also been proved that the problem of determining whether the knapsack problem has exactly one solution is indeed NP-complete. You can do these things in two steps. First you try to prove that this particular knapsack has just one solution. Then given that, you can proceed to find that solution. The first problem is NP-complete. This does not contradict the statement that the next problem (namely finding the solution) may not be NP-complete.

Professor Massey - Is there any other contribution on this point?

Professor Wolf - You mentioned that the knapsack problem is NP-complete. Does this have any meaning in cryptography? Is this the worst case?

Dr. Andelman - It was shown at least in some very degenerate examples that you can have an NP-complete problem which is trivial to solve from the cryptographic point of view.

Professor Massey - Was this not Lempel's example [11]? Possibly Professor Ziv can tell us about this. As I understand Dr. Lempel has developed an example which showed it was simple to break a PKC system due to Even and Yacobi that was equivalent to solving an NP-complete problem. It was the first time that such a thing was proven.

Professor Ziv - This is correct. You can work that out in this way: you successively learn on your way, and provided that you do it properly, you will actually converge to the solution.

Lempel, Even and Yacobi demonstrated a cipher the breaking of which is NP-complete, even under chosen plain-text approach. However, given enough known plain text, breaking the cipher reduces, with probability approaching unity, to the simple problem of solving n independent linear equations in n unknowns, where n is the number of key bits [11].

Professor Massey - I think that what you do essentially is to create additional linear equations as you go through each step. Then with probability one, after a short time you have enough of these equations to solve the problem and to break the key. At least in that one example it would appear that designing a cryptographic system based on an NP-complete problem is a dangerous thing to do.

Professor Wolf - I would like to hear a comment of Professor McEliece as to whether his scheme is an NP-complete problem of decoding a linear code.

Professor McEliece - What you say is true. It is probably a bad risk to base a cryptosystem on an NP-complete problem, but it is still a worse risk to base a cryptosystem on a problem of polynomial complexity. So it is not that NP-completeness is something undesirable - it may be necessary but it is certainly not sufficient to be secure.

Professor Massey - One problem with NP-completeness is that it is a worst case measure and does not guarantee that you have a good average computation, but on the other hand it seems not to imply that you need to have a bad average computation. It may be that among NP-complete problems there are some that would make good cryptographic problems, while others would not.

Professor McEliece - If you have good algorithms, they certainly will be NP-complete.

Professor Massey - I call Dr. Johannesson, our next panel member, to make his statement.

Dr. Johannesson - First of all I wish to say that my interest in this field is rather amateurish - I am not a cryptographer. Here I will use some ideas from sequential decoding to attack a problem which I encountered a few years ago - it was a logarithm problem in a PKD. Together with Tore Herlestam I developed a heuristic algorithm for finding logarithms over $GF(2^P)$ [12]. Since the arithmetic in $GF(p)$ is not very well suited to most computers a variant is used by MITRE Corp [13], viz. $GF(2^{127})$. As an illustration I shall show how a logarithm can be determined with a little example in $GF(2^7)$, where you can easily do the calculations by hand.

The question is what is the logarithm of some polynomial, say

$$f(t) = t^6 + t^5 + t^3 + t + 1.$$

Obviously we have to find the exponent L: $f(t) = t^L$.

We use the primitive polynomial in this field:

$$p(t) = t^7 + t + 1$$

and we can easily calculate the inverse of t:

$$0 = t^7 + t + 1 \rightarrow t(t^6 + 1) = 1 \rightarrow t^6 + 1 = t^{-1}.$$

Let $L = L_0 + L_1 2 + \ldots + L_6 2^6$ be the binary representation of $L = \log f(t)$. The idea is that if $L_r = 1$ then by multiplying by t^{-2^r} we eliminate the (r + 1)st binary digit in L, which means that the Hamming weight of the logarithm has been reduced. We now take for granted that the Hamming weights (W_H) of $\log f(t)$ and $f(t)$ are dependent and that instead of reducing $W_H(\log f(t))$ we might try to reduce $W_H(f(t))$ just as well. Unfortunately the inequality $W_H(t^{-2^r}f(t)) < W_H(f(t))$ is not generally true, but we can iterate the process and hope that the inequality will be satisfied within a limited number of steps. Now let us tabulate the products of t^{-1}, t^{-2}, t^{-4},....times $f(t)$ and obtain some new polynomials (Table 1):

r	$t^{-2^r}f(t)$	$W_H(t^{-2^r}(t))$
0	$t^{-1}f(t) = t^6 + t^5 + t^4 + t^2$	4
1	$t^{-2}f(t) = t^5 + t^4 + t^3 + t$	4
2	$t^{-4}f(t) = t^6 + t3 + t^2 + t + 1$	5
3	$t^{-8}f(t) = t^5 + t^3 + t^2$	3
4	$t^{-16}f(t) = t^6 + t^5 + t^4 + t + 1$	3
5	$t^{-32}f(t) = t^3 + t^2$	2 ←
6	$t^{-64}f(t) = t^5 + t^4 + t^3 + 1$	4

Table 1

If we arrive at a new polynomial for which the Hamming weight W_H is equal to one, we know the logarithm and stop the procedure. Otherwise, we take the polynomial corresponding to the smallest Hamming weight (namely $t^3 + t^2$) and iterate the procedure (see Table 2):

r	$t^{-2^r}(t^3 + t^2)$		$W_H(t^{-2^r}(t^3 + t^2)$
0	$t^{-1}(t^3 + t^2) = t^2 + t$		2
1	$t^{-2}(t^3 + t^2) = t + 1$		2
2	$t^{-4}(t^3 + t^2) = t^5$	STOP →	1
3	$t^{-8}(t^3 + t^2) = t$		1
4	$t^{-16}(t^3 + t^2) = t^6 + t^5 + t^4 + t^3 + t^2 + t$		6
5	$t^{-32}(t^3 + t^2) = t^5 + t^4 + t^3 + 1$		4
6	$t^{-64}(t^3 + t^2) = t^6 + t^5 + t^3 + t^2$		4

Table 2

Thus we start with $(t^3 + t^2)$ and we do the same thing all over. As soon as we reach the Hamming weight 1, we stop. At this stage we notice that $\log(t^{-4}(t^3 + t^2)) = 5$ and, hence, that $\log(t^3 + t^2) = 9$. Now we can go back to the previous calculation and since $\log(t^{-32}(f(t))) = \log(t^3 + t^2)$ we deduce that if we add $9 + 32 = 41$, we obtain: $\log(t^6 + t^5 + t^3 + t + 1) = 41$.

This works beautifully for such a small field, but you cannot do it as easily for larger fields. We have gone further and tried a non-trivial case of $GF(2^{31})$. The method works, but now we have not only to multiply by t^{-1}, t^{-2}, t^{-4}, ..., but also we have to square. In this way we are rotating the exponent. Certain positions could be more sensitive than others to the multiplication by t^{-2^r}. We computed the logarithms for all polynomials of Hamming weights two and three. With this database we can then use Hamming weight < 3 as a stop condition and do a table look-up before back tracking as described before. Using this strategy we computed the logarithms over $GF(2^{31})$ for 1000 randomly chosen polynomials, on the average, after 85 iterations. Then what happens if you go to still larger fields, such as $GF(2^{127})$? I do not know yet.

I think that it is a fact when we look back at the history of cryptography, that heuristic attacks have been quite successful, though we at that time did not know why they worked. If you play around a little with polynomials, you can hopefully get some intuitive feeling for what you should do next. Although there is no theory at present, probably after a few years someone may find a reason for the success.

I will conclude with a general remark on PKC. What I find surprising is that people believe that we can hide a small amount of information (the secret key) and then use it to encrypt an infinite amount of information. It seems to be an information theoretic perpetuum mobile!

<u>Professor Massey</u> - Thank you Dr. Johannesson. I think that actually you have two parts to your case. One is the insecurity and difficulty of obtaining logarithms in $GF(2^n)$, and the general attack against PKC, in that you really suggest for any PKC you should have a limit on the amount of data which has to be encyphered. So, does anyone like to debate with Dr. Johannesson either of these points? I might add: 'of these unreasonable points'. Since no one responds to my invitation, I will have to say something.

In connection with the question of the logarithm in $GF(2^n)$, maybe we can go a little into the background of this problem. In general, if you take the logarithm in a finite field, say $GF(q)$, and I am neutral here as to whether q is a prime p or a power of some prime like 2^n, you first consider the multiplicative group of this field and you know that this is a group of the order $q-1$.

Take an element α that is primitive in that group. This means that
α is a primitive (q-1)-th root of unity. Then we can write any
element in this group $\beta = \alpha^n$. The finding of that n is the problem
of taking a logarithm. We know, since the work of Pohlig and
Hellman [14], that when q-1 has as factors all small numbers, then
it is easy to take logarithms. This is the reason for their
suggestion that when you use GF(p), then p-1 should have a large
prime factor. Of course p-1 is going to be even, so we should
really write p-1 = 2p', where p' could be prime. Dr. Johannesson
has pointed out already that you would do better to work in GF(p),
rather than say work in $GF(2^{127})$, where $2^{127}-1$ is prime, even
though there is no small prime factor here. I will agree with him
that the only reason that we feel we can do this is that we have a
better feeling of how to do arithmetic in the field GF(2P) than
we do when we meet GF(p), where p is a large prime. It seems to me
that our experience of working with numbers in that form is more
unfamiliar, so we have a kind of in-built feeling that it must be
more difficult to take logarithms in GF(p). However, I do not see
any theoretical reason for this and I guess that you could not
offer any theoretical reason either. Could you?

Dr. Johannesson - Yes, you are right, I cannot. You have made an
important point. You can design an algorithm which works without
theory and you do not know why it works. This is typical in
cryptography.

Professor Massey - I might mention that there is another group who
are following your work, namely the group with Dr. Ian Blake in
Canada. He and his friends are working on $GF(2^{127})$. I talked to
him recently on the phone and he said that now he has calculated
about 6000 of the 10000 special relations that he is going to need
to take a logarithm in that field. He is hoping to obtain an
algorithm with lots of storage and pre-calculated functions that
may take a logarithm in a few hours on one computer. He expressed
the opinion that the next candidate, namely $2^{521} -1$ is
unassailable. There are some numbers in between that might be
interesting, namely $2^{241} -1$ (although this is not a prime), which
has a prime factor of the size about 2^{216}, and that might be a good
number.

Dr. Johannesson - I would like to make one more point. In our case
we have used trinomials as primitive polynomials. At Sandia Labs
they did investigations [15] which showed that when there are more
coefficients in a polynomial that are not zero, then it is more
tough to get the algorithm to work. I do not know why this is so.
I took once a worst case polynomial and tried it. Then I was
not able to find a Hamming weight of 1 in a reasonable time.
However, I could solve some equations for polynomials with Hamming
weight 2, to create a data base of that Hamming weight.

Professor Massey - Why did you not start at the beginning by
changing bases?

Dr. Johannesson - But how? As far as I know this is a difficult problem.

Professor Massey - We should now let other people talk, especially since we have only five minutes before the break. We have a question from Dr. Gopinath.

Dr. Gopinath - Could you say in more detail why is it better to have a system with a single key. Is this true? You said that the information about the key leaks through the data. On the other hand, it seems like that if you go to large fields (similar to the ones we have been looking at here), then in order to make them effective you need a lot of data.

Dr. Johannesson - If you have a small number of keys, then there is a structure between the keys and the encrypted data. The more you are using the system, the more information about the key will leak through the data.

Professor Massey - Professor Ziv has told us about a case where a key has leaked out, or rather where the deciphering algorithm has leaked out through the public algorithm, given enough uses.

Dr. Heegard - I have heard that a system with a lot of keys has been established in conventional cryptographic systems.

Professor Massey - If I could re-phrase this point. Do you see the PKS as really only PKD for conventional cryptosystems?

Professor Ingemarsson - Just to clarify this point again. Here there are two areas. Either you can use a PKD which was the idea of Diffie and Hellman, or you can use PKE as was suggested by Rivest, Shamir and Adleman. My conclusion is that PKD might work for specific applications, but that PKE for data transmission has serious drawbacks when it comes to practical use. These are two different instances.

Professor Massey - Our next speaker is Mr. Wright, who is a 'pro'.

Mr. Wright - I shall tell you of my experience in actually using cryptography in connection with the project 'Universe' which was discussed already during another panel discusussion [16]. The basic situation is that we are trying to connect local area networks of various sizes throughout the U.K., e.g. Cambridge University, University of Loughborough, Marconi Research Centre, etc., via the OTS satellite. Once you have a satellite in a system you basically broadcast and you worry about the security. A typical connection is Marconi and Logica (in London), two partners in the security experiment. At Marconi we are using hardware 'Data Encryption Standard' (DES) [17] for the basic data encryption, whereas Logica have implemented DES in software. Amazingly we both agree with each other. We are using PKC for the key distribution.

Obviously, since there are only two of us, in this system the key distribution poses no problems. We need more experience in using a key distribution system in earnest. If you have a secret key system, then it is well known that the number of keys you need grows as the square of number of users. Thus, for N users you need N(N-1)/2 keys. One way to reduce this number is to have some trusted key distribution centre (KDC), so that you give only one secret key to each user at the KDC. Then you have problems with authentication and everyone has to be able to trust the centre. That is why we really need to have PKC. There is the Diffie-Hellman method [1] which was already discussed here. We are using RSA PKC to encrypt the key. The system we are looking at is as follows: A wants to contact B. A devises a DES key and PKC encrypts under B's public key. He can also sign with his private key for authentication purposes. Then A makes his connection to B. When A has successfully connected, the first piece of data that is passed is the so-called session key, the DES key encrypted under PKC. Suppose now that we wish to send a 512 bit message encrypted by RSA [4]. We are using RSA scheme since the trapdoor knapsack is not very secure at the moment. To implement this we put it on an 8-bit micro (INTEL 8080 MDS). In fact it took us hours to encrypt or to decrypt the key. One case took five hours, another three hours. That was not very practical. What we had to do was to introduce an extra layer of keys. We call it a master key which we change periodically using PKC. Then the session keys are encrypted under DES using this so-called master key.

This is about all I wish to say. The message is that we still need PKC for key changes and that we have in fact successfully demonstrated the system of key changes and also the DES encryption for a large amount of data via a satellite.

Professor Massey - Thank you Mr. Wright. Do we have questions for the speaker?

Professor Cattermole - One of the points made earlier on was that the more data you send using a particular key, then you weaken the system more and more, and you are likely to be discovered. If you are using a key which by itself is a short sequence of data, do you tend to overcome a risk like that?

Mr. Wright - That is quite true.

Professor Massey - When I discussed with Mr. Wright what he was going to be saying today, the thing that surprised me most is not that it took five hours for encypherment by RSA on an 8-bit microprocessor, but that you could do it at all. This seems quite extraordinary for the size of integers that one is dealing within RSA systems. Can you tell us what size are p and q in your system?

Mr. Wright - I cannot say precisely. Our colleagues at Logica wrote that piece of software.

Professor Massey - The general recommendation is about 100 to 200 decimal digits, primes naturally.

Dr. Heegard - Is your communication link 2-way and in that case why did you not use the Diffie-Hellman algorithm?

Mr. Wright - Yes, it is 2-way, but with the Diffie-Hellman you need to pass messages in two directions to exchange the keys, whereas with the RSA you just encrypt it. One person just dreams up a key and starts a session, then it is encrypted and sent over the satellite link, so you only need one satellite hop.

Dr. Goodman - I think that they have used in Logica a 200 decimal digit prime, a pretty enormous number to multiply on an 8-bit chip.

Professor Golomb - I would like to ask the people who are interested in the RSA algorithm if they have given any thought to the relative merit of doing ordinary arithmetic as the RSA scheme originally proposed, versus doing it in finite field arithmetic. In particular, if you look at, say, multiplying two 100-bit ordinary primes, versus multiplying two 99-degree irreducible polynomials over $GF(2)$, the arithmetic over $GF(2)$ is simply an easier arithmetic to implement. Therefore the question remains: Is there any differential advantage in the cryptographic value of doing it in one way or another. Considering my own method [18] that enables you to separate two factors of various polynomials, if the two factors are irreducible of the same degree and if they each have the same primitivity and the same trace, I am not aware of any wedge you can use to drive between them that makes the problem any easier than searching extensively for irreducible factors of that degree. It seems to me that the problem of factoring in ordinary arithmetic is harder. My question is: Is anybody interested in this?

Dr. Andelman - I have looked at this problem. Using polynomials, it is very simple to cryptoanalyse the RSA scheme. There is a paper by Reed [19] on this problem.

Professor McEliece - If I understand Professor Golomb's question properly, it seems that the problem of factoring polynomials in a field is, because of Berlekamp's algorithm [22], a very simple problem which would render the "finite field RSA" to be very weak.

Professor Massey - We now turn our attention to another speaker, namely Professor Blom from Sweden.

Dr. Blom - I will discuss a problem similar to that presented by Mr. Wright but perhaps with a few different aspects. PKC systems promised to be a solution for handling large networks in secrecy. Anyone can join such a group of users to get secure communication through the algorithm used in such a network. The original idea was like this (see Figure 5). Each user U

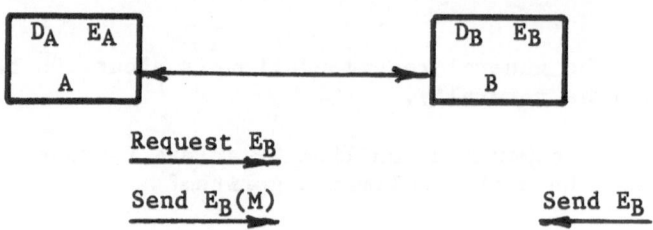

Figure 5.

PKC for Two Users

invents an encryption key E_U and a decryption key D_U and nobody knows the public keys in advance. Here A sends a message to B over the network asking B to send his public key E_B and then B sends E_B back. Then user A uses that encryption key to send the message $E_B(M)$ back to B again. In a network we could of course have a wire tapper between A and B. Let us see how such a threat situation develops (see Figure 6). Now, when A asks B to send his secret encryption key E_B, this message gets intercepted by the wire-tapper W and W sends his public key E_W back to A. He also sends a request to B, to send his encryption key. Thus, when A encrypts the message

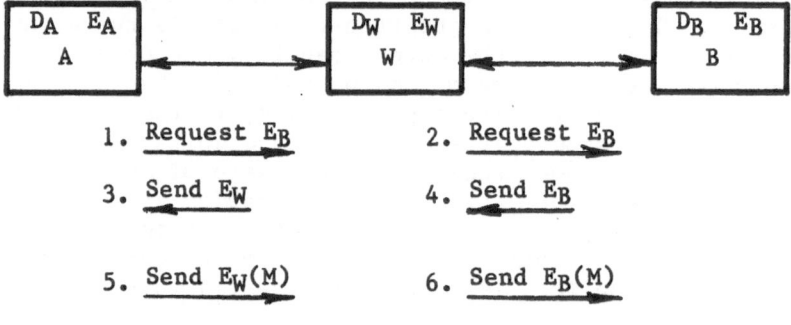

Figure 6.

PKC with Wire Tapper for Two Users

to B he is really using the encryption key of the wire-tapper and A sends the encrypted message $E_W(M)$. The wire-tapper then decrypts the message and re-encrypts it using E_B before he passes it over to B. In this way A and B will not notice that they have a wire-tapper between them. One has to do something to remedy such a situation and to counteract the threat.

A and B could have a common secret key that they could share and use to authenticate without doubt that it is really A that sends a message to B, or that B sends a message to A. They could use such a key in a conventional two-way crypto-system. It need not be a PKC – they could use such a secret key right away.

What happens if there are more users in the network? One should then devise a scheme with a trusted party to guarantee that A gets the right copy of E_B, the encryption key of B. This is usually done in the following way (See Figure 7). We have a server here, denoted S, and he has a secret decryption key D_S and a public encryption key E_S. It is assumed that all the users in the network knows his public encryption key E_S. Now when A wants to send a message to B, he asks B for encryption key E_B. Then B sends back his encryption key E_B, together with the signature of this key, that is the message $D_S(B,E_B)$. Since A knows the encryption key, he can now verify that E_B is indeed the secret key of B. The signature $D_S(B,E_B)$ is thus used to show that only B will be able to decipher this message. This is practical if you can distribute the encryption key of the server to all the users. But you also have to generate the signature of the encryption key of each user. To get the authentic signature, you have to maintain a secure channel from each user up to the server. How do we get such a secure channel? Well, to get a secure channel some secret knowledge has to be shared between two communicating parties. This secret

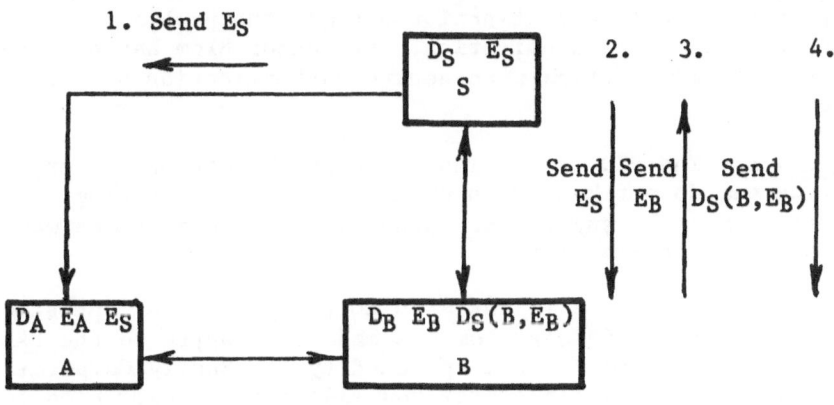

5. Request E_B

6. Send E_B and $D_S(B,E_B)$

7. Send $E_B(M)$

Figure 7.

PKC with Server

information could either be used to authentiate the transmission in a PKC or as a key in a conventional crypto system. Both ways one can obtain a secure channel. But as you always need secret information, why not use it as a key in a conventional crypto

496

system and let the server play the role of a key distribution center. My conclusion then is that in many cases you are better off if you use conventional crypto systems.

Professor Massey - Thank you Professor Blom. I thought that your presentation would be a commercial for PKC. Do we have any questions?

Dr. Heegard - Do you avoid the problem if you publish these keys and send copies to everybody in the network?

Dr. Blom - Still you have to devise a procedure to change the public key. You can, as a user, wish to change your key to another one and then you have to maintain a secure channel to the key distribution centre.

Professor Massey - Then you have to give everybody a copy of your key; you have to display it on a wall and to publish it everywhere. How would you know that this was an authentic signature, and not belonging to someone else? One thing has to be made clear in PKC, namely that there is a distinction between the problem of secrecy and the problem of authentication. Professor Blom has pointed out that it is really difficult to secure authentication by public key methods.

Dr. Goodman - The PKC system implies a prior authentication, otherwise you do not know with whom you are communicating. There does not seem to be any way of authentication in a secure way without secret information.

Professor Ingemarsson - I just wish to point out that the same conclusions that Professor Blom has made also apply to the PKD systems. For example, in the PKD used by the Sperry Corporation for a radio communication system you also have a prior need to know to whom you are talking, and if you have this prior information you share some secret with the person you are talking to. You might as well use that secret as a master key to encrypt a session key, and you might as well use conventional encryption. So the same conclusions are also valid for PKD.

Professor Farrell - I am confused now, for Dr. Harari seemed to be saying that a PKC system does provide a solution to the authentication problem.

Dr. Harari - No, it is not a 100 percent satisfactory solution; it provides merely a path to such a solution.

Professor Massey - I would like to inject a little to this statement that secrecy is required. Really what is required is the integrity of a PKC system. You must be sure that what you get is what was purported to be, and not that no one else is going to see it. To me it is an easier problem to deliver money securely to a bank in a truck, than to deliver it secretly. You do not care whether people watch the truck drivers get out and bring money. For the authentication problem it seems we need that kind of advance secure communication.

Dr. Blom - But you do not have any trucks in computer networks.

Professor Messerschmitt - That is the problem and what you should do is to safeguard your computer networks from confusion. In this case it is the bills, the money itself that is the communication, and it would be a nuisance if you could not count the money that easily.

Professor Massey - That is a good point. It seems that in many situations life suffers from cryptography. It is easier to certify that something is authentic than it is to do something secretly.

Professor Messerschmitt - That is probably true. When people are printing money, you can easily examine it, though you have to do it carefully, for these printers have an expensive enough process to produce those bills, so it is difficult for someone else with limited resources to reproduce that printing process.

Professor Massey - Is that not precisely what PKC does, to make something difficult for others to copy? So that it needs someone very clever to pay for the price of it.

Professor Messerschmitt - That case is a little different in that you are making it expensive to do the encoding, and then you can just examine whether the encoding was done. It is not using higher mathematics, just the very much greater financial resources of the government relative to the counterfeiter.

Professor Ingemarsson - I have found I think a nice way to formulate all these problems. In any secure communication system you will at one time at least need a secure channel. You can use it for prior authentication. Similarly as when a new employee joins a company. He goes there in person, so that his face becomes familiar to everyone else; what he really does is to use a secure channel. The same applies to Dr. Heegard's comment, when he claims that a wide publication of a public key would help. Indeed it would if you regard that as a secure channel.

Professor Massey - Now we shall listen to Professor Ziv, of the Technion in Israel.

Professor Ziv - What I want to do is to search and to speculate on the existence of one-way functions, which are very essential for signatures and for authentication. The idea is that a customer comes with a card bearing his name and a "signature" on the other side of it, which is usually a very long sequence. He has another sequence somewhere in his mind and he feeds this sequence into a

498

small computer. This computer then generates a sequence which
should be the same as the one on the other side of the card. The
idea is to choose a function for this process which is not easily
inverted. In other words, we are trying to make it very difficult
to regenerate his secret memorised signature from the printed
signature which is in the public domain (i.e. on the card). If he
loses his card, nothing happens at all. It is very interesting to
find out whether such functions do exist or do not, and this is one
aspect that I am interested in. Few years ago I had formulated
together with Dr. A. Lempel some convictions, namely that one-way
functions do exist, but now I am not so sure that the results that
we have derived are strong enough to be truly meaningful as you
will see later.

So, coming back to one way
functions, we all sort of feel
that they do exist because we
are well aware of the existence
of one-way devices.

I will give two examples:
One is a diode shunted by a
resistor. If you go one way it
is easy, but it is difficult
for the current to go the other
way. There is another device,
that face in example 2. Just
turn it around and see!

Example 1   Example 2

So there should be such mathematical entities. I would prefer
to confine the discussion to one-to-one functions in a finite field
by limiting myself to $GF(2^n)$. This is both important and
convenient. The first question is, are there 'hard' functions
(forget for the moment the one-way functions) that is, functions
which are hard to compute? The good news is that we know the
answer to this. One way to compute the complexity is to count the
number n of field operations (e.g. $+$, $-$, $\times$) which are needed to
compute such a function. In 'hard' functions the number of
computation steps grows exponentially with n. This is sometimes
called the 'network complexity', and you do not limit yourself in
choosing this rather simple complexity measure when the network
complexity is at least proportional to n^3 or more. The standard
Turing machine complexity, in the sense of time-space complexity,
has then the same complexity. Hence it does not cost you anything
to limit yourself to this particular simple model. Surely
enough we can now establish the happy fact that actually most
functions are hard, so at least you do not have to worry about
that. Most functions are at least exponentially hard [21].

However this is not what we really need. What we need is a
one-to-one function over $GF(2^n)$ which is hard one way and easy in
the other direction. If you want to define this mathematically,
you look for a function F such that:

$$\frac{\text{complexity } [F(.)]}{\text{complexity } [F^{-1}(.)]} \xrightarrow[n \to \infty]{} \infty$$

i.e., this ratio should go to infinity for <u>most</u> inputs $x \in GF(2^n)$,
as $n \to \infty$. Do such functions exist? Yes ! One of them is x^3. You
may not like this, but in one direction its complexity grows as n,
and in the other as n log n. So the definition is good, but in the
hard direction we should like to have the complexity grow at least
exponentially. The very fact that easy functions are so rare,
because most functions are hard, throws some light on the situation
that it is not so easy to find one-way functions. This is the only
comment I wish to make.

<u>Professor Steiglitz</u> - There is a result by G. Brassard, Fortune and
Hopcroft [20] that goes essentially like this: If there is a
one-to-one onto mapping that is polynomial complexity in one
direction, and NP hard in the other, then NP = Co-NP. So either
there are no good one-way functions in this sense, or the unlikely
result that NP = Co-NP is true.

<u>Profesor Ziv</u> - I am aware of this result. This was our starting
point. What I was trying to avoid is exactly that 'either - or'.
Everything is hanging on the fact that something is not quite
clear. You cannot really verify an 'either-or' condition. The
point that I want to stress again is the fact that most functions
are hard, yet it is very difficult to find one-way functions
because easy-functions are very rare.

<u>Professor McEliece</u> - Why have we this result the cube root of x
has the complexity of n log n?

<u>Professor Ziv</u> - It is simple to show that it is easy to compute x^3.
The inverse is slightly more difficult. You can establish it by
considering that you need logs to evaluate roots. There is an
available published report on this subject [21].

<u>Professor Massey</u> - I take it that the good news is relative to the
listener of what was said. For a long time we had only bad news!
It seems that bad news for someone might be good news for another.
When Professor Ziv told me here about his contribution to this
discussion, I was quite excited about this, because here is at
least someone who has made a rigorous attack on this problem. That
is why, though he thought himself to be a 'contra', I put him here
as a 'pro'.

I do not know whether our last speaker is a 'contra' or a
'pro'. He will have to tell you this himself. Let us hear Dr.
Goodman.

<u>Dr. Goodman</u> — I want to say that my position on this matter is categorically and unequivocally on the fence. I want now to go back to my Public Key Broadcast System which I presented already in my lecture here [6]. What I am trying to do is to broadcast a session key out to users and I am going to use a PKC system to do this. So I must therefore be a 'pro' at least so far! My objective is to devise a system such that the sender encrypts the data to produce a single cryptogram which is sent out to many users. Only the intended users can decrypt correctly the cryptogram to get the session key. The diagram in Figure 8 shows how we can go about it by introducing redundancy. Here only the first two bits of the message form the session key, and this ensures that receivers 1 and 2 get the correct key when they decrypt the cryptogram, but that the receiver 3 does not. Let us look now at this in terms of the trapdoor–knapsack. Consider the notation:

x_{ij} = bit i of message to user j,

a_{ij} = i-th knapsack vector of user j (Public Key),

C_j = cryptogram to user j.

We have:

$$C_1 = \sum_{i=1}^{n} a_{i1}\, x_{i1} \quad \text{and} \quad C_2 = \sum_{i=1}^{n} a_{i2}\, x_{i2}$$

What we actually want is to maintain $C_1 = C_2$ for the first p bits of the message. Therefore:

$$C = \sum_{i=1}^{p} a_{i1}\, x_{i1} + \sum_{i=p+1}^{n} a_{i1}\, x_{i1}$$

and also:

$$C = \sum_{i=1}^{p} a_{i2}\, x_{i2} + \sum_{i=p+1}^{n} a_{i2}\, x_{i2}.$$

Hence we obtain:

$$\sum_{i=1}^{2n-p} a_k\, x_k = 0, \quad \text{where}$$

MESSAGE–CRYPTOGRAM MAPPINGS

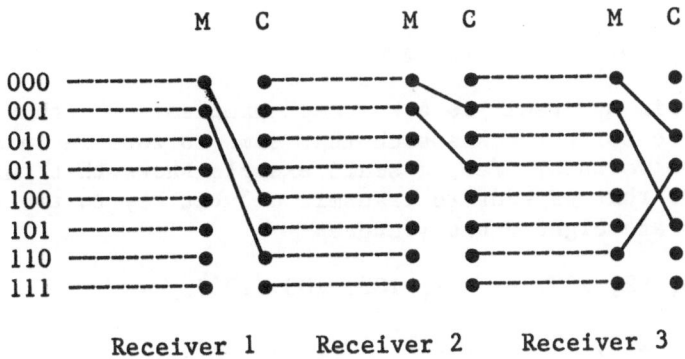

Receiver 1   Receiver 2   Receiver 3

SYSTEM

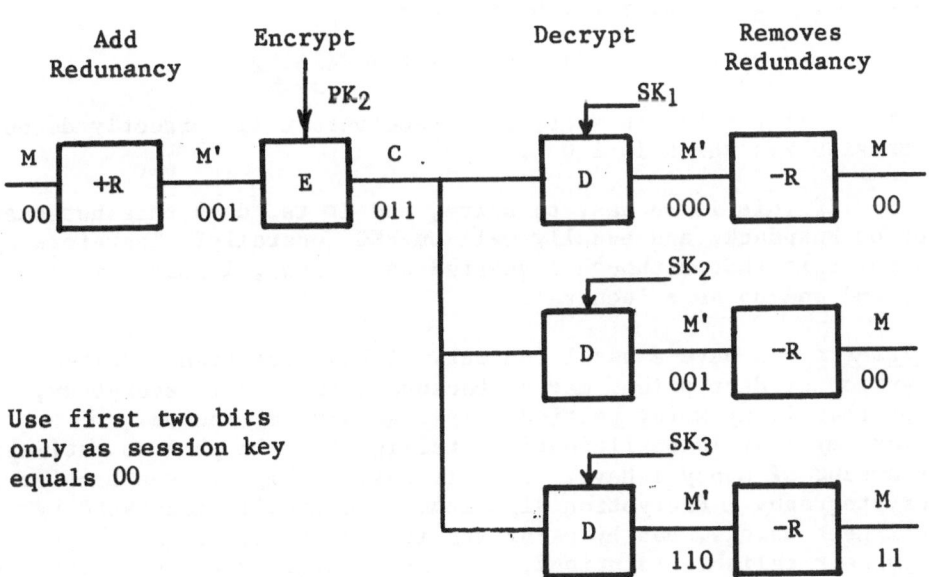

Figure 8. Public Key Broadcast

$x_k = x_{k1}$ for $k = 1$ to n,

$x_k = x_{k-n-p}$ for $k = n+1$ to $2n-p$

$a_k = a_{k1} - a_{k2}$ for $k = 1$ to p

$a_k = a_{k1}$ for $k = p+1$ to n

$a_k = a_{k-n+p}$ for $k = n+1$ to $2n-p$.

However, this by itself is a knapsack problem! In other words, we want to solve the knapsack that sums to zero in this case. Can we solve this? Yes, a small example shows that this can be done. Assume that we want to transmit a 3-bit key to two users. The public keys are eight 8-bit vectors:

$a_{11} = (123, 92, 233, 61, 11, 188, 103, 134)$

$a_{12} = (132, 210, 177, 50, 201, 107, 88, 54)$

To solve this for a 3-bit key, we need to solve the $2n-p = 13$ bit knapsack which sums to zero. This is easily done by some manipulation. For example, if we choose:

$x_{11} = (1\ 0\ 1,\ 1\ 0\ 0\ 1\ 0)$

$x_{12} = (1\ 0\ 1,\ 1\ 0\ 1\ 0\ 1)$

then we have: $C_1 = 123 + 233\ 61 + 103 = 520$

$C_2 = 132 + 177 + 50 + 107 + 54 = 520.$

Thus, if $C = 520$ is sent, both receivers will correctly decode the session key which is 1 0 1.

So, if this is so easy to solve, and it is, does this not cast doubt on knapsack, and equally well on PKC generally? Therefore my main point is that although I started as a 'pro', I must jump the fence and end up as a 'contra'.

Dr. Trzaska - I have a small comment. I have not been involved in encryption or decryption, may be because I am open to everybody, except that is my wife, particularly regarding my income. However, my lady is sufficiently intelligent to be able to estimate what amount of money I have. Here is some analogy to the problem of cryptography. Encryption algorithms are usually generated by intelligent beings, not by using statistics, or by trying to obtain 100 percent reliable solutions, but rather by iterations. Consider for example the case presented by Professor Ingemarsson. He needs 10^5 years to decrypt something. Yet we know that Sherlock Holmes decrypted something while smoking a single pipe! I consider encrypting as an a-posteriori process, for it exists in a human brain, and I suppose it could be solved by telepathy. In the U.S.A. a big effort is made to investigate extra-sensory perception. By telepathy, or by other similar way, a coded message could be decyphered even before it was created.

Professor Massey - The question seems to be how to guard a crypto-system against a telepathic attack! I would like now to open the floor for general discussion. You can question any of the speakers, and the speakers can possibly have a word or two to give second thoughts to propositions they have made earlier.

Dr. Andelman - I want to make a general comment on the strength of cryptosystems. Everybody knows that substitutions and permutations are weak systems, yet you can still build very strong systems out of them. Take for example the DES [14]. This is just a smart concatenation and combination of individually weak systems. I want to extend this concept to PKC. People look at PKC systems as simple tractable problems and suggest that you should use them as they are. As an example, take the knapsack. Everybody makes a big fuss that it was broken. In fact, the iterative knapsack was not broken. So my simplistic suggestion is to use PKC systems as building blocks, to build more powerful systems by combining them. Look at DES + PKC. This is not necessarily a right mixture, but you can use the first one to encrypt the key and then use the other system which I claim might be stronger.

Professor Massey - Are you suggesting that by iterating simple PKC systems one might build strong cryptosystems!

Dr. Andelman - I am sure you can do it, to design really strong systems.

Professor Bhargava - Mr. Chairman, I wonder whether you can tell us what is your position here? Are you a 'pro' or a 'contra'?

Professor Massey - I am not sitting on a fence. To describe my position, I would like to quote a black joke which appeared in the 'Newsweek' magazine at the time of the Vietnam war. An army general said: "It's not much of a war, but it's the only war we have!". I feel the same way about the PKC systems. What is the alternative? The only alternative is the classical cryptography and its history has been very non-glorious. The successes have been on the side of crypto-analysts, and not on the side of cryptosystem designers. Dr. Andelman has said that the DES system is secure. I do not know that it is secure. Are people, and the same people, spending the same amount of time to break DES, as they are spending on breaking RSA, or knapsack? Maybe someone else can say something about this? In the case of most classical systems of which I am aware, they depend for their security on iterations of several operations which appear to be very confusing but are not probably secure. There is no alternative, as I see, to PKC. We simply must use PKC. However, we are beginning, as Professor Ziv has said, to get some mathematical results to show that one might be able to prove that some PKC ystems will be secure. This was a good start, to show some asymmetry in functions.

<u>Profesor McEliece</u> - Just to follow what Dr. Goodman has said, for his problem was very interesting. We have heard a lot about digital signatures and digital keys. At the California Institute of Technology we have a sort of master key system which consists of a hierarchy of keys. The president has a key which opens every door, and a new graduate student has a key which opens one or two doors only. I think in a broadcast or in a network situation it is a very interesting combinatorial mathematical problem to find an analogue of this 'master key' system. If you have a message to which everyone is entitled to listen, there need to be no protection. Then you may have some other material which may need a lot of protection. You can imagine some sort of a tree with secret sharing levels. That is a problem one needs to think about.

<u>Professor Wolf</u> - I have <u>an</u> answer to this question, but it is not <u>the</u> answer. Suppose we have eight users as shown in Figure 9, numbered from 1 to 8. What you do is to layer them into sets starting from the bottom. Two possibilities occur to you immediately. The first is to give each user 1,2,....,8 a different key. Then to transmit to the set of all users requires eight transmissions.

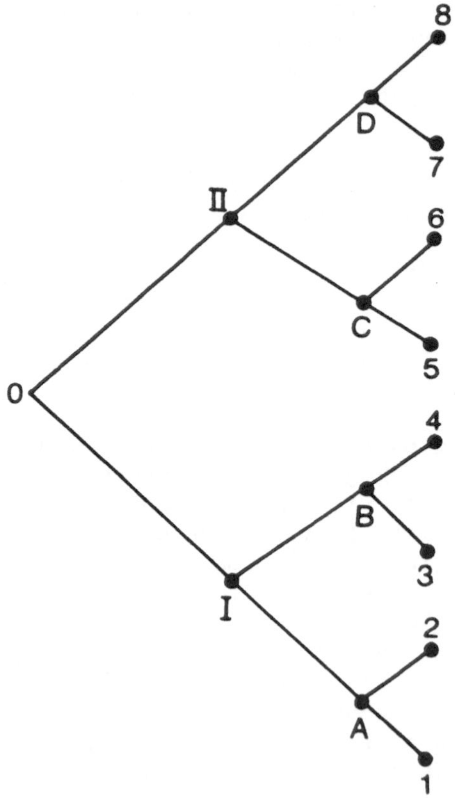

Another approach is to have a key for each set. Now a given user needs to store many keys but no more than one transmission is required to transmit to any set.

Somewhere in between these two possibilities is the following.

You assign a key to nodes 1, A, 3, I, 5, C, 7 and II. Then no user need store more than three keys and no more than three transmissions are required to transmit to any set. So there are several approaches which need to be studied.

Figure 9

Dr. Goodman – This is a problem which I tried to point out in my lecture [6]. If you can use somehow the structure of the network to distribute the keys, you can possibly arrange that distribution without having the number of keys increasing dramatically. This number is going to increase with the complexity of a network, but it need not increase so rapidly. In the particular case of a ring, when such an economy attempt was made, it was difficult to make the ring secure because there could always be some people who will cheat in this. In some cases you need not send the broadcast key in multiple packets, but you could just whizz round the ring. There are ring systems that you can buy in the U.K. for local networking where Professor McEliece's idea would be quite useful, and that is the problem they would need to solve.

Professor Ingemarsson – There is also such a problem in PKD systems. Consider for instance a military conversation when say station A lifts his receiver and asks for station B. Then a random number is generated by station A and this is sent to station B. When B lifts his receiver, he sends his random number to A. Both stations have a common α [see Equation (1)] of the PKC system of theirs. They raise their random numbers to powers x and y respectively (naturally modulo p everything) and what they obtain then is α^{xy} mod p. Then they have a common key which is used for encryption. I can imagine an operational situation when A calls B and initially they talk freely, that is openly. They both authenticate each other, knowing each other's voice, or by some administrative procedure. From then on they both turn to a secret communciation.

Professor Massey – I know of a military system (at least back in the dark ages) when they had a problem with a voice scrambling system. The commanders never liked to use it because they could not recognise each other's voice, so they did not trust that they really had the right speaker at the other end, even though they could have been reassured by conventional cryptography that they had the right speaker. Professor Ingermarsson's idea is an interesting case for the use of PKD systemns, which might permit one to utilise a human kind of authentication in this awesome voice scrambling system.

Professor Ingemarsson – In the particular case mentioned by me the encryption is used on a digitised speech – they have digital telephony, and not a voice scrambling system. Such scrambling systems should be avoided by all means.

Professor Massey – I disagree with you. In Switzerland we have a voice scrambling system which appears to be relatively good.

Professor Messerschmitt – I should mention in this discussion that optical disc storage technology is coming along well to the point that we can store there an incredible number of bits in a permanent storage mechanism. I wonder if this does not suggest that the one-time pad is now a much more practical cryptography system.

<u>Professor Massey</u> - Indeed, it is interesting, for in a binary system you could simply take a sequence, add it modulo 2 to your plain text, and as long as that sequence also exists at the receiving end, it can be subtracted modulo 2, and you can show that this would result in an economic and perfectly secure crypto-system (the one-time pad).

<u>Dr. Harari</u> - There is one thing about sequences that have been used in cryptography, that they are long and complex. If we can discover an algorithm that can generate a primitive polynomial of high degree with few coefficients, it would be interesting if we can implement this on a shift register. This would have an advantage over all the DES and RSA of little computer requirements for a crypto-system. People tend to forget that in all the modern cryptographic systems you need a lot of computing for a little bit of crypto. A system, which can be proved to be safe and which would have good acceptance levels with fewer computations per bit, would indeed be very welcome.

<u>Professor Massey</u> - I would like to tell you a little story. I recently visited a Swiss company which produces cryptographic equipment. They told me that it is not so important whether a system is secure, as it is whether their customers believe it to be secure. It is rather difficult to sell cryptographic equipment to someone who is suspicious of that equipment. They regard the present debate on possible security or lack of security of PKC systems as one of the possible reasons for dooming the future of PKC systems in a commercial market. People simply are too unsure about them, or have no reason to be more secure than with conventional systems. If they <u>believe</u> that a system if secure, you can sell it to them.

<u>Professor Golomb</u> - Relative to the comment which Professor Messerschmitt has made on the availability of a new method of storage of data, this could help in the key distribution problem and thus ease up the pressure on the efficiency of PKC systems. You could argue at least that one can distribute more keys each time such a distribution has to be made anew.

<u>Professor Messerschmitt</u> - My point was that if you only have to distribute one-time pads perhaps once in ten years, the physical distribution problems for the more secure one-time pad system may be no more difficult than distributing keys.

<u>Professor Massey</u> - You might have a problem in generating continuously those random numbers.

<u>Professor Messerschmitt</u> - I would just put a resistor and a Gaussian noise generator with a threshold device to do it. Then I would store the lot on two discs, one for the sender and one for the receiver.

<u>Professor Wolf</u> – I have heard that people who actually generate random sequences do not like to do this by using physical mechanisms, but rather they would prefer to use mathematical techniques, even though there is no good mathematics to do this.

<u>Professor Cover</u> – What possible mathematical mechanism can you have that has a sufficient number of degrees of freedom to generate such random data streams?

<u>Professor Wolf</u> – I have no answer to that, but I do know that these people do not like for instance to use a gas tube to do it, since they say that such a device, or a mechanical device could have a bias.

<u>Dr. Heegard</u> – Maybe they could use π! Then they could expand this to many places and use these digits.

<u>Professor Cover</u> – You certainly would not use π! How does Professor Ingemarsson generate his random sequences?

<u>Professor Ingemarsson</u> – I am not sure! A user might pick on his birthdate, although this would not please the military people, for they are very careful not to use human beings for this, for these are very non-random.

<u>Professor Steiglitz</u> – Concerning the problem of generating random numbers from physical devices, one problem you run into is that you would like to get numbers that are esentially independent. In order to do so, when you use finite bandwidth devices, you have to wait a certain amount of time. You must be sure to separate these signals both in time and in space! If you have to fill up a video-disc, this might take a long time indeed. My second comment follows Dr. Trzaska on defeating telepathy. I think you can turn that around and say that perhaps encryption systems are a good tool to use to investigate the subject of extra-sensory perception. You can simply ask someone who is clairvoyant to decrypt a message!

<u>Professor Massey</u> – On this we have to finish this discussion, by thanking all the speakers (Applause.)

References

[1] W. Diffie and M.E. Hellman, "New Directions in Cryptography",
     IEEE Trans. on Inf. Theory, IT-22, No. 6, November 1976,
     pp 644-654.

[2] A. Shamir, "A Polynomial Time Algorithm for Breaking
     Merkle-Hellman Cryptosystems", Internal Report (1983), Applied
     Mathematics, The Weizmann Institute, Rehovat, Israel.

[3] R.C. Merkle and M.E. Hellman, "Hiding Information and
     Signatures in Trapdoor Knapsacks", IEEE Trans. on Inf. Theory.
     IT-24, No. 5, September 1978, pp 525-530.

[4] R.L. Rivest, A. Shamir and L. Adleman, "A Method for
     Obtaining Digital Signatures and Public-Key Cryptosystems",
     ACM Comm. 21, No. 2, February 1978, pp 120-126.

[5] S. Harari, "The Seal method in Transmission and Storage" (in
     this volume).

[6] R. Goodman, "Processing Techniques in Public Key
     Cryptosystems" (in this volume).

[7] R.J. McEliece, "A Public Key System based on Algebraic Coding
     Theory", Jet Propulsion Laboratory, Pasadena, DSN Progress
     Report, 1978.

[8] C.P. Schnorr (Fachbereich Mathematik, Universitaet Frankfurt)
     and H.W. Lenstra (Mathematich Instituut, Universiteit
     Amsterdam), "A Monte Carlo Factoring Algorithm with Finite
     Storage", Internal report, May 1982 (revised in September
     1982).

[9] M.O. Rabin, "Digitalised Signatures and Public-Key Functions
     as Intractable as Factorisation", MIT, Laboratory for Computer
     Science, Report MIT/LCS/TR-212, January 1979.

[10] A.M. Odlyzko, "Cryptoanalytic Attacks on the Multiplicative
     Knapsack Cryptosystem and on Shamir's Fast Signature System",
     IEEE International Symposium on Information Theory, St.
     Jovite, Canada, September 26th -30th, 1983.

[11] A. Lempel, "Cryptology in Transition", Computing Surveys, Vol.
     11, No. 4, pp 285-303, December 1979.

[12] T. Herlestam and R. Johannesson, "On Computing Logarithms over
     GF(2P)". BIT, Bind 21, Hefte nr. 3, 1981, pp. 326-334.

[13] S. Berkovits, J. Kowalchuk and B. Schanning, "Implementing
     Public Key Scheme", IEEE Commun. Mag., 17, May 1979, pp.2-3.

[14] S.C. Pohlig and M.E. Hellman, "An Improved Algorithm for
Computing Logarithms over GF(p) and its Cryptographic
Significance", IEEE Trans. on Inf. Theory, IT-24, January
1978, pp 106-110.

[15] R.F. Brickell, and J.H. Moore, "Some Remarks on the
Herlestam-Johannesson algorithm for Computing Logarithms over
GF(2P)". Advances in Cryprography: Proceedings of Crypto '82.
Plenum Publishing Corporation.

[16] Panel Discussion on "Layered Protocols and Local Area
Networks" (in this volume).

[17] National Bureau of Standards. "Data Encryption Standard", NBS
FIPS Pub. 46, January 15th, 1977.

[18] S.W. Golomb, Shift Register Sequences, Holden-Day Inc., San
Francisco 1957. Revised edition: Aegean Park Press, Laguna
Hills, CA., 1982.

[19] D.W. Kravitz and I.S. Reed, "Extension of RSA
Crypto-Structure: A Galois Approach", Electronics Letters,
18th march 1982, Vol. 13, No. 6, pp 255-256.

[20] G. Brassard, S. Fortune and J. Hopcroft, "A Note on
Cryptography and NP CoNP-P", Technical Report TR-338,
Department of Computer Science, Cornell University, Ithaca,
N.Y., 1978.

See also:
G. Brassard, "A Note on the Complexity of Cryptography", IEEE
Transactions on Information Theory, Vol. IT-25, March 1979,
pp 232-233.

[21] A. Lempel and J. Ziv, "On the Complexity of Functions and the
Existence of On-Way Functions in Finite Fields",
Technion-I.I.T., Haifa, Israel (1980).

See also:
A. Lempel, G. Seroussi and J. Ziv, "On the Power of
Straight-Line Computations in Finite Fields", IEEE
Transactions on Information Theory, Vol. IT-28, November 1982,
pp.875-879.

[22] E.R. Berlekamp, Algebraic Coding Theory, New York,
McGraw-Hill, 1966 (see Chapter 6).

SEMANTIC-SYNTACTIC APPROACH TO PATTERN ANALYSIS AND IMAGE PROCESSING[†]

K. S. Fu

School of Electrical Engineering
Purdue University
W. Lafayette, Indiana 47907

ABSTRACT

Many mathematical methods have been proposed for solving pattern analysis and image processing problems. They can be grouped into two major approaches, decision-theoretic or statistcal .approach and structural or syntactic approach. From the point of view of pattern representation and description, we can discuss pattern analysis and image processing in terms of a unified semantic-syntactic approach. This paper will describe the basic formulation of this approach and discuss a major result in syntax-semantics tradeoff. Recognition methods based on this proposed approach and their implementation will also be discussed.

1. INTRODUCTION

Many mathematical methods have been proposed for solving pattern analysis and image processing problems [1]. They can be grouped into two major approaches, decision-theoretic or statistical approach and structural or syntactic approach [1-6]. From the point of view of pattern representation or description, we can discuss pattern recognition in terms of single-entity representation versus multiple-entity representation, and suggest a combined semantic-syntactic approach on the basis of using attributed languages.

Consider a m-class pattern analysis problem. When we consider each pattern as a single entity we can use a set of n characteristic measure-

[†]This work was supported by the NSF Grant ECS 81-19886.

ments (features) to represent each pattern under study. In such a case, each pattern is represented by a n-dimensional feature vector and the classification of patterns can be accomplished by applying various techniques in discriminant analysis and statistical decision theory. Such an approach is often called decision-theoretic or statistical approach [1-3]. However, when the patterns under study are very complex or when the number of pattern classes m is very large (for example, in fingerprint identification or scene analysis problem) the number of features n required for analysis could also become very large. Consequently, the classical decision-theoretic approach often becomes ineffective or computationally infeasible in solving this kind of problems. One way to approach this kind of problems is to represent a complex pattern by its simpler subpatterns and hope that we can treat each simpler subpattern as a single entity and use decision-theoretic methods [4,5]. Of course, the relations among subpatterns must be taken into consideration. On the other hand, if each subpattern is again very complex, we may have to represent each subpattern by even simpler subpatterns, until we are sure that the simplest subpatterns, called "pattern primitives," can be easily treated by simple decision-theoretic methods. By doing so, the feature extraction and selection problem (for pattern primitives) will also be much simplified. Graphically, we can express such an approach in terms of a hierarchical tree structure as shown in Fig. 1.

At least two problems occur when we employ this approach for pattern representation. The first problem is the selection of subpatterns (and primitives). We may want to decompose each pattern into subpatterns and primitives based on some prespecified simple relations among subpatterns, for example, sampling of a waveform [7-9] or regular decomposition of an image pattern [10]. On the other hand, the patterns under study may naturally contain subpatterns which can be extracted using a

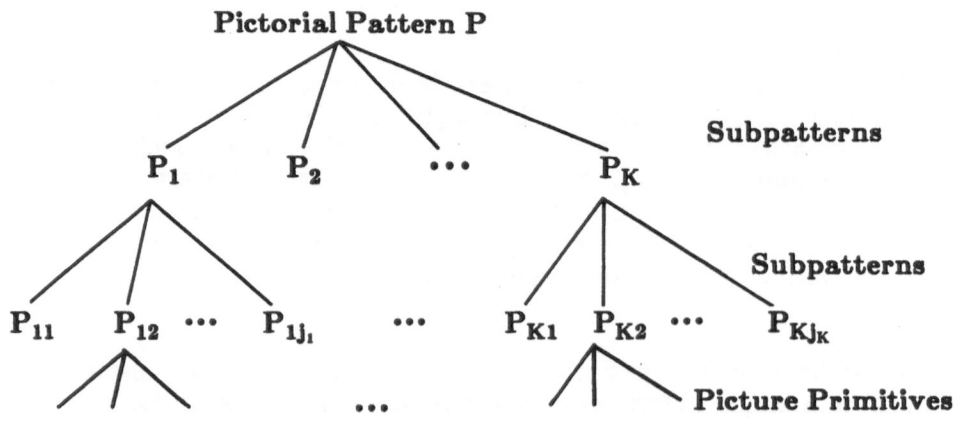

Fig. 1 Hierarchical Structural Representation of a Pattern.

waveform or image segmentation technique [11]. In the latter case, each subpattern may be easily interpretable and physically meaningful [12-13]. Once a decomposition or segmentation process is carried out, the pattern primitives are extracted and treated as single entities. Different kinds of primitive may require different sets of features for characterization [7, 12]. Primitives of the same kind can certainly be analyzed in terms of the same set of features [8,9,15].

The second problem is the additional analysis required when we use the hierarchical structure representation. If we are limited to use only one-dimensional string representation of patterns, results from formal language theory can be directly applied [16]. The problem here is how to handle the situation of context-sensitive languages. For the description of high dimensional patterns, trees and relational graphs are often more effective [5]. Efficient procedures for the analysis of tree and graph structures are thus required. Recent studies have shown that a powerful representation of patterns can be achieved by using attributes (semantic information) as well as structural (syntactic) information [17, 18]. In this paper, a brief review of these results is given. Problems for further investigation along this direction are also discussed.

2. BASIC FORMULATION

In the decision-theoretic approach, a pattern is represented by a feature (or attribute) vector

$$\underline{X} = [x_1, x_2, ..., x_n]$$

For a pattern S consisting of two subpatterns X and Y with relation R between them, we can write

$$S \to XRY \tag{1}$$

Generally speaking, X, Y and relation R can be described by their feature or attribute vectors, that is,

$$A(X) = \underline{X} = [x_1, x_2, ..., x_n]$$

$$A(Y) = \underline{Y} = [y_1, y_2, ..., y_\ell]$$

and $\quad A(R) = [r_1, r_2, ..., r_k]$

where $A(X)$ denotes the attribute vector of X. From (1), the attribute vector of S can be expressed as a function of the attributes of X, Y and R

$$A(S) = \phi(A(X), A(Y), A(R)) \tag{2}$$

For a special case, when R is a "left-right concatenation" relation, we can write

$$S \to XY \tag{3}$$

and $\quad A(S) = \psi(A(X), A(Y))$

Example 1: Let X be a straight line segment with attributes ℓ_X and θ_X where ℓ_X is the length of X and θ_X is the orientation (angle with respect to the horizontal axis). Similarly, we can use the attributes ℓ_Y and θ_Y to describe straight line segment Y. After we connect Y to X to form a pattern S, the length and the orientation of S can be easily calculated by

$$\ell_S = \ell_X + \ell_Y \tag{5}$$

$$\theta_S = \tan^{-1} \left[\frac{\ell_X \sin \theta_X + \ell_Y \sin \theta_Y}{\ell_X \cos \theta_X + \ell_Y \cos \theta_Y} \right] \tag{6}$$

where $A(X) = (\ell_X, \theta_X)$, $A_Y = (\ell_Y, \theta_Y)$ and $A(S) = (\ell_s, \theta_s)$.

Example 2: Attributed Shape Grammar [19]

Consider that a line pattern can be decomposed into a sequence of primitives, each is a curve segment. Let the attribute vector of curve segment C be

$$A(C) = (\vec{C}, L, \Phi, Z)$$

where $\vec{C}$ is the vector length, L the total length, Φ the total angle change, and Z a symmetry measure of C. Let the attribute of the concatenation relation CAT between two adjacent curve segments be the angle "a" between them. Thus, a pattern N consisting of two curve segments C_1 and C_2 concatenated through an angle a can be expressed as

$$N \rightarrow C_1 \text{ CAT } C_2 \tag{7}$$

and

$$A(N) = A(C_1) \oplus_a A(C_2) \tag{8}$$

where $A(C_1) \oplus_a A(C_2)$ denotes the following computation of attributes:

$$\vec{C}_N = \vec{C}_1 + \vec{C}_2 \tag{9}$$

$$L_N = L_1 + L_2 \tag{10}$$

$$\Phi_N = \Phi_1 + a + \Phi_2 \tag{11}$$

and

$$Z_N = Z_1 + Z_2 + \frac{1}{2} [(\Phi_1 + a)L_2 - (\Phi_2 + a)L_1] \tag{12}$$

Similarly, if a pattern S consists of two subpatterns N_1 and N_2, then

$$S \rightarrow N_1 \text{ CAT } N_2$$

and

$$A(S) = A(N_1) \oplus_a A(N_2)$$

The operation $\oplus_a$ is associative. That is, if

$$S \rightarrow N_1 \text{ CAT}_1 N_2 \text{ CAT}_2 N_3$$

then

$$A(S) = A(N_1) \oplus_{a_1} A(N_2) \oplus_{a_2} A(N_3)$$

$$= [A(N_1) \oplus_{a_1} A(N_2)] \oplus_{a_2} A(N_3) \tag{13}$$

$$= A(N_1) \oplus_{a_1} [A(N_2) \oplus_{a_2} A(N_3)]$$

It is not difficult to see that such a representation is quite flexible and powerful in describing object contours for shape analysis. With the introduction of attributes, only one kind of primitive (curve segment) is required.

Example 3: The PDL (Picture Description Language) proposed by Shaw [20] can be extended to include attributes as follows:

Let the concatenation relations of +, -, x and * in PDL be represented by CAT $(+,\phi)$, CAT $(-,\phi)$, CAT (x,ϕ) and CAT $(*,\phi)$ respectively. In addition to the coordinates of head and tail we can introduce other attributes to the primitives such as shape and texture features. For example, the following primitive X can be described by the shape of the primitive as well as the locations of its head and tail.

Thus the subpath N

(1) $N \to X+Y$ represents
$\qquad$ where $A(N) = \psi_+(A(X), A(Y), \phi)$

(2) $N \to X-Y$
$\qquad$ where $A(N) = \psi_-(A(X), A(Y), \phi)$

(3) $N \to XxY$
$\qquad A(N) = \psi_x(A(X), A(Y), \phi)$

(4) $N \to X*Y$
$\qquad A(N) = \psi_*(A(X), A(Y), \phi)$
$\qquad$ where $X \to X_1+X_2$

In terms of the hierarchical structure representation shown in Fig. 1, we can describe a pattern structure by a set of structural composition or syntax rules [5]. In general, the addition of attributes to a set of structural rules makes the pattern description more effective and flexible. Such an approach of using syntactic (structural) as well as semantic (attribute) information of patterns can be formulated in terms of attributed grammars [21, 22].

An attributed (string) grammar is a four-tuple
$\qquad G = (V_N, V_T, P, S)$

where V_N is a set of nonterminals, V_T a set of terminals, and $S \epsilon V_N$ the start symbol. For each $X \epsilon (V_N \cup V_T)$, there exists a finite set of attributes $A(X)$, and attribute x_i of $A(X)$ has a set, either finite or infinite, of possible values D_{x_i}. P is a set of production rules each of which consists of two parts, a syntactic or structural rule and a semantic or attribute rule. The syntactic rule is of context-free form

$$N \rightarrow \alpha \ , \quad N \ \epsilon \ V_N \ \text{and} \ \alpha \ \epsilon (V_N \cup V_T)^+$$

Let $\alpha = C_1 C_2 \cdots C_k$, $C_i \ \epsilon \ (V_N \cup V_T)$ for $1 \leq i \leq k$. The semantic rule is a mapping

$$f: D_{C_1} \times D_{C_2} \times \cdots \times D_{C_k} \rightarrow D_N \tag{14}$$

or we can write that

$$A(N) = f(A(C_1), A(C_2),...,A(C_k)) \tag{15}$$

where $A(N)$, $A(C_1)$,...,$A(C_k)$ are the attribute sets or attribute vectors of N, $C_1, \ldots ,$ and C_k respectively.

The mapping f can be a closed-form function, that is, the attributes of N can be expressed functionally in terms of the attributes of $C_1, C_2, \ldots ,$ and C_k. f can also be a computation algorithm which takes the attribute values of $C_1, C_2, \ldots ,$ and C_k and any other available information as input and the attribute values of N as output. We can also include the case that (15) represents a logical predicate or functional constraint to indicate the applicability of a certain syntactic rule.

The above definition of attributed grammar follows Knuth's formalism closely [21]; two kinds of attributes are included in the semantic rules, inherited attributes and synthesized attributes. The former are those aspects of meaning (attributes) coming from the context of a phrase (sub-pattern) in a string (pattern), whereas the latter are those aspects of meaning which are built up from the basic vocabulary (primitives) within the phrase. In syntactic pattern analysis, if a top-down parsing is employed to analyze pattern structures, inherited attributes are more convenient to use because they can be computed in a top-down fashion, starting from the start symbol S of the grammar. On the other hand, if a bottom-up parsing is preferred, then synthesized attributes should be used, which are computed in a bottom-up fashion.

It is noted that by deleting all semantic rules, an attributed grammar is reduced to a context-free grammar. However, if each pattern X is treated as a single entity or as a primitive and not decomposed into sub-patterns, X will be characterized only by its attribute vector $A(X) = \underline{X} = [x_1, x_2,...,x_n]$ and there is no need of structural or syntactic rules. The conventional m-class pattern classification problem using decision-theoretic approach can be expressed as follows:

Syntactic rule	*Semantic rule*
$S_1 \rightarrow X$	$D_1(X) = \underset{1 \leq i \leq m}{\text{Max}} \{D_i(X)\}$

$$S_2 \rightarrow X \qquad\qquad D_2(X) = \underset{1 \leq i \leq m}{\text{Max}} \{D_i(X)\}$$

$$\vdots$$

$$S_m \rightarrow X \qquad\qquad D_m(X) = \underset{1 \leq i \leq m}{\text{Max}} \{D_i(X)\}$$

where $D_i(X)$ is the discriminant function for class i. In this expression, each semantic rule is used as a semantic constraint to indicate the applicability of the corresponding syntactic rule. That is, when $D_j(X) = \text{Max} \{D_i(X)\}$, $1 \leq i \leq m$, the syntactic rule $S_j \rightarrow X$ is applied, which means that X is from class j.

3. SYNTAX-SEMANTICS TRADE-OFF

In addition to provide a more precise and flexible description of patterns, the use of attributes can also reflect the trade-off between syntactic and semantic complexities. That is, semantic (attribute) information can also be used to compensate the low syntactic complexity in pattern description. The following examples illustrate this idea.

Example 4: It is known that $L = \{a^n b^n c^n \mid n=1,2,....,\}$, which describes a set of triangles with different sizes, is a context-sensitive language $L(G_1)$ with production rules

$$
\begin{aligned}
G_1: \quad & S \rightarrow aSBA \qquad bB \rightarrow bb \\
& S \rightarrow aBA \qquad bA \rightarrow bc \\
& AB \rightarrow BA \qquad cA \rightarrow cc \\
& aB \rightarrow ab
\end{aligned}
$$

We can introduce the following attributes to the primitives and relation:

i) the length of primitive -- ℓ_a, ℓ_b, ℓ_c
ii) the angle of the left-right concatenation -- θ

Let CAT(a,b) represent the relation "concatenation" between a and b. Using PDL notations, we can represent the left-right cancatention between a and b with angle θ as $CAT(a,b) = (+, \theta)$. An attributed context-free grammar G_2 can be constructed to generate L.

G_2:	Syntactic rule	Semantic rule
(1)	$S \rightarrow ABC$	$CAT(S,A) = CAT(c,a) = (+, 120°)$
		$\ell_S = \ell_A + \ell_B + \ell_C$
(2)	$A \rightarrow aA$	$CAT(A,B) = CAT(a,b) = (+, 120°)$
		$\ell_A = \ell_a + \ell_A$

(3)	$B \rightarrow bB$	$CAT(B,C) = CAT(b,c) = (+, 120°)$ $\ell_B = \ell_b + \ell_B$
(4)	$C \rightarrow cC$	$CAT(C,A) = CAT(c,a) = (+, 120°)$ $\ell_C = \ell_c + \ell_C$
(5)	$A \rightarrow a$	$CAT(a,b) = (+, 120°)$ $\ell_A = \ell_a = 1$
(6)	$B \rightarrow b$	$CAT(b,c) = (+, 120°)$ $\ell_B = \ell_b = 1$
(7)	$C \rightarrow c$	$\ell_C = \ell_c = 1$

It is noticed that in G_2 the attribute rule associated with each syntactic rule not only calculates the length of each side of the triangles generated but also imposes the constraint about the next production rule to be used in the generation. For example, after the production rule (1) is applied $CAT(S,A)$ indicates that S must be concatenated to A. This means that the next production rule applied must be (2) or (5). Similarly, after (2) is applied, the next production rule selected must be (3), etc. Replacing the attribute or semantic rules in G_2 by the constraints of selecting the next production rule, we obtain precisely the following context-free programmed grammar G_2'.

G_2':	Core Production	Success field	Failure field
(1)	$S \rightarrow ABC$	$\{2,5\}$	ϕ
(2)	$A \rightarrow aB$	$\{3\}$	ϕ
(3)	$B \rightarrow bB$	$\{4\}$	ϕ
(4)	$C \rightarrow cC$	$\{2,5\}$	ϕ
(5)	$A \rightarrow a$	$\{6\}$	ϕ
(6)	$B \rightarrow b$	$\{7\}$	ϕ
(7)	$C \rightarrow c$	ϕ	ϕ

It is noted that G_2' can be considered as a context-free (core) grammar with a control diagram [23] shown in Fig. 2. Such a control diagram can in turn be interpreted as the semantic constraint for the application of syntactic rules. It is also possible to construct an attributed finite-state grammar G_3 to generate L.

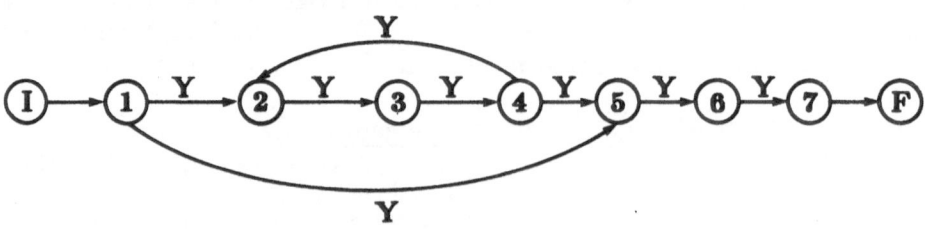

Fig. 2 Control Diagram of G_2'.

G_3:

	Syntactic rule	Semantic rule
(1)	$S \rightarrow aA$	$\overline{CAT(a,b)} = (+, 120°)$
		$\ell_S = \ell_a + \ell_A$
(2)	$A \rightarrow bB$	$CAT(b,c) = (+, 120°)$
		$\ell_A = \ell_b + \ell_B$
(3)	$B \rightarrow c$	$CAT(c,a) = (+, 120°)$
		$\ell_B = \ell_c$
		$\ell_a = \ell_b = \ell_c = n, \; n = 1, 2, \ldots$

Example 5: Chromosome Classification

1) Conventional syntactic approach - with the following primitives as terminals [24]:

$$V_{TM} = \{\underset{a}{\cap}, \underset{b}{|}, \underset{c}{\cup}, \underset{d}{\}}\}$$

three kinds of chromosomes--median, submedian, and acrocentric--are segmented accordingly, and are shown in Fig. 3 together with their string representations.

The nonattributed grammars to characterize the three kinds of chromosome patterns could be as follows [5]:

$G_{median} = (V_{NM}, V_{TM}, P_M, S)$

$V_{NM} = \{S, A, B, D, H, J, E, F\}, V_{TM} =$ as above

$P_M: S \rightarrow AA, \; D \rightarrow FDE, \; H \rightarrow a$

$\quad A \rightarrow cB, \; D \rightarrow d, \; J \rightarrow a$

$\quad B \rightarrow FBE, \; F \rightarrow b, \; E \rightarrow b$

$\quad B \rightarrow HDJ$

$G_{submedian} = (V_{NS}, V_{TS}, P_S, S), V_{TS} = V_{TM}$

$V_{NS} = \{S, A, B, D, H, J, E, F, W, G, R, L, M, N\}$

$P_S: S \rightarrow AA, \; D \rightarrow FDE, \; G \rightarrow FG, \; L \rightarrow HNJ$

$\quad A \rightarrow cM, \; D \rightarrow FG, \; W \rightarrow WE, \; R \rightarrow HNJ$

$\quad B \rightarrow FBE, \; D \rightarrow WE, \; F \rightarrow b, \; G \rightarrow d$

$\quad B \rightarrow FL, \; L \rightarrow FL, \; E \rightarrow b, \; W \rightarrow d$

$\quad B \rightarrow RE, \; R \rightarrow RE, \; H \rightarrow a, \; N \rightarrow FDE$

$\quad M \rightarrow FBE, \; J \rightarrow a$

$G_{acrocentric} = (V_{NA}, V_{TA}, P_A, S), V_{TA} = V_{TM}$

$V_{NA} = \{A, B, D, H, J, E, F, L, R, W, G\}$

$P_A: S \rightarrow AA, \; D \rightarrow FG, \; G \rightarrow FG, \; R \rightarrow HDJ$

$\quad A \rightarrow cB, \; D \rightarrow WE, \; W \rightarrow WE, \; G \rightarrow d$

$\quad B \rightarrow FL, \; L \rightarrow FL, \; L \rightarrow HDJ, \; W \rightarrow d$

$\quad B \rightarrow RE, \; R \rightarrow RE, \; H \rightarrow a, \; E \rightarrow b$

$\quad J \rightarrow a, \; F \rightarrow b.$

A close observation of the three kinds of chromosome shows that they can be easily classified by measuring the lengths of their arm pairs. The length of an arm pair is represented by the average of the lengths of the two arms on either side of a chromosome. If the length of the left arm pair is approximately equal to that of the right arm pair, then the pattern is a median chromosome. If they differ significantly, then it is a submedian chromosome, or if the length of one arm pair is almost zero, then it is an acrocentric chromosome. However, since conventional grammars such as those shown above cannot incorporate attributes (numerical data) within their production rules, the classification consequently can only rely on symbolic syntax analysis which, when made to take care of numerical information contained in the input patterns, is usually not very effective and efficient.

2) Attributed grammar with synthesized attributes for chromosomes [17]. We now show that an attributed grammar with synthesized attributes can be used for chromosome description and classification. First, we remove the fixed-length restriction from terminal b, and let its length be an attribute. For example in Fig. 3 (a) we consider the curve between terminal d and terminal a as a single terminal b. The resulting segmentations and string representations for the chromosomes are shown in Fig. 4. Note that the string representation of a median and a submedian chromo-

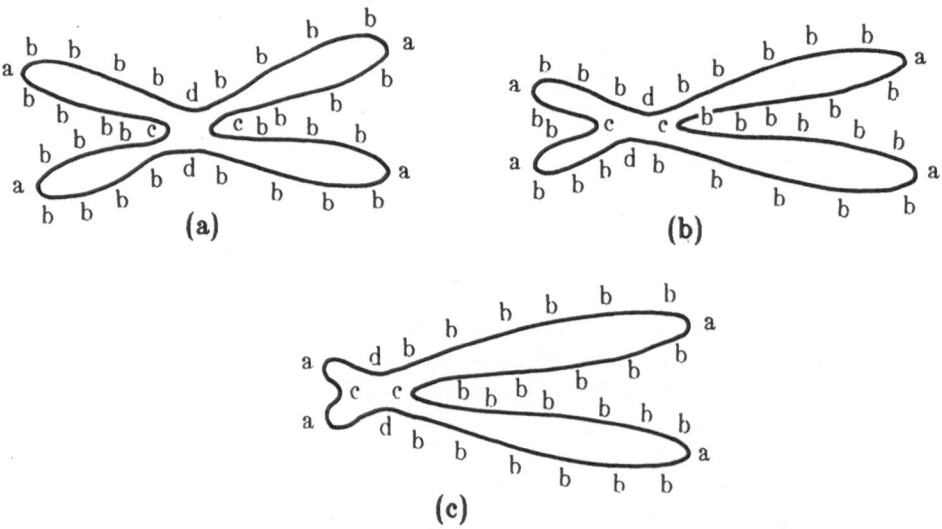

Fig. 3 Three classes of chromosomes. (a) Median chromosome z_M = cbbbabbbbbdbbbbabbbcbbbdbbbbabbb. (b) Submedian chromosome z_S = cbabbbdbbbbbabbbbcbbbbabbbbbdbbbab. (c) Acrocentric chromosome z_A = cadbbbbbbabbbbbbcbbbbbabbbbbbbdac.

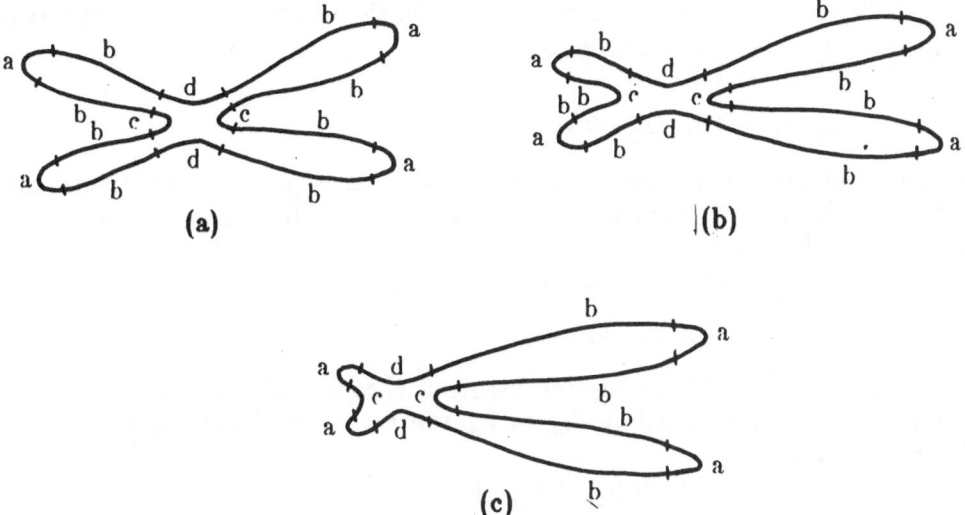

Fig. 4 Three classes of chromosomes. (a) Median chromosome z_M = dbbabcbabdbabcbab. (b) Submedian chromosome z_S = dbabcbabdbabcbab. (c) Acrocentric chromosome z_A = dbabcbab-daca.

somes have become identical, but this is not a problem since our discrimination between them will rely on the difference of their synthesized attributes. The grammar is given as follows in which superscripts of nonterminals are used just for the purpose of discriminating identical nonterminals on the right-hand side of a syntactic rules.

A chromosome grammar with synthesized attributes:
$$G_S = (V_T, V_N, P, S)$$

$V_T = V_{TM}$ as specified previously,

$V_N = \{S, Q_1, Q_2, R_1, M_1, M_2, A, B, C, D\}$

with attribute sets as follows:
  a) $A(A) = A(C) = A(D) = A(M_2) = \phi$,
  b) $A(X) = \{l_X\}$, for $X = B, M_1, R_1, R_2, Q_1, Q_2$,
  c) $A(S) = \{l_{S1}, l_{S2}\}$.

P: 1) $S \rightarrow Q_1^1 Q_1^2$; $l_{S1} = l_{Q_1^1}$, $l_{S2} = l_{Q_1^2}$

   2) $S \rightarrow Q_1 Q_2$; $l_{S1} = l_{Q_1}$, $l_{S2} = l_{Q_2}$

   3) $S \rightarrow Q_2 Q_1$; $l_{S1} = l_{Q_2}$, $l_{S2} = l_{Q_1}$

where nonterminal S represents the whole pattern, Q_1 or Q_2 represents an arm pair connected to a primitive d, attribute l_{S1} or l_{S2} is an arm pair length in the chromosome, and l_{Q_1} or l_{Q_2} is the average length of the two arms in the left or right arm pair:

4) $Q_1 \rightarrow DR_1; l_{Q_1} = l_{R_1}$

5) $Q_1 \rightarrow DR_2; l_{Q_2} = l_{R_2}$

where nonterminal R_1 or R_2 represents an arm pair, D represents the primitive d, and attribute l_{R_1} or l_{R_2} is the average length of the two arms in left or right arm pair.

6) $R_1 \rightarrow M_1^1 C M_1^2, l_{R_1} = (l_{M_1^1} + l_{M_1^2})/2$

7) $R_2 \rightarrow M_2 C M_2; l_{R_2} = 0$

where nonterminal M_1 or M_2 represents a chromosome arm, C represents the primitive c, and attribute l_{M_1} or l_{M_2} represents the length of arm M_1 or M_2 respectively.

8) $M_1 \rightarrow B^1 A B^2; l_{M_1} = (l_{B^1} + l_{B^2})/2$

9) $M_2 \rightarrow A$

10) $A \rightarrow a$

11) $B \rightarrow b; l_B = l_b$

12) $C \rightarrow c$

13) $D \rightarrow d$

where nonterminals A, B, C, D represent primitives a, b, c, d, respectively and attribute l_B or l_b is the length of primitive b.

From Example 4 and Example 5, it can be seen that with respect to the generation of a language for pattern description there exists a trade-off between the syntactic complexity and the semantic complexity [18, 25]. It is possible to construct a context-free or finite-state grammar with semantic or attribute information equivalent to a context-sensitive grammar without attributes. Thus, attributed finite-state grammar has been proposed as a normal form for pattern description [26]. In such a case, the analysis and recognition of patterns can be accomplished by a finite-state automation with associated semantic or attribute computations. Of course, in practical applications, the balance between syntactic and semantic complexities will probably also be affected by the actual implementation considerations.

A block diagram for pattern analysis using the semantic-syntactic approach is shown in Fig. 5.

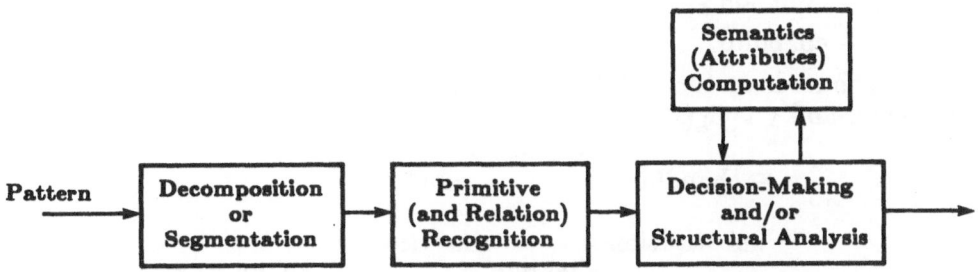

Fig. 5 Block Diagram of a Semantic-syntactic Pattern Analysis System

4. DISTANCE MEASURES BETWEEN ATTRIBUTED STRINGS

Because of its simplicity in computation, the use of distance measures for the classification of syntactic patterns has received increasing attention recently [5]. For non-attributed strings, the Levenshtein distance between two strings is defined as the minimum number of substitutions, deletions and insertions required to transform one string to the other. Weighted Levenshtein distance and weighted distance have also been suggested and applied to practical problems [5]. For attributed strings, only the following two distance measures have recently been proposed. Their utility to practical applications still needs to be examined. 1) Let x and y be two attributed strings

$$x = a_1 a_2 \text{ --- } a_n$$
$$y = b_1 b_2 \text{ --- } b_m$$

The distance between x and y is defined as

$$d_1(x,y) = \alpha d_L(x,y) + \beta d_A(x',x'')$$

where α and β are two weighting coefficients, $d_L(x,y)$ is the Levenshtein distance between x and y. $d_A(x',x'')$ is the attribute distance between x and y after the syntactic errors (or symbolic differences) between the two strings are eliminated. Let

$$x' = a_1' a_2' \cdots a_k' (= x'' = a_1'' a_2'' \cdots a_k'')$$

Then $d_A(x',x'')$ could be defined as

$$d_A(x',x'') = \sum_{i=1}^{k} w_i \, d(A(a_i'), A(a_i''))$$

where $A(a_i')$ and $A(a_i'')$ are the attributed vectors of a_i' and a_i'' respectively. It is noted that when $k = 1$,

$$d_A(x',x'') = w_1 d(A(a_1'), A(a_1''))$$

which could be the Euclidean (or weighted Euclidean) distance between the attributed vectors $A(a_1')$ and $A(a_1'')$.

Example 6: Consider two strings
  x = cbbabbdb

  y = cbabdbb

with attributed vectors A(a), A(b) and A(c). The Levenshtein distance between x and y is $d_L(x,y) = 3$ since

$$\text{substitution}$$
  y = cbabdbb   cbab$\underline{b}$bb

  · substitution   insertion
        cbab$\underline{bd}$b   c$\underline{b}$babbbd = x.

The underlined symbol indicates where the error occurs. After eliminating all the symbolic errors, we obtain
  $x^{'}$ = cbabb = $x^{''}$

However $A(x^{'}) \neq A(x^{''})$. Let $x^{'} = a_1^{'} a_2^{'} a_3^{'} a_4^{'} a_5^{'}$ and $x^{''} = a_1^{''} a_2^{''} a_3^{''} a_4^{''} a_5^{''}$. Thus

$$d_A(x^{'},x^{''}) = \sum_{i=1}^{5} w_i d(A(a_i^{'}), A(a_i^{''}))$$

2) In [26], a distance measure between two attributed strings x and y is defined as

$$d_2(x,y) = d_{syn}(x,y) + d_{sem}(x,y)$$

where the syntactic distance between x and y is defined as $d_{syn}(x,y) = \alpha \big| N_1 - N_2 \big|$, α is a constant, N_1 is the number of productions used to generate x, and N_2 is the number of productions to generate y. The semantic distance between x and y is defined as

$$d_{sem}(x,y) = \underset{N}{\text{Min}} \{ \sum_{i=1}^{N} d(A(a_i^{'}),A(a_i^{''})) \}$$

where N is the length of the shorter string in {x,y}. It is noted that, in order to calculate $d_2(x,y)$ the knowledge of the grammar generating x and y is required. Also, in calculating $d_{sem}(x,y)$, all the substrings with length N of the longer string in {x,y} need to be explored.

5. DESCRIPTION OF HIGH-DIMENSIONAL PATTERNS

Besides the use of PDL, trees and relational graphs have been proposed for the description of high-dimensional patterns [4,5,13,14]. When we use trees or graphs for pattern description, the nodes usually represent subpatterns and the branch between two nodes represents the relation between the two corresponding subpatterns. Needless to say, both subpatterns and relations can have attributes. The following examples illustrate the use of attributed trees and attributed relational graphs for pattern description.

Example 7: Let a be a straight line segment with attributes l_a and θ_a where l_a is the length of a and θ_a is the angle of a with respect to the horizontal axis and $-30° \leq \theta_a \leq 30°$. That is to say that a is a more or less horizontal straight line segment with length l_a. Similarly, let b be a more or less vertical straight line segment ($240° \leq \theta_b \leq 300°$) with length l_b. We can then use the tree in Fig. 6(b) to represent the character E shown in Fig. 6(a). It is interesting to see that when $l_{a_3} = 0$ the character becomes ⌐ and primitive a_2 is essentially deleted. On the other hand, when $l_{a_3} = 0$, the character becomes F. The deletion of a_2 or a_3 causes a structural change. That is, in this case, character E is changed to ⌐ or F. Small changes in the values of l_a, θ_a, l_b and θ_b will not change the structure of pattern E, and can be used to describe the effect of noise and distortions on E. (See Fig. 6(c)).

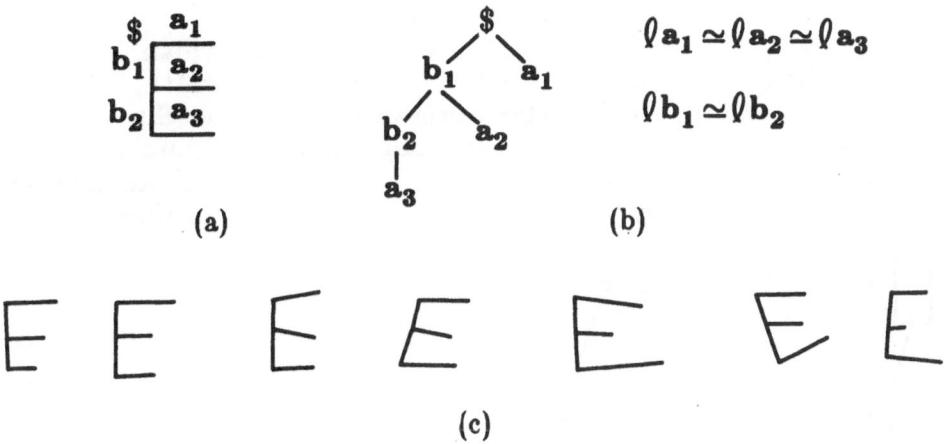

(a)　　　　　　　　　　　　(b)

(c)

Fig. 6　(a) Character E and (b) its tree representation (c) Samples of distorted E.

It is noted that when the pattern structure is preserved, only semantic deformation or substitution errors are allowed. When the pattern structure is not preserved, both semantic and syntactic errors can occur. A significant change of semantic information could result in a syntactic (deletion or insertion) error. On the other hand, a substitution could be interpreted as a deletion error followed by an insertion error. Syntax-directed translation schemata has been proposed as a pattern deformation model for substitution, deletion and insertion errors [27,28]. When patterns are characterized by an attributed grammar, extension of the syntax-directed translation model has recently been made by including semantic as well as syntactic deformations [26].

Example 8: In 3D solid modeling for CAD/CAM, a manufactured product

526

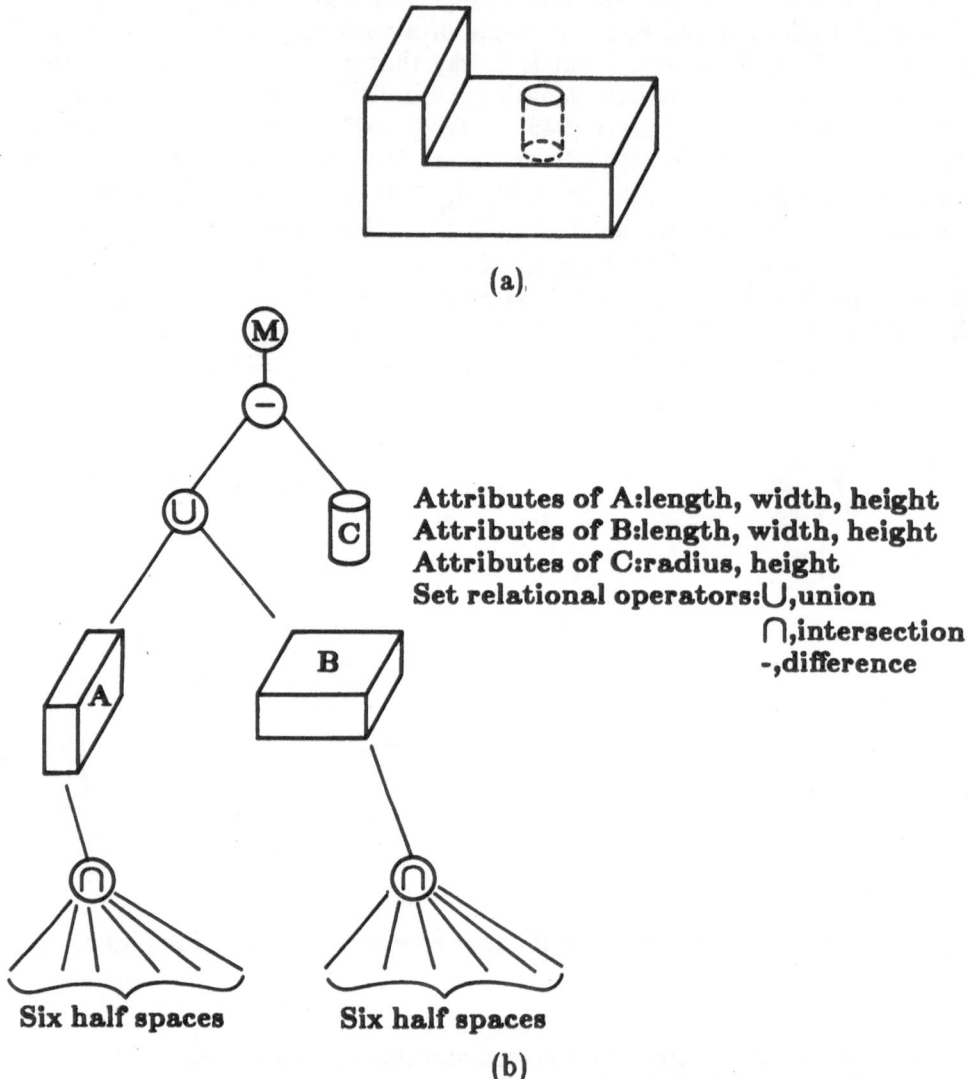

(a)

Attributes of A:length, width, height
Attributes of B:length, width, height
Attributes of C:radius, height
Set relational operators:∪,union
∩,intersection
-,difference

Six half spaces   Six half spaces

(b)

Fig. 7   (a) Product M, and (b) its relational graph representation.

or machine part is usually represented syntactically [29]. An example is given in Fig. 7 in terms of a relational graph representation. Obviously, changes of size and/or other parameters describing the product can be made by varying the attributes of the components (primitives and subpatterns) of the product (pattern) in the graph representation.

Example 9: A 3D scene shown in Fig. 8(a) can be described by the attributed relational graph shown in Fig. 8(b). It is noted that by varying the attributes of subpatterns and relations, a variety of similar scenes can be described.

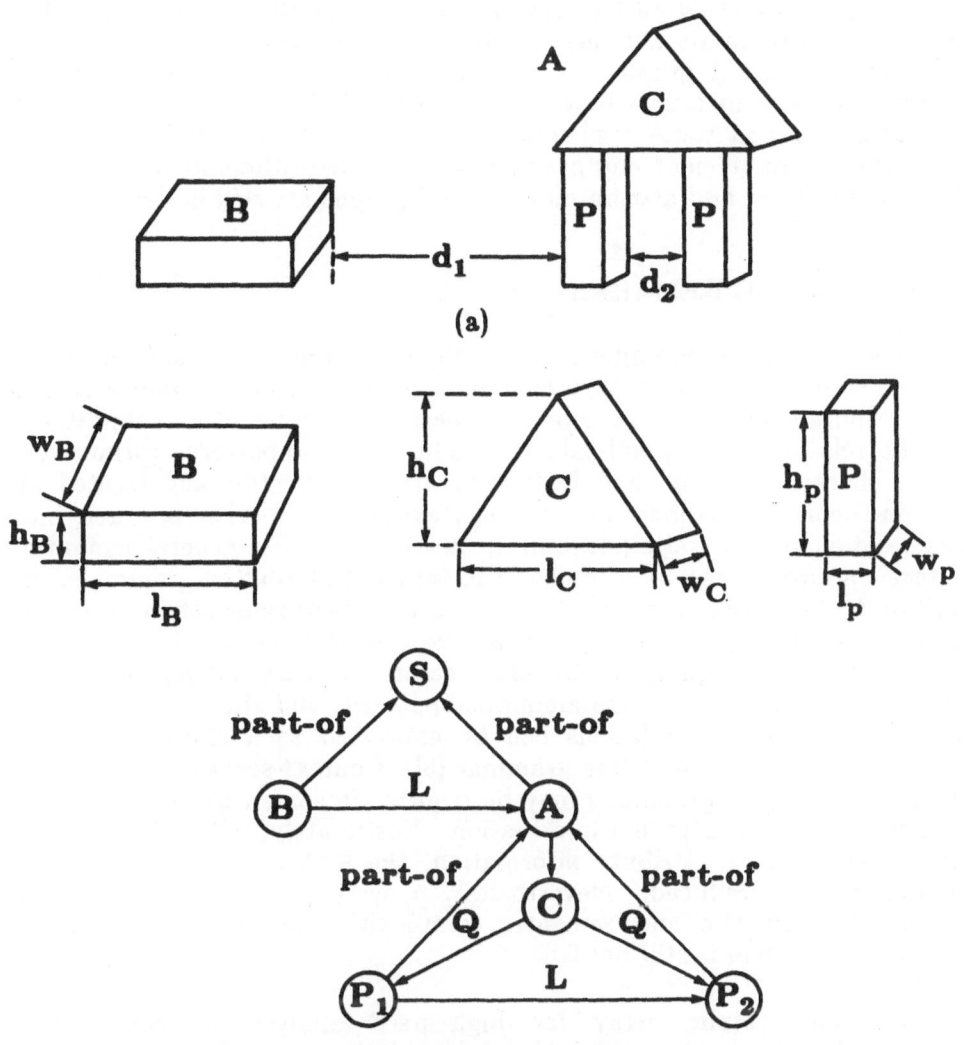

Attributes of brick B: l_B, w_B, h_B
Attributes of trianglular prism C: l_C, w_C, h_C
Attributes of pillars P_1 and P_2: l_p, w_p, h_p
Attributes of relation "Left-of" L: distance d
Relation Q: supported-by

(b)

Fig. 8   (a) Scene A, and (b) its attributed relational graph representation.

Analysis of attributed tree grammars and some special classes of attributed graph grammars has been studied recently [30-32]. Algorithms for attributed relational graph and subgraph matching have been proposed [33-35]. Distance measures between two attributed relational graphs have been proposed and some applications demonstrated [36,37]. There is no doubt that more efficient and practically useful algorithms for the analysis of attributed trees and attributed relational graphs are still in demand.

6. CONCLUDING REMARKS

When the patterns under study are quite complex or when the structural or contextual information contained in the patterns is important, it is often more effective to describe each pattern in terms of its subpatterns and the relations among subpatterns. The basic subpatterns (primitives) can be characterized by their feature or attribute vectors and treated by decision-theoretic methods, and the contextual information is taken into consideration by syntactic (structural) methods. Such a general semantic-syntactic approach can be formulated in terms of attributed grammars. In addition to the representation of a pattern by a feature or attribute vector (as in the decision-theoretic approach), we can describe a pattern using a string, a tree, or a graph of attribute vectors. Contextual information is characterized by the relations among subpatterns and the grammar rules. Markovian contextual relations can be expressed as a stochastic finite-state or stochastic context-free grammar [5]. Context-sensitive or stochastic context-sensitive grammars may be required in some cases to characterize more complex contextual information. Fortunately, with the introduction of semantic or attribute information, the high syntactic complexity required can be reduced. Nevertheless, a systematic method to select appropriate syntactic and semantic complexities for a specific pattern recognition problem is still needed.

A VLSI systolic array for high-speed analysis of context-free languages has been proposed [40]. the analysis is based on the Cocke-Kasami-Younger algorithm. This pipelined triangular array, constructed of $n(n+1)/2$ processing cells, can be used in syntactic pattern analysis. Each cell has two unidirectional data channels and one control line along each direction. Data appear as strings of symbols flowing through the recognition matrix from left to right and bottom to top. This two-dimensional array can recognize any input string of length n in 2n time units. This context-free language analyzer and its extension to analyze finite-state languates are described in more details in [40]. VLSI architectures for high-speed analysis of context-free languages using Earley's algorithm and its error-correcting parser have recently been proposed [41,42]. A special purpose VLSI processor which consists of three systolic arrays has been designed for fast classification of seismic waveforms [43].

It is interesting to notice that, with the formulation introduced in this paper, relationships between pattern analysis and (rule-based) artificial intelligence can be established. The rules used in AI systems are, in general, of the form [38, 39]: IF E (evidence) THEN H (conclusion). Referring to Section 2, we can express the conventional one-dimensional two-class pattern classification problem as follows: Let n = 1, m = 2

E: E_1 $x_1 > t$ (threshold), or $D_1(x_1) > D_2(x_1)$

   E_2 $x_1 < t$, or $D_1(x_1) < D_2(x_1)$

H: H_1 x_1 is from class 1

   H_2 x_1 is from class 2

The decision-theoretic classification rule can then be stated as

IF   E is E_1 THEN   H is H_1

IF   E is E_2 THEN   H is H_2

It is certainly possible to extend this example to the case where n > 1 and m > 2. For the case with uncertainty, the classification rule can be expressed as

IF   E is E_1 THEN   H is H_1 with $P(H_1 | E_1)$

where $P(H_1 | E_1) = \dfrac{P(E_1 | H_1)P(H_1)}{P(E_1)}$.

In the case of a pattern S consisting of two subpatterns X and Y related by R, as shown in (1), we can express the recognition rule as

IF X and Y, and X and Y are related by R THEN S

In order to include uncertainty, we can also express the recognition rule as

IF X, Y and R

THEN S with $P(S | X,Y,R)$

REFERENCES

1. K. S. Fu, and A. Rosenfeld, "Pattern Recognition and Image Processing," *IEEE Trans. Computers*, Vol. C-25, Dec. 1976.
2. K. S. Fu, *Digital Pattern Recognition*, Springer-Verlag, Second Edition, 1980.
3. J. Kittler, K. S. Fu and L. F. Pau, ed., *Pattern Recognition: Theory and Applications*, D. Reidel Publishing Co., 1982.
4. T. Pavlidis, *Structural Pattern Recognition*, Springer-Verlag, 1977.
5. K. S. Fu, *Syntactic Pattern Recognition and Applications*, Prentice-Hall, 1982.
6. R.C. Gonzalez and M. G. Thomason, *Syntactic Pattern Recognition - An Introduction*, Addison-Wesley, 1978.
7. R. Bonamini, R. DeMori, A. Lettera, R. Roggers and E. Sandretto,

530

"An Electrocardiographic Signal Understanding System", in *Pattern Recognition: Theory and Applications, ed.* by J. Kittler, K. S. Fu and L. F. Pau, D. Reidal Publ., 1982.

8. D. A. Giese, J. R. Boure and J. W. Ward, "Syntax Analysis of Electroenncephalogram," *IEEE Trans. Syst., Man, Cybern.*, Vol. SMC-9, August 1979.

9. H. H. Liu and K. S. Fu, "A Syntactic Approach to Seismic Pattern Recognition," *IEEE Trans. Patt. Anal. Mach. Intel.*, Vol. PAMI-4, March 1982.

10. B. Moayer and K. S. Fu, "A Syntactic Approach to Fingerprint Pattern Recognition," *Pattern Recognition*, Vol. 7, 1975, pp. 210-233.

11. K. S. Fu and J. Mui, "A Survey on Image Segmentation," *Pattern Recognition*, Vol. 13, 1981, pp. 3-16.

12. G. Stockman, L. N. Kanal and M. C. Kyle, "Structural Pattern Recognition of Carotid Pulse Waves Using a General Waveform Parsing System," *Comm. ACM*, Vol. 19, Dec. 1976.

13. P. H. Winston, "Learning Structural Descriptions from Examples," in *The Psychology of Computer Vision*, ed. by P. H. Winston, McGraw-Hill, 1975.

14. A. Rosenfeld, "Picture Processing: 1980", *Computer Graphics and Image Processing*, Vol. 16, May 1981.

15. J. L. Mundy and R. E. Joynson, "Automatic Visual Inspection using Syntactic Analysis," Proc. 1977 IEEE Computer Society Conference on Pattern Recognition and Image Processing, June 6-8, Troy, N.Y.

16. A. V. Aho and J. D. Ullman, *The Theory of Parsing, Translation and Compiling*, Prentice-Hall, 1972.

17. W. H. Tsai and K. S. Fu, "A Syntactic-Statistical Approach to Recognition of Industrial Parts," Proc. 5th Int'l. Conf. Pattern Recognition, Dec. 1-4, 1980, Miami Beach, FL.

18. K. S. Fu, "Recent Advances in Syntactic Pattern Recognition," NSF Workshop on Structural and Syntactic Pattern Recognition, June 22-24, 1981, Saratoga Springs, N.Y.

19. K. C. You and K. S. Fu, "A Syntactic Approach to Shape Recognition Using Attributed Grammars," *IEEE Trans. Syst., Man, Cybern.*, Vol. SMC-9, June 1979.

20. A. C. Shaw, "A Formal Picture Description Scheme as a Basis for Picture Processing Systems," *Information and Control*, Vol. 14, 1969, pp. 9-52.

21. D. E. Knuth, "Semantics of Context-Free Languages," *J. Math. Syst. Theory*, Vol. 2, 1968, pp. 127-146.

22. P. M. Lewis, D. J. Rosenkrantz and R. E. Stearns, *Compiler Design Theory*, Addison-Wesley, 1976.

23. H. Bunke, "Programmed Graph Grammars," in *Graph Grammar and Application to Computer Science and Biology*, ed. by V. Claus, H. Ehrig and G. Rosenberg, Springer-Verlag, 1979.

24. K. S. Fu, "Syntactic Models for Image Analysis" in *Modelle und*

Strukturen, ed. by B. Radig, Springer-Verlag, 1981.

25. A. Pyster and H. W. Buttleman, "Semantic-Syntax-Directed Translation," *Information and Control*, Vol. 39, 1978, pp. 320-361.

26. J. W. Tai and K. S. Fu, "Semantic Syntax-Directed Translation for Pictorial Pattern Recognition," Proc. 6th International Conference on Pattern Recognition, Oct. 19-22, 1982, Munich, Germany.

27. T. I. Fan and K. S. Fu., "A Syntactic Approach to Time-Varying Image Analysis," *Computer Graphics and Image Processing*, Vol. 11, 1979, pp. 138-149.

28. M. G. Thomason, "Stochastic Syntax-Directed Translation Schemata for Correction of Errors in Context-Free Languages," *IEEE Trans. Computers*, Vol. C-24, Dec. 1975.

29. K. S. Fu, "Attributed Grammars for Pattern Recognition--A General (Syntactic-Semantic) Approach," Proc. 1982 PRIP Conference, June 14-17, Las Vegas.

30. Q.Y. Shi and K. S. Fu, "Parsing and Translation of (Attributed) Expansive Graph Languages for Scene Analysis," Proc. 6th International Conference on Pattern Recognition, Oct. 19--22, 1982, Munich, Germany.

31. Q. Y. Shi and K. S. Fu, "Efficient Error-Correcting Parsing for (Attributed and Stochastic) Tree Grammars," *Information Sciences*, Vol. 26, 1982.

32. A. Sanfeliu and K. S. Fu, "Tree-Graph Grammars for Pattern Recognition," Proc. 2nd International Workshop on Graph Grammars, Oct. 4-8, 1982, Osnabruck, Germany.

33. W. H. Tsai and K. S. Fu, "Error-Correcting Isomorphisms of Attributed Relational Graphs for Pattern Analysis," *IEEE Trans. System, Man, and Cybernetics*, Vol. SMC-9, Dec. 1979.

34. W.II. Tsai and K. S. Fu, "Subgraph Error-Correcting Isomorphisms for Syntactis Pattern Recognition," *IEEE Trans. System, Man, and Cybernetics*, Vol. SMC-13, no. 1, 1983.

35. L. Shapiro and R. M. Haralick, "Organization of Relational Models for Scene Analysis," Dept. of Computer Science, VPI & SU, September 1981.

36. A Sanfeliu and K. S. Fu, "A Distance Measure Between Attributed Relational Graphs for Pattern Recognition," Proc. 6th International Conference on Pattern Recognition, Oct. 19-22, 1982, Munich, Germany.

37. A. Sanfeliu, K. S. Fu and J. M. S. Prewitt, "An Application of a Distance Measure Between Graphs to the Analysis of Muscle Tissue Patterns," NSF Workshop on Structural and Syntactic Pattern Recognition, June 22-24, 1981.

38. R. O. Duda, P. E. Hart and N. J. Nilsson, "Subjective Bayesian Methods for Rule-Based Inference Systems," Proc. 1976 National Comuter Conference.

39. D. A. Waterman and F. Hayes-Roth, ed., *Pattern-Directed Inference*

532

Systems, Academic Press, 1978.

40. K. H. Chu and K. S. Fu, "VLSI Architectures for High-Speed Regcognition of General Context-Free and Finite-State Languates," Proc. 9th International Symposium on Computer Architectures, April 1982, Austin, Texas.

41. Y. T. Chiang and K. S. Fu, "A VLSI Architecture for Fast Context-Free Languate Recognition (Earley's Algorithm)," Proc. 3rd International Conference on Distributed Computing Systems, Oct. 18-22, 1982, Miami, Florida.

42. Y. T. Chiang and K. S. Fu, "Parallel Processing and VLSI Architectures for Syntactic Pattern Recognition and Image Analysis," Tech. Rept. TR-EE 83-4, Purdue University, Jan. 1983.

43. H. H. Liu and K. S. Fu, "VLSI Systolic Processor for Fast Seismic Classification," Proc. 1983 International Symposium on VLSI Technology, Systems, and Applications, March 30 - April 1, Taipei, Taiwan.

CLUSTERING TENDENCY PROBLEM IN PATTERN ANALYSIS*

by

Erdal Panayırcı[†]

Faculty of Electrical Engineering, Technical University of Istanbul

and

Marmara Research Institute
Istanbul, Turkey

ABSTRACT. Determining the structure of multi-dimensional patterns is an important problem in exploratory pattern analysis. Clustering methods have been used extensively for this purpose. However, clustering algorithms will locate and specify clusters in data even non are present. It is therefore appropriate to measure the clustering tendency or randomness of a pattern set before subjecting it to a clustering algorithms. We survey the work that has been done in developing measures of clustering tendency. Methods based on quadrats, spatial model fitting, distribution of interpoint distances are reviewed with special attention to distance-based methods that use sampling origins. One of the goals of this study is to assess their applicability to high dimensional patterns. The successes and failures of these methods are discussed as well as suggestions and directions for future study.

---

* Research supported in part by NSF Grant ECS-8007106.

† E.Panayırcı was a Visiting Professor in the Department of Computer Science, Michigan State University, under Fulbright-Hays and NATO Senior Scientific Programs.

534

1. INTRODUCTION

The goal of Pattern Analysis is to understand the structure of a given pattern set. We consider pattern sets in which n objects, called "patterns", are given in d-dimensional space, d<<n, whose axes represent the measurements made on each pattern and are called "features". The word "structure" is difficult to define in a useful way for such a pattern set but the key factors are the interactions among patterns and among features and sample size considerations. Cluster analysis has been used extensively for this purpose by organizing the patterns into natural disjoint groups, called clusters. A wide collection of clustering algorithms is known, each operating under various criteria. Some surveys include Anderberg [1], Everitt [2], and Hartigan [3]. Computationally, clustering algorithms are relatively expensive, and their run time increases greatly with an increase in the number of patterns as well as the number of features. The most serious drawback of these algorithms is that they will find clusters even if the pattern set is entirely random according to various definitions which we explore at length. See Dubes and Jain [4] for some examples of this phenomenon. It is therefore appropriate to measure the "clustering tendency" or "randomness" of a pattern set before subjecting it to a clustering algorithm. The issue of clustering tendency, and cluster validity, has virtually been ignored in the literature and this makes it difficult to interpret the results of a clustering algorithm [5].

A type of pattern set, which is of interest here, can be presented to a clustering algorithm in the form of "pattern matrix". A pattern matrix is an N x d matrix whose rows denote patterns, or points in a d-dimensional apace, and whose columns consist of measurements made on each pattern. This paper examines techniques that assess the general nature of the spatial arrangement of the patterns and determine which, if any, of the following descriptions fits the pattern matrix.

i. The patterns are arranged randomly;

ii. The patterns are aggregated, or clustered;

iii. The patterns are regularly spaced.

This problem of examining the global nature of the spatial arrangement of patterns will be termed the clustering tendency problem We wish to pose the clustering tendency problem in a hypothesis testing framework. Our null hypothesis will always be that the given pattern set is random. If the null hypothesis fails to be rejected at some significance level than we will say that the pattern set does not exhibit a tendency to cluster and it is not worth while to apply a clustering algorithm to this pattern set. A number of techniques for the analysis and modeling of two-dimensional patterns is avaible in the literature [1,6,7,8]. The development of these

techniques was motivated by a need to interpret and analyze large
amounts of two-dimensional data collected in various ecological and
socio-geographic studies. These tests have not been tried out on
high-dimensional data and it is not known whether a direct generalization (to d>2 dimensions) of these statistics would lead to useful
tests.

The main purpose of this paper, is to survey the work that has
been done in developing measures of clustering tendency, to see
whether some of the test statistics known to be powerful in two-
dimensions can be generalized to d>2 dimensions, and to discuss the
successes and failures of these methods as well as suggestions and
directions for future study.

The next section introduces spatial point processes which are
used to give a formal definition of our concept of "random",
"clustered" and "regular" structure. In Section 3, the sampling
window problem is introduced and discussed. In Section 4, most of
the existing statistical methods for clustering tendency are described,
some of them are generalized to $d(d>2)$ dimensions, and the computational feasibility of the methods, including computation of size and
power of the tests are considered. Finally, in Section 5, the main
conclusions, reached, are summarized and directions for further
research are given.

2. SPATIAL POINT PROCESSES

Spatial point processes are stochastic models of discrete events,
or patterns, represented as points in a d-dimensional pattern space.
Our framework for studying the clustering tendency problem is based
on spatial point processes and follows the work of Hammersley [9]
and Ripley [7]. Formally, a spatial point process is a stochastic
process whose parameter set is contained in a multidimensional space.
The term "stochastic point process" usually refers to a stochastic
process with a one-dimensional parameter set, such as an interval
of the real line. Many applications of stochastic point processes
have been studied and much theory has been developed for this case
[10,11,12]. By contrast, relatively little is known about spatial
point processes.

A spatial point process is desribed by a family of random
variables, $\{Z(A)\}$, where $Z(A) < \infty$ is the number of patterns in the
bounded Borel set A, which is a subset of the pattern space, X,
contained in d-dimensional Euclidean space. The distributions of
these random variables are governed by the intensity of the process
generating the patterns, or the variation with position in the expected number of patterns per unit volume, and the interactions among
the patterns and among the features.

i. Poisson Process

The simples and most important case of spatial point process is Poisson process. A Poisson process has a constant intensity, λ patterns per unit volume, and,

$$\Pr\,[\,Z(A) = k\,] = \frac{[\lambda\nu(A)]^k}{k\,!}\,\exp\,[-\lambda\nu(A)\,]\quad\text{for}\quad k = 1,2,\dots$$

where $\nu(A)$ is the Lebesque measure of A, which can usually be taken as the volume of A. In addition, $Z(A_1)$ and $Z(A_2)$ are independent random variables if A_1 and A_2 are disjoint sets. Poisson processes will serve as models of randomness. In fact, our tests of clustering tendency will involve testing the null hypothesis (H_o) that the given pattern set is a sample from a Poisson process. Figure 1, shows a realization of a Poisson process.

ii. Clustered Processes

Among a large number of models for clustering structures, the Neyman-Scott processes are perhaps the most well known, [13], [14]. A Poisson "parent" process having intensity λ is first observed to obtain cluster centers, in a sampling window $EC\chi$. The clusters themselves are generated by the "daughter" process which consists of a radial distribution function, usually taken to be multivariate Guassian, and a integer poisson distribution having intersity $\mu(\mu\gg\lambda)$ which controls the number of points per cluster. Figure 3 shows a realization of a Neyman-Scott process. Classical cluster analysis thus attempts to estimate the position of the cluster centers and identify the members of the daughter process.

When the clusters are tight, well separated, and reasonable dense (extreme clustering case), a simple clustering model, called the modified Thomas process is preffered. The steps for generating a sample from the modified Thomas process is listed below.

a) Generate a sample from a Poisson process in d-dimensions with intensity λ. Each point serves as a cluster center.

b) Associate a radius, r, with each cluster center. The radii are samples of independent, identically distributed expo- nential random variables. Each radius defines a sphere around a cluster center.

c) Generate N independent and identically distributed patterns uniformly inside each sphere. The values of N are samples of independent, identically distributed Poisson random variables with intensity μ.

iii. Hardcore Process

The next general class of model for the multidimensional patterns is the so-called "hardcore" or "inhibitory" models. These models all have the feature that no two points of the process are

within a distance 2r of each other, where r is a fixed positive
number called the "interaction radius". In some sense, hardcore
models represent a regular structure which we may characterize as
anticlustering or repulsion among patterns. There are a number of
models for generating hardcore processes. The details appear in
Matern [15] and Ripley [7]. The simple sequential inhibition (SSI)
process is a more reasonable model of regularity in real data.
Imagine each pattern being encased in a small d-dimensional sphere.
The spheres are "hard" in that they can not overlap or intersect.
To generate a realization of an SSI process, the spheres are placed
sequentially into a sampling window, which ensures that no two spheres
overlap. The process terminates when a predetermined number of patterns have been inserted, or the sampling window is full. The essential parameter of an SSI process is the packing density, which can
be expressed as follows when the volume of the sampling window is one.

$$\rho = \lambda A r^d$$

Here, r is the radius of each hard sphere, λ is the expected number
of patterns per unit volume, and A is the volume of a unit-radius,
d-dimensional hypersphere,

$$A = \pi^{d/2} / \Gamma(\frac{d}{2} + 1) .$$

The maximum packing density when d=2 has been shown to be $\pi/\sqrt{12}$, or
about 0.9069, which is achieved by arranging the circles in a triangular lattice. Maximum packing densities are not known exactly for
d larger than 2 but it is clear that the maximum ρ will decrease
rapidly as d increases. Note that an idealization of a regular
arrangement of patterns can be obtained when we place each pattern
at the vertices of a d-dimensional lattice with side length a. All
positions in the lattice are assumed to be filled with single patterns. Then, the appearance of a hardcore realization is similar to
a noisy lattice as shown in Figure 2.

3. SAMPLING WINDOW

Our mathematical model for the generation of patterns is a
multidimensional spatial point process. We imagine a probabilistic
mechanizm scattering patterns throughout d-dimensional Euclidean
space. We observe the process through a bounded, d-dimensional sampling window, which is a frame around the data. The choice of a sampling window or bounded set over which the patterns are assumed to
be dispersed is critical. In a data collection problem, the sampling
window is usually imposed by the environment. For example, the forest
boundary specifies the sampling window when we are studying the
spatial distribution of trees. On the other hand, a given set of
patterns can be bounded by a hypercube and also by a hypersphere.
The convex hull of the patterns will probably occupy only a small

portion of the volume of the enclosing hypersphere or hypercube. The choice of a hypercube as a sampling window may make the patters appear to be a single tight cluster in a large box and distort the structure of any micro clusters in the data itself. We beleive that this problem of defining a sampling window to be intractable without contextual information. Techniques proposed by Grenander [16] describe the estimation of the convex hull of a scatter of points (patterns), but even then, there are statistical assumptions made about the form of the pattern set.

In pratice, the sampling window that we use will depend on the method of analysis to be adapted for the problem of clustering tendency. As it will be seen later, in the case of direct distributional test, we fit a hypershere to the pattern set by translating the sample mean to 0, whitening (i.e. decorrelate the coordinates and scale the coordinates so that they have sample variance 1), and then dividing all coordinates by the maximum norm of the whitened vectors to obtain a final scaling within the unit sphere. In other contexts, notably the distance methods, we use a rectangular sampling frame. This is accomplished by decorrelating the pattern set and then finding the dimensions of the best fitting hyper-rectangle by taking the maximum and minimum in each coordinates. There will inevitably be some open space between the convex hull of the given pattern set and the window selected used to enclose it. This space or volume increases greatly with the dimension, so in some cases it may be appropriate to consider a rectangle or sphere inside the extend of the pattern set.

4. TESTS FOR CLUSTERING TENDENCY

A test for clustering tendency based on a set of N patterns (points) in a d-dimensional pattern space can be stated as a statistical test of the following null hypothesis.

H_o: The patterns are generated by a Poisson process with intensity λ patterns per unit volume.

We seek test statistics whose distributions are known, at least asymptotically, under H_o, which have large power against various alternatives, and which do not depend on λ.

The size of a statistic is the probability of rejecting H_o when H_o is active and is fixed a-priori. The power is the probability of rejecting H_o when some alternative hypothesis is active. In general, it is extremely difficult to derive distributions for test statistics under realistic alternative hypothesis.

We attempt to review almost all tests that can be used to assess clustering tendency. The richest source of such test is the ecological literature. The occurence of plant species, trees, moon craters, or contagious organizms are observed. We exploit tests for structure

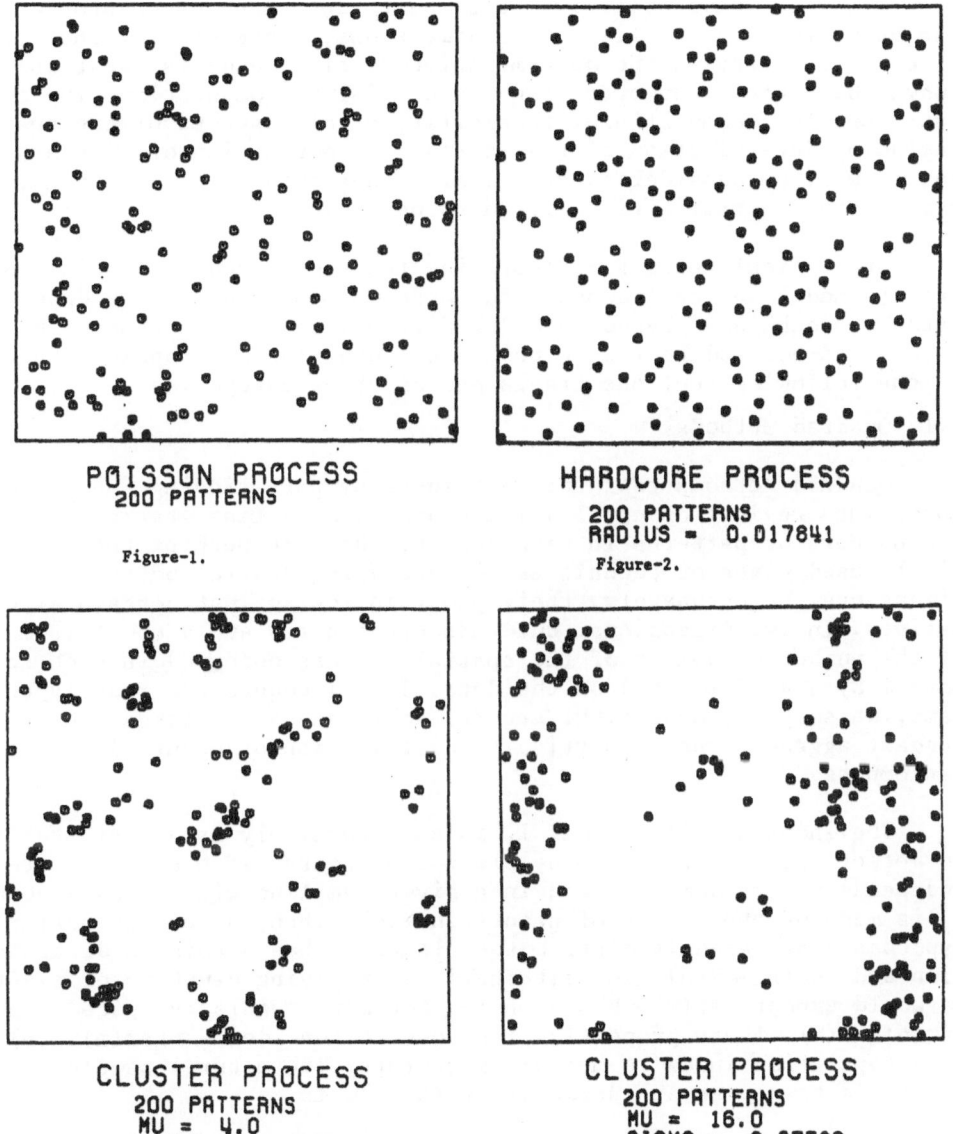

POISSON PROCESS
200 PATTERNS

Figure-1.

HARDCORE PROCESS
200 PATTERNS
RADIUS = 0.017841

Figure-2.

CLUSTER PROCESS
200 PATTERNS
MU = 4.0
SIGMA = 0.02500

CLUSTER PROCESS
200 PATTERNS
MU = 16.0
SIGMA = 0.07500

Figure-3.

but are not normally interested in the estimation of parameters, which is primary importance in Ecology. Almost all Ecological applications are pharased in two dimensions, for obvious reasons, while pattern analysis requires d-dimensional tests.

Thus, our main objectives will be to determine which procedures can be usefully generalized to d-dimensions, state clearly the form of each test, review the determination of size, and list what is known about power. An overriding issue will be computation. The computation. The computational feasibility of all aspects of the test, including determination of the statistic, computation of size and power, will be considered. Practical issues such as edge effects, sample size, normalization will also be included.

Statistical tests for clustering tendency based on spatial point process model can be grouped into three main categories: Quadrad Methods, Techniques Based on Model Fitting, and Distance-based Method Several ad-hoc and heuristic procedures can also be proposed. Each of the following sections treats one of these categories.

4.1. Quadrad Methods

Quadrad Methods test for randomness by partitioning the pattern space into regions of equal volumes and forming test statistics from the numbers of patterns in each region. They are perhaps the most widely used class of techniques for analyzing spatial pattern in two dimensions. Unfortunately, their power is low against certain alternatives. In two dimensions, quadrads are squares and a count is made of the number of points of the spatial process observed in each square formed by a grid imposed on the data. If all counts are roughly the same, we suspect regularity; several small and a few large counts suggest aggregations; a particular distribution of counts indicates randomness.

The index of Dispersion, I, is an extensively used test statisti to detect spatial structure by plant ecologists. If N quadrats are independently randomly placed in a given sampling window and a count, n, is made of the number of plants in each, then, under the null hypothesis of randomly distributed plants, n has a Poisson distribution and $Var(n) = E(n)$. For alternatives involving clusters of plants we would expect $Var(n) > E(n)$, while for more regularly spaced plants $Var(n) < E(n)$. These properties lead one to consider $w = Var(n)/E(n)$ as a population index to measure structure. The natural sample statistic to use as an estimator of $(N-1)w$ is

$$I = \text{Index of Dispersion} = [\sum_{i=1}^{N} (n_i - \bar{n})^2] / \bar{n}$$

It has been shown that under the null hypothesis I has an approximate Chi-square distribution having $N-1$ degrees of fredom which enables tests of randomness to be performed. Distributions of dispersion statistics under alternative hypotheses of clustering and regularity have also determined [17].

Cross [18] has analyzed quadrat-based tests for clustering tendency. The major defect of such tests is their inability to detect and test spatial arrangements at more than one scale, set by the quadrat mesh. The Greg-Smith [19] approach attempts to remedy this deficiency. The other serious drawback in extending quadrats to general spaces is the selection of quadrat size. Nonrandomness may not be detected unless the number of patterns and the number of quadrats are large. We normally work with sparse data, such as a few hundred patterns, in pattern analysis, so the standart chi-square tests for evaluating the quadrat counts would be statistically useless.

4.2. Techniques Based on Model Fitting

Ripley [7] has provided a complete and mathematically sound treatment of spatial point processes that is applicable to pattern analysis. Ripley's underlying assumption is that the patterns lie in a known, homogeneous region of the pattern space called the sampling window $E \subset X$. Homogeneity requires that the process generating patterns in E has no preferred direction over another and is capable of producing patterns throughout the entire pattern space. Many of Ripley's models are inherently two-dimensional and are applied to ecological problems where X is obviously two-dimensional.

The most promising source of test for the clustering tendency appears to be the technique of spatial modelling developed by Ripley. Ripley's procedure for modelling a set of patterns requires estimating two parameters, namely, λ and $K(t)$. The intensity λ, is the expected number of patterns per unit volume, so

$$E [Z(A)] = \lambda \nu (A)$$

where $\nu(A)$ can be taken as the volume $A \subset X$. The parameter $K(t)$ involves second moment information defined by $E[Z(A) Z(B)]$, where A, B $\in X$, are two bounded measureble sets. Ripley than points out that

i. $\lambda K(t)$ is the expected number of patterns within a spherical neighborhood of radius t centered at an arbitrary pattern but not including the centre pattern

ii. $\lambda^2 K(t)$ is the expected number of ordered pairs of (distinct) patterns within distance t of each each other.

Both parameters λ and $K(t)$ are invariant under translations and rotations of the pattern space. Specifying λ and $K(t)$ creates a family of models. In two dimensions, it can be shown that the Poisson process is specified by its intensity λ and $K(t) = \pi t^2$.

For the Neyman–Scott cluster process, $\lambda = \alpha E(N)$ and

$$K(t) = 1 + \frac{\alpha}{\lambda^2} \, E\{N(N-1)\}F(t)$$

where α is the intensity of the Poisson process describing the parent process and the daughter process is described by a random number of independent, identically distributed points. $F(t)$ denotes the distribution function of the distance between two arbitrary points of the daughter process.

For the hard-core process defined by Matern [15], in which a Poisson process of intensity α is first sampled and then, any point within 2R of any other is deleted whether or not this has already been deleted, we have:

$$\lambda = \alpha \exp(-4\pi\alpha R^2)$$

and

$$K(t) = \begin{cases} 0 & t \le 2R \\[2ex] \dfrac{2\pi}{\alpha^2} \displaystyle\int_0^t t' \exp\{u(t')\}dt' & t \ge 2R \end{cases}$$

where $u(t)$ denotes the total area of two discs, having equal radii, 2R, whose centers are t units apart.

Ripley's procedure for modelling a set of patterns requires estimating the intensity, λ, and the second-order function $K(t)$. The functional form of $K(t)$ is particularly important in determining an appropriate model. If the intensity λ is known, Ripley proposes the following unbiased estimator for $K(t)$, in two-dimensions. Extentions to multi-dimensional spaces are discussed at the end of the section.

$$\hat{K}(t) = \Sigma \, k(\underline{x},\underline{y})/\lambda^2 \nu(E)$$

where $k(\underline{x},\underline{y})$ is the reciprocal of the portion of the perimeter of the circle centered at $\underline{x}$ and passing through $\underline{y}$ which is within the sampling window. The sum is over all distinct pairs of patterns $(\underline{x},\underline{y})$ within distance t of each other. $\nu(E)$ denote the area of the sampling window. When λ is unknown, it can be estimated by simply counting the total number of patterns, n, in the sampling window divided by the area. That is, $\hat{\lambda} = n/\nu(E)$. The estimator for $K(t)$ is then obtained by replacing λ with $\hat{\lambda}$ as follows.

$$\hat{K}(t) = \Sigma \, k(\underline{x}, \underline{y})\nu(E)/n^2$$

Ripley points out that this estimator has a small bias when t is small, and calculates this bias exactly for a Poission process, [7].

The actual process of fitting a model to a given set of n patterns starts with the computation of $\hat{\lambda}$ and $\hat{K}(t)$. The main strategy is to simulate the proposed model, compute several estimates of K(t) on a Monte Carlo basis, and see if $\hat{K}(t)$ for the given pattern set lies within the average $\hat{K}(t)$ for the simulations. Note that K(t) need not be known for the process, although other parameters of the process are required to define the model being simulated. If K(t) is known, to within a few parameters, it can also be plotted and compared to $\hat{K}(t)$.

The technique of fitting a model to a given set of n patterns provides source of tests for the clustering tendency problem. If the null hypothesis is that the patterns were generated by a Poisson process, the intensity, λ, is the only parameter and can be estimated by $\hat{\lambda}$. The poisson process with intensity $\hat{\lambda}$ is then simulated m times. The values of $\hat{K}(t)$ for these simulated realizations are observed and the upper and lower envelopes are plotted over the m samples. We can define the acceptance region of a test by requiring $\hat{K}$ for the given pattern set to be within the envelope of $\hat{K}$ for the simulations througout this range. Thus, if the observed $\hat{K}(t)$ stays inside envelope for any t, then we accept the null hypothesis of Poisson process.

The number of candidate models avaible for a set of patterns is quite large. One usually starts with the simples model, namely, the Poisson spatial point process. If this randomness model is rejected, other models can be tested in order of complexity. The form of K(t), estimated from the given set of patterns, can suggest an appropriate model. For example, if K(t) indicates that all pairs of patterns are at least a distance r apart, a hard-core model with inhibition distance r might be suitable. No procedures exist for ordering the various models to be tested. It is quite possible that a satisfactory model cannot be found in a reasonable amount of computer time.

Computational realities are key considerations in extending Ripley's procedures to high dimensions (d>3). In principle, Ripley's models are applicable to any bounded set in d-dimensional space. However, an unreasonable amount of computation is required when K(t) is estimated by known methods and there is a considerable computational burden in doing Monte Carlo simulations to test the goodness of the fit even in the low dimensional cases (d<4). A fundamental theoretical problem is the adequacy of models based on second order statistics in describing multidimensional phenomena. Procedures based on high-order statistics might be needed to fit a variety of models to multidimensional data.

A fundamental problem in using the models proposed above, such as the Neyman-Scott model, is the large number of parameters involved in each model. The number of cluster centers, the number of patterns per cluster, and the parameters of parent and daughter processes must all be hypothesized or estimated. In practice, one needs information

about some of these parameters from prior studies. Testing the global
fit of these models is especially troublesome because the form of the
model must be specified before observing the data. One practical
suggestion is to project a subset of the patterns to two dimensions
with a standart technique [20] and infer some of the gross parameters
from the projection.

4.3. Distance Based Methods

Distance-based methods present a rich source of potential tests
for clustering tendency. A great deal of work has been reported in
the literature on distance-based tests. Most of these tests were first
suggested in the ecological literature, where two-dimensional spatial
point processes provide reasonable models for the growth of trees in
a forest, the scattering of plants in a field, or the spread of a
contagious organizm. We consider several d-dimensional generalizations of these tests in the contex of clustering tendency.

Perhaps the most obvious test involving only interpattern distances is based on the distribution of these distances. Section 4.3.1
defines this test and explains why it is not a good indicator of
clustering tendency. The number of "small" interpattern distances,
on the other hand, leads to a promising test for clustering dendency,
as developed in Section 4.3.2. The tests based on the near-neighbor
concept is treated in Section 4.3.3.

4.3.1. Tests Based on Interpattern Distance Distribution

This method is based on the knowledge of the distribution of
distance, r, between two points chosen at random in a bounded set
in d-dimensions. The most obvious test for clustering tendency from
interpattern distances alone would compare the empirical, or, observed,
distribution of interpattern distances to the theoretical distribution
under a randomness hypothesis. For example, suppose two points are
chosen at random (independently and uniformly) from a unit-diameter
sphere in d-dimensions. The density function for the Euclidean distance between these points is know, Hammersley [9], Lord [21],
Alagar [22], Solomon [23] to be

$$f(r) = 2^d \, r^{d-1} \, I_{1-r^2}(d + \frac{1}{2}, \frac{1}{2}) \quad \text{for} \quad 0 \le r \le 1$$

where $I_x(p,q)$ is the incomplete Beta function. The null hypothesis
for a test of clustering tendency based on this distribution would be:

H$_o$: The patterns are drawn independently from a uniform distribution in a hypersphere.

The test statistic would be either a chi-square or a Kolmogorov-
Simirnov statistic, (see Conover [24]) which compare the theoretical
distribution to the observed distribution. The test itself would
accept H$_o$ for suitably small values of the test statistic.

A similar approach can be taken by using the distribution of distance of two points placed randomly in a hypercube of side L. The exact distribution of interpoint distance, r, between two points chosen at random in a square was given by Bartlett [25], in two dimensions and by Panayirci [26] in three dimensions. Panayirci [26] has also derived the asymptotic distribution of r for d>3.

Cross [18] explains the difficulties inherent in this procedure. First is the problem of estimating the interpattern distribution. At most, n/2 of the n(n-1)/2 distinct interpattern distances are independent and the only independent observations are required in all statistical tests for quantitatively comparing two distributions, such as the Kolmogorov-Smirnov or chisquare tests. Cross [18] suggests observing ten sets of n/2 independent interpattern distances and accepting H_o if all ten sets are accepted as matching the known distribution. This assumes that the patterns have been scaled to a unit-diameter hypersphere. He reports on several simulations which show this test to be so sensitive to data scaling as to be impractical. For example, whitening, or de-correlating patterns generated uniformly from a unit-diameter hypersphere lowered the number of acceptance of H_o, as compared to raw data. The null hypothesis was soundly rejected when patterns were obtained uniformly over a hypercube instead of a hypersphere, with and without whitening. Similary, H_O was usually rejected when the patterns came from a hyperellipse rather than a hypercube.

Based on the above results we conclude that the direct distributional test has limited applicability to the problem of clustering tendency unless the sampling frame (bounded region) is known. If the boundaries of the sampling frame are known, then we can apply the exact distribution that we need to perform the test. However, for an arbitrary shape we may have serious technical difficulties in determining the exact distribution. If the shape is unknown, however, forcing the data into a sphere or a cube will distort the interpoint distances and result in spurious rejections of the null hypothesis when it is in fact true for the orginal shape.

4.3.2. Tests Based on Small Interpattern Distances

In this section, we will focus on statistics based on small inter pattern distances, and certain limiting distributions will be given which lead to approximations for sizes and power of tests for clustering tendency.

Let n patterns $\underline{x}_1$, $\underline{x}_2$,...,$\underline{x}_n$ be given in d-dimensions. We will use the counting function of interpoint distances to test for spatial randomness, i.e., the function

$$Y_n(r) = \sum_{1 \leq i \leq j \leq n} \sum I[d(\underline{x}_i, \underline{x}_j) \leq r]$$

where $d(\underline{x}_i, \underline{x}_j)$ is d-dimensional Euclidean distance between the patterns $\underline{x}_i$ and $\underline{x}_j$ and $I[.]$ is the indicator function, $I[z]=1$ if and only if z is "true". Thus, $Y_n(r)$ is the number of pairs of patterns closer than r, counting each pair once. The advantages of such a statistic are listed below.

i. It uses only distances which are invariant under Euclidian motions;

ii. Only the given patterns are required by the statistic, thus eliminating the need to generate sampling origins [27].

iii. Its similarity to Ripley's K(t) function implies that it measures the second order properties of the sample.

The main disadvantage of the counting function is that the distribution of $Y_n(r)$ has not been derived analytically, even in the relatively simple case of n patterns randomly placed in a d-dimensional hypersphere (unless n=2). Recently, however, some results have been published on the asymptotic properties of the counting function as n increases without bound [28-29]. The main results reported in this section are based on these asymptotic properties.

A model for the joint density function of n patterns over a sampling window D_n was proposed by Strauss [30] and is given below.

$$f_n(\underline{x}_1, \underline{x}_2, \ldots, \underline{x}_n) = \frac{1}{c(n,\nu)} \exp[\nu y_n(r_1)] \quad \text{if}$$

$$x_i \in D_n \quad \text{for all i and } y_n(r_o)=0.$$

where, $y_n(r)$ is the observed value of the counting function, r_o is the minimum distance allowed between patterns $(r_o>0)$, called the inhibition distance, r_1 is the radius of influence, $r_o<r_1$, ν is the clustering parameter, and $c(n,\nu)$ is a normalizing constant.

Under this model, one can generate many spatial point processes over the bounded region D_n to describe random as well as cluster and regular hypotheses. Some of them are given below, following the categorization of Kryscio, Saunders, and Funk [31]

Case 1: $\nu = 0$ ... Random positioning of patterns:

    a) $r_o=0$ ... Uniform density over $D_n(H_o)$

    b) $r_o>0$ ... Matern's hardcore process

Case 2: $\nu > 0$ ... Clustering

    a) $r_o=0$ ... Strauss's clustering, (see Kelly and Ripley[41])

    b) $r_o>0$ ... Saunder, Kryscio, and Funk's hardcore Strauss clustering

Case 3: $\nu < 0$ ... Repulsion or Regularity.

The number of small interpoint distances determines the distribution of the patterns under the models described above. Therefore, this number captured in $Y_n(r)$, appears to be an attractive statistic for measuring clustering tendency. Several alternative hypothesis can be generated in terms of the parameters, ν and r_o. In all cases, the null, or randomness hypothesis is as follows.

$$H_o: \nu = 0 \quad \text{and} \quad r_o = 0 \text{ (case 1-a)}$$

The actual distribution of $Y_n(r)$ has not been derived and is not likely to be found in closed form even for simple D_n. Saunders and Funk [28] have proposed using the limiting distribution of $Y_n(r)$ as n grows large to approximate the distribution of $Y_n(r)$ for finite n for purposes of computing size and power.

Saunders and Funk [28] proved that under some "sparseness" conditions, $Y_n(r)$ has limiting poisson distributions with mean λr^d under H_o. Here, the positive constant λ comes out of the sparseness conditions. The limiting distribution of $Y_n(r)$ is also Poisson under the alternatives represented by cases 1b-3 so the size and power of the test statistic $Y_n(r)$ can be found from a table of the Poisson distribution and the threshold of the test can be fixed. Since these limiting results rely upon large n and sparseness, we must know suitable ranges for n and r under which this approximation is reasonable.

As a concluding remark we note that, the model suggests an attractive statistic for measuring clustering tendency based on small interpattern distances. This statistic reduces the clustering tendency problem from d to one dimension since they require only the number of close pairs, between small interpattern distances. Under conditions of sparseness, the asymtotic distribution of the test statistic is known whenever the underlying density can be specified. Our simulation show a good fit when r is small for moderate sample sizes in two dimensions. Two crucial factors that must be considered when using this statistic are dimensionality and sampling window. The power studies showed that the statistic, $Y_n(r)$, is not useful in spaces dimensionality exceeding five. The size and shape of the sampling window had dramatic effects on the distribution of statistic. Effective and practical strategies for choosing a sampling window need to be deviced and evaluated befor the test suggested here can be widely applied in pattern analysis.

4.3.3. Tests Based on Near-Neighbor Distances

Tests based on small interpattern distances have low power against clustered alternatives. For example, suppose the patterns are organized into well defined clusters but distributed uniformly over each cluster. The empirical distribution of small interpattern distances should be very close to that experienced when the patterns are uniformly distributed over the entire sampling window. The tests proposed in this section overcome this difficulty by using "sampling origins" and do not depend on the intensity of the Poisson process assumed under H_o. These sampling origins are placed in the previously idenditified sampling window according to a Poisson distribution, which is to say uniform.

Near-neighbor distance based methods present a rich source of potential tests for clustering tendency. They are computationally attractive and have been extensively applied to forestry and plant ecology in two dimensions. The performances of several statistics, including the Hopkins [32], Holgate [33], T-square [34] and Eberhard [35] statistics have been compared but no one test has consistently dominated the others [36]. The T-square statistics are difficult to compute in d-dimensions [18] while little is known about the theoretical properties of the Eberhard statistic. Cross [18] states them in d-dimensions and compares their performances under H_o. Cox and Lewis [27] recently proposed a test statistic having a known distribution under the randomness hypothesis, as well as under alternative of clustering and regularity. They successfully applied the test to various real and artifical data sets in two dimensions and demonstrated its superiority over Holgate's statistic. Panayirci and Dubes [37-38] examined a d-dimensional extention of the Cox-Lewis statistic and investigated its power as a function of dimensionality in discriminating among random, aggregated, and regular arrangements of patterns in d-dimensions. We now give the details of two of the most powerful distance-based methods, namely, Hopkins' test and Cox-Lewis test.

Hopkins' Method [32] for assesing randomness is one of the earliest methods and involves the distribution of "nearest neighbor" distances in a Poisson process. It can be easily shown that the probability density of the distance from any pattern to its first nearest neighbor pattern is given by,

$$f(r) = \lambda r^{d-1} d \exp(-\lambda r^d \nu) \qquad (1)$$

where ν is the volume of a d-dimensional sphere of radius 1, ν being given by $\pi^{d/2}/\Gamma(\frac{d}{2}+1)$.

Hopkins Method

To define the statistics of interest, we adapt the following notation. The n patterns, or locations observed, are d-dimensional vectors denoted $\underline{x}_1, \underline{x}_2, \ldots, \underline{x}_n$. The m sampling origins or "posts", randomly selected in the sampling window $E \subset X$, are denoted by the d-dimensional vectors, $\underline{y}_1, \underline{y}_2, \ldots, \underline{y}_m$. Let u_i be distance from $\underline{y}_i$ to its nearest pattern, and let w_i be a random sample of m near neighbor

distances of n patterns for $i = 1, 2, \ldots, m$. We require $m \ll n$ because
we assume that origin-to-pattern nearest neighbor distances $\{u_i\}$ are
represented by independent random variables. That is, no pattern
should be the neighbor of more than one origin. Ripley [7] suggests
$m < 0.1n$. Under the null hypothesis of Poisson, $\{u_i\}$ and $\{w_i\}$ have an
identical distributions given by Equation (1). The Hopkins[i] statistic
is defined as,

$$h = \frac{\Sigma u_i^d}{\Sigma (u_i^d + w_i^d)}$$

where all sums are over the range $1, 2, \ldots, m$. Under the null hypothesis, the distances from origins to nearest patterns should, on
the average, be the same as the interpattern nearest neighbor distances so h should be about 1/2. If the patterns aggrageted, or
clustered into tight balls, the origin-to-pattern nearest neighbor
distances should, on the average, be larger than the interpattern
nearest neighbor distances so h should be larger than 1/2. By similar
reasoning, h is expected to be less than 1/2 for regularly spaced
patterns. Thus we can form one-sided or two-sided statistical test
of H_o. Assuming that all 2m random variables representing nearest
neighbor distances are statistically independent, the Hopkins statistic can be shown to have a Beta distribution with parameters (m, m).
The mean and variance of h under H_o are 1/2 and 1/4(2m+1), respectively.

Cox-Lewis Method

We now propose a new test for clustering tendency based on a
d-dimensional version of the Cox and Lewis statistic, that has a
uniform distribution under H_o. The distribution of the statistic
under regularity and clustering models are also given and the powers
of some tests based on this statistic are established.

The Cox-Lewis statistic is defined in terms of the pairs of
distances $\{u_i, v_i\}$, $i = 1, 2, \ldots, m$. Here, u_i denotes the distance from
the i th sampling origin to its nearest pattern and v_i denotes the
distance from that pattern to its nearest pattern. We again assume
randomly located origins and $m \ll n$ for statistical independence.
The intuitive idea behind the Cox-Lewis statistic is demonstrated
in Figure 4. If u_i, the distance from sampling origin i to its near
neighbor, is consistently smaller than v_i, the interpattern near-
neighbor distance for pattern i, we suspect a regular arrangement
of patterns. If the reverse holds, we suspect clustering, If neither
distance dominates the other, randomness is suggested. A means for
capturing this idea in a statistic is described below in d-dimensions.

Cox-Lewis and consequently Cormack [39] showed that $(2U_i/V_i)^2$
has a uniform distribution when condition on the event $2U_i < V_i$, no

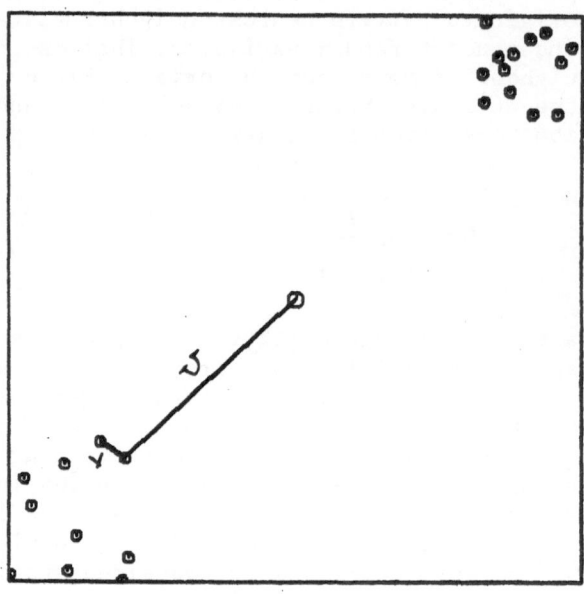

(a) Random Data

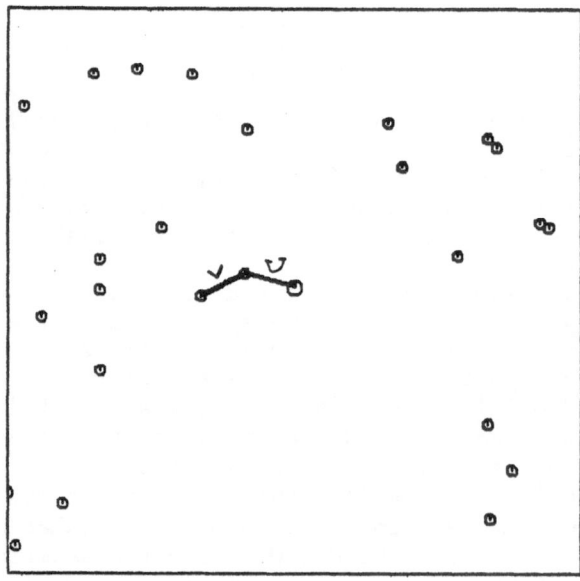

(b) Clustered Data

Figure-4. Demonstration of Cox and Lewis Statistic.

matter what the spatial distribution of patterns. The random variables U_i and V_i represent the measurements u_i and v_i, respectively. Thus, pairs of measurements satisfying this condition cannot contribute to an assessment of clustering tendency and can be ignored, and the information about spatial patterns resides essentially in the remaining m'<m pairs of measurements. Relabelling the sampling origins so near-neighbor distances can be expressed as follows

$$\{u_i, v_i, \ i = 1,2,\ldots,m' \quad \text{where} \quad v_i < 2u_i\} \ .$$

For simplicity of notation, we replace m' by m; thus, m will hence forth denote the number of sampling origins actually used so that $2u_i/v_i > 1$ for $i = 1,2,\ldots,m$. When m is small, the pairs of random variables U_i, V_i, representing the observations $\{u_i, v_i\}$, can be taken to be independent and identically distributed. For convenience, the index i is dropped and the test for clustering tendency will be based on the random variable R (for each i) defined below [38].

$$R = g(B)/p_1 [A+g(B)]$$

where

$$g(B) = 2^d A \ Sin^d(\tfrac{B}{2}) - \tfrac{A'}{d}[(d-1)2^d I(\tfrac{\pi-B}{2}) Sin^d(\tfrac{B}{2}) + (d-1) \ I(B) - Sin^{d-1}(B)]$$

and

$$A = \pi^{d/2}/\Gamma(\tfrac{d}{2} + 1); \quad A' = \pi^{\frac{d-1}{d}} /\Gamma(\tfrac{d+1}{2})$$

$$B = 2 \ arc \ Sin \ [V/2U]$$

$$I(\gamma) = \int_0^{\gamma} [Sin\theta]^{d-2} \ d\theta \ , \quad p_1 = Pr[2U>V]$$

Panayirci [40] proved that under randomness, $p_1 = 1-2^{-d}$ and the random variables $\{R_i\}$, $i = 1,2,\ldots,m$, are independent and identically distributed with uniform distributions over the unit interval so the testing schemes defined for two dimensions can also be applied in d-dimensions.

Panayirci and Dubes [40] have also derived the distribution of R, under an idealization of a regular arragement of patterns in which all patterns occur at the vertices of a d-dimensional lattice with side length a, and under a modified Thomas process, which is a simple model for the extreme clustering [36]. The powers of Neyman-Pearson tests of hypotheses based on the average Cox-Lewis statistic,

$$\bar{R} = \frac{1}{m} \sum_{i=1}^{m} R_i$$

are derived and compared. It is observed that the power is a unimodal function of dimensionality in the test of lattice regularity with the minimum occuring at 12 dimensions.

The power of the Cox-Lewis statistic is also examined under hard-core regularity and under Neyman-Scott clustering with Monte Carlo simulations. The Cox-Lewis statistic leads to one-sided tests for regularity having reasonable power and provides a sharper discrimination between random and clustered data than other statistics. The choice of sampling window is a critical factor. The Cox-Lewis statistic shows great promise for assesing the gross structure of pattern set.

5. CONCLUSIONS AND DIRECTIONS FOR FURTHER RESEARCH

In this paper, we have studied in some detail several potential methods for measuring clustering tendency, namely, the quadrad-based methods, the methods based on model fitting and the distance methods based on interpoint distances and near-neighbor distances.

The quadrat methods seem to be the weakest and of little utility in attacking the clustering tendency problem. In principle, these methods can be used for high-dimensional pattern sets. Note that since we must count the number of patterns falling in each quadrat, we would seem to require a large number of quadrats and consequently a large memory space in high dimensions. Thus, the selection of quadrat size presents a serious drawback to extending quadrats to general patter spaces. Nonrandomness may not be detected unless the number of patterns and the number of quadrats are large. Even if the number of patterns is large, the quadrat size will affect the threshold for accepting the randomness hypothesis.

The most promising source of tests for clustering tendency appears to be the technique of spatial modelling disscussed in Section 3, and we feel that the main trust of future research should be in the development of multi-dimensional versions of Ripley's spatial modelling approach using K(t) function. The use of K(t) is predicted on the fact that the second moment information is enough to distinguish the processes in two dimensions. Higher order covariance structures may be needed to distinguish the various processes in dimensions greater than two. In addition, significant computationa problems are present in high dimensions both in simulation and in computing K(t).

Among the distance-based tests, the direct distributional test discussed in Section 4.3.1 seems to be able to provide a good test of randomness, but it is very sensitive to the shape of the sampling window. It is extremely difficult to compute the distribution of the distance between two points placed at random in an arbitrary shape in all dimensions. On the other hand, the tests based on small interpoint distances suggest an attractive statistic for measuring clusterin tendency. The problem is reduced from d to one dimensions by this statistic, since it requires only the number of close pairs between small inter-pattern distances. Two crucial factors that must be considered when using this test are dimensionality and sampling window. The power studies showed that the test statistic is not useful in spaces dimensionality greater than five and has low power against clustered alternatives. The size and shape of the sampling window have dramatic effects on the distribution of statistics. More research is needed for effective and practical strategies in choosing a sampling window before the test suggested here can be widely applied in pattern analysis.

Near-neighbor distance based methods present a rich source of potential tests for clustering tendency. In this paper, we investigated two of the most powerful methods, namely, the Hopkins' method and the Cox-Lewis method. These methods were extended to multidimensional patterns and their powers as a function of dimensionality in discriminating among random, aggregated, and regular arragements of patterns were examined. The d-dimensional Hopkins and Cox-Lewis statistics were defined and their distribution under a randomness hypothesis of a Poisson spatial point process were given. The powers of the Cox-Lewis statistic and the Hopkins statistic were also examined under hard-core regularity and under Neyman-Scott clustering with Monte Carlo simulations. The Cox-Lewis method was found to be powerful and provided a sharper discrimination against both hard-core and cluster alternatives than other statistics. However, the choice of sampling window was again a critical factor.

Finally we note that further study is needed for the following:

i. Techniques for determining the sampling window.

ii. Investigation of other distance methods such as those discussed by Diggle [36].

iii. The influence of both the number of patterns and the number of sampling origins on the distance method statistics.

iv. The extension of the work of Ripley [7] to high dimensions.

v. Further research on the distribution of small interpattern distances.

6. ACKNOWLEDGMENS

The author is deeply indebted to Professors Richard C. Dubes and Anil K. Jain of the Computer Science Department, Michigan State University, U.S.A., for innumerable simulating and encouraging discussions on the research and the results presented in this paper.

554

REFERENCES

[1]   Anderberg, M.R., *Cluster Analysis for Applications*, Academic Press, New York, 1973.

[2]   Everitt, B., *Cluster Analysis*, John Wiley and Sons, New York, 1974.

[3]   Hartigan, J.A., *Clustering Algorithms*, John Wiley and Sons, New York, 1975.

[4]   Dubes R.C., and Anil, A.K., "Clustering Methodologies in Exploratory Data Analysis", *Advances in Computers*, V.19, pp. 113-228, Acedemic Press, New York, 1980.

[5]   Dubes, R.C., and Anil, A.K., "Validity Studies in Clustering Methodologies", *Pattern Recognition*, V.11, pp. 235-254, 1979.

[6]   Pielou, E.C., *An Introduction to Mathematical Ecology*, John Wiley and Sons, New York, 1963.

[7]   Ripley, B.D., "Modelling Spatial Patterns (with discussion)", *Journal of Royal Statistical Society*, Series B 39, 172, 1977.

[8]   Rogers, A., *Statistical Analysis of Spatial Dispersion*, Pion, London, 1974.

[9]   Hammersley, J.M., "Stochastic Models for the Distribution of Particles in Space", *Supply Advances in Apply Probability*, pp. 47-68, 1972.

[10] Daley, D.T., and Vere-Jones, D., "A Summary of the Theory of Point Process", in Stochastic Point Processes (P.A.W.Lewis, Ed.) John Wiley and Sons, N.Y., pp. 299-383, 1972.

[11] Snyder, D.L., *Random Point Process*, John Wiley and Sons, N.Y. 1975.

[12] Srinivasan, S.K., *Stochastic Point Process and Their Application* Griffin Statistical Monograph No.34, Hafner Press, N.Y., 1974.

[13] Neyman, J. and Scott, E.L., "Processes of Clustering and Applications", in *Stochastic Point Processes* (P.A.W.Lewis, Ed.) , John Wiley and Sons, N.Y., pp. 646-681, 1972.

[14] Neyman, J., "On a New Class of Contagious Distributions Applicable in Bacteriology and Epidermiology", *Annals of Mathematical Statistic*, Vol. 18, pp.35-37, 1939.

[15] Matern, B., "Spatial Variations", *Medd. Fran Statens Skogforskningsinstitut*. Vol. 49, pp. 1-144, 1960.

[16] Grenander, U., "Statistical Geometry, a Tool for Pattern Analysis", *Bulletion of American Math. Soc.*, Vol. 79, pp. 829-856, 1973.

[17] Rogers, A., *Statistical Analysis of Spatial Dispersion*, Pion, London, 1974.

[18] Cross, G.R., "Some Approaches to Measuring Clustering Tendency", Technical Report TR-80-03, Department of Computer Science, Michigan State University, 1980.

[19] Greig-Smith, P., *Quantitative Plant Ecology,* 2nd ed., Butterworths, London, 1964.

[20] Biswas, G., Jain, A.K., and Dubes R.C., "Evaluation of Projection Algorithms", Technical Report TR-79-03, Computer Science Department, Michigan State University, 1979.

[21] Lord, R.D., "The Distribution of Distances in a Hypersphere", *Annals of Mathematical Statistics,* Vol. 25, pp. 794-798, 1954.

[22] Alagar, V.S., "The Distribution of the Distance Between Random Points", *Journal of Applied Probability,* Vol.B., pp. 558-566, 1976.

[23] Solomon, H., *Geometric Probability,* SIAM, Philadelphia, 1978.

[24] Conover, W.J., *Practical Nonparametric Statistics,* John Wiley and Sons, New York, 1971.

[25] Bartlett, M.S., "The Spectral Analysis of Two-Dimensional Point Processes", *Biometrica,* Vol. 51, pp. 299-311, 1964.

[26] Panayirci, E., "Some New Results on the Topological Dimensionality of Signal Sets", *Bulletin of Technical University of Istanbul,* Vol. 35, No. 1, pp. 119-141, 1982.

[27] Cox, T.F., and Lewis, T., "A Conditional Distance Ratio Method for Analyzing Spatial Patterns", *Biometrica,* Vol. 63, pp.483-491, 1976.

[28] Saunders, R., and Funk, G.M., "Poisson Limits for a clustering Model of Strauss", *Journal Applied Probability,* Vol. 14, pp. 795-805, 1977.

[29] Silverman, B. and Brown, T., "Short Distances, Flat Triangles and Poisson Limits", *Journal Applied Probability,* Vol. 15, pp. 815-825, 1978.

[30] Strauss, D.J., "A Model for Clustering", *Biometrica,* Vol. 62, pp. 467-472, 1975.

[31] Kryscio, R.J., Saunders, R., and Funk, G.M., "A Note on Tests for Randomness in Sparse Spatial Patterns", (Under review by *Biometrics,* 1980.

[32] Hopkins, B., with an appendix by J. Skellam, "A New Method for Determining the Type of Distribution of Plant Individuals", *Annals of Botany,* Vol. 18, pp. 213-226, 1954.

[33] Holgate, P., "Tests of Randomness Based on Distance Methods", *Biometrica,* Vol. 52, pp. 345-353, 1965.

[34] Besag, J.E., and Gleaves, J.T., "On the Detection of Spatial Pattern in Plant Communities", *Bulletin of the International Statistical Institute,* Vol. 45, pp. 153-158, 1973.

[35] Eberhardt, L.L., "Some Developments in Distance Sampling", *Biometrics,* Vol. 23, pp. 207-216, 1967.

[36] Diggle, P.J., Besag, J., and Gleaves, J.T., "Statistical Analysis of Spatial Point Patterns by Means of Distance Methods' *Biometrics,* Vol. 32, pp. 659-667, 1967.

[37] Panayirci, E., and Dubes, R.C., "A New Statistic for Assesing Gross Structure of Multidimensional Patterns", Technical Report, TR-81-04, Department of Computer Science, Michigan State University, East Lansing, MI, 48823, USA, 1980.

[38] Panayirci, E., and Dubes, R.C., "A Test for Multidimensional Clustering Tendency", *Pattern Recognition,* Vol. 16, No. 4, pp. 433-444, 1983.

[39] Cormack, R.M., "The Invariance of Cox-Lewis's Statistic for the Analysis of Spatial Data", *Biometrica,* Vol. 64, pp. 143-144, 1977.

[40] Panayirci, E., "Generalization of the Cox-Lewis Method to d(d>2) Dimensions", Technical Report, TR-03. Marmara Research Institute, Gebze, Turkey, 1983, (Also Submitted to *Biometrica*).

[41] Kelly, F.P., and Ripley, B.D., "A Note on Strauss's Model for Clustering", *Biometrica,* Vol. 63, pp. 357-360, 1977.

A VIEW OF PATTERN RECOGNITION IN RELATION TO ERROR CONTROL CODING

M.C. Fairhurst   P.G. Farrell

The Electronics Laboratories   The Electrical Engineering Dept.
The University   The University
Canterbury, Kent   Manchester
CT2 7NT, U.K.   M13 9PL, U.K.

1.   BACKGROUND

It is clear that there exist some obvious parallels between
techniques employed in the solution of pattern recognition
problems and those underlying error-control coding. For example,
the most obvious areas of similarity might be grouped as follows:-

$$\begin{array}{ll}
\text{pattern recognition} & \equiv \text{error correction} \\
\text{pattern detection} & \equiv \text{error detection} \\
\text{feature extraction} & \equiv \text{encoding} \\
\text{classification} & \equiv \text{decoding}
\end{array}$$

It is appropriate, therefore, that this paper should begin
with a brief review of the pattern recognition field, both to
illustrate the range of approaches possible and to clarify the
terminology so that these suggested parallels might be explored
and developed in more detail. The review which follows is not
intended to be exhaustive, but to lead the newcomer to the field
to an understanding of the problems and the methodology by
sketching selected and contrasting techniques.
    The paper perhaps raises more questions than it answers but
should provoke discussion and indicate directions for future more
detailed study.

2.   INTRODUCTION TO PATTERN RECOGNITION

The term *pattern recognition* describes a process which allows
the naming of a general category of objects in response to data

which forms a specific "pattern" example, this process assuming
the existence of pattern *classes* or categories defined by identi-
fiable common attributes among the members of class.

Even more fundamentally, a *pattern* is defined by a pattern
vector $\underline{X} = \{x_1, x_2 \ldots x_N\}$ where individual elements correspond
to some measurable feature of the pattern, the existence of which
implies a requirement for data sensing and extraction of appropriate information for pattern definition. Likewise a pattern
feature may be defined as a measurable property of a pattern which
is selected to contribute to the identification/recognition
process. Since real-world data is often not available in a form
for optimally efficient processing, patterns may be transformed in
a variety of ways before the features are extracted and the
recognition procedure initiated, thereby introducing a *preprocessing* operation. The pattern recognition process may therefore be
summarised in its simplest form in the traditional schematic shown
in Fig. 1.

Approaches to pattern recognition may assume a variety of
forms, and the brief review provided here will begin from the
viewpoint of classical Bayesian decision theory. This both
characterises some of the fundamentally important issues and
provides a reference point for other techniques.

3.   THE BAYES CLASSIFIER

Consider a simple pattern environment in which there may
exist two pattern classes C_1 and C_2 with a priori probabilities of
occurrence $p(C_1)$ and $p(C_2)$ respectively. If it is assumed that an
observation on a pattern generates a measurable feature x, then
$p(x/C_j)$ is the class conditional probability density function for
x, given that the pattern belongs to class C_j.

The pattern recognition problem in this simple situation may
then be regarded as the determination of the probability that the
pattern belongs to class C_j, given that it has the observed
measured feature x. In other words, a recognition system seeks to
assign an unknown pattern to the class C_j such that $p(C_j/x)$ is a
maximum.

Thus, the decision rule adopted might be:

Assign pattern to C_1 if $p(C_1/x) > p(C_2/x)$

Assign pattern to C_2 otherwise.

Bayes rule allows this decision rule to be made practically
viable, since

$$p(C_j/x) = \frac{p(x/C_j) \; p(C_j)}{p(x)}$$

where
$$p(x) = \sum_{j=1,2} p(x/C_j)\, p(C_j)$$

Since $p(x)$ is here just a scale factor, the decision rule may be written as:

Assign to C_1 if $p(x/C_1)\, p(C_1) > p(x/C_2)\, p(C_2)$

Assign to C_2 otherwise.

It is easily shown that this decision rule minimises the average probability of classification error, and the Bayes classifier is therefore a very useful performance reference.

Although this type of approach to classification is highly attractive it requires that the class probabilities and class conditional distributions are known, and these criteria are very infrequently satisfied in practice. These values must therefore be estimated on the basis of data available in a *training set* of known-identity example patterns. Inevitably for practically feasible computation these estimates are based on assumptions (such as the likely *form* of the underlying distributions, for example) and thus a Bayesian error rate represents in practice a lower bound on error probability.

As an example, consider a classification problem in which patterns are characterised by a feature vector $\underline{X}$, and assume that a priori class probabilities are equal. In a particular problem where, for example, typical values for $\underline{X}$ represent slightly distorted versions of some class-prototype vector $\underline{X}^*$, then it might be appropriate to make the assumption that the class conditional densities can be approximated by multivariate normal distributions with equal covariance matrices.

Thus $p(\underline{X}/C_i)$ may be approximated by the expression:

$$p(\underline{X}/C_i) = \frac{1}{(2\pi)^{N/2} |\Phi_i|^{\frac{1}{2}}} \exp\left| - \frac{1}{2}(\underline{X}-\underline{\mu}_i)\ \Phi_i^{-1}(\underline{X}-\underline{\mu}_i) \right|$$

where
Φ_i = covariance matrix for the ith class

$\underline{\mu}_i$ = vector of means for all N components of $\underline{X}$

$|\Phi_i|$ = determinant of Φ_i

Φ^{-1} = inverse of Φ

The covariance matrix and mean vector can be computed for each class, and a decision function ψ_{ij} derived such that

$$\psi_{ij} = \ln \frac{p(\underline{X}/C_i)}{p(\underline{X}/C_j)}$$

ψ_{ij} is computed for all $i,j (i \neq j)$ and $\underline{X}$ assigned to C_k such that ψ_{kj} is maximum.

4. LINEAR DECISION FUNCTIONS

In adopting a Bayes classifier approach it is assumed that information is available about the underlying probability distributions of pattern classes occurring in the operating environment of the classifier, and training samples are used to estimate parameter values.

An alternative approach is to make no assumptions about pattern statistics but rather to assume a specified form for the decision functions. In particular, the assumption of a linear decision function allows conceptual simplicity and mathematical tractability.

The general form of a linear decision function $g(\underline{X})$ may be written as

$$g(\underline{X}) = \underline{W}\ \underline{X}$$

where $\underline{X} = \{x_1,\ x_2\ \dots\ x_N,\ 1\}$

is the pattern vector augmented by the additional unit component for mathematical convenience (but without disturbing its geometrical properties in feature space).

$$\underline{W} = (w_1,\ w_2\ \dots\ w_N,\ w_o)$$

is a coefficient or *weight* vector.

Several classifier configurations are possible using linear decision functions and may best be described geometrically in feature space. An example for the three-class case (shown for a 2-dimensional feature space for convenience) is illustrated in Fig. 2, in which classification is achieved on the basis of a number of pairwise class discriminations. A decision function $g_{ij}(\underline{X})$ is used to distinguish between the ith and jth classes and the decision rule adopted for this configuration is that $\underline{X}$ is assigned to class C_i such that $g_{ij}(\underline{X}) > 0$ for all $j \neq i$.

An important aspect of this pattern recognition scheme is that appropriate values for the components of the weight vector must be derived. This is generally achieved by implementing an incremental updating procedure, utilising available training samples, which converges to provide a suitable set of weight values. Such a scheme is successful only when data is appropriately distributed in feature space, and may often require a sacrifice in performance in return for analytical simplicity.

5.   CLUSTERING TECHNIQUES

Clustering techniques provide an intuitively appealing approach to pattern classification by considering directly distance relations among patterns. This type of classifier assumes no knowledge of the statistical feature distributions but requires instead a large number of training samples. The assumption on which this approach is based is that patterns which are separated by a small "distance" in feature space are likely to belong to the same class, while large "distances" between patterns suggest membership of different classes. The distance measure adopted may be Euclidean distance or some other appropriate metric, and its choice is left to the discretion of the classifier designer for a particular application.

In the simplest case, assume that the only information available about a set of pattern classes is a knowledge of the existence of a set $\{S\}$ of labelled patterns, where

$$\{S\} = \{(\underline{X}_1,\ C_1)\ \ldots\ (\underline{X}_M,\ C_M)\}$$

where C_i represents the class to which the training sample $\underline{X}_i$ belongs.

In response to a pattern $\underline{X}$ of unknown identity the classifier seeks to find the element $(\underline{X}',\ C')$ of the set $\{S\}$ such that the distance between $\underline{X}$ and $\underline{X}'$ is minimised. The pattern $\underline{X}$ is then assigned to the class C' of its *nearest neighbour* (NN). In other words, the decision rule adopted is

Assign $\underline{X}$ to C' if   .

$$\delta(\underline{X},\ \underline{X}') = \min_{i} \delta(\underline{X},\ \underline{X}_i)$$

where δ is the distance metric chosen.

This decision rule is known as the 1-NN rule, and may in practice be extended to a generalised K-NN rule, where the classifier searches for the K nearest neighbours to $\underline{X}$ from $\{S\}$, and assigns $\underline{X}$ to the class which is most represented in the labels of these neighbours.

A further interesting feature of clustering algorithms is that they easily allow classification on an unsupervised basis. Confronted by a set of patterns of unknown identity, patterns are examined in turn and a pattern is assigned to a cluster if its distance with respect to any currently available prototype is less than some predefined threshold value. A pattern falling outside this threshold boundary in feature space is used to define a new cluster centre, and so on. This technique is also used in the vector quantisation process for voice, picture, and other source coding tasks.

6. SYNTACTIC PATTERN RECOGNITION

In recent years much interest has developed in a fundamentally different approach to the pattern recognition problem. This syntactic approach focusses on the inherent structure of patterns and, using concepts from formal language theory, seeks to generate an essentially linguistic description of what constitutes a pattern class.

The general approach is to decompose a pattern into its basic primitive elements which represent allowable symbols in some grammar or set of syntax rules which permit the generation of meaningful sentences. This grammar provides the set of rules which governs the composition of primitives which define a pattern. The recognition process in this scheme is achieved by parsing a "sentence" to determine whether or not it is syntactically allowable with respect to the specified grammar, and hence whether a pattern may be regarded as belonging to some predefined class of patterns.

This approach is the subject of another separate contribution and consequently will not be discussed further here.

7. THE BINARY N-TUPLE CLASSIFIER

The n-tuple method of pattern recognition has a long history and represents a conceptually straightforward technique which can be implemented in a variety of forms. The method may be regarded as representing a (computationally simple) compromise between assuming statistical independence of pattern features and seeking a complete knowledge of the underlying statistical feature distributions.

The basic configuration of an n-tuple classifier scheme is illustrated in Fig. 3. The method employs binary feature sampling with memory cell arrays, where feature representation corresponds to the ordered grouping of binarised pattern points sampled by memory cells which associate feature occurrences with pattern classes by storing information describing the appearance of the possible feature-defining binary groups over a number of labelled training patterns.

Each possible recognition class C_k is associated with a processing network (the C(k)-network) comprising an array of memory cells, where the input vector of the ith cell is an n-tuple denoted by

$$\underline{x}_i^{(k)} = (x_{i,1}^{(k)}, x_{i,2}^{(k)} \ldots\ldots x_{i,n}^{(k)})$$

where $\underline{x}_i^{(k)}$ is a subset of $\underline{X}$, an input pattern.

The C(k)-network is trained with a number of patterns from class C_k, when the ith memory cell is forced to store a 1 at an address defined by the vector $\underline{x}_i^{(k)}$. After training, in response to an unknown pattern $\underline{X}$ the kth network computes a decision function $f^k(\underline{X})$, the numerical summation of binary cell outputs, each of which signifies the occurrence or non-occurrence of the currently sampled n-tuple $\underline{x}_j^k$ in at least one training pattern of the class C_k, and $\underline{X}$ is assigned to class C_k such that $f^k(\underline{X}) > f^\ell(\underline{X})$ for all $\ell \neq k$. In other words, $\underline{X}$ is assigned to the kth class such that the number of n-tuples in $\underline{X}$ with a state corresponding to that found in at least one training pattern in the kth class is greatest.

The major parameters which determine the performance of the classifier are the number of binary pattern points sampled by a memory cell (n-tuple size), the way in which these points are selected and grouped (the feature mapping) and the training set size. These parameters are all inter-related, but the relation between n-tuple size and recognition performance is significant, and an optimal value is often sought empirically for a given data set. Where pattern classes are particularly well-defined a relatively small sample size can yield acceptable performance, while a classifier operating on more diversely specified data might typically require an n-tuple size of 12 or greater for optimal performance. As n-tuple size increases, however, storage requirements increase dramatically and hence a compromise is often sought between attaining an acceptable performance level and keeping total storage requirements within reasonable bounds. One possible trade-off is to use a multi-layered n-tuple classifier.

The n-tuple technique for pattern recognition will be discussed further below.

8. PATTERN RECOGNITION AND ERROR CONTROL CODING

The primary aim of this paper is to provide some background material prerequisite to further exploration and development of the parallels indicated in the interaction between the techniques of pattern recognition and those of error control coding. Thus further discussion of potentially fruitful areas for specific consideration is indicated. The following discussion will centre on the n-tuple approach to classification in raising specific issues for study for, though pattern recognition techniques in general deserve attention in this context, the n-tuple approach is of particular interest since implementation of an n-tuple system is possible in a variety of configurations where the possibility of trade-off in hardware/software, speed/cost and performance/cost factors, provides an immense degree of flexibility of choice for the system designer. Furthermore, by its inherent structural tendency towards parallelism this type of classifier allows the probability of implementation for very fast operation, and decreasing unit memory costs offer obvious advantages for n-tuple systems.

It is clear that there exists a parallel between a binary code word which may be (randomly) distorted through noise and the

concept of, say, visual patterns coded as binary vectors where a
pattern class can be regarded as an "archetypal" (class-defining)
vector with noisy or distorted versions of that pattern representing individual instances within that class.
it is now possible to point towards some areas of investigation
which might bring together the concepts of pattern recognition and
error coding.

a) Minimum distance decoding

(i) Random errors

The primary objective of minimum distance decoding is to
determine for a noise-distorted "archetype" (the transmitted code
word) the form of the archetype of which the received data is most
likely to be a distorted version. This parallels very closely the
situation illustrated by the example of, say, alphanumeric
character recognition, or of a Bayes Classifier in general. Here,
an unknown pattern may be regarded as a noise-distorted version of
a pattern representing a well-formed, stylised character, where
each pattern class would be associated with one (or possible more)
such archetypes.

Thus it is possible to consider a system which utilises a
classifier, trained with appropriate examples of received
(identified) code words, to map from the set of received data
blocks (representing unknown patterns) to code words corresponding
to pattern class archetypes. Particularly important problems for
investigation would be related to the selection of allowable code
words at the source and the possibility of localization in
possible error distributions with a view to n-tuple selection.

(ii) Burst errors

For burst error correction a similar scheme might be envisaged,
with n-tuples now formed from an ordered selection of data bits to
minimise the spread of values in the decision function for received
data patterns defining a particular code word. In addition, the
selection of n-tuple size could be an important factor in relation
to expected burst length.

b) Sequential decoding

As indicated above, in a conventional n-tuple classifier
configuration the expected improvement in performance for increasing n-tuple size can only be achieved at the expense of a
significant increase in the storage requirements of a practical
implementation. Recent research has shown, however, that a two-
level configuration of processing cells can give substantially
improved performance levels without such escalating storage costs.

In essence the configuration assumes two levels of processing.
The first employs a conventional single layer system (with modest

n-tuple size) from which the output is not a unique classification decision but a set of "guesses" which are passed to a second decision layer. This process enables the error rate in the first layer to be kept very low. The second processing layer then reaches a classification decision over a much reduced number of classes by performing a sequence of highly efficient pairwise class dichotomies. Such a procedure seems to have some resemblance to the process of sequential decoding, and suggests another area for further study.

c) Soft decision decoding

Most practical problems in pattern recognition result in the design of a classifier with an error rate which is acceptable in a specific context. In practice it is possible to avoid making a classification decision in a marginal case (with the attendant high risk of error) by "rejecting" an input pattern. This idea of a *rejection class*, defined in terms of conditions imposed on the classifier decision functions can often allow significant reductions in the overall error rate, though at the expense of retaining many unclassifiable samples. However, in situations where errors are regarded as extremely costly this approach is very useful, since there remains the possibility of processing the rejected patterns in an alternative way, perhaps employing more extensive or time-consuming computation.

In the n-tuple classifier this concept is easily introduced by defining a *confidence level* θ, such that a decision is only made if the value of the largest decision function exceeds that of the next largest by an amount θ. Thus a pattern $\underline{X}$ is assigned to the class C_k such that $f^k(\underline{X}) > f^{\ell}(\underline{X}) + \theta$ for all $\overline{\ell} \neq k$. If no such k can be found then $\underline{X}$ is assigned to a rejection class. Fig. 4 shows the general effect of varying θ in a typical classifier, where in this example the data is derived from an experiment in alphanumeric character recognition.

The concept of a rejection class, coupled with some scheme such as the two-level classifier configuration of the previous section suggests a relation with the concept of soft-decision decoding, which permits some flexibility in the interpretation of received data and its decoding.

d) Feature weighting

In the conventional n-tuple classifier all features carry the same notional weight in terms of their contribution to the decision function. Thus n-tuple features which occur infrequently over a set of training samples are just as important in the determination of a classification decision as features occurring frequently and, as a result, a relatively small proportion of unrepresentative samples in a training set can in some cases significantly degrade recognition performance.

An alternative approach is to utilise a configuration which weights n-tuple features in $\underline{X}$ in relation to their frequency of occurrence in training set samples. This scheme, for little extra implementation expense, can improve classifier performance significantly in some situations.

The concepts underlying the frequency-weighted n-tuple classifier seem to parallel the underlying features of "priority coding" schemes, where, in order to take into account the relative degree of importance attached to certain subsets of data bits in the codeword, the coding scheme introduces areas of high integrity in the received data. This non-uniform distribution of confidence in received data suggests a motivation for exploring the potential of a feature weighted classifier structure in implementing the decoding process. These concepts also have obvious application to unequal error protection codes (in which certain digits in a word are better protected than others), and asymmetric error protection codes (in which certain symbols, when they occur (ONES, say) are better protected than others).

9.   CONCLUSION

The areas to which specific attention has been drawn suggest some possible ways in which techniques of pattern recognition might be brought to bear on problems of error control coding. Clearly these suggestions have outlined specific problem areas and a specific approach to the implementation of a classification scheme, and it is emphasised that these are intended to be merely illustrative of more general and wide-ranging considerations, although a more detailed example can be found in the work of Wride. Optimism about the potential for fruitful interaction between these two disciplines can ultimately only be justified if thorough and imaginative studies are undertaken, and it is towards that objective that this paper seeks to motivate the reader.

BIBLIOGRAPHY

B. G. Batchelor (Ed.), "Pattern Recognition: Ideas in Practice", Plenum Press, 1978.

W. W. Bledsoe and I. Browning, "Pattern Recognition and Reading by Machine", Proc. EJCC, 1959, pp. 225-236.

P.A. Devijver, P.A. and J. Kittler, "Pattern Recognition - A Statistical Approach", Prentice-Hall, 1982.

M. C. Fairhurst and T. J. Stonham, "A Classification System for Alphanumeric Characters Based on Learning Network Techniques", Digital Processes, No. 2, 1976, pp. 321-339.

M. C. Fairhurst and M. Mattoso Maia, "A Two-layer Memory Network Architecture for a Pattern Classifier", Pattern Recognition Letters, No. 1, 1983, pp. 267-271.

M. C. Fairhurst and S. R. Kormilo, "Some Economic Considerations in the Design of an Optimal N-tuple Pattern Classifier", Digital Processes, No. 3, 1977, pp. 321-329.

P. G. Farrell, "A Survey of Error-Control Codes", Algebraic Coding Theory and Applications, Longo (Ed.), CISM No. 258, Springer-Verlag 1979.

D. J. Hand, "Discrimination and Classification", Wiley, 1981.

L. Kanal, "Patterns in Pattern Recognition", IEEE Trans. Inform. Theory, IT-20, 1974, pp. 697-722.

R. J. McEliece, "The Theory of Information and Coding", Addison-Wesley, 1977.

J.T. Tou and R. C. Gonzalez, "Pattern Recognition Principles", Addison-Wesley, 1974.

J. R. Ullmann, "Pattern Recognition", Butterworth, 1974.

A. Wride, "The Application of Pattern Recognition Techniques to the Decoding Problem", M.Sc. Dissertation, University of Kent, 1971.

MARKOV SOURCE MODELING OF TEXT GENERATION

Frederick Jelinek

Continuous Speech Recognition Group

IBM T.J. Watson Research Center

1. *Introduction*

A language model is a conceptual device which, given a string of past words, provides an estimate of the probability that any given word from an allowed vocabulary will follow the string. In speech recognition, a language model is used to direct the hypothesis search for the sentence that was spoken. In fact, its probability assignments are a factor in evaluating the likelihood that any word string was uttered, and in ordering the word string hypotheses to be examined (1). Language models are also useful in text encoding for compression or transmission, in character recognition of printed or handwritten text, etc. Ideally, the probability assigned by a language model to any word should depend on the entire past sequence of words. However, it is practically impossible to implement such a dependence: 1) The probability estimate involved would have to be extracted from a giant sample of training text containing billions of words; 2) If the probabilities were extractable, sufficient space in computer memory could not be found to store them.

It must be remembered that in most languages, as the vocabulary increases, an ever increasing majority of words contained in it are nouns and verbs. In any sentence, when a noun is to be predicted, almost any noun in the vocabulary is possible. Thus the conditional probability

$$p\ (w_k/w_{k-1},...,w_1) \tag{1}$$

of most words w_k will be positive when the past $w_{k-1},...,w_1$ can be followed by either a noun or a verb. Hence the number of non-zero probabilities (1) would be of the order of M^k which is huge for even moderate values of M and k. Therefore, for reasons of estimation as well as estimate storage, past strings on which the prediction is based must be partitioned into a manageable number of appropriately chosen equivalence classes. Such equivalence classes might be grammatical or semantic. A particularly simple expedient is a n-gram classification which states that all past strings ending in the same substring of n-1 words are equivalent. The Continuous Speech Recognition project, of which I am a member, has been using successfully a three-gram approach in its recognition experiments. Thus the prediction is based on an estimate of the probability $p(w_1/w_2,w_1)$ of the occurrence of word w_3 given that it is preceded by words w_2 and w_1, i.e.,

$$p(w_1,w_2,...,w_n) \cong p(w_1,w_2)\prod_{k=3}^{n}p(w_k/w_{k-1},w_{k-2}) \tag{2}$$

Even this truncated estimate cannot be obtained crudely by simply counting up relative frequencies of trigrams in a given text. In fact, when we divided a large text of laser patent descriptions into two subsets of 300000 and 1500000 words, we found that 23% of trigrams occurring in the first subset have never taken place in the second, larger one. Thus a language model based on a relative frequency trigram model would declare as impossible much of actually occurring text. It follows that it is necessary to smooth out the trigram estimates in some way. One possibility is through linear interpolation. Let f be the relative frequency function,

$$f(w_k/w_{k-1},...,w_1) \overset{\Delta}{=} \frac{N(w_1,w_2,...,w_{k-1},w_k)}{N(w_1,w_2,...,w_{k-1})} \tag{3}$$

where $N(w_1,...,w_k)$ denotes the number of occurrences of the string $w_1,...,w_k$ in the text. Then we estimate the required probability by the formula

$$\hat{P}_1(w_3/w_2,w_1) = \lambda_1 f(w_3/w_2 w_1) + \lambda_2 f(w_3/w_2) + \lambda_3 f(w_3) \tag{4}$$

where the weights $\lambda_1, \lambda_2, \lambda_3$ are an appropriately chosen probability distribution. How this choice is made is discussed in Section 5.

Before discussing the quality of the interpolated estimator (4), let us point out one of its advantages. Consider a vocabulary of, say, 5000 words. Formula (4) then serves to estimate $(5000)^3 = 1.25 \times 10^{11}$ different parameters. But any text of N words used to estimate relative frequencies can contain at most N different trigrams and bigrams, so the number of parameters that must be stored will not exceed $2N + 5002$. In fact, an office correspondence data base used at IBM contains only 395000 and 162000 distinct trigrams and bigrams, respectively. Thus the required storage is 5.62×10^5 rather than 1.25×10^{11} numbers.

How can one judge the quality of the estimate (4)? Obviously by using it to (over-) estimate the entropy H of some related test text, distinct from the one used to calculate the relative frequencies. The desired estimate formula is

$$\hat{H} = -\sum f^*(w_1 w_2 w_3) \log \hat{P}_1(w_3/w_2 w_1) \tag{5}$$

where f^* corresponds to the trigram relative frequencies of the test text, and the sum is over all trigrams occurring in it (the test text is a sample of one that should be compressed or transmitted). In order to tell us about the quality of (4), the result (5) must be compared to

572

some reliable estimate of entropy of the test text. Although Shannon [2] in his classical paper, and Cover [3] and others later have developed techniques to estimate entropies of languages (using human subjects as their language models!), the numerical values estimated by them are questionable, since they have been obtained on the basis of too little data which moreover was not necessarily typical. In any case, their estimates would not apply to specific sub-corpora of English, such as office correspondence.

There is a strong suspicion that language model (4) needs a lot of strengthening, at least when the relative frequencies involved are based on less than, say, 50 million words of text. In particular, when the bigram w_1, w_2 occurred only once or twice then the trigram component of the estimator (4) is very unreliable. This can be somewhat remedied by making the weights λ depend on the number of observations of the conditioning bigram w_1, w_2 and unigram w_2, but this will not solve the problem. For instance, if neither the sequence *bought old* nor the word *old* was ever followed in the training data base by the word *terminals*, then (4) results in the prediction

$$\hat{P}_1(terminals/old, bought) = \lambda_3 f(terminals) \tag{6}$$

But the relative frequency of *terminals*, or of other plural nouns that might be expected to follow *bought old*, will undoubtedly be low in comparison with the relative frequency of totally unsuitable words like *the*, *into*, *is*, etc. This fact suggests adding a grammatical component $q((w_3/g(w_2), g(w_1))$ to (4), where $g(w)$ is the part-of-speech [noun, verb, adjective, ...] that w assumed in the running text. The new estimate then becomes

$$\hat{P}_2(w_3/w_2 w_1) = \lambda_1 f(w_3/w_2 w_1) + \lambda_2 f(w_3/w_2) + \lambda_3 f(w_3) + \lambda_4 q(w_3/g(w_2)g(w_2)) \tag{7}$$

Two straightforward ways to obtain the probability q are

$$q(w_3/g(w_2) = g_2, g(w_1) = g_1) = f(w_3/g_2 g_1) \tag{8}$$

and

$$q(w_3/g(w_2) = g_2, g(w_1) = g_1) = \sum_{g_3} f(w_3/g_3) f(g_3/g_2 g_1) \tag{9}$$

They both involve annotating the training text by parts of speech and then collecting relative frequencies. Since English words can have many parts of speech (e.g. *light* is a noun, verb, adjective, or adverb) annotation cannot be done by dictionary lookup. One must either employ humans (thus limiting severely the amount of training text), or devise an automatic approach involving a grammar or some other suitable mechanism.

We have already observed that grammar could be used not just to assign parts of speech to words, but to place a past word string $w_1, w_2, ..., w_{k-1}$ into a grammatical equivalence class. The latter would consist of all strings that are identically characterized by some appropriate grammar. If ϕ is such a grammatical characterization function, relative frequency counts $f(w_k/\phi(w_{k-1}, ..., w_1))$ could be obtained from the training text and used as one of several linearly weighted estimators contributing to the estimate of the desired prediction probability $p(w_k/w_{k-1}, ..., w_1)$. In fact, $f(w_k/\phi(w_{k-1}, ..., w_1))$ could simply be added to the trigram estimators of (7) as a fifth component.

Grammatical analysis has much wider application than as a language model component. It is usable in all sorts of text processing such as text understanding or translation. But analysis is very difficult in practice because grammars are notoriously ambiguous and usually require hand-tailoring based on many pragmatic compromises. One way to proceed might be to assign probabilities to all analyses of the given sentence and choose the one with the highest value. The approach introduced in Section 7 assigns such probabilities based on

574

statistical parameters of the underlying production model whose values were optimally adjusted relative to a training text.

We introduced three problems: determination of optimal interpolation weights λ_i, automatic annotation of text by parts of speech, and grammatical analysis of text. The solution of the first two is based on an algorithm for estimating parameters of Markov sources. Its generalization leads to the solution of the last problem.

2. *Markov Sources*

By a Markov source, is meant a collection of states connected to one another by transitions which produce symbols from a finite alphabet. Each transition t from a state s has associated with it a probability $q_s(t)$ which is the probability that t will be chosen next when s is reached. From the states of a Markov source we choose one state as the intial state and one state as the final state. The Markov source then assigns probabilities to all strings of transitions from the initial state to the final state. Fig. 1 show an example of a Markov source.

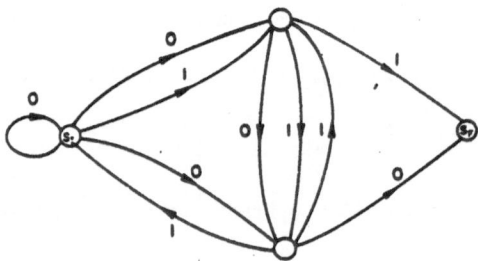

Fig. 1. A Markov source

A Markov source is defined more formally as follows. Let $\mathscr{S}$ be a finite set of states, $\mathscr{T}$ a finite set of transitions, and $\mathscr{A}$ a finite alphabet. Two elements of $\mathscr{S}$, s_I and s_F are distinguished as initial and final states, respectively. The structure of a Markov source is a 1-1

mapping $M:\mathcal{T}\rightarrow\mathcal{S}\times\mathcal{A}\times\mathcal{S}$. If $M(t) = (l,a,r)$ then l is called the predecessor state of t, a the

output symbol associated with t, and r the sucessor state of t. Thus $l = L(t)$, $a = A(t)$, and

$r = R(t)$, where L, A, and R are *predecessor*, *label*, and *successor* transition functions,

respectively.

The statistical parameters of a Markov source are the probabilities $q_s(t), s\in\mathcal{S}-\{s_F\}, t\in\mathcal{T}$,

chosen such that

$$q_s(t) = 0 \quad \text{if} \quad s\neq L(t)$$

and

$$\sum_t q_s(t) = 1, \quad s\in\mathcal{S}-\{s_F\}. \tag{10}$$

A string of n transitions (*) $\mathbf{t}_1^n$ for which $L(t_1) = s_I$ is called a path; if $R(t_n) = s_F$, then it

is a complete path. The probability of a path $\mathbf{t}_1^n$ is given by

$$p(\mathbf{t}_1^n) = q_{s_I}(t_1)\prod_{i=2}^{n} q_{R(t_{i-1})}(t_i). \tag{11}$$

Associated with path $\mathbf{t}_1^n$ is an output symbol string $\mathbf{a}_1^n = A(\mathbf{t}_1^n)$. A particular output

string $\mathbf{a}_1^n$ may in general arise from more than one path. Thus, the probability $P(\mathbf{a}_1^n)$ is given

by

$$P(\mathbf{a}_1^n) = \sum_{\mathbf{t}_1^n} P(\mathbf{t}_1^n)\delta(A(\mathbf{t}_1^n),\mathbf{a}_1^n) \tag{12}$$

where

$$\delta(a,b) = \begin{cases} 1 & \text{if} \quad a = b \\ 0 & \text{otherwise} \end{cases}$$

(*) $\mathbf{t}_1^n$ is a short-hand notation for the concatenation (string) of the symbols $t_1, t_2, ..., t_n$.

Strings are indicated in boldface throughout.

In practice it is useful to allow transitions which produce no output. These null transitions are represented diagrammatically by interrupted lines (see Fig. 2). Rather than deal with null transitions directly, it is convenient to associate with them the distinguished letter ϕ. We then add to the Markov source a filter (see Fig. 3) which removes ϕ, transforming the output sequence $\mathbf{a}_1^n$ into an observed sequence $\mathbf{b}_1^m$ where $b_i \in \mathscr{B} = \mathscr{A} - \{\phi\}$. Sources will not be allowed to have closed circuits of null transitions (although this restriction could be eased).

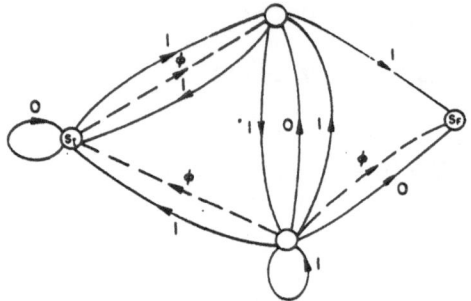

Fig. 2. A Markov source with null transitions.

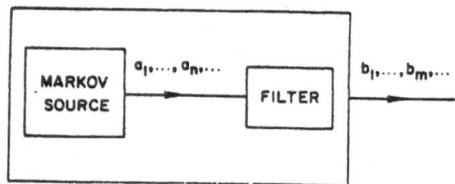

Fig. 3. A filtered Markov source.

3. The Viterbi Algorithm

Given an observed sequence $\mathbf{b}_1^m$, generated by a Markov source, it is of interest to find the most likely path of transitions $\mathbf{t}_1^n (m \leq n)$ that could have caused the former. This is best accomplished by the well known Viterbi Algorithm [4,5] described below.

Let $\tau_k(s)$ be the most probable path to state s which produces output $\mathbf{b}_1^k$. Let $V_k(s) = P(\tau_k(s))$ denote the probability of the path $\tau_k(s)$. We wish to determine $\tau_m(s_F)$, the most probable path into the final state s_F.

Because of the Markov nature of the process, $\tau_k(s)$ can be shown to be an extension of $\tau_{k-1}(s')$ for some s'. Therefore, $\tau_k(s)$ and $V_k(s)$ can be computed recursively from $\tau_{k-1}(s)$ and $V_{k-1}(s)$ starting with the boundary conditions $V_0(s_I) = 1$ and $\tau_0(s_I) = $ *the null string*. Let $C(s,a) = \{t \mid R(t) = s, A(t) = a\}$. Then

$$V_k(s) = \max \{ \max_{t \in C(s,b_k)} V_{k-1}(L(t))q_{L(t)}(t), \ \max_{t \in C(s,\phi)} V_k(L(t))q_{L(t)}(t) \}. \tag{13}$$

If the maximizing transition t is in $C(s, b_k)$ then $\tau_k(s) = \tau_{k-1}(L(t)) \circ t$; otherwise t must be in $C(s, \phi)$ and $\tau_k(s) = \tau_k(L(t)) \circ t$ [$\circ$denotes concatenation]. Note that in (13) $V_k(s)$ depends on $V_k(L(t))$ for $t \in C(s, \phi)$. $V_k(L(t))$ must therefore be computed before $V_k(s)$. Because closed circuits of null loops are not allowed, it is possible to order the states $s_1, s_2, s_3, \ldots$, such that $t \in C(s_k, \phi)$ and $L(t) = s_j$ only if $j < k$. If we then compute $V_k(s_1), V_k(s_2)$, etc., in sequence, the necessary values of $V_k(L(t))$ will always be available when required.

4. *Automatic Estimation of Markov Source Parameters from Data*

Given a Markov Source with a transition structure $\mathscr{T} \to \mathscr{S} \times \mathscr{A} \times \mathscr{S}$. We wish to estimate the transition probabilities $q_s(t)$ most likely to have caused a fixed observed output sequence $\mathbf{b}_1^m$.

Let $P_i(t, \mathbf{b}_1^m)$ be the joint probability that $\mathbf{b}_1^m$ is observed at the output of a filtered Markov source and that the i th output b_i *spans* t, that is, either $A(t) = b_i$ or $A(t) = \phi$ and the last output generated previous to the transition t was b_{i-1}.

578

The *count*

$$c(t, \mathbf{b}_1^m) \overset{\Delta}{=} \sum_{i=1}^{m} P_i(t, \mathbf{b}_1^m) / P(\mathbf{b}_1^m) \qquad (14)$$

is the Bayes *a posteriori* estimate of the number of times that the transition is used when the string $\mathbf{b}_1^m$ is produced. If the counts are normalized so that the total count for transitions from a given state is 1, then it is reasonable to expect that the resulting relative frequency

$$f_s(t, \mathbf{b}_1^m) \overset{\Delta}{=} \frac{c(t, \mathbf{b}_1^m) \delta(s, L(t))}{\sum_{t'} c(t', \mathbf{b}_1^m) \delta(s, L(t'))} \qquad (15)$$

will approach the transition probability $q_s(t)$ as m increases.

This suggests the following iterative procedure for obtaining estimates of $q_s(t)$.

1) Make initial guesses $q_s^0(t)$ of the values of $q_s(t)$.

2) Set $j = 0$.

3) Compute $P_i(t, \mathbf{b}_1^m)$ for all i and t based on $q_s^j(t)$.

4) Compute $f_s(t, \mathbf{b}_1^m)$ and obtain new estimates $q_s^{j+1}(t) = f_s(t, \mathbf{b}_1^m)$.

5) set $j = j + 1$.

6) Repeat from 3.

To apply this procedure, we need a simple method for computing $P_i(t, \mathbf{b}_1^m)$. Now $P_i(t, \mathbf{b}_1^m)$ is just the probability that a string of transitions ending in $L(t)$ will produce the observed sequence $\mathbf{b}_1^{i-1}$, times the probability that transition t will be taken once $L(t)$ is reached, times the probability that a string of transitions starting with $R(t)$ will produce the remainder of the observed sequence. If $A(t) = \phi$, then the remainder of the observed sequence is $\mathbf{b}_i^m$. If $A(t) \neq \phi$ then, of course, $A(t) = b_i$ and the remainder of the observed sequence is $\mathbf{b}_{i+1}^m$. Thus if $\alpha_i(s)$ denotes the probability of producing the observed sequence $\mathbf{b}_1^i$ by a string of transitions starting from the state s_I and ending in s, and $\beta_i(s)$ denotes the probability of producing the observed sequence $\mathbf{b}_i^m$, by a string of transitions starting from the state s, and ending in s_F then

$$P_i(t,\mathbf{b}_1^m) = \begin{cases} \alpha_{i-1}(L(t))q_{L(t)}(t)\beta_i(R(t)) & \text{if } A(t) = \phi \\ \alpha_{i-1}(L(t))q_{L(t)}(t)\beta_{i+1}(R(t)) & \text{if } A(t) = b_i \end{cases} \tag{16}$$

The probabilities $\alpha_i(s)$ satisfy the equation

$$\alpha_0(s) = \delta(s, s_I) + \sum_t \alpha_0(L(t))\gamma(t,s,\phi) \tag{17a}$$

$$\alpha_i(s) = \sum_t \alpha_{i-1}(L(t))\gamma(t,s,b_i) + \sum_t \alpha_i(L(t))\gamma(t,s,\phi) \quad i \geq 1 \tag{17b}$$

where

$$\gamma(t,s,a) = q_{L(t)}(t)\delta(R(t),s)\delta(A(t),a). \tag{18}$$

As with the Viterbi algorithm described in Section 3, the absence of null circuits guarantees that the states can be ordered so that $\alpha_i(s_j)$ may be determined from $\alpha_{i-1}(s), s \in \mathscr{S}$, and $\alpha_i(s_k)$, $k<j$.

The probabilities $\beta_i(s)$ satisfy the equations

$$\beta_m(s_F) = 1 \tag{19a}$$

$$\beta_i(s) = \sum_t \beta_i(R(t))\zeta(t,s,\phi) + \sum_t \beta_{i+1}(R(t))\zeta(t,s,b_i) \quad i \leq m, s \neq s_F \tag{19b}$$

where $\beta_{m+1}(s) = 0$ and

$$\zeta(t,s,a) = q_{L(t)}(t)\delta(L(t),s)\delta(A(t),a). \tag{20}$$

Step 3) of the iterative procedure above then consists of computing $\alpha_i(s)$ for all s and i in a forward pass over the data, $\beta_i(s)$ for all s and i in a backward pass over the data, and finally $P_i(t,\mathbf{b}_1^m)$ from (16). We refer to the iterative procedure together with the method described for computing $P_i(t,\mathbf{b}_1^m)$ as the Forward-Backward Algorithm.

The probability $P(\mathbf{b}_1^m)$ of the observed sequence $\mathbf{b}_1^m$ is a function of the probabilities $q_s(t)$. To display this dependence explicitly, we write $P(\mathbf{b}_1^m; q_s(t))$.

Baum [6] has proven that $P(\mathbf{b}_1^m; q_s^{j+1}(t)) \geq P(\mathbf{b}_1^m; q_s^j(t))$ with equality only if $q_s^j(t)$ is a stationary point (extremum or inflexion point) of $P(\mathbf{b}_1^m; \cdot)$. This result also holds if the transition distributions of some of the states are known and hence held fixed during the estimation procedure.

5. *Interpolation of Probabilities*

The problem treated in this section is the evaluation of interpolation weights λ such as those occurring in (4) and (7). We will treat (7) first in a slightly generalized way, the rest will follow easily. (7) can be rewritten as

$$
\begin{aligned}
\hat{P}_2(w_3/w_2 w_1) = &\lambda_1 p(w_3/\phi_1(w_1,w_2)) + \lambda_2 p(w_3/\phi_2(w_1,w_2)) \\
&+ \lambda_3 p(w_3/\phi_2(w_1,w_2)) + \lambda_4 p(w_3/\phi_2(w_1,w_2))
\end{aligned}
\tag{21}
$$

where ϕ_i are equivalence functions and p are probability distributions over the words w_3. For instance, in (7) ϕ_2 partitions the set of pairs w_1,w_2 into subsets whose elements have the same w_2 value, and ϕ_4 partitions the set of pairs w_1,w_2 according to the part of speech pair to which they correspond.

Now $\hat{P}_2$ can be thought of as arising from a Markov source having 6 states that generates outputs w_3 (see Figure 4). Starting in the initial state 0 one can reach four different next states (1,2,3, or 4) via null transitions whose probabilities are $\lambda_1, \lambda_2, \lambda_3$, and λ_4, respectively. From each of these latter states there are exactly M transitions into the final state 5, each labelled with a different word w_3 of the allowed vocabulary of size M. The transition out of state i labelled with the word w_3 is taken with (known, fixed) probability $p(w_3/\phi_i(w_1,w_2))$.

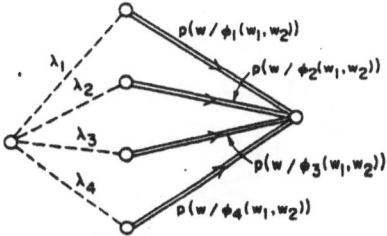

Fig. 4. A section of the interpolated trigram language model corresponding to the state determined by the word pair w_1, w_2.

Thus the problem of obtaining best interpolation weights λ_i is equivalent to estimating Markov source parameters from data, so the procedure of the Forward-Backward algorithm of the preceding section applies.

Note that if the previously fixed probabilites $p(\cdot/\phi_i)$ are themselves estimated from data (as they would be if they were relative frequencies), then that basic data must be disjoint from the data used to estimate λ's [5].

It can be shown that for all the cases discussed in the present section, the Forward-Backward algorithm leads to the unique maximum likelihood (for the observed data) interpolated weight solution as long as the inital weights λ_i are all chosen to be positive.

6. *Part of Speech Annotation of Text*

In this section we wish to introduce an automatic mechanism determing the part of speech (POS) classes g of words of some given text w_1^n. In practice it is convenient to have a larger POS alphabet g than the 8 basic parts of speech taught in high schools. In our work in the Continuous Speech Recognition project we use 40 parts of speech.

It is possible to take the view that the text $\mathbf{w}_1^n$ is generated by the following probabilistic mechanism:

$$p(\mathbf{w}_1^n) = \sum_{\mathbf{g}_1^n} \prod_{i=1}^{n} k(w_i/g_i)h(g_i/g_{i-1},g_{i-2}) \qquad (22)$$

where the sum is over all possible part of speech strings $\mathbf{g}_1^n$, and $g_{-1} = g_0 = TRM$, the conventional start-of-sentence part of speech.

Formula (22) can be interpreted from a Markov Source point of view. The states of the source are specified by pairs g'',g' and transitions to at most $||G||$ different states g', g are possible. There are at most M transitions between states g'', g' and g', g. They are labelled with words w. The relevant transition probability is

$$k(w/g) \quad h(g/g', g'') \qquad (23)$$

We can thus use training text and the forward-backward algorithm to try to learn the source statistical parameters k and h. Having learned these, we can use the Viterbi Algorithm of Section 3 to find the most likely sequence of Markov Source states corresponding to the training text. If, as a result of this process, the word w_i is determined to correspond to the state g_{i-1}, g_i, then it is labelled by the part of speech g_i.

What should be the initial values of k and h for the start of the forward-backward process? Experience shows that the following are adequate:

$$h(g/g', g'') = \frac{1}{||G||} \quad \text{for all} \quad g,g',g'' \in G \qquad (24)$$

$$k(w/g) = \frac{k^*(g/w)f(w)}{\sum_{w'} k^*(g/w')f(w')} \qquad (25)$$

In (25), $f(w)$ is the relative frequency of the word w in the text. If $r(w)$ is the number of different parts of speech that w can have (as allowed, say, by some dictionary) then

$$k^*(g/w) = \begin{cases} 0 & \text{if } g \text{ is not a part of speech allowed to } w \\ \dfrac{1}{r(w)} & \text{otherwise} \end{cases} \tag{26}$$

The annotation of the 5000 word vocabulary office correspondence text worked on by the Continuous Speech Recognition project was carried out with 97% accuracy relative to annotation by experts. This exceeds all published annotation results obtained by any alternate method.

7. *Automatic Extraction of Context Free Production Rule Statistics from Training Text.*

A context free grammar is a mechanism capable of generating strings of *terminal symbols*. When applied to natural languages (such as English), the symbols would be words and the strings involved would be sentences. The formal definition [8] involves a set of *terminal symbols* $\mathcal{T} = \{a,b,...\}$, a set of *phrase markers* $\mathcal{N} = \{A,B,...\}$ including a special distinguished phrase marker S (the sentence start symbol) and a set of phrase marker *rewrite (production) rules* $\mathcal{R}$ of two forms: $A \to BC$ and $A \to x$ where $x \in \mathcal{T}$ and $A,B,C, \in \mathcal{N}$. (Without loss of generality, we are restricting our attention to the so called *Chomsky normal form* of the rules.) The production process starts with the rewriting of the sentence symbol S by some allowed rule and continues until all phrase markers have been replaced by terminals.

The above is not a full formal definition, but the following example will clarify the intuitive idea. Let $\mathcal{T} = \{a,b,c\}, \mathcal{N} = \{S,A,B,C\}$ and let $\mathcal{R}$ consist of the set of rules

584

$$S \rightarrow A\,B \quad A \rightarrow a$$
$$S \rightarrow B\,B \quad B \rightarrow b$$
$$A \rightarrow A\,C \quad C \rightarrow c \tag{27}$$
$$A \rightarrow C\,S$$
$$B \rightarrow B\,A$$
$$C \rightarrow C\,A$$

Then the following left-to-right successive application of rewrite rules, $S \rightarrow AB$, $A \rightarrow CS$, $C \rightarrow c$, $S \rightarrow AB$, $A \rightarrow a$, $B \rightarrow b$, $B \rightarrow BA$, $B \rightarrow b$, $A \rightarrow AC$, $A \rightarrow a$, $C \rightarrow c$, will produce the sentence $c\,a\,b\,b\,a\,c$. The production process can be represented by the *parse tree* of Fig. 5.

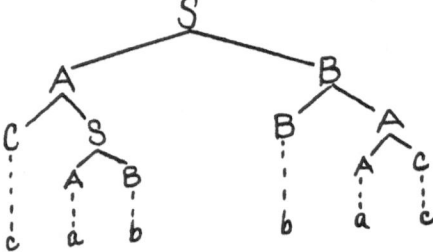

Fig. 5. Parse tree of sentence $c\,a\,b\,b\,a\,c$.

The set of sentences that can be produced by the context free grammar is the *language* of the grammar, and it is obvious that the number of sentences in the language of (27) is infinite. One additional condition must be imposed on the rules: no non-terminal can be introduced which through the rewriting process cannot lead to a string of terminals only.

Grammars are in general *ambiguous* in that different production processes may result in the same terminal string. (27) is ambiguous because the production process $S \rightarrow AB$, $A \rightarrow CS$, $C \rightarrow c$, $A \rightarrow a$, $S \rightarrow BB$, $B \rightarrow b$, $B \rightarrow b$, $B \rightarrow AC$, $A \rightarrow a$, $C \rightarrow c$, leading to the parse of Fig 6. results in the same string as that of Fig. 5.

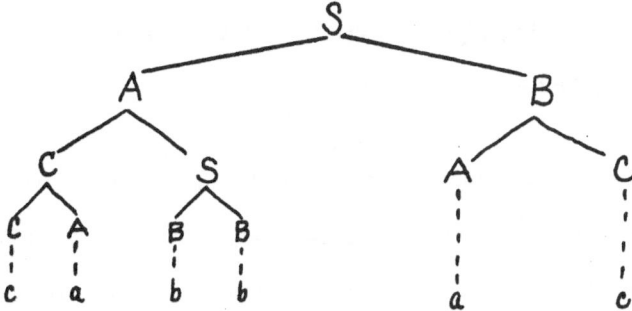

Fig 6. Alternate parse tree of sentence c a b b a c.

The productions $\mathscr{R}$ may have probabilities associated with them. These may be unknown and one may wish to estimate them from a sample of text (presumably) generated by the grammar. If the latter is unambiguous, the estimation problem is easy. One simply generates the unique parse leading to the text sample, and by straightforward counting determines the relative frequency with which one or another rule is used to rewrite each particular phrase marker.

If the grammar is ambiguous, however, then a different method must be used. This case is the interesting one, if only because the knowledge of the production probabilities allows us to assign probabilities to each of the possible parses and thus to determine the most probable one. The extraction of the needed probabilities that cover the general case is called the outside inside (O/I) algorithm [9] and is a generalization of the FB algorithm described in Section 4.

8. *The Outside/Inside Algorithm*

A sentence y_1, y_2, ..., y_T is generated by a context free grammar (CFG). Let X (i, j) denote the phase marker *dominating* y_i, ..., y_j. Thus y_i, ..., y_j are the leaves of the subtree whose root node is $X(i,j)$. As an example, in Figure 5, $X(2,3) = S$ and $X(4,6) = B$.

The parse tree generation can be fully described by the sequence

$$\Big\{\{X(i_1,i_1), i_1 \epsilon [1, T]\};\{X(i_2,i_2+1), i_2 \epsilon [1,T-1]\};\{X(i_3,i_3+2), i_3 \epsilon [1, T-2]\};$$
$$\dots ;\{X(i_{T-1}, i_{T-1}+T-2), i_{T-1} \epsilon [1,2]\}; X(1,T)\Big\} \tag{28}$$

where $X(1, T) = S$ (the unique sentence marker) and $[i,j]$ denotes the integer sequence $i, i+1,\dots, j$. In above, we set $X(i,j) = \phi$ if no phrase marker spans exactly the interval $[i, j]$. Thus, the tree of Fig 6 producing the sentence $c\ a\ b\ b\ a\ c$ is described by $[C,A,B,B,A,C;C,\phi,S,\phi,B;\phi,\phi,\phi,\phi;A,\phi,\phi;\phi,\phi;S]$, and that of Fig 5 is $[C,A,B,B,A,C;\phi,S,\phi,\phi,A;A,\phi,\phi,B;\phi,\phi,\phi;\phi,\phi;S]$.

We would like to develop an algorithm, similar to the FB algorithm, suitable for estimation of probabilities $P_\alpha(\alpha \rightarrow \beta\gamma)$ of re-write rules. Here, α,β,γ are phrase markers. To simplify things we will eliminate rewrite rules of the form $A \rightarrow \alpha$, by introducing the empty terminal symbol ϕ and using rules of the form $A \rightarrow \alpha\phi$ instead.

Assume that k sentences were generated, the ith of length T_i being $y_{V_{i-1}+1}, \dots, y_{V_i}$, where $V_i = \sum\limits_{j=1}^{i} T_j$, $V_o = 0$. Then the update on the probability $P_\alpha(\alpha \rightarrow \beta\gamma)$ (analogous to the update (15) of the FB algorithm) will be computed by the formula

$$P'_\alpha(\alpha \rightarrow \beta\gamma) = \frac{C(\alpha,\beta,\gamma)}{\sum\limits_{\beta\gamma} C(\alpha,\beta,\gamma)} \tag{29}$$

where the contents of the counter $C(\alpha,\beta,\gamma)$ are given by

$$C(\alpha,\beta,\gamma) = \sum\limits_{i=1}^{k} \sum\limits_{r=V_{i-1}+1}^{V_i-1} \sum\limits_{t=r+1}^{V_i} \sum\limits_{s=r}^{t-1} P\{X(r,t) = \alpha,\ X(r,s) = \beta,\ X(s+1,t) = \gamma\ /\ y_1, y_2, \dots y_{V_k}\} \tag{30}$$

Of course, the probabilities in (30) are computed on the basis of the pre-update values of $P_\alpha(\alpha \rightarrow \beta,\gamma)$. Since sentences are assumed generated independently of each other, we see from

(30) that we are really interested in probabilities

$$P\{X(r,t) = \alpha, \ X(r,s) = \beta, \ X(s + 1,t) = \gamma \ / \ y_1, \ldots y_T\} \qquad (31)$$

for all $1 \leq s \leq r < t \leq T$ and all observed sentences $y_1, \ldots, y_T$. But

$$P\{X(r,t) = \alpha, X(r,s) = \beta, X(s + 1,t) = \gamma, \ y_1, \ldots, y_T\} = \qquad (32)$$

$$P_\beta\{y_r, \ldots y_s\} \ P_\gamma\{y_{s+1}, \ldots y_t\} \ P_\alpha\{\alpha \to \beta\gamma\} \ P^*\{y_1, \ldots, y_{r-1}, y_{t+1}, \ldots y_T, \alpha\}$$

where $e(r,s/\beta) \overset{\Delta}{=} P_\beta \{y_r, \ldots y_s\}$ is the probability that the terminal string $y_s, \ldots y_r$ is generated from a phrase marker β, $a(\beta,\gamma/\alpha) \overset{\Delta}{=} P_\alpha \{\alpha \to \beta\gamma\}$ is the probability that the rule $\alpha \to \beta, \gamma$ is used to re-write α, and $f(\alpha,r,t) \overset{\Delta}{=} P^* \{y_1, \ldots y_{r-1}, y_{t+1}, \ldots y_t, \alpha\}$ is the probability that in the re-write process, α is generated and that the strings *not* dominated by it are $y_1, \ldots, y_{r-1}$ on the left and $y_{t+1}, \ldots y_T$ on the right.

The underlying production probabilites $P_\alpha\{\alpha \to \beta\gamma\}$ are given (originally by an arbitrary guess) at every step of the iteration process and the other probabilities of (32) must somehow be computable from these production probabilities. Since

$$e(r,t/\alpha) = \sum_{\beta,\gamma} \sum_{s=r}^{t-1} a(\beta,\gamma/\alpha) \ e(r,s/\beta) \ e(s + 1, t/\gamma), \qquad (33)$$

one can proceed recursively, by first computing $e(s,s/\alpha)$ for all $\alpha, s = 1, \ldots, T$ then $e(s, s + 1/\alpha)$ for all $\alpha, s = 1, \ldots, T - 1$ etc., until $e(s,s + T-2/\alpha)$ for all $\alpha, s = 1,2$ and finally $e(1,T/S)$. This is best done in the bottom-up parsing process described in Section I of the Appendix.

Next,

$$f(\beta,r,s) = \sum_{\alpha,\gamma} \left[\sum_{t=s+1}^{T} f(\alpha,r,t) \; a(\beta,\gamma/\alpha) \; e(s+1,t/\gamma) \right.$$
$$\left. \sum_{q=1}^{r-1} f(\alpha,q,s) \; a(\gamma,\beta/\alpha) \; e(q,r-1/\gamma) \right] \qquad (34)$$

(34) is clearly a top-to-down process to be performed after all the values of $e(r,t/\alpha)$ have been computed. Proceed as follows: The initial condition is

$$f(\alpha,1,T) = \begin{cases} 1 & \alpha = S \\ 0 & \text{otherwise} \end{cases} \qquad (35)$$

Then

$$f(\beta,r,T) = \sum_{\alpha,\gamma} \sum_{q=1}^{r-1} f(\alpha,q,T) \; a(\gamma,\beta/\alpha) \; e(q,r-1/\gamma) \qquad (36)$$

are first computed for all β and $r = 2$; next for all β and r = 3, until at last for all β and r=T.

Next compute

$$f(\beta,r,T-1) = \sum_{\alpha,\gamma} \left[f(\alpha,r,T) \; a(\beta,\gamma/\alpha) \; e(T,T/\gamma) \right.$$
$$\left. + \sum_{q=1}^{r-1} f(\alpha,q,T-1) \; a(\gamma,\beta/\alpha) \; e(q,r-1/\gamma) \right] \qquad (37)$$

for all β and r=1, next all β and r=2, . . . , finally for all β and r=T-1., etc. Thus (37) is computed in the order $s = T$, $T - 1, . . ,1$; $r = 1,2, . . . , s$; all β, with the last index varying fastest and the first slowest. The top-down process generating the f-values is described in Section II of the Appendix

Since $P\{y_1, . . . ,y_T\} = e(1,T/S)$, it follows that the counter $c(\alpha,\beta,\gamma)$ providing the *update* for $a(\beta,\gamma/\alpha)$ will for the sentence $y_1, \ldots y_T$ be increased by the contribution

$$\left[\sum_{r=1}^{T-1} \sum_{t=r+1}^{T} \sum_{s=r}^{t-1} e(r,s/\beta) \; e(s + 1,t/\gamma) \; f(\alpha,r,t) \right] \frac{a(\beta,\gamma/\alpha)}{e(1,T/S)} \tag{38}$$

The O/I algorithm then proceeds by iterative steps. One starts with a guess of production probabilities $P_\alpha(\alpha \rightarrow \beta\gamma)$, carries out the counter accumulation process (38) over the sentences of the training text, updates the production probabilities according to (29), and proceeds with the next iteration. The whole proceedure should be started with a bottom-up parse of Section I of the Appendix establishing all the possible analyses of the training sentences, and saving these for further iterations. Section III of the Appendix addresses the need for normalization of parameter values during the process. Section IV introduces a Viterbi-type algorithm determining the most likely parse of a sentence and its probability.

REFERENCES

1) L.R. Bahl, F. Jelinek, and R.L. Mercer, "A Maximum Likelihood Approach To Continuous Speech Recognition," *IEEE Transactions on Pattern Analysis and Machine Intelligence,* Vol. 5, March 1983.

2) C.E. Shannon, "Predictions and Entropy of Printed English," *Bell Systems Technical Journal,* pp. 50-64, January 1951.

3) T.M. Cover, and R.C. King, "A Convergent Gambling Estimate of the Entropy of English," *IEEE Transactions on Information Theory,* Vol. IT-24, No. 4, July 1978.

4) A.J. Viterbi, "Error Bounds for Convolutional Codes and an Asymptotically Optimal Decoding Algorithm," *IEEE Transactions on Information Theory,* Vol. 13, No. 4, pp. 260-269, (April 1967)

5) G.F. Forney, Jr., "The Viterbi Algorithm," *Proceedings IEEE,* Vol. 61, pp. 268-278, March 1973.

6) L.E. Baum, "An Inequality and Associated Maximation Technique in Statistical Estimation of Probabilistic Functions of Markov Processes," *Inequalities,* Vol. 3, pp. 1-8, 1972

7) F. Jelinek and R.L. Mercer, "Interpolated Estimation of Markov Source Parameters From Sparse Data," *Proceedings Workshop Pattern Recognition In Practice,* May 21-23, 1980, Amsterdam, the Netherlands: North Holland.

8) J.E. Hopcroft and J.D. Ullman, "Introduction to Automata Theory, Languages, and Computation," Addison-Wesley, 1979.

9) J.K. Baker, "Trainable Grammars for Speech Recognition," *Proceedings of the Spring Conference of the Acoustical Society* of America, 1979.

APPENDIX

I. *Bottom-up Parsing Algorithm Establishing Values* $e(r,t/\alpha)$

String $y_1, y_2, \ldots, y_T$ is to be parsed, assuming CNF. Parsing proceeds by index from left to right.

1. List all rules $A_i \rightarrow y_1$ $i = 1, \ldots, k(1)$, and establish entries $[1, A_i, 1, e]$. Call them old and place in stack 1 in alphabetical order of A_i. Let $e = P(A_i \rightarrow y_1)$.

2. List all rules $B_j \rightarrow y_2$ $j = 1, \ldots, k(2)$, and arrange entries $[2, B_j, 2, e]$ into a new stack $[(e = P(B_j \rightarrow y_2)]$. For each entry in the new stack see if there exist A_i in the old stack 1 and a "parent" C having production $C \rightarrow A_i B_j$ if so, create a *new* entry $[1, C, 2, e_c]$ with pointers to $[1, A_i, 1, e_A]$ and $[2, B_j, 2, e_B]$ and place in new stack 2 in alphabetical order of C. Also, $e_c = e_A \cdot e_B \cdot a(A,B/C)$. When for B_j all A_i in old stack 1 have been examined for sibling character, $[2, B_j, 2, e_B]$ is removed from new stack and placed in old stack 2 below the $[1, C, 2, e_c]$ entries, and in alphabetical order of B. When placing an entry into a stack, the latter must be examined for a possible duplicate. If such exists, the new entry is consolidated with the duplicate. This means that the pointer pair of the new entry is added to the pointer pair of the old and that the old value of e_c is updated by adding to it the corresponding value of the new entry.

3. The General Step

    List all rules $D_l \rightarrow y_m$ $l = 1, \ldots, k(m)$, establish entries $[m, D_e, m, e](e = P(D_l \rightarrow y_m))$ and place in new stack in alphabetical order of D_l. In the following process, the new stack will in general contain entries $[i, D, m, e_D]$, $i \leq m$. Take the top entry $[i, D, m, e_D]$ of new stack and examine old stack $i - 1$ to see if it contains entry $[j, E, i - 1, e_E]$ for which F exists such that $F \rightarrow ED$. If so, create the entry $[j, F, m, e_F]$, $e_F = e_E \cdot e_D \cdot a(E,D/F)$, and make it point to $[j, E, i - 1, e_E]$ and

$[i, D, m, e_D]$. If $j>1$ place $[j, F, m, e_f]$ in the new stack, otherwise in the old stack m. When the top of the new stack $[i, D, m, e_D]$ has been examined for all potential siblings in the old stack i-1, it is removed from the new stack and placed in the old stack m, and examination of the current top of the new stack begins. When the new stack is exhausted, the next step involving $G \rightarrow y_{m+1}$ begins. As mentioned above, new stack entries are consolidated with duplicate old ones. Also, stack m is maintained in the order of the first two indices of $[i, D, m, e_D]$, with i varying slower than D. This is done not just to facilitate the consolidation process, but also the top-down generation of f-values (see next section).

4.   When the process terminates, if the top entry of stack T is not $[1, S, T, e_s]$ (we assume that S stands for the "top" letter of the alphabet) then the sentence did not parse. All entries in all stacks that are not ultimately chained to that entry could now be purged, because they do not contribute to any valid parse.

The above generation of the stacks specifying all possible parses needs to be carried out in its entirety only in the first iteration of the update process. In all succeeding iterations one keeps the stacks as they are and only re-computes the e-values of their entries. In this way one avoids the unnecessary computation of entries that are not chained to the $[1, S, T, e_s]$ entry and thus do not contribute to any complete parse of the output string $y_1, ..., y_T$.

II. *Top-down Process Computing $f(\alpha,r,t)$ values and contributing to counters) $C(\alpha,\beta,\gamma)$*

Assume that all entries of all stacks have the form $[i, A, j, e_A, f_A]$ where initially $f_A = 0$. The algorithm will make use of the stacks created in the bottom up process, will proceed from right to left, and will take advantage of the stack ordering established in the previous section.

1.   Set $f_s = 1$ in the entry $[1, S, T, e_S, f_S]$

2. In all entry pairs $[1, \beta, r, e_\beta, f_\beta]$ and $[r + 1, \gamma, T, e_\gamma, f_\gamma]$ pointed to from the entry $[1, S, T, e_s, f_s]$ set $f_\beta = f_S \, a(\beta, \gamma/S)e_\gamma$ and $f_\gamma = f_S \, a(\beta, \gamma/S)e_\beta$. Note that this fixes forever the f_γ values of entries $[2, \gamma, T, e_\gamma, f_\gamma]$

3. Purge all entries $[2, \alpha, T, e_\alpha, f_\alpha]$ such that $f_\alpha = 0$. If a remaining entry for which $f_\alpha > 0$ points to the pair $[2, \beta, r, e_\beta, f_\beta]$ and $[r + 1, \gamma, T, e_\gamma, f_\gamma]$ then update $f_\beta = f_\beta + f_\alpha \, a(\beta, \gamma/\alpha)e_\gamma$ and $f_\gamma = f_\gamma + f_\alpha \, a(\beta, \gamma/\alpha)e_\beta$. This fixes forever values of entries $[3, \gamma, T, e_\gamma, f_\gamma]$.

4. Proceed as in 3 by purging and "extending" entries $[r, \alpha, T, e_\alpha, f_\alpha]$, $f_\alpha > 0$, for $r = 3, 4, \ldots, T$.

5. Purge all entries $[1, \alpha, T - 1, e_\alpha, f_\alpha]$ such that $f_\alpha = 0$. If a remaining such entry points to a pair $[1, \beta, r, e_\beta, f_\beta]$ and $[r + 1, \gamma, T - 1, e_\gamma, f_\gamma]$ then update $f_\beta = f_\beta + f_\alpha \, a(\beta, \gamma/\alpha)e_\gamma$ and $f_\gamma = f_\gamma + f_\alpha a(\beta, \gamma/\alpha)e_\beta$

6. Continue in this vain down the stack T-1, then down stack T-2 starting with $[1, \alpha, T - 2, e_\alpha, f_\alpha]$ entries, etc., until treating stack 2. At that moment all f_α values in all stacks are fixed, and the stacks contain only entries that contribute to some complete valid parse of $y_1, y_2, \ldots, y_T$.

With the f-values fixed one is ready to contribute to the counters $c(\alpha, \beta, \gamma)$. One simply goes through all entries $[r, \alpha, t, e_\alpha, f_\alpha]$ of all stacks, finds all entry pairs $[r, \beta, s, e_\beta, f_\beta]$ and $[s + 1, \gamma, t, e_\gamma, f_\gamma]$ to which those entries point, and contributes to the counter $c(\alpha, \beta, \gamma)$ the count $e_\beta \cdot e_\gamma \cdot f_\alpha$.

III. Normalization

The sequential computation of e and f values, if carried out interactively as indicated in the preceding sections, might lead to underflows. In fact, the value of the probability $e(r, s/\beta)$

will decrease as the span length $s - r$ gets larger. Similarly, $f(\alpha,r,t)$ decreases with increasing span lengths r and $T - t$. To avoid numerical problems during the calculation, some normalization is necessary.

Let $n(i)$, $i = 1, ..., T$ be a set of normalizing factors whose appropriate value will be discussed below. Define

$$e^*(r,s/\beta) \triangleq e(r,s/\beta)[\prod_{j=r}^{s} n(j)] \qquad (A-1)$$

$$f^*(\alpha,r,s) \triangleq f(\alpha,r,s)[\prod_{j=1}^{r-1} n(j)][\prod_{i=t+1}^{T} n(j)] \qquad (A-2)$$

Then it is easy to check that equations (33) and (34) hold with e^* substituted for e and f^* substituted for f.

We can thus carry out the computation of the preceding two sections so as to obtain the normalized quantities e^* and f^*. Of course, the values of the normalizing factors $n(i)$, to be worth anything, must be functions of the computation results themselves, the aim being to keep e^* and f^* within acceptable numerical range. We will go directly to the general step of Section II.

Assume that $n(i)$ and $e^*(j,i/\beta)$ are correctly determined for all β, $j \leq i$, and $i = 1, 2, ..., m - 1$. That means that the first $m - 1$ "old" stacks are set. Then carry out step 3 of Section II to obtain a *preliminary* old stack m with entries $[i, D, m, e_D^+]$, where the values for the leaves of the tree from which we start are

$$e^+(m,m/D) = e(m,m/D) = P(D \rightarrow y_m) \qquad (A-3)$$

Note that the process obtains

$$e^+(r,m/\alpha) = \sum_{\beta,\gamma} \sum_{s=r}^{m-1} a(\beta,\gamma/\alpha) \; e^*(r,s/\beta) \; e^+(s + 1,m/\gamma) \qquad (A-4)$$

in the sequence $r = m-1, m-2, ..., 1$. Therefore, if (A-1) holds, then because of the intitial choice (13), we get

$$e^+(i,m/\alpha) = e(i,m/\alpha)[\prod_{j=i}^{m-1} n(j)] \qquad (A-5)$$

Thus, knowing the quantities (A-5) we can now choose the normalizing value $n(m)$ so as to keep

$$e^*(i,m/\alpha) = n(m) \; e^+(i,m/\alpha) \qquad (A-6)$$

within appropriate numerical bounds and satisfy (A-1). Thus the *final* old stack m is obtained by multiplying the e^+ values in the preliminary old stack by the constant $n(m)$. A good choice for $n(m)$ is

$$\frac{1}{n(m)} = \max_{i,\alpha} [e^+(i,m/\alpha)] \qquad (A-7)$$

resulting in

$$e^*(i,m/\alpha) \le 1 \qquad (A-8)$$

with equality for at least one i,α combination.

Note that (A-7) can be re-written as

$$[\prod_{j=i}^{m} n(j)]^{-1} = \max_{i,\alpha} \{[\prod_{j=i}^{i-1} n(j)]^{-1} e(i,m/\alpha)\} \qquad (A-9)$$

Thus $[\prod_{j=i}^{m} n(j)]^{-1}$ is equal to the probability of producing $y_1, ..., y_m$ by the highest probability decomposition of the type $P(y_1, ..., y_{i_1}, / \; \alpha_i \;) \; P(y_{i_{1+1}}, ... , y_2/\alpha_2) ...P(\; y_{i_g+1}, ..., y_m/\alpha_x)$ when it is not taken into account whether α_1, ..., α_k can actually be produced as an initial substring of phrase markers stemming from S. Hence the choice (A-7) is very appropriate to the observed string y_1, ..., y_T.

By the time the f^*– evaluation of Section III begins, all $n(i)$ values are determined, and all stacks contain e^* values. Setting $f^*_S = 1$ for entry $[1,S,T,e^*_S f^*_S]$ in step 1, and carrying out the procedure exactly as described in Section III will result in the computation of f^* values satisfying (A-2). Since that computation consists of carrying out (37) with properly normalized values e^*, the obtained f^* values will automatically fall within tolerable limits. In fact, if the values $n(m)$ are selected by rule (A-7), and

$$f(\alpha,r,s) \;=\; P^*\{y_1, ..., y_{r-1}, y_{t+1}, ..., y_T, \alpha\} \qquad\qquad (A-10)$$

then $f^*(\alpha,r,s)$ can be interpreted as the probability (A-10) normalized by the maximum product of probabilities with which the strings y_1, ..., y_{r-1} and y_{t+1}, ..., y_T can be generated.

At the end of the computation process for the given sentence, the counter $c(\alpha,\beta,\gamma)$ should again be increased by the quantity (38) modified by replacing e with e^* and f with f^* This comes about because the normalizing factor $\prod_{i=1}^{T} n(i)$ cancels out in the modified (starred) expression (38) so that the latter is exactly equal to the unmodified one.

IV. *A Viterbi-type Algorithm for Probabilistic Context Free Grammars*

Once the production probabilities of the context free grammars are known, it is of interest to determine the most likely parse accounting for any given output string y_1, ..., y_T. The way to get this parse, and its probabilities, is based on the following observations.

Consider a phrase marker B and the substring y_r, ..., y_s generated from it. Let $\mathcal{T}^*$ be that subtree stemming from B which denotes the most probable production of y_r, ..., y_s from B (there may be many ways in which the substring can be produced starting with B). If the most probable production of the entire string $y_1, ..., y_T$ contains the marker B spanning $y_r, ..., y_s$, then in that production the subtree stemming from B must be $\mathcal{T}$ *. (The probability of any production of $y_1, ..., y_T$ containing a subtree $\mathcal{T}$ stemming from B can be only increased by replacing $\mathcal{T}$ with $\mathcal{T}$ *, because of the context free nature of the grammar).

Therefore, to obtain the most likely parse of the string $y_1, ..., y_T$ one need only to change the meaning of the notion of *consolidation* in the bottom-up parse of Section II. Namely, when a new entry, say $[j, F, m, e'_F]$ is about to be placed into the stack, one must again examine the latter for a possible duplicate, $[j, F, m, e_F.]$ If it exists and $e'_F \le e_F$ then the new entry is dropped. Otherwise the old entry is simply replaced by the new one (including its pointers). In this way, from each entry $[j, F, m, e_F]$ there will lead only one pointer pair to entries $[j, E, i-1, e_E]$ and $[i, D, m, e_D]$. When the entry $[1, S, T, e_s]$ is finally created, and one parse stemming from it will be present in the stacks, and that parse will be the most probable one.

Part 5.

THE NATIONAL PHYSICAL LABORATORY SESSION

Organised and Chaired by:

E.L. Albasiny
National Physical Laboratory
Middlesex, U.K.

TESTING PROTOCOL IMPLEMENTATIONS

D. RAYNER

Protocol Standards Group
Division of Information Technology and Computing
National Physical Laboratory
Teddington, Middx., UK

ABSTRACT

A large amount of effort is being put into developing international standards for protocols and services in support of Open Systems Interconnection (OSI). All this activity will, however, be in vain unless products are produced in conformance with these standards. In order to give users confidence in the products they buy, implementations need to be tested objectively using a standard suite of internationally agreed tests. In the absence of agreed objective tests, users can gain confidence from testing by a trusted independent assessment centre.

NPL is currently developing suitable techniques for objective or independent testing. The architectural basis which has been established for this work is presented, giving the context within which NPL's current testing system can be discussed. A brief indication is given of the relationship between this and the work of our European collaborators, in France and Germany, and that of others in the USA and Canada.

1. INTRODUCTION

The objective of OSI (10) will not be achieved unless products are produced in conformance with OSI standards. There is therefore a growing interest in establishing centres that can test for this conformance. For high-level protocols, such testing can be carried out remotely via, for example, a public data network. These testing centres need to be seen to be impartial and trustworthy. Their credibility will be crucial to their success.

Conformance testing on its own, however, is not enough. Products also need to be tested to see that they will meet the user's requirements. In particular, the range of options supported, the performance (throughput and response times), and the robustness (the degree of recovery from error situations) all need to be measured. This combination of tests and measurements is termed 'assessment'.

Assessment Centres can also provide subsets of an assessment test suite tailored to meet buyers' needs for acceptance testing. Implementors can be helped if Assessment Centres provide development aids to assist them in debugging their systems.

In order to maintain credibility Assessment Centres may provide an arbitration service. Since testing, however thorough, can only detect the presence of errors not their absence, situations could arise in which two implementations that received favourable assessment reports fail to interwork properly. An arbitration service could investigate and determine the cause of the problem.

Assessment reports will not, therefore, be able simply to state that a given implementation has passed or failed. Instead, they will list the tests performed and the results obtained. They will also give the period over which the tests were run and details of the environment in which the implementation was tested. They will state the functional range over which the implementation was found to conform to the specification, and give the likely implications of its range, performance and robustness for potential applications.

1.1 The UK Position

NPL is developing testing techniques for protocol implementation assessment. The aim is to make them applicable to the eventual international standards. In the meantime, protocols already in use in the UK provide the basis for their development. Initial work has been based on the 'Network Service and Protocol over X.25' (11). This is used on British Telecom's Packet Switched Service (PSS) and other X.25 (5) UK networks. The current emphasis at NPL is on tests for conformance and measuring functional range.

NPL will not, however, run a testing service. Instead, the National Computing Centre (NCC) in Manchester has been selected to run a pilot Assessment Centre based on NPL's system, and including use of testing tools developed by other groups. It will start in summer 1983, funded by the UK Department of Industry.

1.2 International Collaboration

Similar work is going on in other establishments around the world. The Commission of the European Communities (CEC) is partially funding collaboration between groups at NPL, in Project RHIN of Agence de l'Informatique (ADI) in Paris, and in Gesellschaft fuer Mathematik und Datenverarbeitung (GMD) in Darmstadt. ADI has adopted a similar approach (1,2,3) to NPL, whereas GMD is giving higher priority to development aids and arbitration (7). In the collaboration, NPL is concentrating on the Network Layer, ADI on Transport, and GMD on Session in the form of CCITT S.62 Teletex protocols (5).

There is also informal collaboration with groups at the USA's National Bureau of Standards (NBS) (13,14) and at the University of Montreal (4,16). NBS is working towards a Certification Center for implementations of the US Federal Protocol Standards which they are specifying. Their approach is similar to that of NPL and ADI. Work at Montreal is more theoretical at present, with useful results in the automatic production of test sequences from a formal protocol specification.

2. COMPARISON OF DIFFERENT PHYSICAL ARCHITECTURES

The physical architecture that is used by an Assessment Centre will depend on:-

(a) the category of testing service required: assessment, development aid, or arbitration;

(b) the type of protocol(s) concerned: above or below the OSI Network Service;

(c) the feasibility of transporting either the test equipment or the system under test (SUT);

(d) economics: the costs of different architectures and how much the clients are willing to pay.

Here we analyse the possibilities according to the category of testing service required.

2.1 Assessment

The combination of hardware and software used by the Assessment Centre to communicate with the client's SUT during testing is called the Active Tester (AT). Figure 1 shows four possible architectures for assessment. The abbreviations used are as follows:-

SUT   System under test (usually at the client's site)
AT   Active Tester (usually at the Assessment Centre site)
NET   Communications medium, e.g. public data network
PTU   Portable testing unit belonging to the Assessment Centre
TDS   Test Driver System at the Assessment Centre site

```
        .----.        .-----.
(a)     | AT |<--->| SUT |
        '----'        '-----'


        .----.    .-----.    .-----.
(b)     | AT |<--|-NET-|-->| SUT |
        '----'    '-----'    '-----'


        .----.    .-----.    .-----.        .-----.
(c)     | AT |<--|-NET-|----|-PTU-|---->| SUT |
        '----'    '-----'    '-----'        '-----'


        .------------------------------.
        |               AT               |
        |.-----.    .-----.    .-----.|    .-----.
(d)     || TDS |<-|-NET-|->| PTU ||<--->| SUT |
        |'-----'    '-----'    '-----'|    '-----'
        '------------------------------'
```

<u>Figure 1. Architectures for assessment</u>

      In all four architectures shown, the AT will be controlled by the Assessment Centre, whereas the SUT will be controlled as necessary by the client.

      Architecture 1(a) can be used when the AT and SUT can be closely coupled together. It enables the AT to control precisely what traffic goes into the SUT and to monitor exactly what traffic emerges. It requires either the AT or the SUT to be transported to the site of the other. It is feasible at present to construct transportable ATs for the assessment only of Physical and Link Layer implementations. Higher layer assessment requires too much backing storage to monitor all the testing traffic required. Similarly, only microprocessor based SUTs are likely to be transportable.

      To overcome the difficulties and costs of transporting equipment, remote testing over some communications medium can be used, as in architecture 1(b). This is very satisfactory for testing protocols above the OSI Network Service, when the underlying service can be expected to be end-to-end. For testing below the OSI Network Service, however, it is impossible to generate a full range of error cases, and remote monitoring of the activity at the interface to the SUT becomes unsatisfactory. This is true even for testing directly over a subnetwork service, such as X.25 (5).

Thus, for testing involving all or part of the Network Layer, neither architecture 1(a) or 1(b) is adequate. The answer is to insert a PTU between the communications medium and the SUT. The PTU can then generate error cases and monitor the interface to the SUT. Architecture 1(c) shows the PTU acting independently of the AT. In this arrangement the PTU is programmed to act in some pre-determined fashion on the traffic flowing through it. It can scan its input from the AT to look for a sequence which triggers error generation, otherwise it just monitors the traffic. It has the advantage of not requiring any modification to the AT software; so it can be plugged in or omitted as desired. Disadvantages of this are that the AT only has very loose control over its behaviour and it requires sizable backing storage to record what it monitors.

These disadvantages of 1(c) are overcome in 1(d), where the PTU is made an integral part of the AT. The AT software is divided between a TDS and the PTU, with a special control protocol operating between the two. This enables errors to be generated with the same control as provided by architecture 1(a), but allows most of the backing storage to be located at the Assessment Centre site. It also allows some of the costs of using a public data network to be overcome. If bulk data needs to be used within a test, it can be generated or absorbed by the PTU under the control of the TDS, with only the occasional control message traversing the network. With 1(c) the generation of bulk data is likely to be easier to achieve than its controlled absorption.

A disadvantage of both 1(c) and 1(d) is the cost of transporting the PTU to the client's site, ensuring that it is correctly connected up, and protecting it against interference from a client out to change the results of the tests to his advantage. It remains to be seen whether the value of more thorough testing is considered to outweigh these extra costs. Otherwise architecture 1(b) will become the dominant architecture for assessment in all layers above the Link Layer.

NPL is currently using architecture 1(b), but has a PTU under development for use in architecture 1(d). ADI, on the other hand, has developed a PTU (3) for use in architecture 1(c).

2.2 Development Aids

The same four architectures can be used for development aids, but with the client controlling both the SUT and the AT. This means that in architectures 1(b), (c) and (d), the control terminal for the AT has to be connected via the communications medium.

If the Assessment Centre provides a consultancy service for debugging implementations, then the AT's control terminal could be at either the client's site or the Assessment Centre, depending on where it is more convenient for the consultant to be during testing.

Architecture 1(a) will apply if the cost of the AT can be kept low enough that users are prepared to buy their own. It could also apply if either the AT or the SUT is transportable. If SUTs can be expected to be transportable, then workshop facilities could be provided at an Assessment Centre. If in its role as part of a development aid, the AT can be put into a PTU, then the Assessment Centre can have several of them available for hire.

The other three architectures apply to the use of an AT developed for assessment purposes, but driven by the client. These can allow clients to try running the assessment tests before a formal independent assessment is made. They can also be used to provide development aids when the costs or lack of transportability rule out 1(a).

Architecture 1(b) can also be used with an AT which only provides simple facilities such as sourcing, sinking and echoing data, call connection and rejection, and recall by the AT. NPL's experience with using such simple remote development aids in the past is that they can provide a useful partner to test with for a short time, but that the implementor soon outgrows such aids. Nevertheless, they will continue to be used because they are quick and cheap to set up.

2.3 Arbitration

Figure 2 gives four possible architectures for use in arbitration testing, involving two SUTs. Since arbitration is primarily passive, the AT of the previous two types of testing is replaced by an Arbiter.

Architecture 2(a) requires transportability of either the arbiter and one SUT or both SUTs as the price to be paid for being able to monitor precisely the activity at the interfaces to both SUTs. If necessary, error situations can be generated by the arbiter for either or both SUTs to react to.

On the other hand, remote arbitration, as in 2(b), is easy to set up, but only provides adequate monitoring and error generation

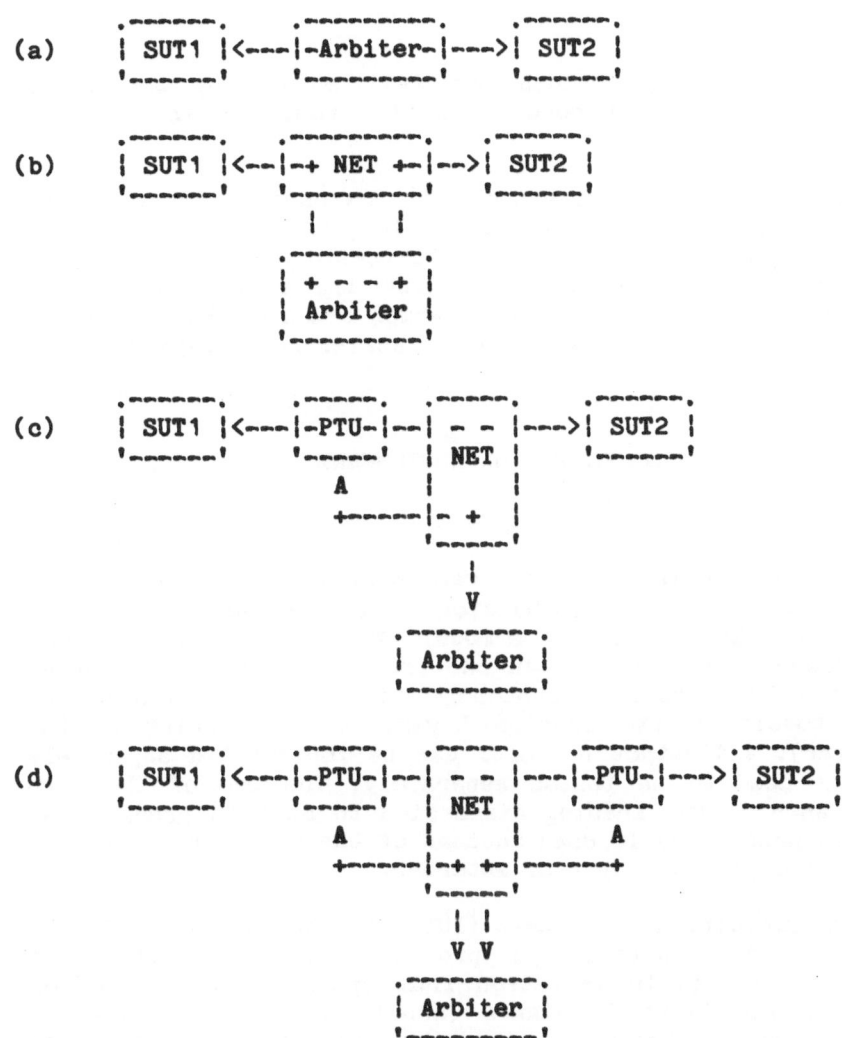

Figure 2. Architectures for arbitration

capabilities for testing above the OSI Network Service. It is possible that problems may arise from the connection crossing the network twice, particularly if the situation to be investigated is time critical.

Architecure 2(c) allows the test connection to cross the network just once, and provides better monitoring and error generation capabilities at the interface to one of the SUTs. This may be adequate if one of the two is much more suspect than the other (e.g. if SUT2 is a Reference Implementation, as discussed below).

Architecture 2(d) provides the full monitoring and error generation capabilities at both clients' sites, but is the most expensive of all.

Any of these four architectures can be used with one of the SUTs replaced by a reference implementation, in which case limited assessment would be possible. The limitation arises from the passive nature of the testing, reducing the control which the Assessment Centre has over the tests which are run. Nevertheless, GMD plans to experiment with such assessment testing using architecture 2(b).

3. GENERAL LOGICAL ARCHITECTURE FOR ASSESSMENT

3.1 The Implementation Under Test

Our logical architecture for assessment is described in terms of OSI (sub)layers. An OSI (sub)layer is a layer or sublayer, as appropriate, of the OSI reference model (10). This is done because some layers may contain more than one protocol, one above another (e.g. Job and File Transfer Protocols in the Application Layer, Teletex Protocols in the Session Layer, or sublayering in the Network Layer). A distinct protocol can be implemented separately and thus may need to be tested separately. The use of the term (sub)layer enables the ensuing discussion to be both precise and concise. The prefix (S) is used instead of the usual prefix (N) to identify a general (sub)layer of interest.

The implementation under test (IUT) will be an implementation of one or more adjacent (sub)layer protocols. It will be chosen so that it can be tested in isolation from application software or protocol implementations in higher (sub)layers. It will have an accessible interface between it and its 'user'. This interface will either correspond to an OSI (sub)layer service boundary or to the user interface to the application process. If no such interface is available, then the implementation is said to be untestable.

3.2 The Need for a Test Responder

Thus the IUT will have an upper interface. The stimuli provided across this interface will affect the protocol behaviour that can be observed externally. If the interface is a realisation of an (S)-service, the (S)-service standard will define the semantics of the activity at each (S)-service access point. This may be realised in any way that the implementor chooses, provided that the semantics are equivalent (i.e. a mapping must be possible between the real interface and the service primitives).

The behaviour of the user of the IUT's upper interface needs to be predictable, so that, as far as possible, the results of tests are dependent only upon the behaviour of the IUT. To this end, the Assessment Centre will define a Test Responder (TR) for the client to incorporate into his system for assessment purposes. If the IUT's upper interface is a human user interface, the behaviour of the TR will have to be carried out manually.

The TR should be as simple as possible to provide the flexibility to enable any desired test to be run. Automatic TRs must be readily portable onto any system, even one with minimal capabilities.

The implementation of an automatic TR will use the local realisation of the service boundary provided by the IUT. Since the details of this interface are implementation-dependent, the TR is defined purely in terms of the service, which provides an abstraction of this interface. It will be left to the implementor to provide an interfacing region between the TR and the IUT. This avoids the need for an Assessment Centre to learn the details of every local interface, and ensures that assessment testing is kept free from local interface considerations.

Algorithms for automatic TRs will be specified (8) in a manner independent of any particular programming language. Nevertheless, to assist clients further, reference TR implementations will be published in a number of popular programming languages. The first reference implementation of the current TR is in Coral 66 (12), a language which is often used in the UK for real-time applications.

The question of how to define TR algorithms for controlling the user interfaces to application processes is for further study. GMD are likely to have to tackle this problem in their testing of Teletex products. For simiplicity, in the rest of the paper, it is assumed that we are discussing the case of an automatic TR interfacing to a service access point.

3.3 General Logical Architecture

Our general logical architecture for assessment testing is shown in Figure 3.

The AT is logically divided into entities which communicate with their peers in the client's system via the appropriate protocol(s). The TR communicates with a Test Driver via a Test Driver-Responder Protocol (TDRP). This is a non-standard protocol, defined specifically for assessment purposes.

610

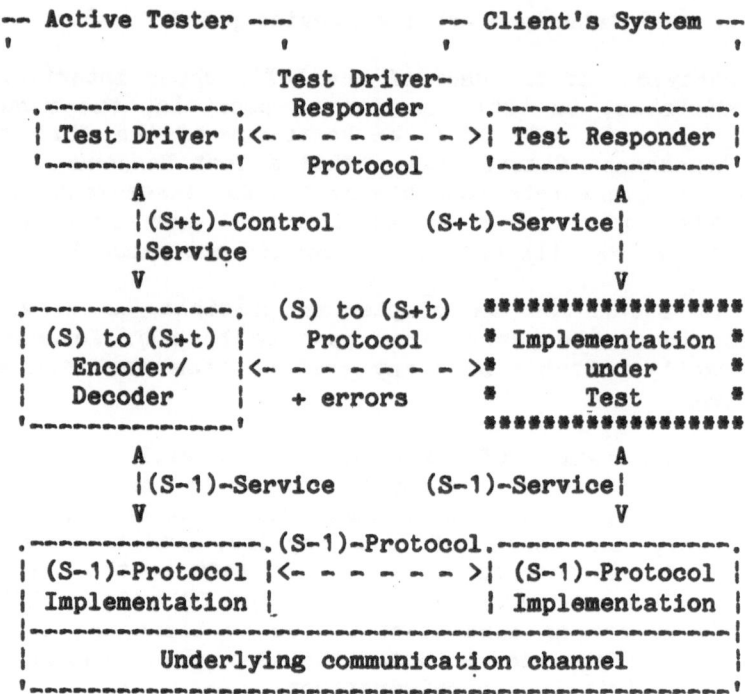

Figure 3. General logical architecture

During testing the IUT is considered to be a 'black-box' containing the appropriate (sub)layer entities within the (sub)layers under test. It will not be possible for the AT to tell whether the implementation is a single process, or a combination of hardware and software such as would be found in an implementation split between a front-end processor and a mainframe. The IUT is simply seen as having two service boundaries, one which should provide the (S+t)-Service and one which should use the (S-1)-Service. The activity across the (S+t)-Service boundary can be recorded by the TR. The activity across the (S-1)-Service can only be observed indirectly from the other end of the communication channel; it is not necessary for this boundary to be accessible within the client's system unless the (S-1)-service provider is to be tested (separately).

Thorough assessment of an IUT requires that it be subjected to error situations as well as normal protocol. It is also desirable that the IUT be subjected to different styles of implementing the same function, where there is real implementor choice. For this reason, it is not adequate to use a reference implementation. Instead, we use an encoder and decoder of both valid and invalid protocol messages. It is operated through a control service which includes the normal (S+t)-Service, but also provides for:-

(a) observation and generation of (S) to (S+t)-Protocol errors;

(b) explicit control over (S) to (S+t)-Protocol mechanisms;

(c) explicit control over the use of the (S-1)-Service when necessary.

The AT's (S-1)-entity is a reference implementation with a minor addition to provide for the generation of some exceptions from within (sub)layer S-1 without violating (S-1)-Protocol.

NBS is using a reference implementation and separate error generator in place of the Encoder/Decoder. ADI is using an error generator combined with an enhanced reference implementation which permits the choice of implementation variants, where there is more than one way of implementing a given protocol function. The advantages and disadvantages of these approaches are being studied in our collaboration.

4. ACTIVE TESTER DEVELOPMENT

The current AT is being developed in stages. The first stage was to develop a manually driven system with tests defined by scenarios (6), similar to those used by NBS (13). This has now been largely superseded by the state table driven system described below. The full development plan is covered elsewhere (15).

4.1 State Table Driven System

This system, illustrated in Figure 4, uses a Test Driver which is table driven. It only handles one connection at a time. Each test is defined by a combination of a state table and a parameter table. The state tables are automatically constructed from a Test Definition Language (TDL) specification. TDL, which is as yet unpublished, is like a simple programming language. For each state, a subset of the possible events have explicit actions and state transitions defined for them. Any other event causes default actions and a default state transition to be executed. Each action's parameters can be taken from the parameter table. This makes it easy to run the same test with, for example, different addresses and timeout values.

In a state table definition of a test, the states are states of the test, which are not necessarily the same as the states of the protocol. The events are of three kinds: messages arriving from the Encoder/Decoder; Test Commands arriving from the

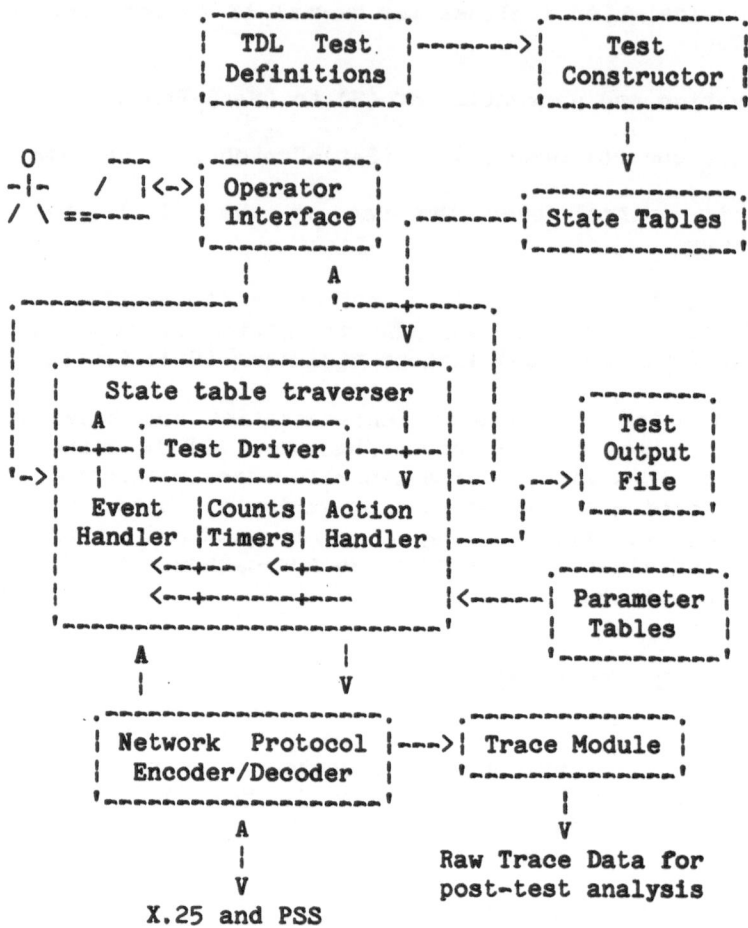

Figure 4.   State table driven system

operator; and internally generated events, such as timeouts and counts exceeding a designated limit. In the case of data arriving from the Encoder/Decoder, it undergoes further decoding by the event handler to separate pure data from TDRP replies. The actions are of four kinds: messages to be sent to the Encoder/Decoder; test output messages to be sent to the operator; internal actions, such as setting timers and limits and incrementing counts; and test output to be deposited in a high-level logging file.

The Test Driver is, therefore, composed of a State Table Traverser, an Event Handler and an Action Handler, with an internal link between the Event and Action Handlers provided both directly and via counts and timers. The direct link is provided by internal actions which are equivalent to Boolean functions whose results are returned as internal events. Internal actions of this kind are, for instance, used to check whether a received TDRP reply is as expected.

The operator specifies which state table and parameter table are to be used for each test. Once a test has been initiated, he receives test output as specified in the test definition, indicating the progress of the test, including the final results. The test definition may allow him to select subtests to be run at particular points in the test. The only other control he has is to abandon a test or subtest, and if he wishes, start another.

All traffic passing through the Encoder/Decoder can be recorded in a raw trace data file. A post-test analysis program can be applied to this file, if required, to determine precisely what happened in a given test. It is expected that this detailed analysis will only be performed when there is some doubt about the result of the test in question.

4.2 Tests being Used

An informal description of a set of tests for the global Network Service over X.25 has been published (9). This identifies seven categories of tests which would be needed for full assessment. Preliminary testing by NPL to date has concentrated on the first two categories: primitive support tests and individual state transitions. The others are parameter variations, various traffic combinations, multiple connections, performance and robustness measurements.

Primitive support tests are designed to establish which service primitives are supported by the IUT. In our current work this involves about 20 individual tests. At the end of this series, a secondary result will be that it will have been established whether or not the TR is working satisfactorily. An example of a TDL definition of such a test is being published elsewhere (6).

Individual state transition tests are the main tests of conformance. They each involve driving the IUT into a given protocol state and then observing its reaction to a specific event. In our current work this involves about 100 individual tests.

5. CONCLUSIONS

The philosophy and architecture for assessment of protocol implementations, as presented in this paper, are applicable to most if not all OSI protocols. Since the interim Network Service being used is similar in many respects to the current ISO Network Service proposals, most of the work now being done at NPL will be easy to adapt to the eventual ISO Network Protocol standards. It could also be adapted to ISO Transport Protocol, but that is already being covered by ADI in France and NBS in the USA.

Internationally, as the number of groups working on third party testing increases, there is the danger of a proliferation of different test suites for the same protocol. There is therefore a need for standardization of sets of objective tests for each OSI protocol, which can be used in first, second or third party testing. This would ensure uniformity of testing, as is achieved in the programming language field by standard compiler validation suites.

ACKNOWLEDGEMENTS

The ideas presented in this paper are the product of the whole NPL Protocol Standards Group, 15 people including past members. This work has been mainly funded by the UK Department of Industry's Electronics and Avionics Requirements Board. The collaborative aspects of the work are being funded by the CEC.

REFERENCES

1. Ansart, J.P., Test and certification of standardised protocols, in: D. Rayner and R.W.S. Hale (eds.), Protocol Testing - Towards Proof?, Volume 2: Testing and Certification (NPL, 1981), 119-126.

2. Ansart, J.P., GENEPI/A - a protocol independent system for testing protocol implementation, in (17), 523-528.

3. Ansart, J.P. and Damidau, J., CERBERE, a tool to keep an eye on high level protocols, in (17), 529-538.

4. Bochmann, G.v. et al., Some experience with the use of formal specifications, in (17), 171-185.

5. Comite Consultatif International Telegraphique et Telephonique, CCITT Yellow Book, output of the VIIth Plenary Assembly of CCITT, held in Geneva on 10-21 November 1980, (ITU, Geneva, 1981).

6. Cowin, G.W., Hale, R.W.S. and Rayner, D., Protocol product testing - some comparisons and lessons, to be published in Proc. Third International Workshop on Protocol Specification, Testing and Verification, to be held in Zurich on 31 May - 2 June 1983, (North Holland, 1983).

7. Faltin, U., Faul, E., Giessler, A., Guenther, I., Orth, W. and Parslow, H., TESDI manual: testing and diagnosis aid for higher level protocols, (IFV-IK-RZ, GMD, Darmstadt, FRG, 1983).

8. Henley, R.F.L. (ed.), Implementation assessment of transport and network services: the test responder specification, prepared by the NPL Protocol Standards Group, NPL Report DNACS 46/81, (July 1981).

9. Henley, R.F.L. and Rayner, D., Implementation assessment of transport and network services: an informal description of tests, NPL DNACS TM 5/81, (July 1981).

10. International Standards Organisation, Data processing - open systems interconnection - basic reference model, ISO/DIS 7498, (April 1982).

11. Linington, P.F. (ed.), A network independent transport service, Annex 1 in particular, prepared by the Study Group 3 of British Telecom's PSS User Forum, SG3 CP(80)2, (NPL, Feb. 1980).

12. Ministry of Defence, Official definition of Coral 66, prepared by the Inter-Establishment Committee on Computer Applications as a language standard for military programming (HMSO, London, 1970).

13. Nightingale, J.S., A benchmark for the implementations of the NBS transport protocol, NBS Draft Report ICST/H.NP-81-20, (NBS, Washington DC, Sept. 1981).

14. Nightingale, J.S., Protocol testing using a reference implementation, in (17), 513-522.

15. Rayner, D., A system for testing protocol implementations, Computer Networks 6 (1982) 383-395.

16. Sarikaya, B. and Bochmann, G.v., Some experience with test sequence generation for protocols, in (17), 555-567.

17. Sunshine, C. (ed.), Protocol specification, testing and verification, (North Holland, 1982).

RESEARCH ON SPEECH RECOGNITION AT NPL

D Schofield

National Physical Laboratory
Teddington, Middlesex, UK

INTRODUCTION

Speech is man's primary means of communication. Direct input
of speech into a machine is therefore an important element in the
range of man-machine interfaces. Advantages include the
possibility of use by untrained people, the potential speed of
interaction (again especially with untrained people) and the ease
with which a person can talk and listen while simultaneously
carrying out some other task.

Speech involves the chain:

BRAIN --> MOUTH --> ACOUSTIC SIGNAL --> EAR --> BRAIN.
   (speaker)   (listener)

Experiments on perception show that the brain seems unable to
process information at more than 50 bits per second (bps). To
represent the acoustic signal itself requires at least 20 000 bps.
A single figure cannot be given for the information content of
speech because it contains information about more than just the
literal meaning, but evidently a major problem of speech
recognition is to remove redundancy. An analysis into phonetic
features is potentially the most effective way of doing this.

The phonetic approach has been a common one, at least in
research. However, the first commercial speech recognition
product, the Threshold recogniser, used a simpler idea of matching
the spectral shape of an utterance to that of a template from the
same speaker. This approach has many limitations (eg it only works
for the person who generated the template) but it is a clear

engineering task, and all subsequent products to date have built
on it. There have been improvements in the way that a speaker's
variable rate of utterance can be allowed for; compensation for
background levels has been added; improved representations have
been used. This approach has produced systems with high
recognition rates (98% or more) under the conditions of the
designs (close-talking microphones, limited vocabulary sizes).
Low-cost VLSI versions are commercially available.

Some commercial products can recognise sequences of words
spoken continuously. This has considerable advantages for a human
operator when compared to the use of isolated words or phrases.
Awkward enunciation is avoided, and the continuous input of digits
and phrases raises the data capture rate up to the level achieved
by keyboards. If the system provides a keyword facility to detect
commands in conversational speech, equipment does not need to be
switched on and off when the user is talking both to it and to
someone else.

Existing commercial devices have turned out badly in one
important respect - relatively few are actually used. Many have
been bought for evaluations and trials, and perhaps the current
generation of trials will produce a definite market. However, the
general opinion seems to be that it is not the performance under
ideal conditions but the performance which can be sustained in
operation which is the weakness. This is due partly to the
inevitable variations in speech and environment, and partly to the
disciplines needed in use.

The programme at NPL

NPL is carrying out research and development into practical
methods for recognition within continuous speech, from varied
speakers, in the presence of background noise. NPL bases its
continuous speech recognition work on the detection of features
which relate directly to the control of the articulatory
mechanism. An example is the presence or absence of voicing, the
excitation of the vocal cavities by the larynx.

Current work uses statistical techniques to analyse
utterances from different speakers and to produce representations
of vocabulary words in terms of phonetic features. During
recognition, pre-processing hardware and front-end software are
used for the identification and classification of features. The
data at this level have been sufficiently reduced to allow
real-time decision making, which compares the list of phonetic
features with phonetic specifications for each acceptable word in
the vocabulary. At the intermediate level, a list of matches with
attached probabilities is stored. The overall match is deduced in
terms of the permitted syntax.

Trial applications at NPL have used context-dependent syntax, for example an avionics system with commands such as:

set height twenty thousand feet.

The keyword SET, imbedded in normal speech, triggers the system and HEIGHT defines a context where a number is required.

The work reported here was carried out inter alia by D.R.Manning, B.E.Pay, R.E.Rengger and J.P.Yardley. This work is funded by the Electronics and Avionics Requirements Board of the UK Department of Industry. During the last couple of years the programme has also been supported by a Speech Recognition Club formed to assist the transfer to industry of the speech recognition technology already developed by NPL.

Why a phonetic approach?

Speech is sampled, say every 10 ms, and the sample or samples are processed in some way (DFT, LPC, hardware) to produce a sequence of frame patterns. The patterns can be analysed in two dimensions - by grouping together similar frames into frame classes, and by segmenting the speech into groups of frames. Either process can be done with a "phonetic" justification. The number of classes to use is a practical choice; it is the segmentation which typifies the feature-based methods. It is these we shall call "phonetic".

Commercial equipment does not segment, but matches whole word patterns. It is now known that this method may also be used for connected speech. A key problem in the matching is the means used to allow for deviations in the rate of speech. The most common solution is to allow "elastic matching" or "time warping" and the commonest applied algorithm uses so-called "dynamic programming". Equally important is of course the set of rules by which similarity of the templates and the utterances is judged.

The strong point of dynamic pattern matching is that the data is handled in a graded manner throughout, as smoothly varying distances etc., and that classification is delayed until the last possible moment. It is the weakness of feature-based methods that, if data is reduced at an intermediate level into a small number of features, any errors introduced have a serious effect which cannot be undone in the subsequent processing. The intermediate classification can in any case only reduce performance. In the laboratory, when both methods have been applied to the same data, the non-segmenting approach produces the best results - eg work at IBM (1) or NPL (below). In the commercial sphere, phonetic methods have not yet demonstrated themselves. Why do workers persist in the segmenting approach?

One attraction is that, by classifying information at a level which is relevant to the listener, a very great data reduction is performed. This simplifies processing and vocabulary representation to the level which could be carried out in present day low-cost equipment. Secondly it provides an obvious starting point for techniques which remove many of the present limitations such as poor discrimination against non-speech sounds and channel limitations. Ultimately, in comparing two elements of speech (utterance and template) it is the perceived or phonetic distance which is relevant. Listeners do seem to categorise at the phoneme level. Some variations in pattern are admissible within a word, some indicate different words. This information must be incorporated in a system for it to be stable in performance between the laboratory and the field, even for a system adapted for an individual speaker. Finally, a phonetic analysis should facilitate mappings between speakers, of the kind humans can quickly make when starting a new conversation.

To achieve these advantages, either the detection of features has to be better than it now is, or means have to be developed to deal with sets of features with attached probabilities. Of course in real speech many features are ambiguous and are resolved at a higher level in the brain, but at present the acoustic performance seems still to need improvement.

WORK AT NPL

NPL has consistently followed the phonetic approach, and has been demonstrating practical phonetic recognisers for the last several years. Our early work (2) assumed that the task of discovering the parameters in speech that conveyed its meaning would be made easier if we studied speech uttered under ideal conditions. The term "ideal" in this context meant the use of a wide bandwidth, high quality studio microphone in a relatively quiet environment, connected to the analysing system by a high fidelity channel.

We have, however, found by experience that this is not the best platform from which to start, as the analysing techniques developed for this special situation cannot always be transferred to more natural conditions. To overcome this problem we now specify that any speech processing procedure must be capable of working properly under the following less than ideal conditions:

Noisy environments - the analysing hardware must function consistently in normal room ambient noise and also when other sounds such as aircraft noise or equipment hum are present. It should be tolerant to signal to noise ratios typically as low as 10 dB.

Low quality channels and microphones - restrictions to the bandwidth of the microphone, or the channel connecting the microphone to the analysing hardware, must not affect the analysing hardware any more than it would a human listening to the same speech. For instance, the public telephone frequency bandwidth of 300 Hz to 3 KHz should be tolerated.

Unconstrained speech - the system must consistently analyse speech whether the words are spoken in isolation, or strung together as is more natural. Variations between speakers with the same general accent should not cause any major differences in the data produced.

The original bias of the NPL work was to analyse speech in terms of its production rather than in terms of its perception - the articulators are more universal than phonetic units such as phonemes. However experiments with human listeners have also been used, both with synthetic stimuli and with speech reconstituted from the analysed features.

The Hardware Preprocessor

The analysis of voiced speech has used essentially information from the spectral shape such as can be resolved in practice by a listener. This information has been extracted from the speech signal by autocorrelation techniques. To do this in real time, it was necessary to construct hardware preprocessors, the most recent being called SID3 (3,4).

SID3 analyses an analogue signal that may contain speech, and produces continuously at its output 64 bits of digital information which represent the way 16 speech parameters vary. It consists of a set of independent hardware modules each coupled to a speech bus. The speech bus is an analogue line giving modules access to the pre-amplified speech signal. Each module is responsible for the measurement of a particular feature or class of features. The digitally-encoded output from each module is gated onto an output highway which may be sampled by the software every 10 ms.

The 16 parameters chosen cover the following aspects:

Speech quality in terms of voiced, frication and quiet.
Resonances in four bands which can be related to the
     frequency and amplitude of the larynx vibration and
     the two lowest formants.
Spectral shape coefficients for higher frequencies.
Overall signal amplitude.

Feature extraction - the original model

The pre-processing software encodes, smooths and segments the samples to form a list of features called the STM (because in our system it is held in a short-term memory). The encoding of each frame is carried out at two levels, the class (eg voiced, fricative, voiced+fricative, nasal, quiet) and the type within each class. In the original work, the determination of voiced types used a partition of the F1-F2 (formant) plane. The vowel map of Peterson and Barney (5) is the best known example. Various boundaries were used, almost always allowing for considerable overlap so that many frames produced multiple type classifications. For fricatives, four types were used, based on a partition of the space generated by SID from the fricative spectrum (height, width and frequency at peak height).

The smoothing process removed short-term transitions, especially class changes, which are not relevant to perception. A moving window established the locally most frequent type, and other types within some threshold were retained. This involves two adjustable parameters, window size and threshold.

Segmentation was based on three main hypotheses: that within a certain maximum period, the order in which phonetic events occur is unimportant; that beyond a certain period, the duration of steady events is unimportant, and that certain class and type transitions are of articulatory importance (ie are segment boundaries). This leads to a representation where segments can have multiple labels, each with an associated "weight" derived from the number of samples in the segment with this character.

String matching - the WISPA system

The first recognition package which was used with SID3 was WISPA (6). WISPA was designed as a research tool to evaluate various strategies for recognition of continuously spoken speech, so it had a large number of system parameters and was made from many linked sub-units. Because of the large reduction in data rate through the use of features, a small computer (an LSI11) was quite adequate as a research tool working on vocabularies of up to 30 words.

WISPA operated in terms of input features; it was used with the features described above. In the training phase, one or more utterances are processed for each word in the vocabulary to make a (possibly composite) template for that word. This consisted of bit arrays with each bit representing the presence or absence of a particular feature in a given position in the word. Three maps were used to hold features of high, medium and low weights, the thresholds being adjustable.

During recognition, each template was compared with the corresponding representation of the input. The computation was essentially a word-spotting one, ie every template was compared with the input starting at every point. The determination of how closely the template matched the input was based on a string-matching model. One string can be derived from another by successively pairing the same elements, by inserting an element, by deleting an element, or by changing one element into another. Penalties are attached to the last three activities. Using dynamic programming, a penalty matrix can be built up where $P(i,j)$ represents the lowest cost of matching the first i elements of the template with the first j elements of the utterance. Suitable minima in $P(l,j)$, where l is the template length, identify the possibility of the occurrence of this word starting at the current starting point.

In this way a spotting matrix can be built up, with columns representing the words in the vocabulary and rows the possible starting times. The matrix elements represent likelihoods and durations. An independent segmentation package was used to derive the most probable utterance by resolving ambiguities and allowing for any overlapping features such as the middle "ss" in "six-seven".

A series of experiments with WISPA gradually refined the performance to the point where it was visibly limited by the adequacy of the input features, at least for words spoken in isolation (this was at 98% recognition for a database of digits, averaged over three speakers). For three-digit strings a much poorer performance of 52% was obtained, indicating remaining weaknesses. However at this time a much more efficent approach to the vocabulary representation was introduced.

Admissible variations within a word

The probability that two spoken words are analysed to be identical is very low indeed. There will always be quantitative and/or qualitative variations and such variations occur quite independently of any external influences. These variations are manifest with a single speaker. Naturally, emotional state, nasal congestion and so on contribute to variations in articulation. However the fundamental reasons for variations in pronunciation are context and motivation.

It is the context within which a word is spoken that determines which parts of that word must be articulated "well" and which parts need not. In achieving the goal of being successfully understood there is no motivation to articulate the whole word precisely.

An example is the word "7" in the context of numbers. The "S" in Seven appears to be an important part of that word. As such from many utterances one finds considerable qualitative stability. It is also a feature in that word which is stressed as a function of time. The duration however is variable, typically in the range 100-250 ms. Thus the "S" in Seven is qualitatively stable and quantitatively unstable.

In the first "e" in Seven we find the reverse situation. It appears we can say Suv'n, Serv'n, Sav'n, Siv'n and so on without changing the perceived meaning. As an unstressed feature it is short and relatively consistent in duration. Thus the "e" in Sev'n is quantitatively stable and qualitatively diverse.

All linguistic information in the language will take on one of the four combinations determined by the fundamental rule of context and motivation. These variations which can be referred to as "admissible variations" operate quite independently of any external influence.

Commonly used algorithmic approaches cannot "know" whether a difference between data and template is admissible or not. Thus the penalty applied is the same in both cases. This has the undesirable effect of closing the penalty differential between "good" and "bad" results. Also, the commonly used technique of basing a word template on a single utterance means that the template itself cannot reflect admissible variations. NPL has subscribed to the view, for many years, that the template should reflect the admissible variations, although this requires deriving templates from many utterances of the same word.

The early work at NPL (pre-WISPA) used manual techniques and produced as many templates (per word in the vocabulary) as were required to encompass the admissible variations. This early work was successful in establishing the feasibility of processing and recognising continuous speech in real time. It was however commercially impracticable, being labour intensive and requiring special knowledge to build a new vocabulary. This system was archived in 1978 to provide an operational demonstration and to study the long term "durability" of the recognition. To date the system's recognition ability has remained consistent and the admissible variations reflected in the templates have clearly contributed to this.

Subsequent automation (in WISPA) highlighted the weakness of having more than one template per vocabulary word. The automatic training program would typically produce as many as 20 to 30 admissible sequences from utterances of the same word. Although this was aggravated to some extent by incompatibility with other parts of the overall system, it became clear that there should be

just one template per vocabulary word which must reflect the
admissible variations.

Sesame

During 1981 a new package known as SESAME was developed. This
used a large number of utterances of a keyword from a number of
speakers to build admissible variations into the templates. This
is a statistical exercise based on the probability of features
occurring and on their duration. By building in the admissible
variations at training time the matching algorithms during
recognition can be simpler and do not require dynamic programming,
leading to a shorter response time than WISPA. The main motive
however is to improve the robustness of recognition.

An attribution map is built in the form of a lookup table
where there are three 16-bit computer words per feature per
feature position. For each feature at every feature position there
is a high, medium and low probability word. Each bit is unique to
a vocabulary word. Thus one ATTMAP can accommodate 16 words. A
typical ATTMAP contains 4200 bytes for words up to approximately
650 ms duration. This area has to be reserved for any vocabulary
size from 1 to 16. When fully utilised for 16 words the ATTMAP is
equivalent to 263 bytes per vocabulary word.

Results using SESAME show improvement over WISPA, especially
for sequences of words. For example, using the digits collected
from three "donors", a fourth person obtained a recognition rate
of 72% for totally correct three-digit phrases. This may be
compared with an average of 74% for the three donors themselves.
It is interesting that some of the donors performed better on the
three-speaker database than they did on their own subset of it.
All speakers had the same general accent.

Here also it was apparent that the real block to progress was
the adequacy of the feature extraction.

Experiments using Template Matching

NPL has looked at a number of the dynamic programming
approaches to continuous recognition. The first was developed by
Sakoe (7) and used in the Nippon Electric Company DP100 equipment.
This matched each pattern against the input in all possible
positions (limited by the permissible lengths of words). It uses
considerable computing power and is implemented by a built-in
vector processor.

Bridle et al (8) describe a simplified method which works by
a single pass through the input and requires less computing,
though still quite a powerful processor is required. Commercial

systems using this approach are available from Logica and from Marconi Space and Defence Systems. The experiments described here have used this approach.

Other groups use broadly similar methods, for example Rabiner's group at Bell Laboratories, Murray Hill New Jersey.

On applying this method at NPL, the initial results were encouraging in that they showed that the SID3 data had the information to enable digit strings to be recognised over a range of parameters. This was with only a single template for each digit, and not a particularly good one in many cases. The same templates worked on the few examples from other speakers which were tried. The corresponding results for WISPA, which are much poorer, therefore indicate problems in classifying the data rather than in the data itself.

Some of the errors made by the dynamic programming approach might have been avoided by using more effective algorithms for dealing with the costs and penalties and by controlling the time-warping in a better manner. However, some seem intrinsic. For example, the permissible degree of lengthening of different components of speech depends on properties of the speech itself. These could be derived from expert knowledge about speech, or automatically by some statistical approach. For a single speaker, the use of multiple templates will help, but to extend the method to cope with many different speakers it seems likely that a more systematic (we would say "phonetically-based") approach will be needed.

Current work on feature extraction

The feature extractor described earlier did not make use of all the information provided by the hardware preprocessor. In particular it made use of formant frequencies but not formant amplitudes. One direction of development has been to try to incorporate all the information. In this approach, the extractor consists of three major parts. After initial smoothing of the data, samples are segmented into periods of stability. Such periods are named as static features and given sub-names where appropriate. Finally the inter-segment zones are evaluated and given dynamic names where they are analysed to indicate more than a simple co-articulation interface.

A second approach being evaluated is a pure segmentation approach, ie it tries to determine significant points which can be boundaries between segments without at this stage classifying or labelling the segments in any way. This work concentrates on descriptions of the acoustic data at a number of levels of segmentation. A software tool which accepts hypotheses for segmentation at any level by building on previous hypotheses and using earlier transformations is being developed.

FUTURE WORK

Pattern-matching devices of the current generation are extremely impressive and it remains to be seen whether they can break through the barrier from evaluation studies to specific uses. The aim of the NPL work is to contribute to the technology for the next generation of recognisers. These will overtake current products not because they perform better under ideal conditions, but because they will greatly widen the area of practical use. Items for work are:

Re-evaluation of the formant-based technique.
Dynamic feature classification.
Feature analysis and adaptive training algorithms.
Collection of data for training from a large population.
Vocabulary library management.
Standard calibration techniques.

REFERENCES

1. Bahl, L.R, Jelinek, F and Mercer, R.L. A Maximum Likelihood approach to Continuous Speech Recognition. IEEE Trans. PAMI-5 (1983), 179-190.

2. Pay, B.E. and Evans, C.R. An approach to the Automatic Recognition of Speech. Int. J. of Man-Machine Studies, 14 (1981) 13-27.

3. Rengger, R.E. and Manning, D.R. A Hardware Preprocessor for use in Speech Recognition: Speech Input Device SID3. NPL Report no. DITC 22/83, May 1983.

4. Schofield, D. and Manning, D.R. A theoretical model for the Speech Input Device SID3, Part 1. Filters. NPL Report no. DITC 20/83, April 1983. Part 2. Correlators. NPL Report no. DITC 21/83, April 1983.

5. Peterson, G.H. and Barney, H.L. Control methods used in a study of the vowels. J. Acoust. Soc. America 24 (1952), 175-184.

6. Yardley, J.P. WISPA: A system for Word Identification in Speech by Phonetic Analysis. PhD Thesis, University of Essex, UK, 1981.

7. Sakoe, H. Two-level DP matching - a Dynamic Programming-based pattern matching algorithm for connected-word recognition. IEEE Trans ASSP-27 (1979) 588-595.

8. Bridle, J.S., Brown, M.D, and Chamberlain, R.M. An algorithm for Connected Word Recognition. Proc.IEEE Conf. ICASSP 82, 2 (1982) 899-902.

A REVIEW OF THE DEVELOPMENT OF STANDARDS FOR DATA ENCIPHERMENT

Wyn L Price

Division of Information Technology and Computing,
National Physical Laboratory, Teddington, UK

INTRODUCTION

It is now almost exactly ten years since the National Bureau
of Standards (NBS), Gaithersburg, Maryland, USA, issued a call for
submission of encipherment algorithms for use in the public domain.
It is therefore appropriate to trace the way in which public domain encipherment standards have developed during this time.

Prior to 1973 there had been little concern outside the military and diplomatic areas for protection of data against illegal
access by clandestine routes, though use of secure storage for
tapes and discs, control of user access to computer facilities by
password, etc., was fairly common and secure operating systems
were being developed. By 'clandestine route' we mean such techniques as line tapping, program bugging, and the like. Though
access to data by regular routes may be well controlled, it is
necessary to consider all possible ways in which data can be com-
promised. Clandestine attacks on data may be passive, where the
data is illegally read, or active, where data is illegally created,
altered or destroyed.

With ever increasing use being made of computer systems for
public and private data processing, it was inevitable that some of
the data handled should be of a sensitive nature; health records,
police files and financial transactions are but three examples.
Though the degree of security surrounding such data in pre-com-
puterised days was often inadequate, computerisation has made it
easier for the intruder to browse and collect information, indeed
systematic monitoring, searches and correlation of data have
become relatively trivial operations. Data in transit between

computer systems may be especially vulnerable when it passes over data communication networks.

During the ten years with which we are concerned, public disquiet about data collection and possible exposure has grown apace and many countries now have data protection legislation either in existence or in advanced stages of preparation. Much of this legislation is mainly concerned with governing the collection and processing of data, but other laws go further to specify the controls on access to data banks that shall be established. We have already suggested that, whilst access by regular routes may be subject to perfectly adequate controls, access by clandestine routes may be overlooked. Unless the latter is successfully prevented, many of the aims of data protection legislation may be thwarted.

One of the most important tools available to the system designer for the protection of data is encipherment. Ciphers and codes are known to have existed for several millenia, but until the advent of the digital computer they were operated either by hand or by fairly primitive machinery, even the electromechanical devices used during the Second World War were relatively unsophisticated, and therefore processing rates tended to be very slow. Since that time encipherment methods have advanced greatly, drawing upon the techniques being developed in digital computing. Many manufacturers offer proprietary encipherment devices, capable of high data rates and often designed particularly for the military and diplomatic markets. Devices from different manufacturers are not usually compatible one with the other and the encipherment algorithms on which they are based are kept secret.

Such encipherment facilities, whilst admirably suited for their purpose, may be inappropriate for general use, particularly when a large community of data users wishes to exchange data securely over a public data network. It is essential that such users employ compatible encipherment equipment and that the protocols by which they control the exchange of enciphered data shall be standardised. This is the nature of the requirement that has led to the development of encipherment standards for use in the public domain.

A STANDARD DATA ENCIPHERMENT ALGORITHM

In May of 1973 NBS published in the US Federal Register an invitation to inventors to submit fully developed encipherment algorithms suitable for protection of data in transit and in storage. Other requirements were that the algorithm should be economically viable, that it should be expressible in LSI hardware and that the details of the algorithm could be made fully public without compromise to its cryptographic strength. The response to this invitation was disappointing; many of the algorithms submitted required further development and none was suitable for the NBS intention. The invitation was reissued in August 1974 and more algorithms were submitted. Of these one, from IBM, was considered worthy of further investigation. The IBM offering was based on an earlier algorithm, called Lucifer [1], which had been developed around 1970 with protection of cash dispensing machine transactions as one of its first applications.

The IBM algorithm met the condition of full publication by virtue of the complexity of the way in which the encipherment key was used to control the transformation of plaintext into ciphertext (and vice versa). The condition of expressibility in LSI hardware, also met by the IBM algorithm, was set in order to avoid placing an excessive load on other components of computer or communication systems. Hardware implementation was also regarded as more easily verifiable and more resistant to malicious interference.

The DES algorithm in encipherment mode accepts plaintext blocks of 64 bits and transforms these into ciphertext blocks of 64 bits under the control of keys which are effectively 56 bits in length; in the Federal standard the key variable is 64 bits long, with the remaining 8 bits allocated to parity, but not otherwise entering into the operation of the algorithm. The algorithm contains sixteen internal cycles or 'rounds' (each with its sub-key of 48 bits derived from the 56 bits of the main key), during which substitution and transposition operations are alternated. In decipherment mode the internal operation is almost identical, but with the 16 sub-keys used in the reverse order.

Following review by the Federal authorities for suitability as a Federal Standard, the details of the algorithm were published in March 1975 as a draft standard, the title chosen being 'Data Encryption Standard' or DES. Publication of the draft standard provoked some criticism from those who considered that the cryptographic strength of the proposed algorithm was insufficient for the purposes envisaged. Appreciable effort was expended on an attempt at systematic cryptanalysis [2], but no significant weakness was identified as a result. Other criticism centered on the possibility of exhaustive search for encipherment keys - testing all keys in succession. Clearly there are far too many possible keys for even the fastest machine conceivable at present to test all of them serially within a time to be of any practical use. On the assumption that a large exhaustive search machine could be built, consisting of a million fast chips operating in parallel on different parts of the key domain, it was claimed [3] that key solutions to known plaintext/ciphertext pairs could be found in less than half a day. However, the assumption is extravagant;

such a machine would be very expensive to make, very expensive to run and very expensive to maintain. No such machine is known to exist at present.

As a result of the criticism levelled at the draft standard NBS held two meetings [4,5] at which technologists and mathematicians were invited to express their views. The outcome of these and other studies led the NBS to go forward and adopt the algorithm as an United States Federal Information Processing Standard, FIPS 46 [6], in November 1976, with a publication date of January 15 1977 and an effective date of July 15 1977. After the effective date Federal agencies were expected to use the DES algorithm to encipher any sensitive data outside the military and diplomatic areas. We shall see that additional standards are necessary before application of the DES can be said to be truly effective; it is doubtful whether the US government agencies found it possible to comply with the requirement to apply the DES in 1977, when none of the associated standards had been defined.

The controversy about the cryptographic strength of the DES continued long after the publication of FIPS 46 and the US government found it necessary to hold a Senate Select Committee to consider whether undue influence had been brought to bear during the specification of the algorithm. The full report of this committee was classified and not published, but a summary [7] was published which gave the standard a clean bill of health and exonerated government agencies from any deliberate attempts to weaken the algorithm.

The algorithm is known to have certain special properties. One of these is that of complementarity - bit inversion of the plaintext and of the encipherment key produces a bit inverted ciphertext. Another special property is that of the so-called 'weak keys', for which all the internal sub-keys have identical values. Because of the identical sub-keys, double encipherment with a weak key reproduces the original plaintext; encipherment and decipherment are exactly equivalent. Four weak keys exist, one of these consists of 56 zero bits and another of 56 'one' bits; the other two consist of equal numbers of ones and zeros arranged so that the sub-keys are all the same. There are also the 12 'semi-weak keys', where for six particular key pairs the sub-keys for one key are exactly the same as the sub-keys for the other key of the pair arranged in the opposite order. If a data block is enciphered with a semi-weak key, then a second encipherment with the other key of the pair restores the plaintext. The semi-weak keys were first discovered by Davies [8].

The existence of weak and semi-weak keys does not necessarily constitute a fundamental weakness of encipherment with the DES algorithm. Prudent users, as a safety precaution, will test randomly generated keys for these values before using them for enciipherment of data and reject any that are found. However, it is unlikely that an intruder could profit by their use even if they are not rejected; the intruder needs to discover that a weak key is in use, which is not necessarily any easier than discovering any other key. The kind of scenario where such profit might be achieved assumes such things as feasibility of resubmission of enciphered data to the DES device (with the same key installed) by the intruder.

Other key values are known to produce certain patterns of repeated sub-keys within the algorithm and there is currently some debate on the subject of alleged weaknesses arising from these; it is a little confusing that the whole class of keys of this type is sometimes called 'weak' as distinct from the two groups of keys we have already mentioned. It has been suggested [9] that there are 25 different categories of such keys (268,419,060 keys in all) and that, given plaintext/ciphertext pairs produced with one of these keys, it is possible to decide to which of these classes the relevant encipherment key belongs. If this is so, then exhaustive search within the class will not take long, about 4 hours is claimed. It is also suggested that a key can be found, given plaintext/ciphertext pairs, in about 8 hours even if it is not in the 'weak' categories; no basis of a method has been cited for this claim. There has been no published description of any way of profiting from properties of any keys in the general weak class other than the 16 special keys already discussed. The claims to which we refer have not yet been substantiated and it remains to be seen whether they can be.

Confidence in the suitability of the DES for its purpose has been expressed by the American National Standards Institute (ANSI) by its adoption as a national standard under the title of 'Data Encryption Algorithm' X3.92-1981 [10], published in July 1981. The responsible technical committee was X3T1, with representatives drawn from industry, commerce, professional institutions and government agencies.

The British Standards Institution (BSI) has been active in the field of data security standards since 1980, when a committee was set up to consider this subject; the designation of this committee is now OIS/21. Its chairman is Donald Davies of the UK National Physical Laboratory, which has played a significant part in the preparation of draft documents for the committee. The main role of the latter has thus far been to provide texts of draft standards to a working group at international level. This was set up by Technical Committee 97 of the International Standards Organisation and is known as Working Group 1 (TC97/WG1). The status of this working group is currently under consideration, a working group directly responsible to TC97 was an unique arrangement, and it may be replaced by a regular sub-committee of TC97.

It is hoped that agreement can be reached through WG1 (or its successor) and TC97 on an international standard for data encipherment. In the meantime a British standard is in preparation.

Implementations of the DES began to appear almost before the official adoption by the US Government. FIPS 46 recognises only hardware implementations and these are available from a range of manufacturers as programmed microprocessors (with low speed performance) or as special purpose LSI (with algorithm execution times as low as 4 microseconds). The ANSI standard recognises both hardware and software implementations (as do the international and British drafts); software implementations are known to have been written and used in various systems.

STANDARDS FOR MODES OF OPERATION OF THE DES

The DES algorithm handles data in 64 bit or 8 byte blocks; used in block mode the DES is said to carry out Electronic Code Book (ECB) encipherment. A message which is longer than 8 bytes can be segmented into blocks, each of which can be separately enciphered and the message sent as a series of individual ciphertext blocks. There are several disadvantages arising if this mode of operation is adopted.

In a text segmented into 8 byte units there is a possibility that identical blocks may occur in different parts of the text. With individual block encipherment the ciphertext corresponding to each identical block will also be identical and a passive observer can detect at least some structure in the enciphered message. This is undesirable, but an active attacker can do much more by storing, processing and retransmitting the message. The attacker can manipulate the enciphered text, causing blocks to be moved about, duplicated or deleted; when the receiver finally gets the altered message and deciphers it there may be nothing to tell that changes have been made. There is therefore a need to establish a chained relationship between successive blocks of ciphertext.

The chained block mode of operation is known as Cipher Block Chaining (CBC), in which, before encipherment, each plaintext block is combined by an exclusive-OR operation with the preceding ciphertext block. At the start of the message the first plaintext block is combined with an 'initialising variable'. The enciphered form of any block depends not only on the corresponding plaintext block, but also upon all the preceding plaintext blocks. If any attempt is made to interfere with the transmitted blocks of such an enciphered message, the effect on the plaintext at the receiver is catastrophic to the sense of the message and should be obvious to the receiver.

Not all applications are well served by a block mode of operation and there is a need for a mode which will deal with smaller units of text, such as bits or characters, in either synchronous or asynchronous transmission; transparency is often a requirement in a data communication system. For this reason the Cipher Feedback (CFB) mode has been developed, in which the units of plaintext are combined (in the manner of a Vernam cipher) by exclusive-OR with bits extracted from the output of a DES device, producing units of ciphertext. The input to the DES device is used as a shift register into which the successive units of ciphertext are fed; initially the input register is set to an initialising variable, which is progressively displaced by the ciphertext. At the receiver the DES device is used in encipherment mode to produce the same pseudo-random bit stream at its output as was used for the exclusive-OR operation of plaintext encipherment. The data unit sizes most likely to be handled in this way are the single bit and the 8 bit byte.

CFB mode of operation (and CBC mode) has the property of some error propagation, so that errors in transmission of ciphertext result in expansion of error in the deciphered plaintext. For some applications (for example, transmission on a noisy channel with appreciable delay, such as a satellite link) this is undesirable and another mode of operation, known as Output Feedback (OFB), has been devised to take account of this. OFB is very like CFB, but, instead of feeding the ciphertext into the input register of the DES device, only the contents of the DES output register are fed back. Output feedback has no error propagation.

Definitions of CBC and CFB modes of operation were published first in early versions of the draft Federal Telecommunications Standard (FTS) 1026, produced by the US National Communications System (NCS) agency, under the auspices of the Federal Telecommunications Standards Committee (FTSC). Progress on FTS 1026 was slow, so the NBS produced a Federal Information Processing Standard, FIPS 81 [11], published in December 1980, which defines all the modes of operation outlined above. In addition FIPS 81 describes a mode of operation for use of the DES to produce authentication of messages.

ANSI has published a national standard for modes of operation, this is 'Modes of Operation for the Data Encryption Algorithm' [12], published in 1983. BSI has a modes of operation standard in preparation. Both of these documents are compatible with FIPS 81.

STANDARDS FOR DES ENCIPHERED COMMUNICATION

It was realised early that implementation of DES encipherment
within a data communication context would require specification of
data formats, methods of key management and so forth. The initial
ideas on these subjects appeared in the early versions of draft
FTS 1026, initially entitled 'Telecommunications: compatibility
requirements for use of the Data Encryption Standard'; a later
title was 'Telecommunications: interoperability and security
requirements for use of the Data Encryption Standard in data
communication systems'. The interoperability requirement aimed to
make compatible all DES equipment procured for Federal use, whilst
the security requirement laid down the following objectives:

a) prevention of disclosure of plaintext messages
b) detection of fraudulent insertion of messages
c) detection of fraudulent deletion of messages
d) detection of fraudulent modification of messages
e) detection of the replay of previously valid messages.

When, in 1977, the first versions of FTS 1026 were prepared,
the aim was to specify methods of using the DES in a compatible
manner in the most general sense. However, it was noted that link
encryption devices were being made that used the DES in a manner
optimised for link applications, and that were incompatible with
synchronous, end-to-end encipherment applications. This caused
several temporary expedients to be proposed. Fortunately the
development of the Open Systems Interconnection (OSI) architectural model [13] of ISO, which began in 1978, went rapidly ahead
and provided a means of achieving a hierarchy of coherent and
compatible data communications protocols. The lower levels of the
OSI structure, physical, link, network and transport levels, have
by 1983 achieved final or almost final definitions. These developments have affected the work on DES-oriented communication
protocols, giving it a firmer basis.

As we have indicated, the early versions of FTS 1026 were
intended to be as general as possible, covering all kinds of communication connections, but it was soon discovered that more specific definitions were called for at the lower layers of OSI. In
June 1981 the draft FTS 1026 was split into two separate documents, draft FTS 1026, covering encipherment at the physical and
link layers, and draft FTS 1025, covering encipherment at the network and transport layers. Since that time the Federal authorities have again altered the scope of FTS 1026, confining it now to
the physical layer only. It seems that work is concentrated temporarily on this document, the other layers to be dealt with later.
Encipherment at the physical layer as defined in draft FTS 1026 is
intended only to protect against passive attack.

Compared with earlier drafts the new version of FTS 1026 is the epitome of simplicity, the technical details extending to just over three pages. In physical layer encipherment, the whole of the physical layer service data unit is enciphered, except for the START and STOP bits during asynchronous operation and parity bits when parity is being restored. An option is provided for restoring asynchronous character parity. The mode of operation specified is 1-bit cipher feedback; this is not described in the standard, instead a reference to FIPS 81 is given. In synchronous transmission once a Clear to Send indication is detected the initialising vector is sent in clear, followed immediately by all the bits of the enciphered physical layer service data unit; the process continues indefinitely, until the disestablishment of the physical connection. In asynchronous operation once a Clear to Send indication is detected the initialising vector is sent, broken into suitable character units and with START and STOP bits added. Enciphered data transmission follows, with each data character enciphered, but with START and STOP bits in clear. The process continues (with the exception of BREAK characters) until the physical connection is broken.

The 1983 draft of FTS 1026 does not specify any form of service message for exchange of keys via the physical connection between the communicating DES devices; the keys are to be loaded either from a keyboard on the DES device or transferred to the DES device from a purpose-built key transport module. Details of this kind of key management are to be found in FTS 1027 [14].

Earlier versions of FTS 1026 had included specification of data formats and two types of service message. Such specifications will still be needed when encipherment at the link and higher layers is defined, and for this reason we mention them here. Plaintext chain formats for protection against passive (reading) attack were defined to include beginning of chain marker, data and end of chain marker, with additional fields for chain identification (CID) and manipulation detection code (MDC) where active attack (alteration or replay) was to be expected. Two types of service messages were defined; plaintext service messages (PTSM) allowed signalling of mismatches of CID and MDC, request for encipherment key and current value of CID; protected service messages (PSM) permitted direct exchange of keys and initialising variables between DES devices and were enciphered to prevent disclosure of these parameters.

ANSI and BSI have also been active in the preparation of drafts for physical layer encipherment standards. The ANSI committee has strong representation from manufacturing interests and it is not surprising to find signs of influence from this quarter. An ANSI draft contains the words "this standard recognizes the existence of several DEA cryptographic devices that predate the standard, and where the operation of this equipment conflicts with the preferred method of operation described, their method of operation is identified as an acceptable alternative". Such a clause does not assist the process of achieving compatibility between equipment purchased from different manufacturers, an aim usually associated with the development of standards. At one stage the differences between the ANSI draft for physical layer encipherment and the Federal draft of FTS 1026 were very marked; fortunately many of these differences have been ironed out and the remaining differences are in respect of the non-preferred options. Both ANSI and Federal drafts cite options for action when a BREAK condition is detected; the ANSI draft cites an option for a bypass mode which is not allowed in the Federal draft.

It should be noted that the draft prepared by BSI for international consideration, whilst in line with the Federal and ANSI work, does not tie itself to the DES algorithm, but allows the possibility of other encipherment algorithms being used within the terms of the standard.

At the link layer the divergences between ANSI and the Federal drafts were at one time even more serious. The Federal draft envisaged encipherment of the Link Service Data Unit, so that the Link header and trailer were sent unenciphered. On the other hand, the ANSI draft preferred encipherment of the Link Service Data Unit together with the fields added by the Link protocol; effectively this meant that the unit of encipherment was the Physical Service Data Unit. ANSI preferred encipherment near the bottom of the Link layer, whilst the Federal authorities preferred it to be near the top. In favour of the ANSI view simplicity of implementation was cited, whilst the Federal approach was considered to be more general.

Work on encipherment enhancement of communication protocols above the physical layer is proceeding slowly; from the work already done it is evident that there will a great deal of hard argument within the standards committees. Fortunately, there is less likelihood of entrenched ideas or existing devices hindering the work at the higher levels.

OTHER ASSOCIATED STANDARDS

A standard for the physical security of DES devices has been published (April 1982, [14]) by the FTSC; this is FTS 1027, entitled 'Telecommunications: general security requirements for equipment using the Data Encryption Standard'. The aims of this standard for DES equipment are:

a) prevention of inadvertent transmission of plaintext
b) prevention of theft, unauthorised use, etc. of the equipment
c) prevention of unauthorised disclosure or alteration of keys
d) provision of interoperability of key loaders and equipment
e) prevention of encryption if a failure of the equipment is
detected.

Amongst other clauses FTS 1027 describes how conforming DES
equipment is required to carry out automatic testing of the cryptographic function. Equipment is required to destroy any stored
keys if an attempt to break in to the equipment enclosure is detected. This demands the creation of a tamper resistant enclosure, which is appropriate for any DES devices not located within
installations that are otherwise physically secured.

Another associated standard is found in the banking domain.
We noted how FIPS 81 includes a definition of the use of the DES
to generate authentication fields on messages. This function of
the DES has been further defined in ANSI standard X9.9-1982 [15],
entitled 'Financial Institution Message Authentication', approved
in April 1982 and developed by ANSI in collaboration with the
American Bankers' Association.

The Federal authorities have issued (August 1982) a draft
FIPS called 'Standard for computer data integrity'; this includes
two applications of the DES, authentication only and authentication with encipherment.

FUTURE DEVELOPMENTS

It should be evident from the brief account we have given of
standards activity in the data security area that, though much
progress has been made, much remains to be done. The rate of
installation of encipherment equipment in the public domain has
disppointed some manufacturers and it is possible that this will
have some effect on the rate at which new standards emerge. It
could be that the demand for security will continue to come mainly
from the financial community, banks and like organisations, with
automatic teller machines, point of sale terminals and home banking developing rapidly. This demand could well be enough to maintain the impetus of the work.

We expect to see progress in the production of encipherment
enhancement standards in the higher levels of OSI. At these
levels it will be necessary to define additional functions, such
those of key distribution centres, essential for key management in
large networks.

It is more than likely that standards will emerge for new

encipherment algorithms, possibly replacing the DES or existing alongside it. It is noteworthy that the NBS has already issued a call (deadline September 1982) for submission of public key crypto-system algorithms for consideration as candidates for a future standard in this new field.

REFERENCES

1. Feistel, H. Cryptography and computer privacy. Scientific American, 228, 5, May 1973, pp. 15 - 23.

2. Hellman, M., R. Merkle, R. Schroeppel, L. Washington, W. Diffie, S. Pohlig & P. Schweitzer. Results of an initial attempt to cryptanalyze the NBS Data Encryption Standard. Information Systems Laboratory Report, Stanford University, November 1976.

3. Diffie, W. & M. Hellman. Exhaustive cryptanalysis of the NBS Data Encryption Standard. Computer, 10, 6, June 1977, pp. 74 - 84.

4. National Bureau of Standards. Report of the Workshop on Estimation of Significant Advances in Computer Technology. NBSIR 76-1189, December 1976.

5. National Bureau of Standards. Report of the Workshop on Cryptography in Support of Computer Security. NBSIR 77-1291, September 1977.

6. National Bureau of Standards. Data Encryption Standard. Federal Information Processing Standard Publication 46, January 15 1977.

7. Senate Select Committee on Intelligence. Reports in 'New York Times', April 13 1978, and 'Computerworld', April 17 1978.

8. Davies, D.W. Some regular properties of the Data Encryption Standard algorithm. NPL Internal Note, 1979.
See also The so-called 'weak' keys of the DES algorithm, NPL Internal Note, 1983.

9. Kolata, G.B. Flaws found in popular code. Science, 219, January 28 1983, pp. 369 - 370.

10. American National Standards Institute. Data Encryption Algorithm. ANSI Standard X3.92-1981.

11. National Bureau of Standards. DES Modes of Operation. Federal Information Processing Standards Publication 81, December 2 1980.

12. American National Standards Institute. Modes of Operation for the Data Encryption Algorithm. ANSI Standard X3.106-1983.

13. International Standards Organisation. Information processing systems - Open systems interconnection - Basic reference model. Draft International Standard 7498, May 22 1982.

14. US General Services Administration. Telecommunications: general security requirements for equipment using the Data Encryption Standard. Federal Telecommunications Standard 1027, April 14 1982.

15. American National Standards Institute. Financial Institution Message Authentication. ANSI Standard X9.9-1982, April 13 1982.

Part 6.

GENERAL ELECTRIC COMPANY RESEARCH LABORATORY SESSION

Organised and Chaired by:

P.V. Collins

G.E.C. Research Laboratories,
Hirst Research Centre, Wembley, U.K.

IMAGE PROCESSING ALGORITHMS

C J Oddy[*] and J R M Mason[+]
[*]GEC Research Laboratories, Marconi Research Centre, Chelmsford, UK
[+]GEC Research Laboratories, Hirst Research Centre, Wembley, UK

1   INTRODUCTION

Since the advent of increased computing power at reduced cost, much interest has been focussed on digital image processing techniques. Many of these have hitherto been too complex to implement in real-time, but with the increasing scale of circuit integration and development of parallel processing architectures they are imminently becoming viable.

This paper describes some of the image analysis techniques developed in GEC Research Laboratories for a variety of applications. The discussion is made with reference to two specific applications, namely synthetic aperture radar (SAR) images for earth resources, and very low bit-rate image storage or transmission for visual communications. These two areas are chosen so as to provide a contrast between images with different properties and to illustrate the effect this has on the optimal choice of analysis procedure.

Figure 1 shows a general image analysis system. Typically the pre-processing stage involves enhancement, noise suppression, or smoothing, the feature extraction stage consists of edge detection, region growing and texture analysis, and the feature description stage utilises chain coding of edges, syntactic description or pattern classification. However, it is clear that these stages entirely depend on the application involved.

SAR images contain high levels of noise, which makes the preprocessing stage very important. For the second image type, considerably more detail can be extracted in the analysis stages. Thus two approaches to segmentation, suitable for the two types of image, are discussed. In addition, for the SAR image analysis, a quantitative assessment procedure involving simulated images, has

646

been developed, and is described.

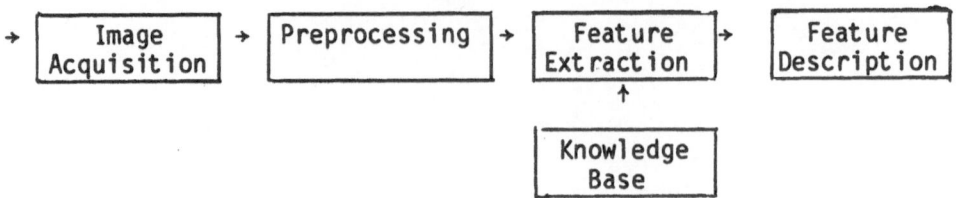

Figure 1

2   AUTOMATIC SEGMENTATION OF SAR IMAGES

2.1   SAR Characteristics

The Segmentation of SAR images presents a particular problem because of the noise or 'speckle' present in the image. This noise is multiplicative (i.e. the standard deviation is proportional to the mean) and is due to the coherent nature of the SAR imaging process. The power, I, of an image pixel from a homogeneous area has an exponential probability distribution of the form:

$$p(I) = \frac{1}{I_0} e^{-I/I_0}$$

Where I_0 is the average power. The speckle variance can be reduced by adding N independent looks to form a 'multi-look' image. The probability density is then given by a Gamma distribution:

$$p(I) = \frac{1}{(N-1)!} \frac{1}{I_0} \left(\frac{I}{I_0}\right)^{N-1} e^{-I/I_0}$$

The mean of this distribution is NI_0 and the variance is $\sqrt{N}I_0$. In addition neighbouring pixels are often correlated because of image over-sampling. For an explanation of SAR imaging see Elachi et al[1].

The greatest problem in the segmentation of SAR images is the reduction of the speckle without too much loss in resolution. A number of schemes for smoothing the image have been designed, and the properties of these are described in the next section. Once the smoothing has been achieved a number of methods are available for segmenting the image and those are described in Section 2.3.

2.2   Smoothing Algorithms

All the smoothing algorithms considered operate on a local window of the image. Many of the simpler smoothing techniques, such as averaging and median filtering, do not perform well on SAR images. It is necessary to look at more complex techniques. Two approaches have been tested. Firstly a number of algorithms have been proposed which are based on the multiplicative properties of the SAR speckle. An attempt is made to estimate the ideal intensity of a pixel, given local estimates for the mean and variance of the distribution (see[2&3]). These algorithms produce an image which is usually very satisfactory but not ideal for subsequent segmentation. This is because the edges tend to become blurred by the smoothing process.

The second approach is to use algorithms which are specifically designed to preserve edges in the image. The algorithm described by Yokoya et al[4] is one example of this. This algorithm preserves, and indeed sharpens, edges but also has a tendency to invent edges where they are not expected. We have produced an extension to Yokoya's algorithm which is designed to deal with multiplicative noise. The ratio of the mean and variance is used to decide whether the local window contains pixels from a homogeneous region or from across a region boundary. Full details of this method are given in Oddy et al[5].

In general we have found that it is best to use a combination of the two approaches. The speckle based smoothing is followed by several iterations of an algorithm which sharpens the edges.

2.3   Segmentation Algorithms

The segmentation has been performed using a 'bonding' scheme. Each pixel is bonded to those of its neighbours which are considered to be in the same region. The bonding is performed according to some local rule. After this operation all the pixels in a region are linked together but not to any pixels in other regions (Figure 2). Various criteria could be used to determine the bonding and two possibilities have been investigated.

Conceptually the simplest criterion for bonding is the intensity difference between neighbouring pixels. For SAR images bonding cannot be performed using an absolute threshold on intensity differences, because of the dependence of the variance on the mean. Instead an adaptive threshold is used, which varies with the mean over a local window. The results are good, except for a tendency to produce 'rogue bonds' across region boundaries at points where the edge is weak or blurred.

A slightly more complex approach is to generate an edge magnitude image using one of the many available edge detection algorithms. Pixels are bonded to their neighbours if the edge magnitude is less than some threshold. Otherwise pixels are bonded to the neighbour with the minimum edge magnitude. Region boundaries are formed where there are local maxima in the edge magnitude image. The threshold used (which represents a decision that no edge is present

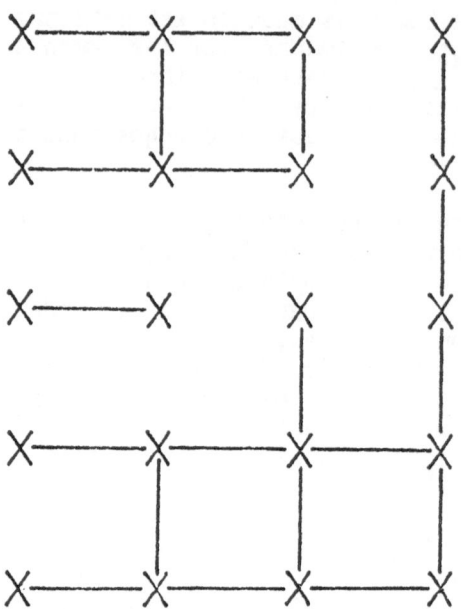

Figure 2 : A bonded image with 3 regions

around that pixel) can be made adaptive in order to give a better performance for SAR images.

2.4   Algorithm Assessment and Optimisation

If an automatic procedure for segmentation of SAR images is to be designed then some method for assessment of possible algorithms must be available. Most papers in the literature only give a visual assessment of the algorithms presented. We have designed a quantitative method in which a simulated image is segmented and a figure of merit is calculated for the result. Test images have been generated with SAR properties and containing regions of known size, shape and radiometric resolution. Each segmentation scheme is performed on the test images.

The bonded test image is input to a boundary tracing program which extracts the chain code for the region boundaries and throws away any edges which are not closed. The image information is stored in a Segmented Image Database. A 'region map' can then be generated from the database. The region map is an image in which each pixel in a region is set to the region identifier. This region map is then compared with an ideal region map (i.e. a perfect segmentation) in order to compute the figure of merit.

The figure of merit is made up of two parts: a detection score and a false alarm score. Each segmented region is classified as a false alarm if less than half the pixels overlap a test region. For the jth segmented region overlapping test region i, the detection score $S_{D_i}^j$ is given by:

$$S_{D_i}^j = 1 - \frac{((S+T)-2V)}{T}$$

$$0 < S_{D_i}^j < 1$$

where S = area of segmented region
T = area of test region
V = area of overlap

The total detection score S_{D_i} for the region is

$$S_{D_i} = \sqrt{\sum_j (S_{D_i}^j)^2}$$

The overall detection score S_D for the image is:

$$S_D = \frac{1}{2N} \left\{ \sum_i S_{D_i} + (\text{no. of regions detected}) \right\}$$

Where N is the total number of test regions.

The false alarm score S_A is given by:

$$S_A = 1 - \frac{(\text{Total area of false alarms})*10}{(\text{image area})}$$

The figure of merit is given by:

$$S_F = \frac{S_A + S_D}{2}$$

Many of the algorithms contain variables (such as thresholds), and the values taken by these are varied to give the best performance for the algorithm. The schemes which perform best on the test images can then be used on real SAR images.

2.5   Results

Slides of some results and examples of figures of merit will be shown.

650

3   EFFICIENT STORAGE AND TRANSMISSION OF DIGITAL IMAGES

3.1   Introduction

A videophone typically transmits visual data sampled at 256x256 pixels/frame, 8 bits/pixel and 25 frames/second, giving a total bandwidth of 12.8 Mb/s. Predictive waveform coding algorithms reduce the redundancy of such an image series so that a data rate of 1-2 Mb/s can be achieved. However, as a digital speech channel has a capacity of 64 kb/s, a useful aim would be to achieve a 64 kb/s view-phone.

As has already been seen, sophisticated coding techniques are not sufficient to produce the reduction factor of 200, and thus an analysis/synthesis method is required. Figure 3 shows a block diagram of the proposed strategy.

The input image is digitised and stored, and then subjected to a segmentation process. The resulting regions are then analysed statistically to determine the nature of their texture, a model of texture is chosen according to the statistics, and the parameters required to describe the chosen texture model are extracted. Thus, the image can now be described purely in terms of the boundaries of the regions, the texture model of each region and the parameters of the texture models, reducing the amount of data used to describe the picture significantly. Further reductions may be made by transmitting just the differences between consecutive frames.

3.2   Segmentation

One of the more popular ways of segmenting an image is to use an approximation of the first derivative of a picture, and take those pixels corresponding to local maxima as probable edge points. In practise it is usually not the local maxima that are chosen but

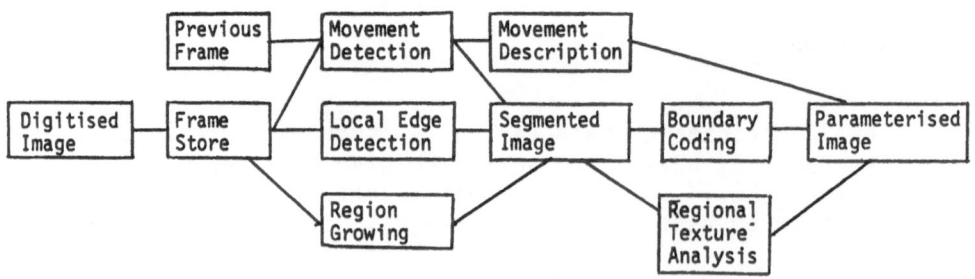

Figure 3

those pixels with gradient values above a given threshold, producing a binary edge array very simply. A typical example of this technique is that suggested by Sobel[6]. This method of edge detection has several disadvantages:

1   Because the local maxima are not necessarily chosen, some edges will not be continuous.

2   The value of the threshold will usually depend on the picture and what is significant in it.

3   Considerable further processing must be used to obtain continuous, single pixel thick edges.

However the gradient operator is generally fairly robust to noise and retains good detail in the image.

An alternative method of edge detection has been suggested by Marr and Hildreth[7] as being a model of human visual processing. The technique involves the finding of the approximate second derivative of the smoothed image and locating those pixels situated at the zero-crossings of the second derivative. The smoothing is performed using a Gaussian filter because it is rotationally symmetric and optimally satisfies the requirements of spatial localisation and frequency localised at the lower end of the spectrum. The second derivative operator used is the laplacian due to its non-directionality. These two operators can be combined into one denoted by $\nabla^2 G$ and called the Marr-Hildreth operator. Figure 4 depicts a three-dimensional plot of the operator for mask width 16. The mask width denotes the radius of the positive part of the masks. There is evidence to suggest that the eye uses various such operators of different sizes, although the method of combination is uncertain.

The advantage of using the Marr-Hildreth operator is that the inclusion of those pixels at zero-crossings of the laplacian (defined as those pixels having the sign of $\nabla^2 G$ different to at least one neighbour) almost always results in connected edge segments. The disadvantages are that the operator is very sensitive to noise and higher mask sizes are unable to cope accurately with edges of high curvature.

Consequently we have two algorithms, both of which have distinct advantages and disadvantages. Thus some sort of 'algorithm engineering' is required to produce a process which combines the strengths of the two approaches.

Experiments suggest that a robust method of combination of the two algorithms is to find the product of the Sobel gradients and the array consisting of the slopes at the zero-crossing pixels and

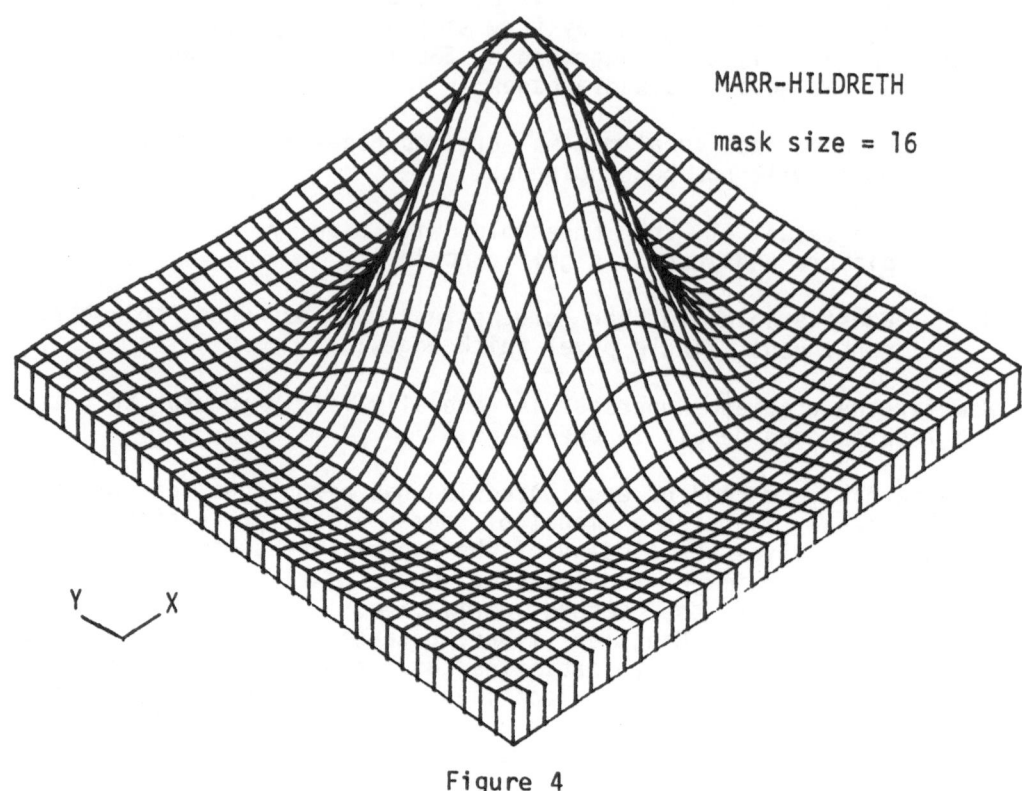

MARR-HILDRETH

mask size = 16

Figure 4

zero elsewhere.   The product then theoretically lies in the range [0,65536], and is treated in two ways.

Firstly it is thresholded at the value 1.   This process obtains a large set of possible edge points (i.e. those points with non-zero gradient from both operators).   Secondly it is thresholded at a higher value of 128, leaving a set of probable edge points.   The set of possible edge points is next cleaned to get a segmentation of the image consisting of single-thickness edge segments and also many   spurious   edge segments.   These   segments   are examined individually to determine if they contain any probable edge points. If no such point is present then the whole segment is discarded as being unlikely.

The above procedure results in a good segmentation of monochrome video images.   Further processes of region merging may be finally applied to remove any remaining spurious regions.

3.3   Texture Analysis

Texture   analysis is a technique that   has been used for various applications, in particular the classification of terrain in Landsat images, and the segmentation of images with textural edges. In this work the texture descriptions are used to effect considerable data reduction, and because of this, the overiding consideration in selection of texture models has been that they should be described using as few parameters as possible.

Having achieved a good segmentation of the image, the texture of each region must be described as efficiently as possible. The strategy employed here is to study the statistics of each region, (first and second order, local and global) and on the basis of the statistics choose from several mathematical models of texture, the one that is most able to synthesise a texture possessing similar qualities to the original. Having selected a model the texture is parametrised accordingly.

To resynthesise the picture the edges of the region, texture model type and parameters are required, generally giving a much lower data rate for each region. Clearly such a resynthesised picture will not be exactly similar to the original but it is hoped that the general impression of the texture will be conveyed. Thus, for the resynthesised image the edges will be fairly accurately portrayed and the texture will be more approximate. This parallels the ability of the human eye to detect edges well but to be less sensitive to texture.

Julesz[8] made a conjecture in 1962 that the human observer could only distinguish textures with differences in second or first order statistics but not higher orders. The conjecture has since been disproved, but the fact that it was made at all, and the length of time for which it stood indicates that this is a good example to follow.

Textural second order statistics may be calculated via the Spatial Grey Level Dependence Matrices (SGLDM).

A spatial grey level dependence matrix is defined on a region R of an image with intensity function $f(x,y)$, $(x,y) \in R$ for a given displacement vector $\underline{a} = (a_x, a_y)$ as the $N_G \times N_G$ matrix $S = (S_{ij})$, where N_G is the number of possible grey levels, and

$$S_{ij} = pr\{f(v,w)=i \text{ and } f(x,y)=j \text{ given that } x= v+a_x \text{ and } y= w+a_y\}$$

These matrices can be then used to provide various textural features[9]. However, due to implementational considerations it is expedient to utilise features that are linear in the values S_{ij}

only, thus foregoing the necessity to store the large SGLDM matrix. The features used include the correlation

$$C(S) = \sum_{i=1}^{N_G} \sum_{j=1}^{N_G} \frac{(i-\mu_x)(j-\mu_y)}{\sigma_x \sigma_y} S_{ij}$$

the local homogeneity

$$L(S) = \sum_{i=1}^{N_G} \sum_{j=1}^{N_G} \frac{1}{1+(i-j)^2} S_{ij}$$

and the inertia

$$I(S) = \sum_{i=1}^{N_G} \sum_{j=1}^{N_G} (i-j)^2 S_{ij}$$

where μ_x and μ_y are the row-sum mean and column-sum mean, and σ_x^2 and σ_y^2 are the row-sum variance and column-sum variance respectively.

These textural features provide information about the relationships between the grey-levels of pixels at a given separation. At larger separations such detail is not required, rather a rough estimate of the correlation is sufficient to enable detection of periodicities. This is performed by calculation of the auto correlation function which is defined in this case as follows:

$$A(i,j) = \frac{\displaystyle\sum_{(x,y),(x+i,y+j)\epsilon R} f(x,y)f(x+i,y+j)}{\displaystyle\sum_{(x,y),(x+i,y+j)\epsilon R} f(x,y)f(x,y)} \quad \text{for } |i|<p, \quad |j|<q$$

The above method of normalisation ensures that the usual tail-off of the autocorrelation function does not occur, so that the autocorrelation of a constant texture is a constant.

Thus, the second order statistics of the unknown texture have been calculated locally, via the SGLDMs, and globally, via the autocorrelation function. These features may now be used to classify the unknown texture, using knowledge of the properties of the textures produced by each model. This can be done in several ways: by constructing a hierarchical decision tree based on the features to produce a quick classification, by a method of analysis-by-synthesis whereby the texture is assumed to be matched to each model in turn, the texture is parametrised to each model and reformed according, the textural features are measured from the reformed texture and a similarity measure calculated between the reformed texture's features and the original texture's features, finally the model producing the highest similarity is chosen as being the best classification.

Typical texture models include a Gaussian random model, a one-dimensional Markov model, a Tesselated model, together with a linearly increasing grey-level weighting function in a specified direction which may be applied to any region.

4   CONCLUSIONS

This paper has discussed image analysis techniques using two contrasting types of image data. Particular aspects that have been studied are preprocessing, segmentation and texture analysis. For both applications a process of 'algorithm engineering' is involved, in order to develop not only the individual algorithms, but also an optimal algorithm package. With synthetic aperture radar images, where noise levels are high, part of this engineering process consists of a quantitative assessment using a figure of merit measure of the algorithms' performance on simulated images.

Some applications of image analysis need not be truly real-time (the SAR application is an example), whereas others (e.g. visual communications) must be real-time. For the latter the implementability of the algorithms is an important aspect, and the systems will require VLSI parallel architectures such as are described in a companion talk in this session.

REFERENCES

1   C Elachi, T Bicknell, R L Jordan and C Wu (1982) 'Spaceborne Synthetic-Aperture Imaging Radars: Applications, Techniques and Technology', Proc. IEEE, Vol. 70, No 10 1174-1209

2   V S Frost, J A Stiles, K S Shanmugam and J C Holtzman (1982) 'A Model for Radar Images and its Application to Adaptive Digital Filtering of Multiplicative Noise', IEEE Trans PAMI, Vol. 4 No 2, 157-165

3   J S Lee (1981) 'Speckle Analysis and Smoothing of Synthetic Aperture Radar', Computer Graphics and Image Processing 17, 24-32

4   N Yokoya, T Kitahashi and K Tanaka (1978) 'Image Segmentation Based on a Concept of Relative Similarity', Proc. 4th ICPR, 645-647

5   C Oddy and A Rye (1983) 'Segmentation of SAR Images Using a Local Similarity Rule', 2nd BPRA Conference on Pattern Recognition (to be published)

6   R O Duda and P E Hart (1973) 'Pattern Classification and Scene Analysis', John Wiley, New York

7   D Marr and E C Hildreth (1980) 'Theory of Edge Detection' Proc. Roy. Soc. B 207, 187-217

8   B Julesz (1962) 'Visual Pattern Discrimination' IRE Trans, Info. Theory, Vol. IT-8, 84-92

9   R M Haralick, K Shanmugam and I Dinstein (1973) 'Textural Features for Image Classification', IEEE Trans. SMC. 3 (6), 610-621

IMAGE PROCESSING SYSTEMS

P V Collins

GEC Research Laboratories, Hirst Research Centre, Wembley, UK

1   INTRODUCTION

This paper describes work on the design and implementation of image processing systems. This work has taken place at the GEC Research Laboratories. The systems are being developed both as specific applications of image processing and to provide a test-bed environment for the development of the techniques themselves. These techniques include both algorithmic aspects and special purpose hardware modules, the latter incorporating novel VLSI architectures. Each of these subjects provide the topics of other papers in this session.

The systems work is based around a highly flexible modular design for an image processing test-bed. Parts of the system can be developed in their own right as applications in the short term. In particular, applications for image archiving and for industrial inspection have already been identified. In the longer term, when suitable VLSI modules become available for inclusion, the aim is to produce a system for real-time image analysis.

The system itself is based around a 68000 microprocessor, interfaced to a suitable amount of image store. Real-time capture and display of image frames or frame sequences may be performed. In addition, software on the 68000 allows in situ implementation of the algorithms. As part of the system, a software environment, containing a prototype image processing language, has been developed so as to allow investigations of the use of image features. Another feature of the system is incorporation of graphics display processing.

Two versions of the system, incorporating some of the modules, will be described. The first is geared towards capture and display of image sequences, and the second towards the analysis of single frames and use of the features obtained. A specific application, namely industrial inspection will then be presented. Finally directions of future work will be discussed.

2   A MODULAR IMAGE/DISPLAY PROCESSOR

The system uses a common memory architecture with the following features:

1   A centralised memory control section providing video rate memory access, processor access, dynamic RAM refresh and vector draw with character generation based around a VLSI GDP chip.

2   Introduction of the basic data item within the memory workspace as being a $(512)^2$X1 segment, this data region being fully relocatable within memory space.

3   Allowing sets of segments to be associated within memory space to form clusters, these being essential to grey scale image processing.

4   Providing a set of common bus structures for all segments realising functions such as video input, video output, processor communication, memory control and cluster control.

The architecture of the system is shown in Figure 1, with particular note drawn to the fact that each segment has the same hardware, the overall system configuration being determined by control software residing on the host CPU.

The system is designed to cope with a maximum of 256 images. Each image can be as simple as a binary (black or white) image or as complex as a full colour (256 'grey' levels per component) three component image. Normal spatial resolution is 512 x 512, but timing strategies have been developed to cope with lower resolutions. Interlaced or non-interlaced scan can also be selected, giving the choice of one 512 line image or two 256 line images. These choices are all software controllable, subject to the limitation of the video input and output bus width. A segment can also be programmed to act as an overlay to be used for superimposing graphics and/or writing onto whatever image is currently being displayed. The way a segment outputs data to the video output bus and selects data from the video input bus are both under software control.

Processor memory accesses to the clusters can be accomplished in two modes:

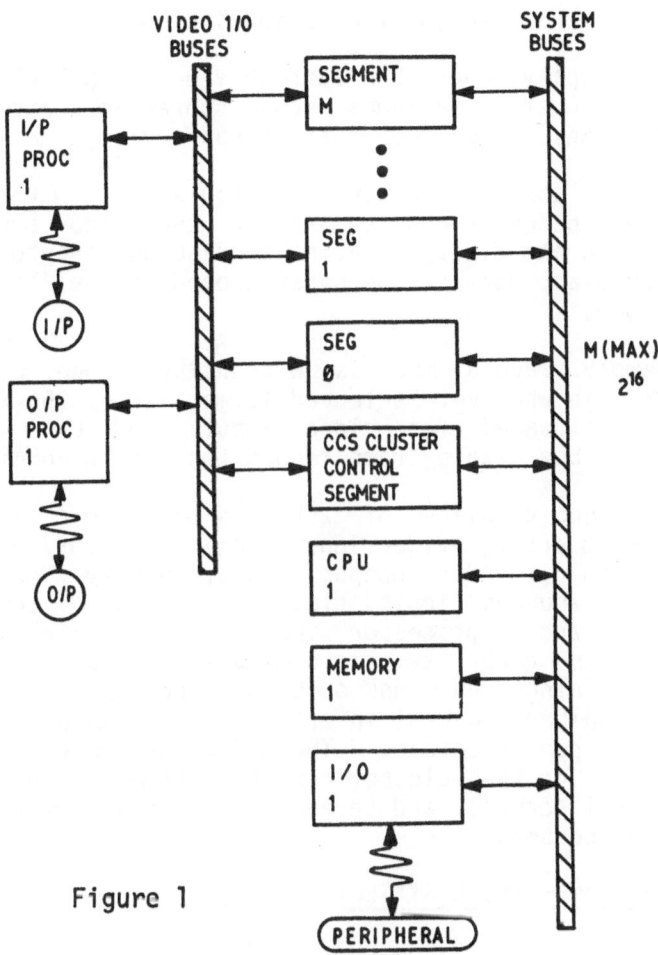

Figure 1

1   Parallel access, where one processor access conveys data to
     eight pixels all on one segment, with each processor data byte
     containing one bit from eight pixels.

2   Pel mapped access, where one processor access again conveys
     data to eight bits, but each processor data byte contains
     information about all of one pixel. In this mode each
     segment's relation (bit plane to bus data bit mapping) to the
     processor data bus is programmable.

     At any one time, one cluster is selected for video output, one
for video input and one for pel mapped processor input/output.
These three functions may be allocated to the same cluster or to
separate ones. Cluster function allocation is controlled by
pointers in the cluster control segment and can be manipulated very
quickly realising for example sequence storage and display in real
time.

The system has two continuous modes of operation:

1   Memory display, where the current video output cluster and any allocated overlay segments output stored image/overlay data to a video monitor (or any other output device).

2   Link mode, where video data is circulated from the video input bus, and hence video camera, to the video output bus via buffers on the cluster control segment and out to the monitor. Overlay plane data can be superimposed on the 'live' image in this mode.

The memory record mode is a transitory one performing one frame record, in whatever selected format (e.g. 256 x 256), into the cluster allocated for video input, and the processor is signalled when the single frame record has been accomplished.

Each segment consists of 32 kbytes of dynamic RAM, organised as a 32K x 8 array, video input logic, vector/character draw interface logic, video output logic, memory control logic, processor input/output logic, cluster addressing logic, segment control logic and a processor interface. The cluster control segment contains global system timing and control strategies as well as containing functions such as vector/character drawing, camera and monitor control interfaces, link mode buffer and an optional light pen interface. The major system control buses are all generated on the cluster control segment. A system of a cluster control segment and several segments is interfaced to the 68000 host processor.

3   SINGLE FRAME VISION SYSTEM[1]

3.1 Hardware

A schematic diagram of this system, when used as a development test-bed, is shown in Figure 2. The host CPU card contains, besides the 68000 processor, 128k RAM, 64k EPROM, two serial I/O channels, 32 parallel I/O lines and a timer. One of the serial interfaces is connected to a VDU, the other is connected to a host machine. Monitor software on the CPU card can connect the two serial channels together such that the VDU will act as a terminal on the host. Software can be developed on the host and then cross-compiled and down line loaded to the 68000 for local execution. A functional version of the software can be programmed into EPROM, making the system independent of the host although the link can be retained for the transfer of image data and programs.

In order to achieve acceptable speed of input, an appropriate frame grabber/store has been developed. This board takes as its input a standard 1V p-p video signal, thus allowing the use of a

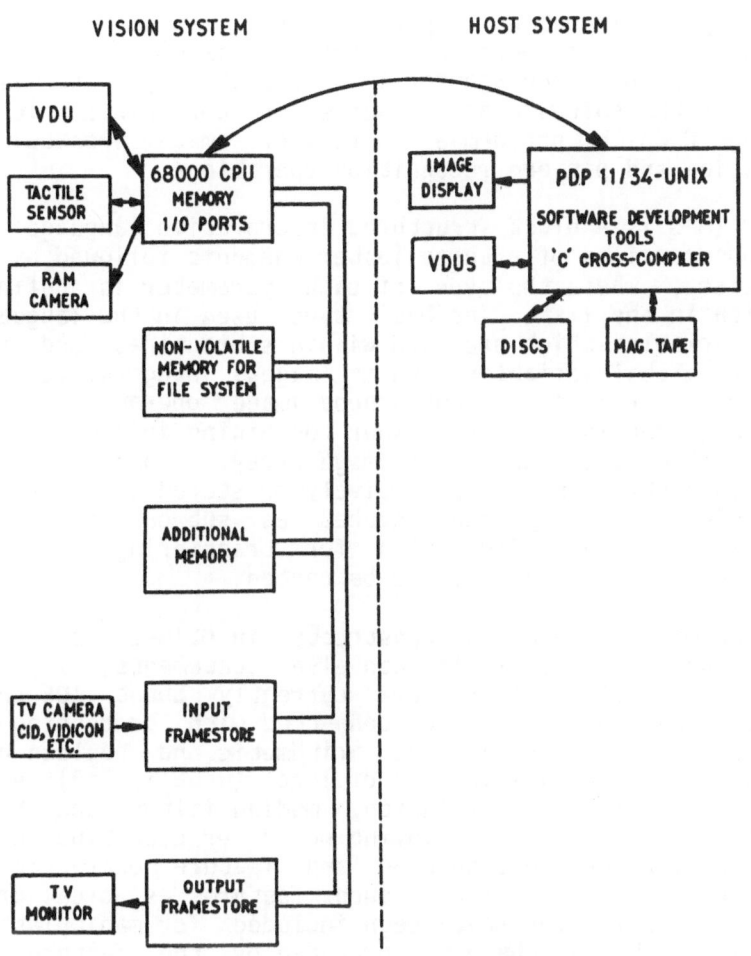

Figure 2 : Hardware schematic

wide range of video cameras or even a video recorder.   The store
area is 256 x 256 x 8, although only 7 bits are required for the
incoming data.   The data can be read off the card directly or, as
an option, a threshold can be set and the data, originally stored
as 7 bits per pixel, can be read out as binary.

The RAM on the CPU card is augmented by a 512k RAM card, a 64k
non volatile RAM card and an additional PROM card.   The non
volatile store is used by the system as a file store.   The use of
this memory eliminates the immediate need to use floppies or a
winchester drive, and consequently allows the equipment to be used
in harsher environments.

3.2 Software

A very important aspect of the system in the software environment which contains a prototype image processing language GLIMPS (the GEC Language for IMage ProceSsing). The main building blocks in the software are a suite of subroutines written in the language C which perform a variety of image processing, feature extraction and pattern recognition operations.

GLIMPS is a block structured interpretive language. The basic commands consists of a three letter mnemonic followed by a list of parameters, where the type of each parameter is defined by its position in the list. The basic types used in the language are the local variables which are used within subroutines, and the various types of global variables such as images, integers, reals, and mask registers for local neighbourhood image operations. Internally each image consists of a header containing information about the image, followed by the actual image array. Commands can either be typed into the system interactively or stored in command files in non volatile memory and invoked as subroutines. The latter mechanism includes facilities for transfering parameters, and allows the subroutine calls to be nested.

The main structuring constructs in GLIMPS are the begin-end block and conditional if-then-else statements, for loops and subroutine calls. There are currently about 105 commands in GLIMPS. Most of these are connected with processing images in various ways. They include arithmetic and boolean operations, together with a rich selection of local (usually 3x3) neighbourhood operations such as convolution, median filter and binary mask matching (suitable for removing noise or detecting edges). In addition various thresholding and feature extraction routines exist, the latter including such features as area or invariant moments. Facilities have been included for manipulating scalar variables such as the data produced by the feature extraction routines. There are also a number of commands for capturing images from a diversity of sources. Finally there are various file handling commands and a simple editor. The file system is very similar to the UNIX structure, with hierarchical directories and devices treated as special files.

4   INDUSTRIAL INSPECTION - AN APPLICATION

The vision system described in the previous section is ideally suited for developing dedicated applications of industrial vision. Thus first experiments are performed to find the most suitable lighting techniques and imaging devices. Next a GLIMPS program is developed to provide the necessary processing. A translator will then take this program and produce a C program with all the appropriate calls to the basic image processing subroutines. If greater speed is required in the final system then various software operations can be replaced with hardware modules such as convolution, binary mask matching and thresholding. Once the C program has been produced it can be linked, loaded and burned into EPROM. This can be run on a system similar to the development system, but without the large memory and time overheads incurred by including the interpreter in the system.

A simple example of such an application is the inspection of contacts. These are produced on an edge strip, with all the contacts perpendicular to the strip. The shape is produced by a multistage press and the inspection task involves checking that three sections of the contact are correctly aligned. A more complex application was developed by using the vision system in association with a robot, in order to establish a simple robot workcell. Three 68000 cards were plugged together with each one running an interpreter based on the original vision system. One 68000 is used as the supervisory processor, one is configured as a vision system and one is used to control the robot. The robot was a Smart Arms 6R/600 research robot. The image processing commands were stripped from GLIMPS and new robot commands added, along with a new data type for the coordinate arrays. This language was installed onto one of the 68000 processors to form the robot controller.

5   FUTURE WORK

This paper has described two versions of image processing system that have been developed, the first suitable for capture and display of image sequences, and the second for the analysis of single frames and use of the features obtained. The overall aim is to work towards real-time image analysis which will find application in a wide variety of fields. In order to achieve this goal, it will be necessary not only to combine appropriate elements of the two versions of vision system (a simple task due to the modular and flexible nature of their design) but also to incorporate many of the developments discussed in the two comparison papers in this session. Thus the VLSI modules will be essential to achieve the required speed of processing, while algorithms and algorithm packages will need to be optimised both with respect to performance and with respect to implementability.

REFERENCE

1   J A Losty and P R Watkins
    Computer vision for industrial applications
    GEC Journal of Science and Technology - to be published

IMAGE PROCESSING WITH VLSI

A G Corry, D K Arvind, G L S Connolly, R R Korya and I N Parker

GEC Research Laboratories, Hirst Research Centre, Wembley, UK

INTRODUCTION

A large variety of tasks in the fields of image processing and numerical computation demand a very high instruction throughput. The figure of 10^9 instructions/sec has for instance been cited as representing a typical computational load for tasks in low-level image processing[1]. Such a high instruction rate cannot be supported by conventional serial (Von Neumann) computer architecture. A close analysis of many image processing algorithms reveals that the same sequence of instructions are normally repeated on every pixel item, pixel by pixel, over the entire image. This inherent parallelism can be exploited to achieve very high computational throughputs. We have been investigating a number of non-Von Neumann architectures which seem to offer a suitable solution[2]. One such architecture - the Single Instruction Multiple Data stream (SIMD) type has shown great promise in the past[3,4,5], but has needed the emergence of VLSI for it to be truly feasible as a cheap and compact system. Such a VLSI system under implementation is the GEC Rectangular Image and Data (GRID) processor system[6]. An alternative approach relies on the identification of common operators, which make up more complex algorithms, and their customizing as special-purpose integrated circuits that operate on an incoming stream of video data.

THE GRID - A PARALLEL PROCESSOR ARRAY

The architecture of the GRID processor system is the well known parallel processor array, shown schematically in Figure 1. It comprises a rectangular array of processing elements (PEs) under common control, all performing the same operations simultaneously

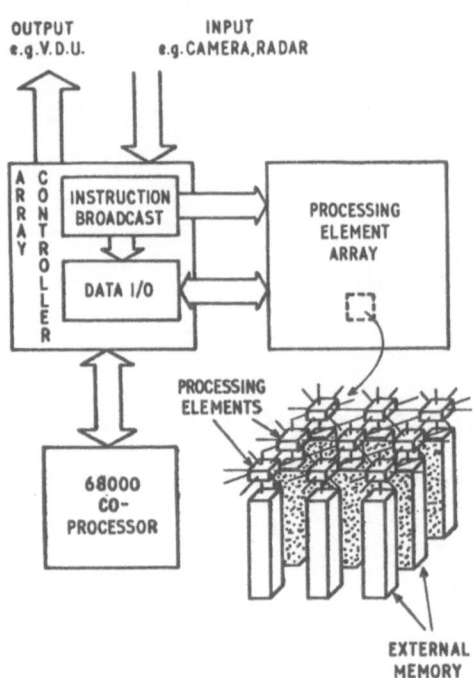

OUTPUT
e.g.V.D.U.

INPUT
e.g.CAMERA,RADAR

Figure 1 : Parallel processor array schematic

on different data items stored in each PE's memory. The interconnections of each processor to its nearest neighbour enable each PE to access the data from its local neighbourhood of processors. A microprogrammed array controller provides for the sequencing and broadcast of both input/output data and instructions to the processor array. The overall system is composed of two distinct processors, a microcomputer and a GRID processor chip based processor array. The microcomputer uses the array as a co-processor executing a high-level language and passing control to the GRID array whenever parallel constructs are encountered.

The major concentration of hardware in the system lies in the processor array. A custom-designed VLSI GRID processor chip renders a cheap and compact implementation of this array. Designed for fabrication in a 2.5 μm bulk CMOS technology, it contains over 50,000 transistors and supports a maximum clock frequency of 10 MHz. Arranged as a rectangularly interconnected array of 64 bit-serial processing elements (PEs), the GRID processor chip includes 32 registers per PE, together with the necessary circuitry for interconnecting many identical chips to construct compact and high performance array processors.

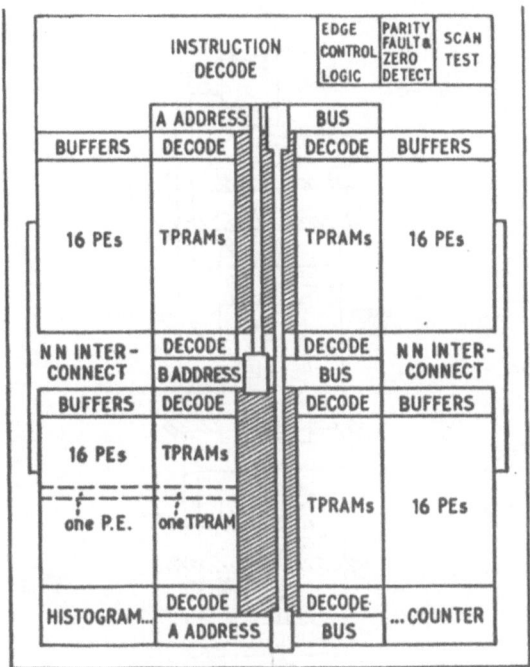

Figure 2 : The GRID chip floor plan

THE GRID PROCESSOR CHIP

The GRID processor chip, a floor plan of which is shown in Figure 2, consists of an 8 x 8 mesh of bit-serial PEs, chip control register, an edge control register, a histogram counter and the peripheral circuitry for instruction decoding, scan-path testing, parity fault and zero detection.

The PE design, as shown in Figure 3 has been optimised to implement image processing and numerical computation tasks efficiently. The bit-serial architecture of the PE allows flexible data formats. Eight-bit, six-bit or even binary wordlength data can be processed without any loss in efficiency.

The two data buses of the PE - the SOURCE bus and the DESTINATION bus, run along the length of the PE. Strung between these two buses are the extended register sets A and B, the ALU, the nearest neighbour switching (NNS) network and the special-purpose one-bit registers.

The extended A,B registers, implemented as a 32 words x 1 bit two-port RAM (TPRAM), allows the storage of intermediate results during grey-level arithmetic operations without recourse to the slower external RAM.

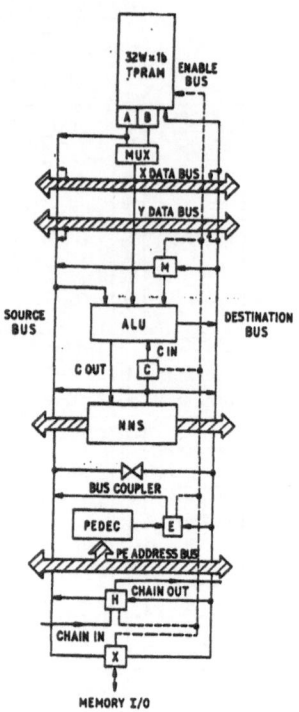

Figure 3 : The GRID PE schematic

The NNS network allows each PE to access the contents of its nearest 3 x 3 neighbourhood of PEs. Through 'corner turning' within the network, diagonal neighbours can be accessed on the square interconnection grid. This has the effect of providing local interconnection to eight neighbours without requiring any extra pins.

The Multiplier (M) register holds the multiplier bit during bit-serial multiplication. It can be gated either with the A or B input to the ALU thus permitting the gated full-add (the basic step of bit-serial multiplication) to be perfomed in one PE cycle.

The Enable (E) register supports conditional instructions at the PE level. When set, all write instructions to the associated PE are enabled. This register is set either as a result of a computation or via the PE addressing network. In the latter case, selected chips, rows, columns or individual PEs can be enabled whilst the remainder of the array is inhibited.

Lastly, all data transfers between the PE and its off-chip RAM are latched in the External Store Register (X). The provision of this port enables external memory expansion to suit particular applications.

GRID SYSTEM ARCHITECTURE

Data Bus Architecture

The most direct path for data transfer between the GRID processor chip and the array controller is via the 'X-' and 'Y-buses'. They lie along the two axes of the array, with one line to each row (X) and each column (Y) of PEs. Rows or columns are selected by the PE addressing circuitry for enabling onto the X-and Y-buses respectively. The buses can be used to broadcast data from the array controller to every column/row on the chip by enabling the whole chip instead of a single column/row. This ability is useful in matrix manipulations for matrix transpositions.

Histogramming

Histogramming is a frequently used image processing operator for obtaining grey-level statistics or for counting the number of edge pixels for perimeter estimation. The H registers of all the 64 PEs are connected as a linear chain which ends in an on-chip counter. By clocking the data down the chain, a histogram count is obtained of all the bits set in the chain. The contents of the counters are then accessed, chip by chip over the array by the array controller, using the X-and Y-buses. Thus the histogram count over the entire image is accummulated at the controller.

Memory organisation

Typically in image processing applications, the input image will be larger than the PE array. The GRID processor system supports two methods for mapping the image onto the processor memory.

(a) Windowed mapping scheme: In this method, the array is effectively stepped over the image and windowed regions (the size of the array) of the image are stored in the external memory. Although simpler to implement, this method has a major disadvantage. During a nearest neighbour transfer operation, some data inevitably 'falls off' the edge of the array.

(b) Pyramidal mapping scheme: This scheme, overcomes the above problem by storing a pxp windowed region of the image in the external memory of each PE. The size of this window is a function of the ratio of the input image size and the array size. Moreover, in a pyramidal mapping scheme each PE can access any of the pixels within a $(2p+1) \times (2p+1)$ neighbourhood of PEs as shown in Figure 4. As a consequence the size of the processor array is proportional to the degree of parallelism necessary to achieve the required computational throughput rather than the size of the image to be processed.

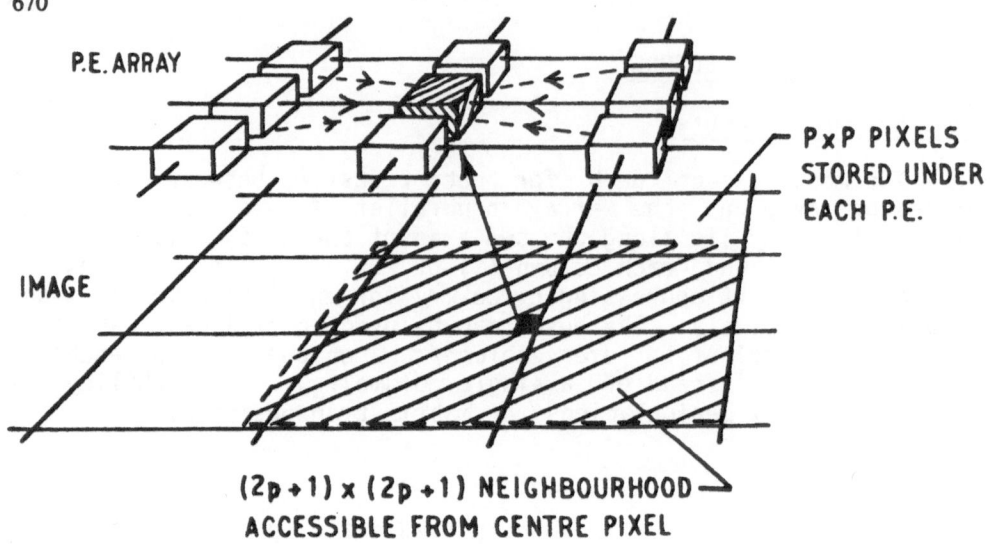

P.E. ARRAY

PxP PIXELS
STORED UNDER
EACH P.E.

IMAGE

(2p+1) x (2p+1) NEIGHBOURHOOD
ACCESSIBLE FROM CENTRE PIXEL

Figure 4 : Accessing a (2p+1)x(2p+1) neighbourhood

GRID 64 - CHIP DESIGN

The technology chosen for the GRID processor chip is the in-house 2.5 μm bulk CMOS process. CMOS is suitable for very large scale integration because of the low power consumption and high driving capability. The design rules have been formulated such that the same mask can be reprocessed in CMOS silicon-on-sapphire (CMOS/SOS). This is particularly attractive for military applications due to the inherent radiation hardness of the CMOS/SOS chips.

Much of the chip design effort was concentrated on the individual PE, to achieve an optimum trade off between silicon area and efficiency. The 'long and thin' shape of the PE was influenced by the pitch of the TPRAM and the need to route the wide control bus efficiently. The orthogonal data and control buses were implemented in a double layer metal process to minimise diffusion delays. Once designed, the PE was stepped across the chip, thus completing a large proportion of the layout task. The peripheral circuitry includes the instruction field demultiplexing and decode, and the histogram counter. These were realised by automatically generated PLAs.

GRID64 is at present at the chip assembly phase, with the major component parts in the process of being interconnected. Other chips in the GRID family (a high speed I/O chip, a VLSI controller) are in various stages of development and are due for inclusion in later generations of GRID systems.

CUSTOMISING ALGORITHMS IN SILICON

A considerable amount of experience in low-level image processing has identified the types of algorithms essential at this level of image understanding[1,7]. The approach here relies on 'factoring-out' the operators necessary to implement these algorithms and integrating these as special purpose/high performance processing elements. A number of potential circuit architectures have been investigated for integrating a variety of kernel functions -convolutions, correlation, spatial filtering -in a form suitable for the support of real-time (TV video rate) computation.

One example of this approach is described in more detail: a two-dimensional median filter. It is a specialised, hardwired functional unit which employs extensive pipelining and can operate directly on the serial scan of an incoming image. There is therefore no requirement to process a certain number of pixels within a fixed period as with the processor array. Architectures under evaluation are based on a design approach termed by Foster and Kung[8] as a 'systolic array'. This type of design is particularly well suited to a VLSI implementation, it being characterized by:

a) the use of only a few types of simple cells;
b) a rectangular data flow and control organization;
c) the use of extensive pipelining and parallelism to achieve a high computational throughput and
d) the multiple use of input data.

The median filter is a tool often used to reduce the effect of discrete impulse noise on an image. In the one-dimensional case the centre pixel in the processing window is replaced by the median of the sequence of pixel values within the window. As an example consider the sequence.

180   190   200   110   120

The median is found by sorting this sequence and choosing the centre value of the sorted sequence. Hence the centre pixel of the processing window (here of length five), 200, is replaced by the median of the sequence, 180. In the two dimensional case a number of geometries are possible. For instance a 5 x 5 square window or the 5 x 5 plus-sign-shaped window depicted in Figure 5.

The basic computational element of the median filter is a sorting network[1]. A nine-input sort array which utilises serial arithmetic is depicted in Figure 6. The array is composed of two types of cell, a delay cell and a sort cell. The delay cell is used for delay equalization. The sort cell consists of a gated

672

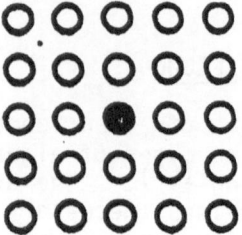

Square shaped

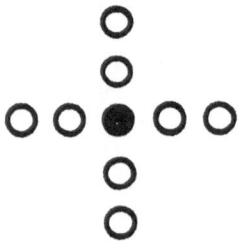

Plus – sign – shaped

Figure 5 : Median filter windows

magnitude comparator and a crosspoint switch. Two serial input
streams enter a sort cell, initially the crosspoint switch is set
in a default direction (i.e. $A_{IN} \rightarrow A_{OUT}$, $B_{IN} \rightarrow B_{OUT}$). Comparison of
the two bit streams will result in either the switch remaining in
its default position or crossing over to its alternative
position (i.e. switching the bit streams $A_{IN} \rightarrow B_{OUT}$, $B_{IN} \rightarrow A_{OUT}$). The
sort array implements a conventional 'bubble-sort' algorithm. Each
sort cell of the array compares two numbers and routes the larger
upward, the smaller downward. A sequence of numbers is grouped in
pairs. Successive application of a column of sort cells will cause
at each stage, larger numbers to 'bubble-up' and the smaller to
'sink down'. The sorting takes place bit serially and is pipelined
at the bit level. The throughput of the median filter is
determined by the delay through a sort cell. Once the pipeline is
full the pixel delay becomes this cell delay times the pixel word
length.

Pixel input (serial)

X_1 X_2 X_3 X_4 X_5 X_6 X_7 X_8 X_9

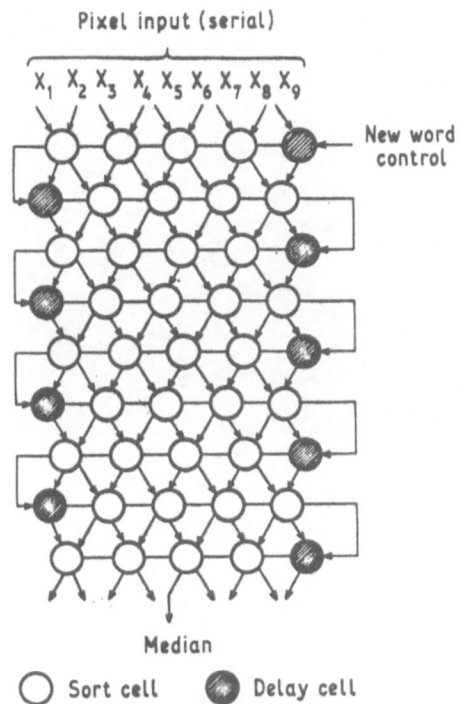

New word
control

Median

◯ Sort cell   ● Delay cell

Figure 6 : 9-input sorting network

In the case of a 5 x 5 plus-sign-shaped median filter the nine pixels of the plus-filter would be sorted with the fifth element of the sorted outputs (i.e. the median of the input sequence) replacing the centre pixel of the plus-window. For this case the sorting network requires only thirty sort cells.   Increasing the number of inputs to the network by either changing the shape of the filter (5 x 5 square) or increasing the size of the filter kernel (7 x 7 plus) results in a rapid growth of the sorting network (e.g. 94 cells for a 5 x 5 square filter). Alternative network structures exist which have a lower growth factor but these suffer from an inherent irregularity in network wiring. These therefore offer little improvement over the more regular 'bubble-sort' when dealing with the sorting of sequences up to a length of twenty-five pixels.

SPATIAL FILTER - CHIP DESIGN

The technology chosen for the Spatial Filter chip was the in-house 3 μm CMOS Silicon-on-Sapphire (CMOS/SOS) process.

A chip photo of the Spatial Filter is shown in Figure 7. It consists of a bubble-sort network capable of sorting twenty-five

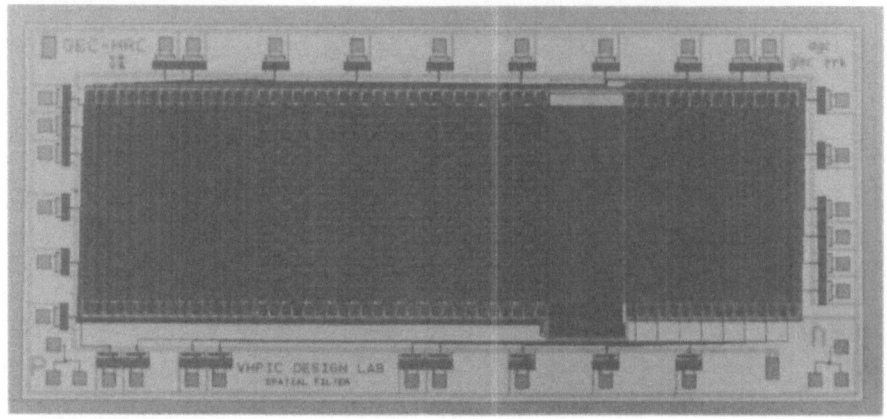

Figure 7 : The spatial filter chip

pixels, together with on-chip circuitry for wordlength programming
pixel storage, serial/parallel and parallel/serial conversion.

The Spatial Filter has been included on a CMOS/SOS
multiproject chip and first samples are expected to provide video
rate computation of the median filter operation.

CONCLUSION

This paper has examined two different design approaches to
computation for low-level image processing. The GRID processor
chip (an array of which forms the heart of the GRID processor
system), and a Spatial Filter chip (a specialised video-rate
computation image processing component) have been described.

The choice of approach in the design of high performance image
processing systems is very much dependent on the nature of
application. One can envisage complex image processing algorithms
being developed on a flexible parallel processor array computer and
then 'programmed' onto silicon as special-purpose devices. Due to
its specialised nature, the latter approach will always out-perform
the processor array on a throughput/chip basis, but once
'programmed' its function is set forever. The special-purpose
device will in general be used to implement solution to problems,
whereas the processor array will be more suitable in applications
where flexibility is demanded.

REFERENCES

1   G R Nudd: 'Image understanding architectures'. Proc Nat Comp
    Conference (1980), 377-390

2   I N Robinson and A G Corry: 'VLSI architectures for low-level
    image processing. Proc IEE Electronic Image Processing
    Conference (1982), UK

3   M J B Duff: 'Review of CLIP4 image processing system'. Proc Nat
    Comp Conference (1978), 1055-1060

4   S F Reddaway: 'DAP - A distributed processor array'. First
    Annual Symposium on Computer Architecture (1973), 61-65

5   K E Batcher: 'Architecture of a massively parallel processor'.
    Proc 7th Annual Symposium on Computer Architecture, 168-173

6   D K Arvind, I N Robinson and I N Parker: 'A VLSI chip for real-
    time image processing'. Proc. 1983 Int Symp on Circuits and
    Systems, USA, pp 405-408

7   K Mori et al: 'Design of local parallel pattern processor for
    image processing'. Proc Nat Comp Conference (1978), 1025-1031

8   M J Foster and H T Kung: 'The design of special-purpose VLSI
    chips'. Computer (1980) 26-40

BEYOND PCM - SPEECH PROCESSING PROSPECTS IN TELECOMMUNICATIONS
APPLICATIONS

T W Chong

GEC Research Laboratories, Hirst Research Centre, Wembley, UK.

ABSTRACT

The advent of VLSI technology is opening up new possibilities
in the realisation of efficient speech processing algorithms
significantly more complex than the universally accepted PCM
method.

At the same time it is stimulating a bewildering range of new
coding techniques. This is bound to have a significant effect on
future communication networks and standards.

Selected illustrations of processing techniques and their
classification are presented, together with key parameters such as
information rate, complexity, delay and quality.

Possible applications in various telecommunications
applications such as modems, integrated digital networks, store and
forward systems are discussed.

1   INTRODUCTION

In principle, the easiest way of representing any bandlimited
signal, such as speech, is to preserve its waveform
characteristics. For instance, PCM, DM (Delta Modulation) and DPCM
are some of the schemes used in the waveform representation of
speech. They are all based on Shannon's sampling theorem which
states that any bandlimited signal can be reconstructed from
samples taken periodically in time if the sampling rate is at
least twice the highest frequency of the signal.

Since the 1960's the only coding technique that has been
defined as as a standard for worldwide use is 64 kb/s PCM (A-or

μ-law). The current trends indicate that with advances both in technology and in signal processing techniques new CCITT standards need to be established to avoid chaos in the network as a result of non-compatible systems being introduced. There are several coding techniques which are capable of giving similar quality to 64 kb/s PCM but at lower data rates (16 kb/s-32 kb/s). The CCITT is, in fact currently finalising a recommendation of an algorithm based on 32 kb/s ADPCM.

Furthermore, it is also believed that the present 64 kb/s PCM may not be the best encoding technique at that rate. This is particularly true when future standard voice terminals are equipped with better types of transducers such as linear microphones. For this type of application a wider bandwidth than the current telephone bandwidth of 3.4 kHz offers a significant improvement in subjective quality. It is envisaged that it would be useful to have in the communications network a means of transmitting/receiving speech at 64 kb/s but with a bandwidth of around 7 kHz. Typical applications would be in loudspeaking telephones, high quality speech on private exchanges, confravision and in later phases of ISDN (Integrated Services Digital Network) development.

There are other more complex schemes which can code speech signals to below 5 kb/s. They are generally known as vocoders, e.g. channel vocoder and LPC (linear predictive coding) vocoder, and can provide quite intelligible speech; however, the quality is unlikely to be acceptable for commercial telephony. Vocoder performance is limited due to several reasons, i) they represent a simple linear model of the voice production mechanism; ii) difficulty in following the changes of the vocal tract and its driving sources faithfully; iii) the process of detecting pitch and the voiced/unvoiced decision is not fully reliable.

It is anticipated that with the progress in VLSI technology it is possible to produce future codecs of reasonable cost and complexity for data rates between 8-16 kb/s. The techniques employed will be a combination of waveform and source coding methods. The quality should be adequate for limited telephony network applications and more robust than the current vocoders.

2   DIGITAL REPRESENTATION OF SPEECH

Speech coding methods fall into two broad classes: waveform and source coding[1,2]. The former generally produce good quality speech at the expense of high data rate. The latter methods on the other hand are capable of coding speech down to a few kbit/s at the expense of compromising quality. At the lowest data rates, speech quality becomes obviously synthetic.

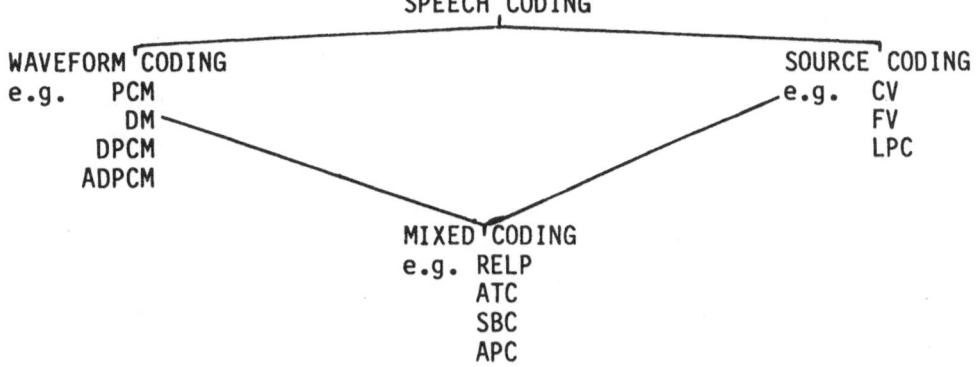

Figure 1 : Coding schemes

In waveform coding, the shape of the analogue signal is preserved by applying the sampling theorem. In the digital telephony network based on PCM the speech signal is sampled at 8 kHz for a signal passband of 300-3400 Hz. A nonlinear quantiser is used (A-or µ-law compading) in which each sample is 8-bit coded resulting in a good signal to quantisation noise ratio over a wide dynamic range of input signal levels, the resolution of low-level signals being equivalent to 12-bit linear quantisation. Techniques such as DPCM and ADPCM are variations of PCM which exploit the correlation between adjacent samples of the input signal in order to achieve a reduction in the number of bits required for each sample for a given quality, as shown in Figure 2.

Using conventional waveform techniques, good quality speech can be achieved with data rates as low as 24 kb/s; any further reduction will result in noticeable degradation.

In source coding the speech production mechanism is modelled with certain assumptions. Firstly, it is assumed that perception is insensitive to the phase content in the signal. In addition, the spectral characteristics corresponding to the shape of the vocal tract are assumed to be stationary for periods of between 10-30 ms duration. In this process of coding speech, the essential parameters which, it is hoped, characterise the speech signal for these short intervals consist of a set of filter coefficients which define the spectral characteristics of the vocal tract, and information regarding the driving sources (vocal folds) as shown in Figure 3. If the vocal folds happen to be vibrating during that interval then information concerning the value of the pitch is required. This corresponds to a voiced sound, and at the decoder end, the filter is excited by a sequence of pulses with a period determined by the pitch value, see Figure 4. On the other hand, if the vocal folds are relaxed then no pitch is detected implying an

a. PCM Coder

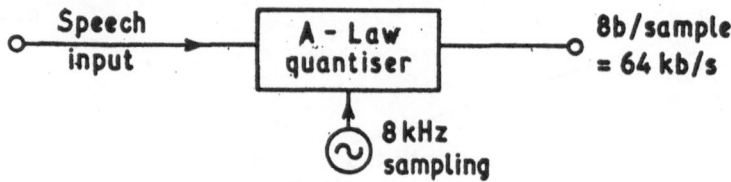

b. DPCM Coder

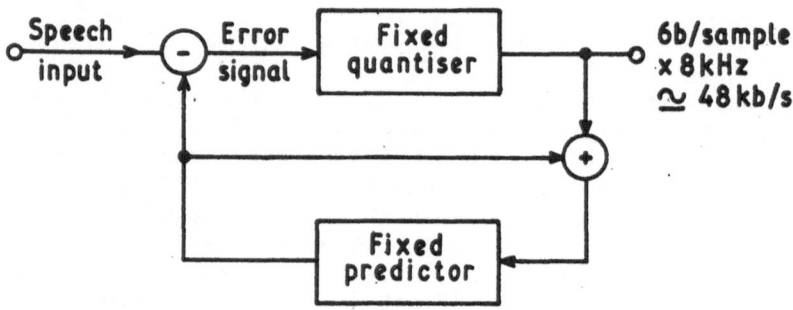

c. ADPCM Coder

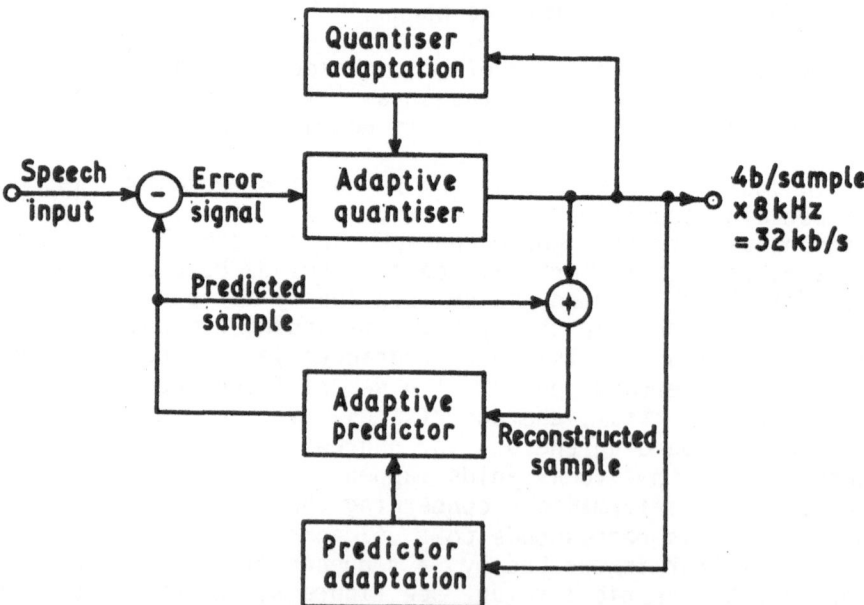

Figure 2 : Derivatives of PCM coding (waveform coding)

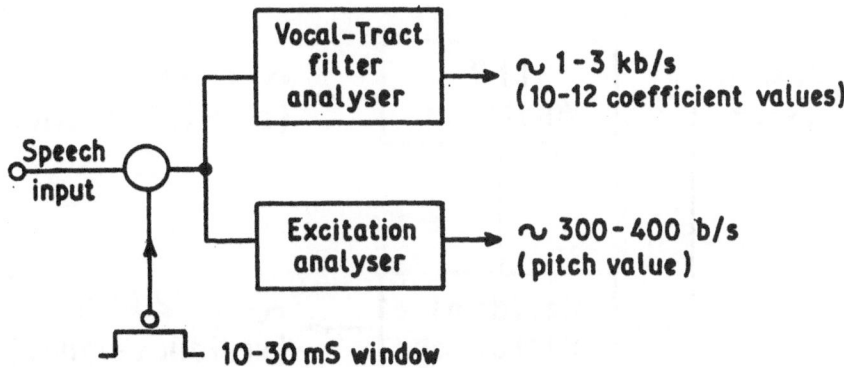

Figure 3 : Basic LPC analysis (source coding)

unvoiced sound, and as such the filter is then excited by a sequence of random pulses with white noise characteristics. Both the driving sources create a wideband excitation of the vocal tract which in turn acts as a linear time-varying filter which imposes its transmission properties on the frequency spectra of the sources.

One of the major drawbacks of source coders is the inability either to extract from the speech signal the pitch value reliably, or to make the right decision as to the voiced or unvoiced nature of the sound. Because of the deficiencies of the simple speech production model the quality of the processed speech can not be improved noticeably by simply allocating more bits to the speech parameters. In fact, a state of diminishing returns is reached for data rates above 6 kb/s.

There is a data rate range extending between 8-16 kb/s in which neither source coding nor waveform coding alone can perform satisfactorily. It is therefore not surprising that the most

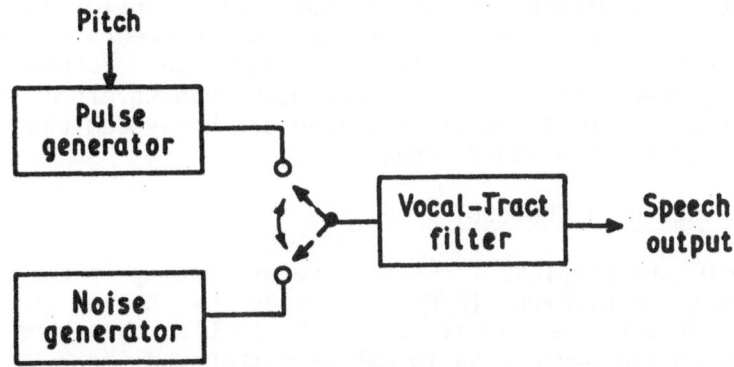

Figure 4 : LPC synthesiser

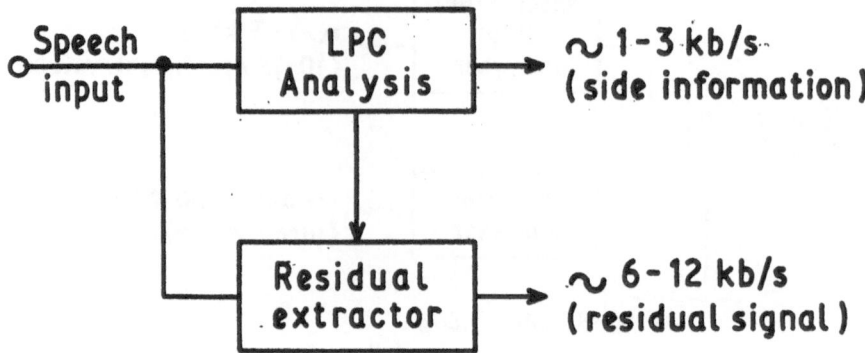

Figure 5 : APC analysis (mixed coding)

suitable coder in this range should involve a combination of the two coding techniques. In this approach source coding techniques are employed to extract the vocal tract characteristics and as a by-product of the process a signal known as the residual is generated, see Figure 5. This contains waveform information concerning the excitation sources as opposed to a single-valued parameter describing the pitch frequency in a source coder. Thus, in principle, if the residual waveform is effectively coded, toll-quality speech can be obtained below 16 kb/s.

Figure 6 summarises how all three classes of coding speech complement each other in terms of achievable quality. Generally speaking, waveform coders are relatively easier to implement in hardware and do not suffer from undesirable delays. For data rates above around 24 kb/s the speech quality should be quite good. Source coders operate on a block-processing basis, and as a result a delay equivalent to a block or more of data is inherent in the system. Quite complex hardware design is involved and the quality that can be achieved so far is not sufficiently adequate for general use in the public network. For mixed coders the problem of delay becomes appreciable only for low data systems. As for complexity, they are harder to implement than waveform coders and depending on the performance required of the system they can also be more complex than source coders.

3 INFLUENCE OF TECHNOLOGY

One of the simplest forms of waveform coding is the well known pulse code modulation (PCM) suggested by Reeves in 1938[3]. However, it was not until the late 1960's that the detailed quantisation law and coding format were standardised by the CCITT. The two logarithmic approximations adopted were μ-law in N America

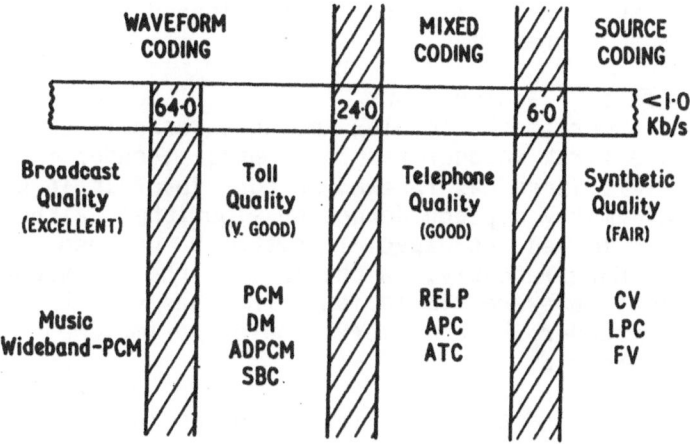

Figure 6 : Speech coding

and A-law in Europe. PCM is established as the worldwide standard for digital telephony.

Any new coding algorithms based on reduced bit-rate coding techniques (<64 kb/s) should be implemented in a format that is compatible with the emerging integrated digital network built around 64 kb/s. Another point that should be borne in mind is that the present telecommunications networks are mostly analogue. Consequently,several coding-decoding processes may be encountered in a subscriber to subscriber connection.

The first PCM codecs in integrated circuit form appeared on the market in the late 1970's. In requiring a mix of analogue and digital circuitry these devices represent a considerable technological milestone. Presently, combined chips in which anti-aliasing filters are incorporated with the PCM codecs are becoming available. Other types of semi-integrated codecs have appeared on the market, such as the CVSD (continuously variable slope deltamod) codec which performs well at 32 kb/s. Unfortunately, in a digital network built with standard 8-bit PCM, transcoding from Delta-Modulation to PCM or vice-versa requires special digital filters.

Currently, the CCITT is set to finalise its recommendations on alternative low bit-rate coding techniques, 32 kb/s ADPCM being the first choice. Several possible algorithms are being evaluated, and undoubtedly technology advances will have an appreciable influence on the final choice, in the sense that more complex algorithms can be considered given that they offer improved performance.

The capability of the codec to provide double the transmission capacity without sacrificing quality noticeably is highly desirable in places such as N America where long-distance transmission capacity is at a premium.

The recent achievements in VLSI technology have also encouraged other more complex methods associated with source coders. The first vocoder was the channel vocoder invented in the late 1930's[5]. There was renewed vocoder interest in the 1960's as a result of improved pitch-detection algorithms, although its use was rather limited to special systems[6] because of cost and complexity. More recently, in the 1970's, the LPC vocoder appeared in the form of software simulation. Another well-known vocoder is the Formant vocoder which is very complex to implement in hardware as a viable product.

Until very recently a large collection of off-the-shelf components was required to implement a typical LPC codec; prototype equipment designed at the Hirst Research Centre used over 200 devices. At the present time the availability of fast digital signal processor devices has enabled a simpler hardware architecture capable of implementing complex algorithms. The general hardware architecture can remain essentially unchanged and yet be capable of performing different coding algorithms by simply modifying the program.

4   APPLICATIONS IN TELECOMMUNICATIONS SYSTEMS

Coding of speech into digital form for communications purposes has several advantages:

a)   robustness of digital transmission
b)   capability of multiplexing different services such as speech and data
c)   secure communication
d)   efficient storage

Several application areas can be identified and these are described in the following sections.

4.1   Integrated Digital Network

In the future digital network, communication between subscribers operates at the basic rate of 64 kb/s and all standard switching nd transmission equipment is designed around it. Lower bit-rate coding such as 32 kb/s ADPCM is currently being standardised by CCITT. This will double the number of subscribers and the quality achieved is claimed to be better than 7-bit PCM. For instance, on a 2Mb/s link, 60 x 32 kb/s channels can be accommodated instead of 30 standard PCM channels[8]. Another application is in the use of 32 kb/s CVSD on leased circuits as proposed by the Mercury System[9].

The IDN is expected to evolve into the Integrated Services Digital Network, ISDN. Here, transmission over local lines at 80 kb/s and 144 kb/s presents severe problems, and the use of 32 kb/s codecs would reduce the required line rate and alleviate the difficulty.

The implementation of 32 kb/s codecs in the telephone network should benefit large countries such as N America where the cost of long-distance transmission is high. Another area is in satellite links such as Intelsat and Eutelsat TDMA systems[10] in order to provide more capacity for data, voice, audio and videoconferencing services.

A study of the application of low bit rate coders to subscriber loops[7] found that 32 kb/s codecs provided a cheaper link than 64 kb/s at distances over several kilometres. As for 16 kb/s codecs, they were not as economical particularly over distances less than several tens of kilometres since the cost of the codec was high and required an echo canceller.

4.2 Coded Speech over Analogue Circuits

Transmitting digital information over an analogue environment presents specific problems. In the case of digitally coded speech signals, several coding-decoding operations between subscribers are possible. In addition, in order to transmit digits in an analogue channel they must be converted into an analogue signal by using a data modem at standard rates of 2.4, 4.8 and 9.6 kb/s. Voice and data can then be mixed into one voice-frequency circuit. They are effective in reducing cost of, say, expensive transatlantic leased circuit connections.

4.3 Voice Announcement Systems

A very appropriate application of low bit rate codecs (2-4 kb/s) is in exchanges such as the Automatic Announcement System[11] of British Telecom's System X. Presently, the system employs 32/48 kb/s waveform coders to compress the speech data. At that rate memory requirements for the announcements are very high and as a result concatenation of individually stored words is employed. An alternative is to use an LPC vocoder operating at, say, 4 kb/s allowing a complete string of words to be stored wherever necessary; in addition, the process of preparing the announcements is far less time-consuming than with the waveform approach. The quality of such a vocoder, in our experience, is quite adequate for this application.

4.4 Voice Store and Forward

Voice store and forward is expected to play a prominent role in the future automated office.   Research has shown that only 30% of all calls made are successful. The 70% failure is normally caused by busy signals, lost calls or the person called not being available.   In addition, only 50% of all calls made require personal interaction.

It is reckoned that with such a facility associated with a digital PABX would result in significantly higher efficiency and productivity in the office.   By leaving messages rather than speaking to the person will also result on the average shorter call lengths.

In such applications good quality speech is required and neither source coders nor conventional waveform coders are suitable.   In the former the quality can be described at most fair, while the latter result in excessive memory capacity.   A mixed coder is most suited since it can provide good quality speech below 16 kb/s.   The cost of the codec should not be of major consideration since only a few are required in a PABX rather than on a per-user basis.

5   REFERENCES

1   J L Flanagan et al : 'Speech Coding', IEEE Trans on Com.,
     $\underline{27}$, No 4, (1979) 710-737.

2   J N Holmes : 'A survey of methods for digitally encoding
     speech signals', The Radio and Elect Eng, Vol 52, No 6,
     (982) 267-276.

3   A H Reeves : French Patent 852183, 1938

4   D J Dooley et al : 'A versatile monolithc PCM codec'. Int
     Zurich Seminar on Digital Communications, Zurich 1976

5   H Dudley : 'Remaking speech'. J Acoust Soc Am., Vol 11,
     (1939) 169-177

6   N G Kingsbury et al : 'A robust channel vocoder for adverse
     environments', ICASSP '80, Denver.

7   M Yoshikawa et al: 'Network applications of new speech
     encoding technologies', GLOBECOM '82, Vol 1, A8.1.1-A8.1.5,
     1982

8   D Cointot et al : 'A 60-channel PCM-ADPCM converter robust
     to channel errors'. GLOBECOM '82, Vol 1, A8.5.1-A8.5.5, 1982

9   D Evans : 'Some practicalities of network planning - a
     pioneering view', Proc of the Int Conf on Sat. and Cable
     TV, CABLE 83, May 1983

10   D McGovern et al 'The European Communications Satellite
     Multi-Service Transponder', British Telecom Eng, Vol 2 (1938)
     32-36

11   D S Cheesman : 'Solid-state voice guidance equipment for
     supplementary telephone services', EUROCON '80, Stuttgart,
     Germany, 1980

List of Delegates

Mr. E. Albasiny, DITC, National Physical Laboratory, Teddington, Middlesex, U.K.

Professor J. Alves, Universidade de Aveiro, Centro de Elec.e Telecom, 3800 Aveiro, PORTUGAL.

Dr. D. Andelman, 23 Ehud St, Haifa 34553, ISRAEL.

Mr. A.J. Anderson, 32 Hill Barn View, Portskewett, GWENT, South Wales, U.K.

Dr. G. Ascheid, Lehrstuhl Fuer Elektrische Regelungs-technik der RWTH Aachen, Templergraben 55, D-5100 Aachen, West Germany.

Dr. M. Beale, The University, Manchester M13 9PL, U.K.

Dr. S. Bellini, Dipartimento di Elettronica, Politechnico di Milano, P.za.L. da Vinci 32, 20133, Milano, ITALY.

Dr. B.K. Bhargava, Concordia University, Room H-915, Dept of Elec. Eng, 1455 de Maisonneuve W., Montreal, CANADA.

Mr. H. Bilgekul, Higher Technological Inst, P.O. Box 95, Magosa, Mersin 10, TURKEY.

Dr. R.E. Blahut, IBM Corp., Bodle Hill Road, Owego, NY 13827, U.S.A.

Dr. R. Blom, Dept of Elec Eng, Linkoping University, S-581 83 Linkoping, SWEDEN.

Mr. J.M. Boucher, 8 Rue Haendel, 29200 BREST, FRANCE.

Mr. G.N. Brown, Dept of Elec Eng Science, University of Essex, Colchester, ESSEX, U.K.

Dr. W. Burger, T.J. Watson Research Centre 31-135, P.O. Box 218, Yorktown Heights, N.Y. 10598, U.S.A.

Professor K.W. Cattermole, "Fairacre", Gandish Road, East Bergholt, COLCHESTER CO7 6TP, U.K.

Mr. D.C. Chaturvedi, ESTEC, Postbus 299, 2200 AG Nordwijk, The Netherlands.

Dr. T.W. Chong, Telecom Research Lab, Hirst Research Centre, East Lane, Wembley, MIDDLESEX, HA9 7PP, U.K.

Mr. I. Cidon, Technion - Dept of Elec Eng, Technion City, Haifa 32000, ISRAEL.

Mr. F.P. Coakley, Dept of Elec Eng Science, University of Essex, COLCHESTER CO4 3SQ, U.K.

Dr. E.T. Cohen, Cyclotomics Inc, 2120 Haste Street, Berkeley CA 94704, U.S.A.

Dr. P.V. Collins, Telecom Research Lab, Hirst Research Centre, East Lane, Wemblet, MIDDLESEX HA9 7PP, U.K.

Dr A.B. Cooper, US Army Ballistic Res. Lab, Aberdeen Proving Ground, MD 21005.

Professor G.R. Cooper, School of Elec Eng, Purdue University, W. Lafayette IN 47907, U.S.A.

Professor T.M. Cover, Dept of Stats and Elec Eng, Stanford University, Stanford CA 94305, U.S.A.

Dr. J.R. Cruz, School of Elec Eng and Comp Science, The University of Oklahoma, Norman OK 73019, U.S.A.

Professor G. Einarsson, Telecommunication Theory, Electrical Eng. Dept, University of Lund, 220 07 LUND, SWEDEN.

Professor A. Ephremides, Elec Eng Dept, University of Maryland, College Park, MD 20742, U.S.A.

Professor P.G. Farrell, Elec Eng Labs, The University, MANCHESTER M13 9PL. U.K.

Professor L.E. Franks, Dept of ECE, University of Massachusetts, Amherst MA 01003, U.S.A.

Dr. V. Freitas, Universidade do Minho Av. Joao XXI, 4700 Braga, PORTUGAL.

Professor K.S. Fu, School of Elec Eng, Purdue University, W. Lafayette IN 47907, U.S.A.

Dr. P. Godlewski, ENST, Dept SYC, 46 Rue Barrault, 75634 Paris Cedex 13, FRANCE.

Professor S.W. Golomb, Prof of Elec Eng and Math, University of Southern California, Powell Hall, University Park, Low Angeles CA 90089-0272, U.S.A.

Dr. R.M.F. Goodman, Dept of Electronic Eng, University of Hull, HULL HU6 7RX, U.K.

Dr. B. Gopinath, Bell Labs., 7C203, Murray Hill NJ 07922, U.S.A.

Professor J.W.R. Griffiths, Electronic and Elec Eng Dept, University of Technology, LOUGHBOROUGH, Leics LE11 3TU. U.K.

Professor D. Haccoun, Ecole Polytechnique de Montreal, Dept of Elec Eng, CP 6079, Succ "A", Montreal H3C 3A7, CANADA.

Mr. E. Haddow, Queen Elizabeth College, University of London, Campden Hill Road, LONDON W8, U.K.

Dr. S. Harari, Universite de Picardie, UER de Mathematiques, 33 rue Saint-Leu 80039-AMIENS CEDEX, FRANCE.

Professor S. Haykin, McMaster University, Comms Research Lab, 1280 Main Street, Hamilton, Ontario 18S 4L7, CANADA.

Dr. C. Heegard, 313 Phillips Hall, Cornell University, Ithaca NY, 14853, U.S.A.

Professor I. Ingemarsson, Dept of Elec Eng, Linkoping University, S-581 83 Linkoping, SWEDEN.

Mr. D. Jancovic, Inst "B Kidric" - Vinca, Dept 270, P.O. Box 522, 11001 BEOGRAD, YUGOSLAVIA.

Dr. F. Jelinek, Manager, Continuous Speech Recog. Gp, IBM Thomas J. Watson Research Center, P.O. Box 218, Yorktown Heights NY 10598, U.S.A.

Dr. R. Johannesson, Dept of Computer Eng, University of Lund, P.O. Box 725, S-220 07 LUND, SWEDEN.

Mr. B.L. Johnson, The MITRE Corp - Mail Stop 142, Burlington Road, Bedford MA 01730, U.S.A.

Dr. G. Kawas-Kaleh, ENST, piece C132, 46 rue Barrault, 75634 Paris Cedex 13, FRANCE.

Dr. A. Leon-Garcia, Dept of Elec Eng, University of Toronto, Toronto, CANADA MS5 1A4.

Dr. E. Lutz, DFVLR, Inst fuer Nachrichtentechnik, 8031 Wessling, West Germany.

Mr. J.R.M. Mason, Telecom Research Lab, Hirst Research Centre, East Lane, Wembley MIDDLESEX, HA9 7PP, U.K.

Professor J.L. Massey, Insat fue Fernmeldetechnik, ETH-Centrum, ZURICH, SWITZERLAND.

Dr. S Matic, Inst "B Kidric"-Vinca, Dept 270, P.O. Box 522, 11001 BEOGRAD, YUGOSLAVIA.

Professor R. McEliece, Dept of Elec Eng 116-81, California Inst of Technology, PASADENA CA 91125, U.S.A.

Professor T.P. McGree, Dept of Elec and Computer Eng, University of Massachusetts, Amherst MA 01003, U.S.A.

Professor D.G. Messerschmitt, 2322 Trotter Way, Walnut Creek, California 94596, U.S.A.

Professor E.C. van der Meulen, Dept of Mathematics, K.U. Leuven, Celestijnenlaan 200 B, B-3030 Leuven, BELGIUM.

Dr. A. Morgul, Bogazici University, Dept of Elec Eng, P K 2 Bebek, INSTANBUL, TURKEY.

Dr. E. Munday, Electronics Labs, University of Kent, Canterbury, KENT CT2 7NT, U.K.

Dr. A. Nugroho, ENST de Bretagne, BP 856, 29279 Brest Cedex, FRANCE.

Mrs. C. Oddy, Marconi Research Centre, CHELMSFORD, U.K.

Mr. M.A. Padlipsky, C/O MITRE Corp., P.O. Box 208, Bedford, MA 01730, U.S.A.

Mr. E.A. Palo, The MITRE Corp - Mailstop E090, Burlington Road, Bedford MA 01730, U.S.A.

Mr. E. Panayirci, ITU Elektrik Fakultesi, Teknik Universite, ISTANBUL, TURKEY.

Mr. I.N. Parker, Telecom Research Lab, Hirst Research Centre, East Lane, Wembley, MIDDLESEX HA9 7PP, U.K.

Dr. B. Pehani, YU 61000 Ljubljana, Trzaska 25, YUGOSLAVIA.

Dr. W.L. Price, National Physical Laboratory, Teddington, MIDDLESEX, U.K.

Dr. J. Quemada, ETSI Telecommunication, Ciudad Universitaria, MADRID 3, SPAIN.

Dr. D. Rayner, DITC, National Physical Laboratory, Teddington, MIDDLESEX, U.K.

Dr. P.J.W. Rayner, Dept of Engineering, Trumpington Street, CAMBRIDGE CB2 1PZ, U.K.

Dr. J.O.'Reilly, Dept of Elec Eng Science, University of Essex, COLCHESTER CO4 3SQ, U.K.

Ms M.J. Rendas, Complexo I do Inic, Instituto Superior Tecnico, Av Rovisco Pais, 1000-LISBOA, PORTUGAL.

Mr. J.R. Rocha, Dept of Elec Eng Science, University of Essex, COLCHESTER CO4 3SQ, U.K.

Dr. Nils Rydbeck, Ericsson Radio Systems AB, 16380 Stockholm, SWEDEN.

Miss M. Sadler, Marconi Research Centre, CHELMSFORD, U.K.

Mr. J. Saras, ETSI Telecomunicacion, Ciudad Universitaria, MADRID 3, SPAIN.

Dr. D. Schofield, National Physical Laboratory, Queens Road, Teddington, MIDDLESEX, U.K.
Dr. G. Seguin, Dept of Elec Eng, Royal Military College, Kingston, ONTARIO, CANADA.
Dr. T. Shepherd, RSRE, St. Andrews Road, MALVERN, WORCS WR14 3PS, U.K.
Dr. J.A.E. Simons, Room B/0301, GCHQ Oakley, Prior Road, CHELTENHAM, Glos, U.K.
Mr. Paul Simpson, RSRE, St. Andrews Road, MALVERN, WORCS WR14 3PS, U.K.
Mr. J.K. Skwirzynski, Marconi Research Centre, CHELMSFORD, U.K.
Professor K. Steiglitz, 130 Longview Drive, Princeton NJ 08540, U.S.A.
Professor C-E Sundberg, Telecommunication Theory, University of Lund, S-22007, LUND, SWEDEN.
Dr. Y. Tanik, Elektrik Muh Bol, ODTU-Ankara, TURKEY.
Professor N. Tepedelenlioglu, Middle East Tech. University, Elec Eng. Dept, Ankara, TURKEY.
Mr. T.C. Tozer, RSRE Sat Comms Centre, Defford, Worcester WR8 9DU, U.K.
Dr. H. Trzaska, Inst of Telecom and Acoustics, Techn. Univ, Wroclaw, Wyspianskiego 27, 50-370 Wroclaw, POLAND.
Ms M.M. Veloso, Instituto Superior Tecnico, CAPS - Complexo I, Av Rovisco Pais, 1, P-1096 LISBOA codex, PORTUGAL.
Mr. B.G. West, Marconi Research Centre, CHELMSFORD, U.K.
Mr. D.L. Whiting, California Inst of Technology 256-80, Pasadena CA 91125, U.S.A.
Professor J.K. Wolf, Dept of Elec and Computer Eng, University of Massachusetts, Amherst MASS 01003, U.S.A.
Dr. K.M. Wong, Dept of Elec and Computer Eng, McMaster University, Hamilton, ONTARIO L8S 4L7, CANADA.
Mr. P.G. Wright, Marconi Research Centre, CHELMSFORD, U.K.
Dr. M.D. Yucel, Middle East Tech. University, Dept of Elec Eng, ANKARA, TURKEY.
Professor J. Ziv, Faculty of Elec Eng, Technion - IIT, Haifa 32000, ISRAEL.